AF836027

Acidophiles

Life in Extremely Acidic Environments

Edited by

Raquel Quatrini

Fundación Ciencia & Vida
Santiago
Chile

and

D. Barrie Johnson

College of Natural Sciences
Bangor University
Bangor
UK

Caister Academic Press

Copyright © 2016

Caister Academic Press
Norfolk, UK

www.caister.com

British Library Cataloguing-in-Publication Data
A catalogue record for this book is available from the British Library

ISBN: 978-1-910190-33-3 (hardback)
ISBN: 978-1-910190-34-0 (ebook)

Description or mention of instrumentation, software, or other products in this book does not imply endorsement by the author or publisher. The author and publisher do not assume responsibility for the validity of any products or procedures mentioned or described in this book or for the consequences of their use.

All rights reserved. No part of this publication may be reproduced, stored in a retrieval system, or transmitted, in any form or by any means, electronic, mechanical, photocopying, recording or otherwise, without the prior permission of the publisher. No claim to original U.S. Government works.

Cover design uses images © D. Barrie Johnson. Top image: Adit in an abandoned copper mine (Cantareras) in the Iberian Pyrite Belt, showing efflorescence of blue copper sulfate on the adit walls and metal-rich waters in the drainage channels. Middle image: An acidic geothermal site in São Miguel (an island in the Azores archipelago) with massive growths of acidophilic microorganisms. Bottom image: An extremely acidic ferruginous (iron-rich) pool in the Rio Tinto, Spain.

Ebooks
Ebooks supplied to individuals are single-user only and must not be reproduced, copied, stored in a retrieval system, or distributed by any means, electronic, mechanical, photocopying, email, internet or otherwise.

Ebooks supplied to academic libraries, corporations, government organizations, public libraries, and school libraries are subject to the terms and conditions specified by the supplier.

Contents

Contributors

Angeles Aguilera
Molecular Evolution Department
Centro de Astrobiología (CSIC-INTA)
Madrid
Spain

aguileraba@cab.inta-csic.es

Rodrigo Almárcegui
Laboratory of Molecular Microbiology and Biotechnology
Department of Biology
Faculty of Sciences
University of Chile
Santiago
Chile

kintaroo@gmail.com

Ricardo Amils
Centro de Biología Molecular Severo Ochoa (CSIC-UAM)
Universidad Autónoma de Madrid
Cantoblanco; and
Department of Planetology and Habitability
Centro de Astrobiología (CSIC-INTA)
Madrid
Spain

ramils@cbm.csic.es

Brett J. Baker
Department of Marine Science
Marine Science Institute
University of Texas
Port Aransas, TX
USA

acidophile@gmail.com

Jillian F. Banfield
Departments of Earth and Planetary Sciences and Environmental Sciences, Policy and Management
University of California Berkeley
Berkeley, CA
USA

jbanfield@berkeley.edu

Violaine Bonnefoy
National Centre for Scientific Research (CNRS)
Bacterial Chemistry Laboratory (LCB)
Mixed Research Unit (UMR) 7283
Aix-Marseilles University
Marseille
France

bonnefoy@imm.cnrs.fr

Juan Pablo Cárdenas
Fundación Ciencia & Vida; and
Facultad de Ciencias Biologicas
Universidad Andres Bello
Santiago
Chile

juan.pablo.cardenas@gmail.com

Mark Dopson
Centre for Ecology and Evolution in Microbial Model Systems (EEMiS)
Linnaeus University
Kalmar
Sweden

mark.dopson@lnu.se

David Fernández-Remolar
Department of Planetology and Habitability
Centro de Astrobiología (CSIC-INTA)
Madrid
Spain

fernandezrd@cab.inta-csic.es

Alonso Ferrer
Department of Biology
Faculty of Chemistry and Biology
University of Santiago
Santiago
Chile

aloferrer@gmail.com

Manuel Ferrer
Department of Applied Biocatalysis
Spanish National Research Council (CSIC)
Institute of Catalysis
Madrid
Spain

mferrer@icp.csic.es

Peter N. Golyshin
School of Biological Sciences
Bangor University
Gwynedd
UK

p.golyshin@bangor.ac.uk

Olga V. Golyshina
School of Biological Sciences
Bangor University
Gwynedd
UK

o.golyshina@bangor.ac.uk

Nicolas Guiliani
Laboratory of Bacterial Communication
Department of Biology
Faculty of Sciences
University of Chile
Santiago
Chile

nguilian@uchile.cl

Susan T.L. Harrison
Centre for Bioprocess Engineering Research (CeBER)
Department of Chemical Engineering
University of Cape Town
Rondebosch
South Africa

sue.harrison@uct.ac.za

Sabrina Hedrich
Federal Institute for Geosciences and Natural Resources (BGR)
Resource Geochemistry
Hannover
Germany

sabrina.hedrich@bgr.de

David S. Holmes
Facultad de Ciencias Biologicas
Universidad Andres Bello; and
Fundación Ciencia & Vida
Santiago
Chile

dsholmes2000@yahoo.com

Carlos A. Jerez
Laboratory of Molecular Microbiology and Biotechnology
Department of Biology
Faculty of Sciences
University of Chile
Santiago
Chile

cjerez@uchile.cl

D. Barrie Johnson
College of Natural Sciences
Bangor University
Bangor
UK

d.b.johnson@bangor.ac.uk

Gloria Levicán
Department of Biology
Faculty of Chemistry and Biology
University of Santiago
Santiago
Chile

gloria.levican@usach.cl

Patricio Martínez
Centre of Proteomics and Metabolomics BioSigma (CPMB)
BioSigma S.A.
Santiago
Chile

pmart007@biosigma.cl

Claudio Navarro
Laboratory of Molecular Microbiology and Biotechnology
Department of Biology
Faculty of Sciences
University of Chile
Santiago
Chile

clnavarrol@gmail.com

Wolfgang Nitschke
National Centre for Scientific Research (CNRS)
Bacterial Chemistry Laboratory (LCB)
Mixed Research Unit (UMR) 7281
Aix-Marseilles University
Marseille
France

nitschke@imm.cnrs.fr

Sanna Olsson
Department of Forest Ecology and Genetics
INIA, Forest Research Centre
Madrid
Spain

sanna.olsson@helsinki.fi

Alvaro Orell
Molecular Microbiology of Extremophiles
Centre for Genomics and Bioinformatics
Faculty of Sciences
University Mayor
Santiago
Chile; and
Prokaryotic Small RNA Biology
Max Planck Institute for Terrestrial Microbiology
Marburg
Germany

alvaro.orell@mpi-marburg.mpg.de

Omar Orellana
Programme of Cell and Molecular Biology
Faculty of Medicine
University of Chile
Santiago
Chile

oorellan@med.uchile.cl

Francisco J. Ossandon
Fundación Ciencia & Vida
Santiago
Chile

fco.j.ossandon@gmail.com

Fernando A. Pagliai
Department of Microbiology and Cell Science
Institute of Food and Agricultural Sciences
University of Florida
Gainesville, FL
USA; and
Laboratory of Molecular Microbiology and Biotechnology
Department of Biology
Faculty of Sciences
University of Chile
Santiago
Chile

fpagliai@gmail.com

Pilar Parada
BioSigma S.A.
Santiago
Chile

pilar.parada@biosigma.cl

Fernando Puente-Sánchez
Molecular Evolution Department
Centro de Astrobiología (CSIC-INTA)
Madrid
Spain

puentesf@cab.inta-csic.es

Raquel Quatrini
Fundación Ciencia & Vida; and
Facultad de Ciencias Biologicas
Universidad Andres Bello
Santiago
Chile

rquatrini@cienciavida.org

Douglas Rawlings
Department of Microbiology
Stellenbosch University
Stellenbosch
South Africa

der@sun.ac.za

Francisco Remonsellez
Laboratorio de Tecnología de Membranas, Biotecnología y Medio Ambiente
Departamento de Ingeniería Química
Universidad Católica del Norte
Antofagasta
Chile

fremonse@ucn.cl

Axel Schippers
Federal Institute for Geosciences and Natural Resources (BGR)
Resource Geochemistry
Hannover
Germany

axel.schippers@bgr.de

Elizabeth L.J. Watkin
School of Biomedical Sciences
Curtin University
Perth
Australia

e.watkin@curtin.edu.au

Carla M. Zammit
School of Earth Sciences
The University of Queensland
Brisbane
Australia

c.zammit1@uq.edu.au

Preface

This textbook provides a comprehensive account of different areas of acidophilic microbiology, ranging from fundamental to applied aspects of knowledge and research. It is subdivided into parts, each dealing with a specific area of the subject. Part II looks at the challenges faced by life forms that grow in extremely acidic environments and how they adapt to meet these challenges. Elevated concentrations of protons/hydronium ions, transition metals and, in many situations, stress induced by high osmotic potentials and redox potentials of the external solutions, are the major challenges faced by acidophilic microorganisms in natural and man-made environments. This part also includes a chapter describing how acidophiles obtain energy and the pathways involved in iron and sulfur oxidation at low pH. Part III includes chapters which describe the physiological and phylogenetic diversities of acidophilic microorganisms – the various genera and species of archaea, bacteria and eukaryotes that have documented thus far. Part IV covers acidophile community dynamics, describing ways in which different species of acidophiles interact with each other in terms of carbon and energy flows, and how these impact the geochemistry of acidic environments, an account of the distribution of acidophiles in both man-made and natural ecosystems, and how quorum sensing is used by acidophiles as a detection mechanism, and how this impacts the formation of biofilms, an important feature of biomining technologies. Part V deals with the various 'omics' that are used to study acidophiles and to answer fundamental question regarding the full breadth of their diversity, metabolisms and the molecular basis of their responses to stress factors. Genomic and metagenomic studies, and what has been and is being learned about acidophiles from proteomic-, mobilomic- and metabolomic-focused research, are described in detail. The final part, Part VI, considers the ways in which acidophiles are used in established and emerging biotechnologies, and describes why these fascinating microorganisms are considered potential candidates for life on other solar bodies, such as Mars, and beyond.

D. Barrie Johnson
Raquel Quatrini

Genus abbreviations

Genus name	Abbreviation
Acidianus	*Ad.*
Acidibacter	*Ab.*
Acidicaldus	*Acd.*
Acidiferrobacter	*Af.*
Acidilobus	*Al.*
Acidimicrobium	*Am.*
Acidiphilium	*A.*
Acidiplasma	*Ap.*
Acidisphaera	*As.*
Aciditerrimonas	*Atn.*
Acidithiobacillus	*At.*
'*Acidithiomicrobium*'	*Atm.*
'*Acidithrix*'	*Atx.*
Acidobacterium	*Abm.*
Acidocella	*Ac.*
Acidomonas	*Amn.*
Aciduliprofundum	*Acp.*
Alicyclobacillus	*Alb.*
Caldisphaera	*C.*
Caldivigra	*Cv.*
Desulfosporosinus	*D.*
Ferrimicrobium	*Fm.*
Ferrithrix	*Fx.*
Ferrobacillus	*F.*
Ferroplasma	*Fp.*
'*Ferrovum*'	*Fv.*
Frateuria	*Ft.*
Hydrogenobaculum	*H.*
Leptospirillum	*L.*
Metallosphaera	*M.*
Methylacidiphilium	*Mt.*
Nitrosotalea	*N.*
Picrophilus	*P.*

Genus name	Abbreviation
Stygiolobus	*Sg.*
Sulfobacillus	*Sb.*
Sulfolobus	*S.*
Sulfurisphaera	*Ss.*
Sulfurococcus	*Sc.*
Thermoplasma	*Tp.*
Thiobacillus	*T.*
Thermogymnomonas	*Tg.*
Thiomonas	*Tm.*
Vulcanisaeta	*V.*

Current Books of Interest

Influenza: Current Research	2016
MALDI-TOF Mass Spectrometry in Microbiology	2016
Aspergillus and *Penicillium* in the Post-genomic Era	2016
The Bacteriocins: Current Knowledge and Future Prospects	2016
Omics in Plant Disease Resistance	2016
Climate Change and Microbial Ecology: Current Research and Future Trends	2016
Biofilms in Bioremediation: Current Research and Emerging Technologies	2016
Microalgae: Current Research and Applications	2016
Gas Plasma Sterilization in Microbiology: Theory, Applications, Pitfalls and New Perspectives	2016
Virus Evolution: Current Research and Future Directions	2016
Arboviruses: Molecular Biology, Evolution and Control	2016
Shigella: Molecular and Cellular Biology	2016
Aquatic Biofilms: Ecology, Water Quality and Wastewater Treatment	2016
Alphaviruses: Current Biology	2016
Thermophilic Microorganisms	2015
Flow Cytometry in Microbiology: Technology and Applications	2015
Probiotics and Prebiotics: Current Research and Future Trends	2015
Epigenetics: Current Research and Emerging Trends	2015
Corynebacterium glutamicum: From Systems Biology to Biotechnological Applications	2015
Advanced Vaccine Research Methods for the Decade of Vaccines	2015
Antifungals: From Genomics to Resistance and the Development of Novel Agents	2015
Bacteria-Plant Interactions: Advanced Research and Future Trends	2015
Aeromonas	2015
Antibiotics: Current Innovations and Future Trends	2015
Leishmania: Current Biology and Control	2015
Acanthamoeba: Biology and Pathogenesis (2nd edition)	2015
Microarrays: Current Technology, Innovations and Applications	2014
Metagenomics of the Microbial Nitrogen Cycle: Theory, Methods and Applications	2014
Pathogenic *Neisseria*: Genomics, Molecular Biology and Disease Intervention	2014
Proteomics: Targeted Technology, Innovations and Applications	2014
Biofuels: From Microbes to Molecules	2014
Human Pathogenic Fungi: Molecular Biology and Pathogenic Mechanisms	2014
Applied RNAi: From Fundamental Research to Therapeutic Applications	2014
Halophiles: Genetics and Genomes	2014

Full details at www.caister.com

Part I

Introduction

Acidophile Microbiology in Space and Time

1

D. Barrie Johnson and Raquel Quatrini

Abstract

The study of extreme acidophiles, broadly defined as microorganisms that grow optimally at pH values below 3, was initiated by the discovery by Waksman and Joffe in the early 1900s of a bacterium that was able to live in the dilute sulfuric acid it generated by oxidizing elemental sulfur. The number of known acidophiles remained relatively small until the second half of the 20th century, but since then has greatly expanded to include representatives of living organisms from within all three domains of life on earth, and notably within many of the major divisions and phyla of Bacteria and Archaea. Environments that are naturally acidic are found throughout the world, and others that are man-made (principally from mining metals and coal) are also widely distributed. These continue to be sites for isolating new species, (and sometimes new genera) which thrive in acidic liquor solutions that contain concentrations of metals and metalloids that are lethal to most life forms. The development and application of molecular techniques and, more recently, next generation sequencing technologies has, as with other areas of biology, revolutionized the study of acidophile microbiology. Not only have these studies provided greater understanding of the diversity of organisms present in extreme acidic environments and aided in the discovery of largely overlooked taxa (such as the ultra-small uncultivated archaea), but have also helped uncover some of the unique adaptations of life forms that live in extremely acidic environments. Thanks to the relatively low biological complexity of these ecosystems, systems-level spatio-temporal studies of model communities have been achieved, laying the foundations for 'multi-omic' exploration of other ecosystems.

This chapter introduces the subject of acidophile microbiology, tracing its origins to the current *status quo*, and provides the reader of this textbook with general information which provides a backdrop to the more specific topics described in subsequent chapters.

Acidophilic microorganisms defined

Acidophiles are broadly defined as life forms that grow preferentially in natural or man-made environments where the pH is well below 7. Together with other categories of 'extremophiles', they have greatly expanded our knowledge of the diversity of life, our understanding on how microorganisms can adapt to seemingly hostile situations, and provided scenarios for the possibility that life forms similar to those that colonize inhospitable niches on Earth, our own planet, may be found on planets and satellites within and outside of our solar system.

While there is no formal definition of what constitutes an acidophile, one proposed by Johnson (2007) has gained general acceptance. That delineated 'extreme acidophiles' as those that have pH optima at 3 or below, and 'moderate acidophiles' as those with pH optima of 3–5, though some species are capable of growing at pH < 3. Acid-tolerant species have pH optima above 5, though they can grow at pH values lower than this. The most acidophilic of all known life forms (the euryarchaeote *Picrophilus*, which currently contains two validated species) has a pH optimum of 0.7 and can grow at pH 0. Given that pH is a logarithmic scale, this means that *Picrophilus* spp. can grow in liquors where the hydronium ion (H_3O^+) concentration is 50–100 times greater than that tolerated even by many extreme acidophiles. The term 'hyper-acidophiles' (analogous to 'hyper-thermophiles' which was introduced when prokaryotes that had temperature optima > 80°C were discovered, some of which grew well above 100°C) is a more appropriate way to describe species that grow optimally at pH < 1, though such prokaryotes appear at present to be very rare.

The same general rule applies with pH as other environmental parameters (temperature, salinity, pressure, etc.) which is that, as the stress factor (in this case, acidity) increases in intensity, so the number of species that can tolerate or grow in such environments decreases. Extreme acidophiles are exclusively microbial. Macroscopic life forms are sometimes found in low-pH environments, such as the angiosperms *Erica* spp. that grow in acidic soils and *Juncus bulbosus* which can colonize the fringes of acidic pit lakes. However, cell membranes of higher plants are susceptible to damage at pH < 4, and the roots of wetland plants such as *Juncus* spp. are generally found in the sediments of acidic ponds and lakes where the pH is significantly greater than that of the water body itself. While it is the case that most extremely acidophilic eukaryotes are unicellular, some multicellular life forms, including rotifers and some crustaceans, can thrive in some low-pH environments. Acidophiles are also widely distributed in the 'tree of life', occurring in all three domains of life, *Bacteria* (Chapter 2), *Archaea* (Chapter 3) and *Eukarya* (Chapter 4), and within multiple phyla in the same domain (Fig. 1.1).

The number of species of extremely acidophilic microorganisms that have been isolated and characterized has increased almost exponentially from just two in the mid-20th century to > 50 in the early 21st century. It is now clear that, as a group, acidophiles are highly heterogeneous from physiological and metabolic perspectives, and can utilize different sources of energy (solar, organic and inorganic chemicals), electron acceptors (oxygen, ferric iron and sulfur) and carbon sources (organic, inorganic, or both) although some metabolic functions seem to be rare (e.g. fermentation) or absent (e.g. dissimilatory nitrate reduction), and chemolithotrophy unusually commonplace amongst acidophilic microorganisms (Johnson and Aguilera, 2015).

The evolution and now routine application of molecular techniques for studying biological systems has resulted in many uncultivated species of microorganisms being detected in extremely low pH, as in other environments. Improved and probably novel cultivation-based approaches and techniques are currently required to catch up and keep pace with what cultivation-free studies are revealing about the biodiversity of extreme environments, including those that are extremely acidic.

Nature, genesis, and global distribution of extremely acidic environments

Extremely acid environments can vary greatly in scale, nature and origin. Gastric acid produced in the lining of the human stomach has a pH of ~1.5 to 3.5 due to the production of hydrochloric acid. In contrast, in both man-made and natural extremely acidic environments, sulfuric acid is usually the mineral acid responsible for the low-pH conditions. In situations where low pH is due to the presence of organic acids, such as humic acids/colloids (as in acid peats) these tend to be moderate (>3) rather than extreme. Notable exceptions are the pH of lemon and lime fruits, which have pH between 2 and 3 due to their elevated concentrations of undissociated citric acid (a tricarboxylic acid with pK_a values of 3.1, 4.8 and 6.4).

Many biological processes, for example fermentation, generate acidity due to the production of small molecular weight aliphatic acids (such as lactic and propionic acids) as metabolic waste products. Acetogenic bacteria have long been used to generate malt and wine vinegars, which have pH values of ~2.5. However, biological processes that generate mineral (inorganic) acids have, at least in theory, greater potential for generating extremely low-pH environments due to the generally lower pK_a values of the acids involved. Nitrification is a two-stage oxidative process that generates nitrous (pK_a 3.4) and nitric (pK_a −1.6) acids. However, the vast majority of prokaryotes that catalyse these dissimilatory reactions (mostly chemolithotrophs) are acid sensitive and only thrive in environments that are self-buffered, such as marine waters and non-acidic soils. In contrast, the dissimilatory oxidation of elemental and reduced forms of sulfur is carried out by prokaryotes that have widely different pH ranges and optima for growth, and includes a number of species of extremely acidophilic bacteria and archaea.

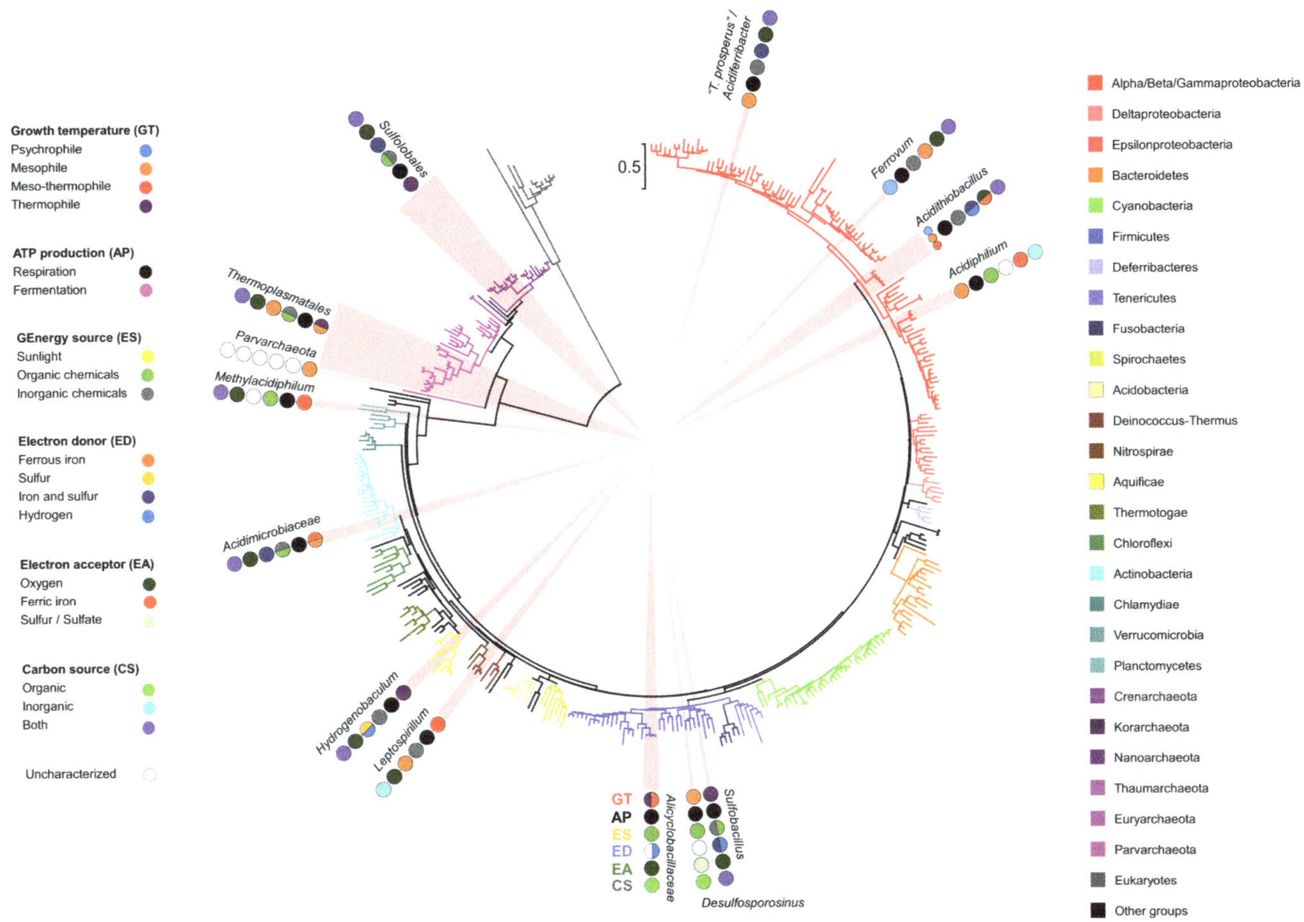

Figure 1.1 Phylogenetic tree of bacterial and archaeal phyla based on the 16S small subunit rRNA gene sequence, constructed as described in Chapter 11. The taxonomic groups are defined by the clade colour index at the right side of the figure and the presence of acidophiles in each clade is indicated in pink. Metabolic properties of each group of acidophiles are defined by the coloured circles outside the tree (coded as in the left side index). The figure was prepared by Juan Pablo Cárdenas (Fundación Ciencia and Vida and Universidad Andres Bello, Santiago, Chile).

Sulfur is one of the more abundant elements in the lithosphere, occurring at ~0.05% on a weight basis. It is also a complex element, and can exist in up to nine different oxidation states ranging from −2 (e.g. in hydrogen sulfide; H_2S) to +6 (e.g. in the sulfate anion; SO_4^{2-}; Steudel, 2000). Catenation (the ability of sulfur atoms to bond to each other to form chains of practically unlimited lengths) is another characteristic of this element. In polythionates (sulfur oxyanions) the sulfur atoms are present in different oxidation states, e.g. in tetrathionate, where the two central sulfur atoms (which are covalently bonded to each other) have oxidation states of zero, while the two terminal sulfur atoms (which are covalently bonded to oxygen as well as to the central sulfur atoms) have oxidation states of +5. Redox transformations of sulfur are common in nature, though mostly this is directed at assimilating the element (e.g. into amino acids, or sulfate esters) and dissimilatory processes (sulfur species being used directly or indirectly as electron donors or acceptors) are confined to some species of prokaryotic microorganisms. The terminal products of sulfur oxidation are the sulfate or bisulfate (HSO_4^-) anions. In situations where dissimilatory sulfate production exceeds the ability of the local environment to counter-balance the net acidity generated by this reaction, pH will decline, and species of prokaryotic microorganisms mediating the reaction will also change from neutrophiles to moderate extremophiles, and ultimately to extreme acidophiles. Sulfuric acid has two dissociation constants, shown in Equation 1.1:

$$H_2SO_4 \underset{pK_a\,-2.8}{\longleftrightarrow} HSO_4^- + H^+ \underset{pK_a\,+1.92}{\longleftrightarrow} SO_4^{2-} + 2H^+ \qquad (1.1)$$

While in most acidic environments associated with sulfur oxidation sulfate is the dominant form, in others (such as water bodies found within the Richmond Mine at Iron Mountain, California) bisulfate is more abundant. In acidic sulfate-rich waters, the sulfate/bisulfate couple can therefore act as an important pH buffer.

Three factors are important in determining whether an extremely acidic environment may be formed as a consequence of dissimilatory sulfur oxidation: (i) reduced forms of sulfur (including elemental sulfur) need to be present in relatively large abundance; (ii) conditions need to be conducive to the (microbial) oxidation of the reduced sulfur, which usually (though not necessarily) requires the presence of oxygen, and (iii) the acidity generated needs to exceed the neutralizing capacity of the niche in which it is occurring. Geothermal sites often fulfil all three criteria. Crystals of prismatic sulfur can form in the vents of volcanoes from the condensation of sulfur dioxide and hydrogen sulfide gases (Equation 1.2):

$$SO_2 + 2H_2S \rightarrow 3S^0 + 2H_2O \qquad (1.2)$$

Solfatara (which derives from the Italian for 'sulfur place') are volcanic and geothermal areas where sulfur gases occur in association with water vapour. Temperatures in solfatara water bodies can approach water boiling point (85° to > 100°C, depending on altitude and the presence of solutes) but tend to cool rapidly as waters flow out from the immediate heat source, giving the opportunity for colonization of the sulfur-rich waters by microorganisms with widely differing temperature ranges. Oxidation of sulfur by archaea and bacteria then generates sulfuric acid (Equation 1.3):

$$2S^0 + 3O_2 + 2H_2O \rightarrow 3H^+ + HSO_4^- + SO_4^{2-} \qquad (1.3)$$

Acidic hot springs are common features within solfatara fields, and thixotropic acidic mud pools form from the destruction of silicates and other minerals by the combined action of extreme acidity and high temperatures. Solfatara can be found around the margins of active and some dormant volcanoes and where the Earth's crust is relatively thin. Examples include Yellowstone National Park (USA), Whakarewarewa (New Zealand), Krisuvik (Iceland), the Kamchatka peninsula (Russia), Sao Michel (Azores), Volcano, Naples and Ischia (all Italy), Djibouti (Africa), the Purico complex (Chile) and some Caribbean islands, such as Montserrat and St. Lucia. Submarine volcanic areas and hydrothermal systems, such as the Mid-Atlantic ridge, the East Pacific Rise, the Guaymas Basin, and active seamounts (e.g. around Tahiti), also discharge large amounts of reduced sulfur from magmatic sources. In contrast to land-based solfatara fields, the acidity generated by sulfur oxidation from submarine discharges is neutralized by the buffering capacity of sea water, though localized areas of (extreme) acidity which act as niches for acidophilic microorganisms can exist close to the venting ducts.

The other major source of reduced sulfur that can act as the point of origin of extremely acidic environments are where sulfide minerals are concentrated, e.g. in metal ore bodies. These may occur as igneous (as in the Iberian Pyrite Belt), sedimentary (e.g. the extensive kupferschiefer deposits in central Europe) or metamorphic (e.g. polymetallic black schist ores in Finland) deposits. Gossans, intensively weathered, oxidized rock strata that are often highly coloured (orange/brown) due to surface layers of ferric iron-rich deposits, are geological features found in many parts of the world. These overlie unweathered rock containing pyrite and other sulfide minerals, and are the primary source of the iron seen at the land surface. Water in the vicinity of, and draining out of, the gossans can be acidic due to the microbially catalysed oxidative dissolution of the sulfidic minerals. Gossans and associated acid rock drainage waters have been reported in northern Greenland (Langdahl and Ingvorsen, 1997) and in Antarctica (Dold *et al.*, 2013). Acid production in such situations is often limited by the physical nature of the unweathered rock, restricting its exposure to both oxygen and water, both of which are necessary for the potentially acid-generating minerals to be degraded. However, when unweathered sulfidic rock strata are mined as ore bodies, the processes involved (excavation, comminution and disposal of reactive mineral wastes) provide conditions that are highly conducive for the activities of lithotrophic ('rock eating') acid-generating acidophiles, and consequently the mining of many metals, and also of coals, which contain variable quantities of iron sulfide minerals and elemental sulfur.

Sulfidic ores are the main sources of many commercially valuable base and semi-precious transition metals, such as copper, zinc, cobalt, nickel and silver, either as single metal sulfides (e.g. millerite; NiS) or in association with other metals (e.g. pentlandite; $(Fe,Ni)_9S_8$). Some metalloids such as arsenic, also form sulfide minerals, such as arsenopyrite (FeAsS) and realgar (As_4S_4). Native fine-grain refractory gold is also often intimately associated with sulfide minerals (arsenopyrite and pyrite, FeS_2), as is the uranium(IV) mineral uraninite (UO_2). Pyrite is the most ubiquitous of all sulfide minerals and is often present in ore bodies in greater quantities than the mineral(s) bearing the target metal(s). Since pyrite (and another commonly encountered iron sulfide, pyrrhotite; $Fe_{(1-x)}S$, where $x = 0–2$) has relatively little commercial value, though it has been, and continues to be, a source of sulfur and sulfuric acid, it often ends up in mineral wastes, such as fine-grain tailings, that are produced during ore processing (Dold, 2010). The oxidative dissolution of pyrite by acidophilic bacteria and its role in generating acidic, metal rich pollution, such as acid mine drainage waters, is among the most intensively studied processes in acidophile microbiology (e.g. Vera *et al.*, 2013). In brief, this involves a primary attack on the mineral by ferric iron, a reaction (Equation 1.4) that is both abiotic and independent of oxygen:

$$FeS_2 + 6Fe^{3+} + 3H_2O \rightarrow 7Fe^{2+} + S_2O_3^{2-} + 6H^+ \quad (1.4)$$

Ferric iron has to be regenerated for this reaction to continue. This is a very slow abiotic reaction at pH <3, though it can be catalysed by numerous species of iron-oxidizing acidophiles in a proton- and oxygen-consuming reaction (Equation 1.5):

$$Fe^{2+} + H^+ + 0.25O_2 \rightarrow Fe^{3+} + 0.5H_2O \quad (1.5)$$

Thiosulfate, generated in Equation 1.4, is unstable in acidic ferric iron-rich liquors and is rapidly transformed to elemental sulfur and various polythionates. These in turn are oxidized by sulfur-oxidizing acidophiles, producing sulfuric acid (e.g. as in Equation 1.3). The reaction summarizing the oxidative dissolution of pyrite at low pH in which all of the products are in their most oxidized forms is therefore:

$$FeS_2 + 3.75O_2 + 0.5H_2O \rightarrow Fe^{3+} + 2SO_4^{2-} + H^+ \quad (1.6)$$

Equation 1.6 indicates that the oxidative dissolution of pyrite is indeed a net acid-generating reaction. However, this is more pronounced when secondary reactions involving the hydrolysis of ferric iron are taken into consideration, such as that shown in Equation 1.7 which depicts the formation of schwertmannite, a common ferric iron mineral that forms in moderately acidic (pH 2 to 4) mine waters:

$$8Fe^{3+} + 14H_2O + SO_4^{2-} \rightarrow Fe_8O_8(SO_4)(OH)_6 + 22H^+ \quad (1.7)$$

Combining Equations 1.6 and 1.7 it can be seen that the microbially mediated transformation of the ferrous iron mineral pyrite to the ferric iron mineral schwertmannite generates 3.75 moles of proton acidity per mole of pyrite oxidized. The hydrolysis and dissolution of ferric iron acts as a powerful pH buffer in oxidized, iron-rich waters, as exemplified in the Rio Tinto in south-west Spain, which maintains a pH of between ~ 2.2 and 2.8 throughout its 100 km length.

Water bodies impacted by mining activities are not only often moderately to extremely acidic, but also typically contain far more elevated concentrations of dissolved transition metals, aluminium and metalloids than non-impacted waters. The former include subterranean water bodies housed within abandoned deep mines (Johnson, 2012), pit lakes that form in opencast voids (e.g. Falagan *et al.*, 2013) and acidic streams than drain mines and mine spoils (e.g. Blowes *et al.*, 2003). Because metal mining was one of the earliest technologies developed by *Homo sapiens* (albeit on small scales in the Iron and Bronze Ages) and, along with coal mining, has been and in many countries remains an important industry, acidic sites caused by this human activity have widespread global distribution.

A brief history of acidophile microbiology

Acidophile microbiology (Fig. 1.2) began with the discovery, by Waksman and Joffe nearly a century ago, of a bacterium that grew on elemental sulfur and which was able to thrive in the dilute sulfuric acid (pH as low as 0.6) that it generated (Waksman and Joffe, 1922). They named this bacterium *Thiobacillus* (*T.*) *thiooxidans*, though it was later reclassified as *Acidithiobacillus* (*At.*) *thiooxidans* by Kelly and Wood (2000) along with other acidophilic species of *Thiobacillus*. Waksman and Joffe (1922) reported that 'sulfur is the all important energy source' for *At. thiooxidans*, and also that is was a strict autotroph and obligate aerobe, and a mesophile.

Some 30 years later, investigations carried out by Arthur Colmer, Kenneth Temple and colleagues in acidic ferruginous waters draining coal mines in eastern USA (Colmer and Hinkle, 1947) led to the isolation of another acidophilic sulfur-oxidizing *Thiobacillus*, though uniquely (at the time) this bacterium could also grow autotrophically using ferrous iron as sole electron donor. Following the description and naming of *Thiobacillus* (later *Acidithiobacillus*) *ferrooxidans* (Temple and Colmer, 1951) other, supposedly related, but novel species of acidophilic iron-oxidizers were isolated from coal mine drainage waters. '*Ferrobacillus ferrooxidans*' was differentiated from *At. ferrooxidans* on the basis that it appeared not to use reduced sulfur as electron donors (Leathen *et al.*, 1956) while '*Ferrobacillus sulfooxidans*' was claimed to grow on elemental sulfur but not on thiosulfate (Kinsel, 1960). Kelly and Tuovinen (1972) later suggested that these claimed differences in physiologies were invalid and that all three were strains of the same species (*T/At. ferrooxidans*). Interestingly, '*Ferrobacillus ferrooxidans*' later assumed the role of type species of *At. ferrooxidans* (ATCC 23270/DSM 14882) though its original description by Leathen *et al.* (1956) (actively motile, pH minimum ~2, and temperature optimum of 20–25°C) is more akin to that of *Acidithiobacillus ferrivorans* (Hallberg *et al.*, 2010) than to *At. ferrooxidans*^{T}, which is non-motile and grows at pH < 1.5 and optimally at ~30°C, suggesting that some cross-over of strains (possibly due to mixed cultures) might have occurred. *At. ferrooxidans* remains by far the most widely studied and reported of all acidophilic bacteria, though iron-oxidizing acidithiobacilli currently includes four classified species.

The 1970s saw some major breakthroughs in isolating new species of acidophiles with contrasting physiological characteristics. These included the first heterotrophic (and moderately thermophilic) acidophile, *Thermoplasma* (*Tp.*) *acidophila* (later renamed *Tp. acidophilum*) which was isolated from a coal refuse pile that had undergone self-heating (Darland *et al.*, 1970). *Tp. acidophilum* was considered at the time to be a *Mycoplasma* (a genus of bacteria that lacks a cell wall) but later shown to be a euryarchaeote, and was therefore was first acidophilic archaea to be formally described. A related genus, *Picrophilus*, contains two species (both heterotrophic) that can grow at pH < 0, making them the most acidophilic of all currently known life forms (Schleper *et al.*, 1995). The first acidophilic bacterium shown to grow on organic carbon was described a few years later (Guay and Silver, 1974).

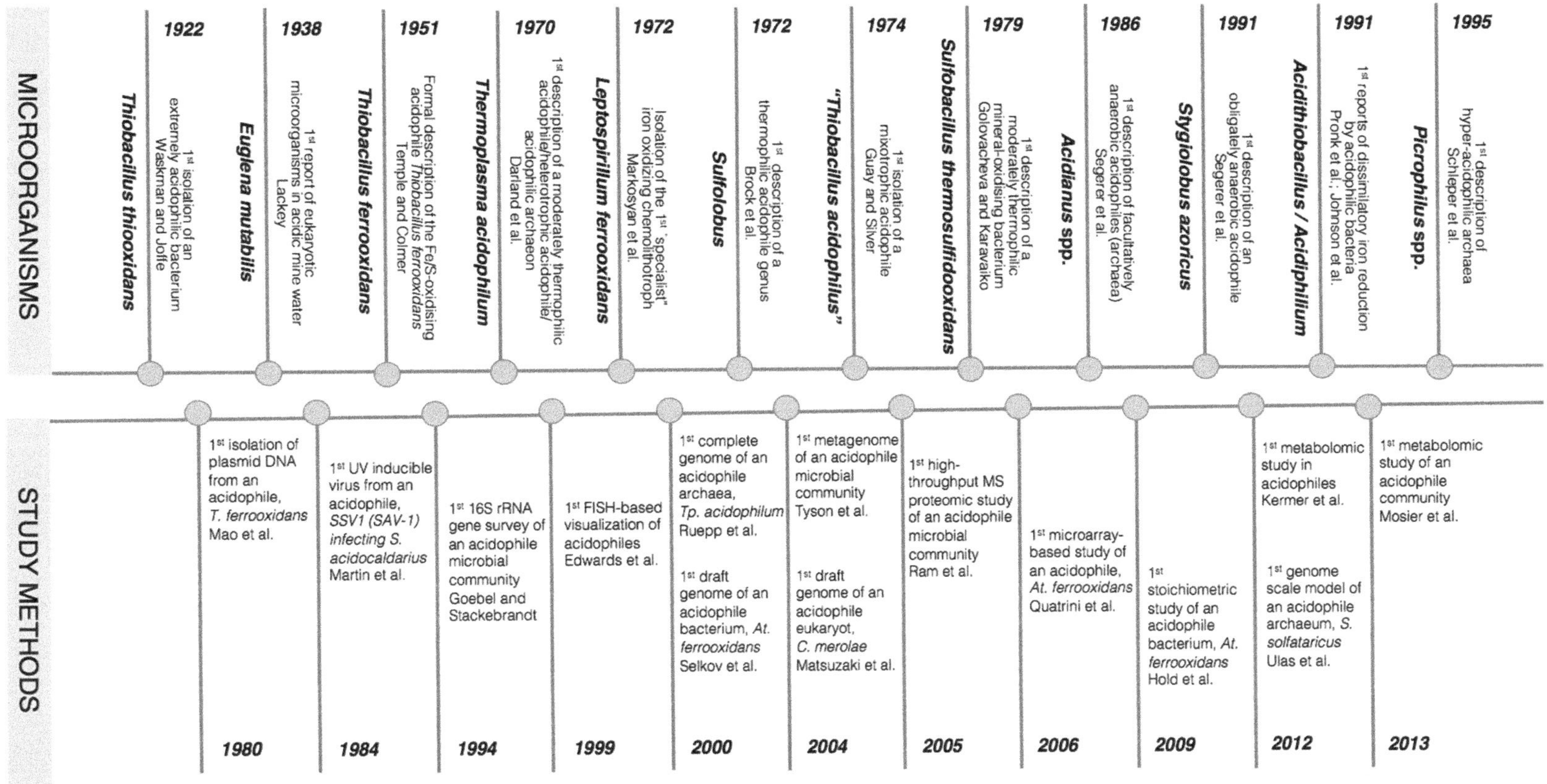

Figure 1.2 Major milestones in the development of acidophilic microbiology.

Named originally as *Thiobacillus acidophilus* since it could grow autotrophically on sulfur, as well as heterotrophically on sugars and some other small molecular weight organic compounds, it was later shown (from its 16S rRNA gene sequence) to be a member of the *Acidiphilium* genus. It remains the sole facultative autotroph in this genus; other classified species, the first of which were described by Arthur Harrison Jr in 1981, are all obligate heterotrophs.

The first extremely thermophilic acidophiles were isolated from Yellowstone National Park, Wyoming, and some other geothermal sites in Italy, Dominica and El Salvador, in the late 1960s and early 1970s. Tom Brock at colleagues described isolates obtained from acidic hot springs and thermal acid soils that grew optimally on sulfur and a variety of simple organic compounds at 70–75°C and at pH 2–3 (Brock *et al.*, 1972). One strain was noted to grow at 85°C. *Sulfolobus* (*S.*) was described as 'a new genus of bacteria' and later confirmed as one of the first genus of the *Crenarchaeota* subdomain. The type strain of first species to be described, *S. acidocaldarius*, is an obligate heterotroph and does not oxidize elemental sulfur, though it was originally described as doing so. The very first description of an extremely thermophilic acidophile had earlier been reported in the 1966 doctoral thesis of James Brierley. This Yellowstone isolate was noted to oxidize ferrous iron and elemental sulfur, and had a lower temperature limit (~70°C) than *S. acidocaldarius* (Brierley and Brierley, 1973), suggesting that it was probably a strain of *S. metallicus*. While all *Sulfolobus* spp. are obligate aerobes, other extremely thermo acidophilic crenarchaeotes include genera such as *Acidianus* spp. that are facultative anaerobes (and oxidize or reduce sulfur) and the obligate anaerobe *Stygiologus azoricus*.

The 1970s also saw the first reports of iron-oxidizing acidophilic bacteria other than *At. ferrooxidans*. A second bacterial genus/species (*Leptospirillum ferrooxidans*) was isolated from a copper mine in Armenia (Markosyan, 1972). *Leptospirillum* spp. form a distinct clade within the *Nitrospira* phylum and appear to use only ferrous iron as an electron donor and are therefore obligate aerobes, since oxygen is the only thermodynamically feasible electron acceptor that can be coupled to ferrous iron oxidation at low pH. *Leptospirillum* spp. in general, and the thermo-tolerant species *L. ferriphilum* in particular (Coram and Rawlings, 2002) are now recognized to be the primary mineral-oxidizing acidophiles in biomining operations, including both heap operations and stirred tanks.

In the 1970s, bacteria that grew autotrophically on both ferrous iron and reduced sulfur but at much higher temperatures (~50°C) than previous chemolithotrophic acidophiles were isolated from mine sites and geothermal areas. These were also originally described as *Thiobacillus* spp. (e.g. Brierley *et al.*, 1978) though many of these isolates were later confirmed to be related to a new genus of Gram-positive spore-forming bacteria, *Sulfobacillus* (Golovacheva and Karavaiko, 1979). One of the notable differences between *Sulfobacillus* spp. and a second genus of moderately thermophilic iron-oxidizing Gram-positive bacteria, *Acidimicrobium* (the earliest isolates of which were also isolated in the 1970s), and *Acidithiobacillus* and *Leptospirillum* spp. is that the former are facultative rather than obligate autotrophs, and their growth is enhanced by organic compounds, such as yeast extract.

Since the 1980s, many new genera and species of acidophilic prokaryotes have been described. These include specialized species that, like *Leptospirillum* spp., have limited metabolic potential and generalist species that, like *Sulfobacillus*, are able to use a wide range of electron donors and acceptors. In addition, species of obligately heterotrophic acidophilic archaea and bacteria have been described that catalyse the dissimilatory oxidation of ferrous iron (e.g. '*Ferroplasma acidarmanus*' and *Ferrimicrobium acidiphilum*) or reduced sulfur (e.g. *Acidicaldus organivorans*) and which, therefore, at least in theory, contribute to mineral dissolution in extremely acidic environments.

Because of the relative ease of observing and, in many cases, identifying them in environmental samples, eukaryotic microorganisms have been known to thrive in many moderately to extremely acidic environments for many years. In a pioneering report published in 1938, James Lackey noted that protozoa and micro-algae were abundant in acidic mine waters that formed in the vicinity of coal mine in eastern USA. He noted that rotifers were present, but prokaryotic blue-green 'algae' were absent, and also that fungi were relatively scarce in the streams he examined. In the same report, Lackey described stones, sticks and leaves in mine water streams being extensively colonized by the phototrophic protozoan *Euglena mutabilis*. *Euglena* spp. in general, and *E. mutabilis* is particular, are now known to be among the most widely distributed acidophilic phototrophs in streams draining metal mines as well as coal mines. Joseph (1953) also reported that euglenoids were abundant in the acid streams he examined, as were fungi (e.g. *Trichoderma*) and the diatom *Navicula viridis*. As with prokaryotic microorganisms, later studies have both confirmed these first reports and expanded the known biodiversity of acidophilic eukaryotes.

The introduction of the polymerase chain reaction and associated DNA- and RNA-based methodologies revolutionized the study of acidophiles as elsewhere in microbiology. In one of the first applications of this developing area of molecular biology, Lane *et al.* (1992) established the phylogenetic relationships between 37 species and strains of iron- and sulfur-oxidizing bacteria, including 19 acidophiles, based on their partial 16S rRNA sequences. They highlighted the important fact that acidophilic prokaryotes have a widespread distribution in the 'tree of life', and also that classification based on physiological trends alone had resulted in some inaccuracies (e.g. '*T. acidophilus*', as referred to above, and '*T. ferrooxidans*' m1, which was noted to be not a *Thiobacillus*/*Acidithiobacillus* sp., and was reclassified almost 20 years later as the new genus/species *Acidiferrobacter thiooxydans* by Hallberg *et al.*, 2011). Classification and identification of acidophiles based on their 16S rRNA (gene) sequences has since become routine, though the use of multiple locus sequence analysis and, increasingly, whole-genome sequences, are emerging as more powerful and accurate approaches to determine the phylogeny of acidophilic microorganisms.

In the 1990s the cloning and sequencing of 16S rRNA genes from pure cultures and environmental samples begun to be used to study the microbial diversity in natural acidic environments, commercial bioleaching processes and acid mine drainages (e.g. Edwards *et al.*, 1999; Goebel and Stackebrandt, 1994). These molecular diversity inventories revealed that the acidic econiches had far less diverse populations than marine, soil and hot spring environments, with both smaller numbers of species and less clonal sequence heterogeneity. Low biodiversity was attributed to the selective nature of acidic biotopes (in terms of pH, redox potential, concentrations of dissolved metals, etc.). Evidence obtained from these and other studies using restriction profiles of 16S rRNA genes and denaturing gradient gel electrophoresis analysis of 5S rRNA genes, revealed that the sequence clones were closely related to known cultured acidophiles and very few if any 'unknown' microorganisms appeared to be present in these acidic econiches. Since then greater understanding of the diversity of organisms present in acidic environments and of their variations in time and space has been obtained through other types of biomolecular studies, including fluorescent *in situ* hybridization (FISH), terminal/restriction enzyme fragment length polymorphism (T-RFLP) analysis, real-time PCR and microarray analysis.

The 2000s saw the advent of next generation sequencing technologies and the radical change in the data generation opportunities brought about by them. The metagenomic analysis of acidic econiches begun with the seminal work of Jill Banfield and colleagues on the microbial communities in AMD in the Richmond Mine tunnels (Tyson *et al.*, 2004). Over the last two decades, and thanks to the relatively low biological and geochemical complexity of this ecosystem, an in-depth characterization and quantitative analysis of the associated microbial communities was achieved. Not only have complete or near-complete genomes of dominant AMD microbes been successfully reconstructed (e.g. Goltsman *et al.*, 2009) but multiple levels of variation between and within the microbial consortia have been revealed and characterized (e.g. Simmons *et al.*, 2008). Community genomics of the Iron Mountain system went far beyond previous gene marker surveys to provide for the first time evidence on the virus–host interactions and the patterns of viral and host distribution over time and space (Andersson and Banfeld, 2008). In addition, several novel taxa were identified and characterized in varying extent thanks to this approach, including a new group of acidophilic nanoarchaea referred to as 'ARMAN' and completely overlooked in other environmental studies before the advent of metagenomics (Baker *et al.*, 2006). Several other metagenomic studies of acidic econiches have followed over the years, including recent integrated community and organism-wide metagenomic and transcriptome analyses. These studies have begun to explain the emerging patterns of variation in the microbial communities controlling geochemical cycling and acid generation in these biotopes, and their correlations with seasonal and spatial fluctuations in geochemical and environmental conditions.

Tandem mass spectrometry-based proteomic (metaproteomic) approaches were also first applied to the study of the natural microbial communities at the Richmond Mine site (Ram *et al.*, 2005). The proteomic research conducted allowed the assessment of the protein inventories of the member organisms in the AMD communities and helped identify relevant differences in physiology of genotypic variants as a function of environment type. These studies were foundational, demonstrating that it was possible to track variations in the proteomic responses of multiple coexisting microorganisms *in situ* and that clues to the functional contribution of the community members could be readily identified (e.g. Wilmes *et al.*, 2009). Also, in the early 2010s, a number of metabolomic studies using stable isotope labelling coupled with targeted or untargeted high-resolution mass spectrometry have been performed on acidophilic communities or acidophile bacterial models, offering further insights into the adaptation strategies and the biology of the organisms that drive geochemical cycles in acidic econiches and related bioprocesses. These studies have enabled, for example, tracking of the carbon flux between a heterotrophic and an autotrophic bacterial species in an artificial mixed culture (Kermer *et al.*, 2012) and provided hints of the metabolites that constitute the sulfur, nitrogen and carbon currencies in natural AMD communities (Moisier *et al.*, 2013).

In parallel, genomic sequencing of a large number of culturable acidophiles covering the three domains of life has provided a robust framework upon which to address fundamental questions related to the biology, the ecology and the evolution of microbial acidophiles. After the public release of the first complete genome of an acidophile archaeon *Thermoplasma acidophilum* (Ruepp *et al.*, 2000), the first draft genome of an acidophile bacteria *At. ferrooxidans* (Selkov *et al.*, 2000) and the first draft genome of an acidophile eukaryote *Cyanidioschyzon merolae* (Matsuzaki *et al.*, 2004) in the early 2000s, a growing number of metabolic reconstructions and physiological studies were performed yielding valuable insight into different aspects of the metabolism and physiology of extreme acidophiles (reviewed in Cárdenas *et al.*, 2010). In the late 2000s and the beginning of the 2010s the first stoichiometric model of an acidophile model bacteria, *At. ferrooxidans*, was obtained (Hold *et al.*, 2009), and the first genome-scale model (GEM) of an acidophile archaea, *Sulfolobus solfataricus*, was reconstructed (Ulas *et al.*, 2012). Use of these models in conjunction with constraint-based methods has begun to allow the simulation of metabolic fluxes induced by different environmental and genetic conditions and the prediction of different cellular metabolic properties. Beyond their relevance for fundamental reasons, both acidithiobacilli bacteria and the sulfolobales archaea are of high interest for industry and biotechnology, and thus these models may prove useful for optimization of biomass production and the generation of relevant enzymes or metabolites.

The application of acidophilic microorganisms in the main biotechnology in which they are involved, the bio-processing of mineral ores and concentrates – generally referred to as biomining – has paralleled that of more fundamental research (Brierley and Brierley, 2013). Although the first purposeful application of mineral-oxidizing bacteria to extract a metal (copper) from a low-grade run-of-mine ore was set up around 15 years after the description of the first iron-oxidizing acidophile (*At. ferrooxidans*), bacterial leaching of metal ores had unknowingly been harnessed as a means of extracting metals since at least the middle ages (Johnson, 2014). Biomining has developed from a relatively basic technology for leaching metals from waste rocks to much more complex engineered bio-heaps and stirred tanks. Although most applications use consortia of mesophilic and thermo-tolerant acidophiles, there has been one successful pilot-scale demonstration using a thermoacidophilic archaeal consortium to bioleach the recalcitrant, but abundant, copper mineral chalcopyrite at ~80°C (Batty and Rorke, 2006). Future developments in biotechnologies that utilize acidophiles will quite probably harness their abilities in areas beyond oxidative mineral oxidation, such as bio-mineralization, organic matter metabolism and reductive bio-processing.

Looking forward: challenges and opportunities

Whilst knowledge of the microbiology of acidic environments has advanced greatly in recent decades, there are still many areas that are poorly understood and questions that remain to be answered. Some are fundamental, such as why most acidophiles are unable to grow in circumneutral pH and, in many cases, mildly acidic waters, and why the upper temperature limit for extremely acidophilic prokaryotes is some 30–40°C lower than those of their neutrophilic counterparts. Cultivation-independent studies have shown that there are unknown species of bacteria and very many unknown species of archaea, living in aerobic and anaerobic low-pH environments (e.g. Baker *et al.*, 2006). The roles of these uncultivated prokaryotes in low-pH environments can only be speculated upon (e.g. Justice *et al.*, 2012). In many cases, the very distant relationships between these uncultivated prokaryotes and characterized species of acidophilic bacteria and archaea (from comparison of their 16S rRNA gene sequences) means that it is not possible to imply or sometimes even give a reasonable guess at the physiological traits of these uncultivated acidophiles. Improved techniques and probably radically novel approaches are needed in order to isolate acidophilic archaea, in particular, from environmental samples, and to cultivate them *in vitro*. This was illustrated in the case of the iron-oxidizing chemolithotroph '*Ferrovum myxofaciens*' (Johnson *et al.*, 2014). Although this bacterium has global distribution in acidic ferruginous waters of pH 2–4, and is frequently the dominant bacterium in large-scale 'acid streamer' growths in these waters, it was not detected until the late 1990s and was not isolated until a modified 'overlay' solid medium was devised, in 2006. Novel acidophilic isolates with hitherto unknown physiological characteristics could open up new opportunities in technologies that use these microorganisms, such as the halotolerant species that are also effective at degrading minerals, which could be harnessed for biomining in areas where only brackish or saline waters are available. There is also the ongoing challenge to isolate and characterize species of hyper-acidophilic prokaryotes, to encompass genera other than *Picrophilus*, and possibly setting new limits for the lowest pH at which organisms can grow.

Genomic analysis of isolates and whole communities, together with other 'omic' approaches, have provided new resources to overcome some of these limitations, whilst providing an in-depth look at the biology of acidophiles and potential for interaction with other microorganisms and with extreme acidic environments. Comprehensive gene lists and catalogues of metabolic pathways have been collected for a number of model acidophiles and acidophilic communities, providing blueprints of their functional potential and task partitioning. Yet, compared with other ecosystems, genomic information of acidophiles is still 'scarce'. Some branches of the tree of life are undersampled (e.g. Johnson *et al.*, 2014), and, perhaps with the exception of *Sulfolobus islandicus* (e.g. Reno *et al.*, 2009) and the group II leptospirilli, few genomes are resolvable at the strain level with the currently existing genomic data. Additional sequencing efforts are required to enable comparative genomic studies that may yield valuable insight into genome architecture, dynamics and evolution. Despite the utility of single cell genomics to study unculturable or 'difficult to culture' microorganisms, no efforts of the kind have been reported in the literature to date, in the case of acidophiles.

Only a few model acidophiles have been thoroughly analysed using metabolic reconstruction and metabolic modelling tools, despite the clear biotechnological applications of many of these microorganisms and an obvious interest in modelling their metabolic fluxes. This is partly due to scarcity of biochemical and metabolic information available on acidophiles in the general and specific literature, compared to well-studied neutrophilic counterparts (e.g. *Escherichia coli*). Much room is still available for proteomic, metabolomic, lipidomic and glycomic analysis of model acidophiles. With the notable exceptions of the Iron Mountain system in the USA (Denef *et al.*, 2010), the Cae Coch abandoned mine system in northern Wales (Kimura *et al.*, 2011) and the Rio Tinto in Spain (Amils *et al.*, 2014), general understanding of the role of prevailing geochemical and environmental factors in shaping diversity patterns in acidophilic communities is still lacking. Further efforts to comprehensively characterize microbial communities inhabiting other acidic econiches are bound to reveal broad trends of microbial distribution and rules of adaptation to acidic environments.

Signatures of acidophilia have been clearly identified in archaea (e.g. high ratio of secondary over primary transport systems; Angelov and Lieb, 2006), some of which are shared with bacteria. In both domains of life, adaptation to extreme acidic conditions is achieved by the concerted action of several mechanisms that contribute to active pH homeostasis through proton exclusion, consumption, exchange and neutralization and by mitigation strategies including cytoplasmic buffering and repair of the damage produced by proton influx (Slonczewski *et al.*, 2009). However, minimal common strategies and specific adaptations of acidophilic bacteria and archaea are poorly explored. More genomes and further comparative genomic analyses are required to search for minimal signatures of acidophilia and to uncover the evolutionary forces acting on microorganisms present in acidic environments. Relevance of these studies extends beyond the understanding of the physiology of extreme acidophiles, touching upon two fields of fundamental interest, the early evolution of life on earth and the study of astrobiology. Further understanding of the transmissible, integrative and translocative elements that populate the genomes of most sequenced acidophiles is also required to understand, the impact of horizontal gene transfer and flexible gene pools on genome evolution and adaptation to extremely acidic econiches. An overwhelming example of the relevant role of horizontal gene transfer in the adaptation to these extreme conditions comes from comparative genomic analysis of the *Thermoplasmatales* and the *Sulfolobales* archaea. *Thermoplasma acidophilum* and *Picrophilus torridus* share approximately 65% of their genes between them, as they do with *S. solfataricus*, a very distant relative thermoacidophile, whereas only 35% of their genes are present in *Pyrococcus furiosus*, a closer relative (Fütterer *et al.*, 2004). Frequent genetic input via horizontal gene transfer during evolution of these archaea seems to have contributed to the evolution of the microorganisms enduring some of most extreme conditions in our planet. From a different perspective, this line of research is likely to be conductive to a more comprehensive understanding of the limitations for genetic manipulation of acidophiles and to open new opportunities for the development of well necessitated tools to transform and further advance functional investigation of acidophiles.

References

Amils, R., Fernández-Remolar, D., and the IPBSL Team (2014). Río Tinto: A geochemical and mineralogical terrestrial analogue of Mars. Life (Basel) *4*, 511–534.

Andersson, A.F., and Banfield, J.F. (2008). Virus population dynamics and acquired virus resistance in natural microbial communities. Science *320*, 1047–1050.

Angelov, A., and Liebl, W. (2006). Insights into extreme thermoacidophily based on genome analysis of *Picrophilus torridus* and other thermoacidophilic archaea. J. Biotechnol. *126*, 3–10.

Baker, B.J., Tyson, G.W., Webb, R.I., Flanagan, J., Hugenholtz, P., Allen, E.E., and Banfield, J.F. (2006). Lineages of acidophilic archaea revealed by community genomic analysis. Science *314*, 1933–1935.

Batty, J.D., and Rorke, G.V. (2006). Development and commercial demonstration of the BioCOP™ thermophile process. Hydrometallurgy *83*, 83–89.

Blowes, D.W., Ptacek, C.J., Jambor, J.L., and Weisener, C.G. (2003). The Geochemistry of Acid Mine Drainage. In Treatise on Geochemistry, Volume 9, Holland, H.D., and Turekian, K.K., eds (Elsevier, Amsterdam, the Netherlands), pp. 149–204.

Brierley, C.L. (2008). How will biomining be applied in future? Trans. Nonferrous Met. Soc. China *18*, 1302–1310.

Brierley, C.L., and Brierley, J.A. (1973). A chemoautotrophic and thermophilic microorganism isolated from an acid hot spring. Can. J. Microbiol. *19*, 183–188.

Brierley, C.L., and Brierley, J.A. (2013). Progress in bioleaching: part B: applications of microbial processes by the minerals industries. Appl. Microbiol. Biotechnol. *97*, 7543–7552.

Brierley, J.A. (1966). Contribution of Chemoautotrophic Bacteria to the Acid Thermal Waters of the Geyser Springs Group in Yellowstone National Park. Ph.D. Thesis, Montana State University, Bozeman, MT, USA.

Brierley, J.A., Norris, P.R., Kelly, D.P., and Le Roux, N.W. (1978). Characteristics of a moderately thermophilic and acidophilic iron-oxidizing *Thiobacillus*. Eur. J. App. Microbiol. *5*, 291–299.

Brock, T.D., Brock, K.M., Belly, R.T., and Weiss, R.L. (1972). *Sulfolobus*: a new genus of sulfur-oxidizing bacteria at low pH and high temperature. Arch. Microbiol. *84*, 54–68.

Cárdenas, J.P., Valdés, J., Quatrini, R., Duarte, F., and Holmes, D.S. (2010). Lessons from the genomes of extremely acidophilic bacteria and archaea with special emphasis on bioleaching microorganisms. Appl. Microbiol. Biot. *88*, 605–620.

Colmer, A.R., and Hinkle, M.E. (1947). The role of microorganisms in acid mine drainage: a preliminary report. Science *106*, 253–256.

Coram, N.J., and Rawlings, D.E. (2002). Molecular relationship between two groups of the genus *Leptospirillum* and the finding that *Leptospirillum ferriphilum* sp. nov. dominates South African commercial biooxidation tanks that operate at 40°C. Appl. Environ. Microbiol. *68*, 838–845.

Darland, G., Brock, T.D., Samsonoff, W., and Conti, S.F. (1970). A thermophilic, acidophilic mycoplasma isolated from a coal refuse pile. Science *170*, 1416–1418.

Denef, V.J., Mueller, R.S., and Banfield, J.F. (2010). AMD biofilms: using model communities to study microbial evolution and ecological complexity in nature. ISME J. *4*, 599–610.

Dold, B. (2010). Basic concepts in environmental geochemistry of sulfidic mine-waste management. In Waste Management, Sunil Kumar, E., ed. (InTech publishers, Rijeka, Croatia), pp. 173–198.

Dold, B., Gonzalez-Toril, E., Aguilera, A., Lopez-Pamo, E., Cisternas, M.E., Bucchi, F., and Amils, R. (2013). Acid rock drainage and rock weathering in Antarctica: important sources for iron cycling in the Southern Ocean. Environ. Sci. Technol. *47*, 6129–6613.

Edwards, K.J., Goebel, B.M., Rodgers, T.M., Schrenk, M.O., Gihring, T.M., Cardona, M.M., Hu, B., McGuire, M.M., Hamers, R.J., Pace, N.R., *et al.* (1999). Geomicrobiology of pyrite (FeS2) dissolution: a case study at Iron Mountain, California. Geomicrobiol. J. *16*, 165–179.

Falagan, C., Sanchez-Espana, J., and Johnson, D.B. (2014). New insights into the biogeochemistry of extremely acidic environments revealed by a combined cultivation-based and culture-independent study of two stratified pit lakes. FEMS Microbiol. Ecol. *87*, 231–243.

Golovacheva, R.S., and Karavaiko, G.I. (1979). *Sulfobacillus* – a new genus of spore-forming thermophilic bacteria. Microbiology (English translation of Mikrobiologiya) *48*, 658–665.

Goltsman, D.S.A., Denef, V.J., Singer, S.W., VerBerkmoes, N.C., Lefsrud, M., Mueller, R.S., Dick, G.J., Sun, C.L., Wheeler, K.E., Zemla, A., *et al.* (2009). Community genomic and proteomic analyses of chemoautotrophic iron oxidizing "*Leptospirillum rubarum*" (group II) and "*Leptospirillum ferrodiazotrophum*" (group III) bacteria in acid mine drainage biofilms. Appl. Environ. Microbiol. *75*, 4599–4615.

Guay, R., and Silver, M. (1974). *Thiobacillus acidophilus* sp. nov.: isolation and some physiological characteristics. Can. J. Microbiol. *21*, 281–288.

Hallberg, K.B., González-Toril, E., and Johnson, D.B. (2010). *Acidithiobacillus ferrivorans* sp. nov.; facultatively anaerobic, psychrotolerant, iron- and sulfur-oxidizing acidophiles isolated from metal mine-impacted environments. Extremophiles *14*, 9–19.

Hallberg, K.B., Hedrich, S., and Johnson, D.B. (2011). *Acidiferribacter thiooxydans,* gen. nov. sp. nov.; an acidophilic, thermo-tolerant, facultatively anaerobic iron- and sulfur-oxidizer of the family *Ectothiorhodospiraceae*. Extremophiles *15*, 271–279.

Harrison, A.P. Jr (1981). *Acidiphilium cryptum* gen. nov., sp. nov., heterotrophic bacterium from acidic mineral environments. Int. J. Sys. Bacteriol. *31*, 327–332.

Hold, C., Andrews, B.A., and Asenjo, J.A. (2009). A stoichiometric model of *Acidithiobacillus ferrooxidans* ATCC 23270 for metabolic flux analysis. Biotechnol. Bioeng. *102*, 1448–1459.

Johnson, D.B. (2007). Physiology and ecology of acidophilic microorganisms. In Physiology and Biochemistry of Extremophiles, Gerday, C., and Glansdorff, N., eds. (American Society of Microbiology Press, Washington, DC, USA), pp. 257–270.

Johnson, D.B. (2012). Geomicrobiology of extremely acidic subsurface environments. FEMS Microbiol. Ecol. *81*, 2–12.

Johnson, D.B. (2014). Biomining – biotechnologies for extracting and recovering metals from ores and waste materials. Curr. Opin. Biotechnol. *30*, 24–31.

Johnson, D.B., and Aguilera, A. (2015). The microbiology of extremely acidic environments. In Manual of Environmental Microbiology, 4th edn, Yates, M.V., Nakatsu, C.H., Miller, R.V., and Pillai, S.D., eds. (American Society of Microbiology Press, Washington, DC, USA), pp. 4.3.1-1–4.3.1-24 (in press).

Johnson, D.B., and McGuiness, S. (1991). Ferric iron reduction by acidophilic heterotrophic bacteria. Appl. Environ. Microbiol. *57*, 207–211.

Johnson, D.B., Hallberg, K.B., and Hedrich, S. (2014). Uncovering a microbial enigma: isolation and characterization of the streamer-generating, iron-oxidizing acidophilic bacterium, "*Ferrovum myxofaciens*". Appl. Environ. Microbiol. *80*, 672–680.

Joseph, J.M. (1953). Microbiological study of acid mine waters; preliminary report. Ohio J. Sci. *53*, 123–127.

Justice, N.B., Pan, C., Mueller, R., Spaulding, S.E., Shah, V., Sun, C.L., Yelton, A.P., Miller, C.S., Thomas, B.C., Shah, M., *et al.* (2012). Heterotrophic archaea contribute to carbón cycling in low-pH suboxic biofilm communities. Appl. Environ. Microbiol. *78*, 8321–8330.

Kelly, D.P., and Tuovinen, O.H. (1972). Recommendation that the names *Ferrobacillus ferrooxidans* Leathen and Braley and *Ferrobacillus sulfooxidans* Kinsel be recognised as synonyms of *Thiobacillus ferrooxidans* Temple and Colmer. Int. J. Syst. Bacteriol. *22*, 170–172.

Kelly, D.P., and Wood, A.P. (2000). Reclassification of some species of *Thiobacillus* to the newly designated genera *Acidithiobacillus* gen. nov., *Halothiobacillus* gen. nov., and *Thermothiobacillus* gen. nov. Int. J. Syst. Evol. Microbiol. *50*, 511–516.

Kermer, R., Hedrich, S., Taubert, M., Baumann, S., Schlömann, M., Johnson, D.B., von Bergen, M., and Seifert, J. (2012). Elucidation of carbon transfer in a mixed culture of *Acidiphilium cryptum* and *Acidithiobacillus ferrooxidans* using protein-based stable isotope probing. JIOMICS 2, 37–45.

Kimura, S., Bryan, C.G., Hallberg, K.B., and Johnson, D.B. (2011). Biodiversity and geochemistry of an extremely acidic, low-temperature subterranean environment sustained by chemolithotrophy. Environ. Microbiol. *13*, 2092–2104.

Kinsel, N. (1960). New sulfur-oxidizing iron bacterium: *Ferrobacillus sulfooxidans*. J. Bacteriol. *80*, 628–632.

Lackey, J.B. (1938). Aquatic life in waters polluted by acid mine waste. Public Health Report *54*, 740–746.

Lane, D.J., Harrison, A.P. Jr, Stahl, D., Pace, B., Giovannoni, S.J., Olsen, G.J., and Pace, N.R. (1992). Evolutionary relationships among sulfur- and iron-oxidizing eubacteria. J. Bacteriol. *174*, 269–278.

Langdahl, B.R., and Ingvorsen, K. (1997). Temperature characteristics of bacterial iron solubilisation and ^{14}C assimilation in naturally exposed sulfide ore material at Citronen Fjord, Greenland (83°N). FEMS Microbiol. Ecol. 23, 275–283.

Leathen, W.W., Kinsel, N.A., and Braley, S.R. (1956). *Ferrobacillus ferrooxidans*: a chemosynthetic autotrophic bacterium. J. Bacteriol. 72, 700–704.

Markosyan, G.E. (1972). A new iron-oxidizing bacterium, *Leptospirillum ferrooxidans* gen. et sp. nov. Biol. J. Armenia. *25*, 26–29.

Matsuzaki, M., Misumi, O., Shin, I.T., Maruyama, S., Takahara, M., Miyagishima, S.Y., Mori, T., Nishida, K., Yagisawa, F., Nishida, K., *et al.* (2004). Genome sequence of the ultrasmall unicellular red alga *Cyanidioschyzon merolae* 10D. Nature *428*, 653–657.

Mosier, A., Justice, N., Bowen, B., Baran, R., Thomas, B., Northern, T., and Banfield, J. (2013). Metabolites associated with adaptation of microorganisms to an acidophilic, metal-rich environment identified by stable-isotope-enabled metabolomics. MBio. *4*, e00484–12.

Quatrini, R., Appia-Ayme C., Denis, Y., Ratouchniak, J., Veloso, F., Valdes, J., Lefimil, C., Silver, S., Roberto, F., Orellana, O., *et al.* (2006). Insights into the iron and sulfur energetic metabolism of *Acidithiobacillus ferrooxidans* by microarray transcriptome profiling. Hydrometallurgy *83*, 263–272.

Ram, R.J., VerBerkmoes, N., Thelen, M.P., Tyson, G.W., Baker, B.J., Shah, M., Blake, R.C. II, Hettich, R., and Banfield, J.F. (2005). Community proteomics of a natural microbial biofilm. Science *308*, 1915–1920.

Reno, M.L., Held, N.L., Fields, C.J., Burke, P.V., and Whitaker, R.J. (2009). Biogeography of the *Sulfolobus islandicus* pan-genome. Proc. Natl. Acad. Sci. U.S.A. *106*, 8605–8610.

Ruepp, A., Graml, W., Santos-Martinez, M.L., Koretke, K.K., Volker, C., Mewes, H.W., Frishman, D., Stocker, S., Lupas, A.N., and Baumeister, W. (2000). The genome sequence of the thermoacidophilic scavenger *Thermoplasma acidophilum*. Nature *407*, 508–513.

Schleper, C., Puehler, G., Holz, I., Gambacorta, A., Janekovic, D., Santarius, U., Klenk, H.-P., and Zillig, W. (1995). *Picrophilus* gen. nov., fam. nov.: a novel aerobic, heterotrophic, thermoacidophilic genus and family comprising archaea capable of growth around pH 0. J. Bacteriol. *177*, 7050–7059.

Selkov, E., Overbeek, R., Kogan, Y., Chu, L., Vonstein, V., Holmes, D., Silver, S., Haselkorn, R., and Fonstein, M. (2000). Functional analysis of gapped microbial genomes: amino acid metabolism of *Thiobacillus ferrooxidans*. Proc. Natl. Acad. Sci. U.S.A. *97*, 3509–3514.

Simmons, S.L., Dibartolo, G., Denef, V.J., Goltsman, D.S., Thelen, M.P., and Banfield, J.F. (2008). Population genomic analysis of strain variation in *Leptospirillum* group II bacteria involved in acid mine drainage formation. PLoS Biol. *6*, e177.

Slonczewski, J.L., Fujisawa, M., Dopson, M., and Krulwich, T.A. (2009). Cytoplasmic pH measurement and homeostasis in bacteria and archaea. Adv. Microb. Physiol. *55*, 1–79.

Steudel, R. (2000). The chemical sulfur cycle. In Environmental Technologies to Treat Sulfur Pollution: Principles and Engineering, Lens, P., and Hulshoff Pol, L., eds (International Association on Water Quality, London, UK), pp. 1–31.

Temple, K.L., and Colmer, A.R. (1951). The autotrophic oxidation of iron by a new bacterium: *Thiobacillus ferrooxidans*. J. Bacteriol. *62*, 605–611.

Tyson, G.W., Chapman, J., Hugenholtz, P., Allen, E.E., Ram, R.J., Richardson, P.M., Solovyev, V.V., Rubin, E.M., Rokhsar, D.S., and Banfield, J.F. (2004). Community structure and metabolism through reconstruction of microbial genomes from the environment. Nature *428*, 37–43.

Ulas, T., Riemer, S.A., Zaparty, M., Siebers, B., and Schomburg, D. (2012). Genome-scale reconstruction and analysis of the metabolic network in the hyperthermophilic archaeon *Sulfolobus solfataricus*. PloS One *7*, e43401.

Vera, M., Schippers, A., and Sand, W. (2013). Progress in bioleaching: fundamentals and mechanisms of bacterial metal sulfide oxidation – part A. Appl. Microbiol. Biotechnol. *97*, 7529–7541.

Waksman, S.A., and Joffe, J.S. (1922). Microorganisms concerned in the oxidation of sulfur in the soil. II. *Thiobacillus thioooxidans*, a new sulfur-oxidizing organism. J. Bacteriol. *7*, 239–256.

Wilmes, P., Remis, J.P., Hwang, M., Auer, M., Thelen, M.P., and Banfield, J.F. (2009). Natural acidophilic biofilm communities reflect distinct organismal and functional organization. ISME J. *3*, 266–270.

Part II

Challenges and Adaptations

Energy Acquisition in Low-pH Environments

2

Wolfgang Nitschke and Violaine Bonnefoy

Abstract

Extreme acidophiles thriving in acidic habitats are characterized by low organic substrate inputs and high concentrations in inorganic substances, in particular sulfur compounds and iron. Their strategies for harvesting energy for growth in these harsh environments are described in this chapter. An overview of the general concepts on bioenergetics is given to highlight the particular challenges faced by extreme acidophiles. Pathways involved in oxidation and reduction of iron and sulfur compounds are presented and discussed. From the available data, it is obvious that, for the same electron donor and acceptor, the metabolic pathways differ sufficiently between different prokaryotes to suggest independent origins followed by diversification and horizontal gene transfers.

Introduction

In contrast to other extremophiles, the habitat wherein extreme acidophiles thrive is often not imposed by the environment but is a result of microbial metabolism itself: the oxidation of inorganic sulfur compounds to sulfuric acid generates the most acidic environments found on our planet. This harsh biotope is not only characterized by its acidic pH, but also by low amounts of organic compounds and by elevated concentrations of soluble metal(oid)s, many of which are highly toxic. Owing to these constraints, the primary producers in acidic ecosystems are microorganisms able to utilize inorganic compounds as carbon (CO_2) and energy sources, i.e. the chemolithoautotrophs. These chemolithoautotrophs provide organic materials (exudates from actively metabolizing and from dead cells) to the mixotrophs (using inorganic compounds as energy source and organic substances as carbon source) and to the heterotrophs (utilizing organic compounds both as carbon and energy sources) that, in turn, deplete these potentially toxic compounds for the chemolithoautotrophs.

A few general concepts in bioenergetics are required to understand energy conversions of acidophiles. The energy sources for bioenergetic mechanisms, i.e. the electron donors and acceptors, can either be environmental chemicals (organic or inorganic) or be derived from light-induced charge separation. The utilization of chemical energy (chemotrophy) involves oxidation-reduction (redox) reactions, that is, the transfer of electrons from an electron donor to an electron acceptor of higher redox potential. This reaction will be exergonic (energy-producing) while the electron transfer from an electron donor to an electron acceptor of lower redox potential will be endergonic (consumption of energy). The greater the difference between reduction potentials between the electron donor and the electron acceptor, the more energy will be released (in exergonic reactions) or required (in endergonic reactions) for each reducing equivalent transferred. This energy is conserved by the formation of high-energy phosphate compounds such as ATP formed from ADP. ATP hydrolysis to adenosine diphosphate (ADP) or adenosine monophosphate (AMP) liberates a large amount of free energy (approximately −12 kcal/mol under typical cellular conditions).

There are two major processes for the conservation of energy: fermentation and respiration. Fermentation is taking place in the absence of an external electron acceptor. In that case, an organic compound generated from catabolism is converted into a simpler compound. ATP is produced by substrate-level phosphorylation through specific enzymatic steps. Only a small amount of energy is released, most of the energy remaining in the fermentation products. Another way to gain energy is through respiration by energy transduction in the cytoplasmic membrane. Electrons are transferred from reducing electron donors to oxidizing electron acceptors. The electron carriers are arranged asymmetrically within the cytoplasmic membrane. A combination of membrane-asymmetric redox reactions (oxidation and reduction taking place on opposite sides of the cytoplasmic membrane) and true proton extrusion outside the cells establishes potential energy in the form of a chemical proton gradient (ΔpH) and a trans-membrane electrical potential ($\Delta\psi$) across the membrane. These forces combine to give the proton motive force (Δp): $\Delta p = \Delta\psi - 2.3(RT/F)\Delta\text{pH}$ (R is the gas constant, T the absolute temperature and F the Faraday constant). The carefully controlled dissipation of this proton gradient releases energy through the ATP synthase, which synthesizes ATP, or through other energy requiring processes. This process is called oxidative phosphorylation.

The respiratory processes of acidophiles are similar to those of neutrophiles. The very large ΔpH – the internal pH is maintained at pH 6.0–7.0 over an outside pH range of 1.0–8.0 (see below and references in Ingledew, 1982) – is compensated by inverting the $\Delta\psi$ from negative inside to positive inside, by active influx systems for cations, e.g. K^+ (Albers *et al.*, 2001). This confers to the acidophiles some protection against cations, in particular to metals that are often present in high concentrations in acidic environments, but also sensitivity to anions, such as nitrate (Alexander *et al.*, 1987). While their biotopes are drastic (low pH and high concentration of metal(oid)s) and scarce in potential organic energy sources, extreme acidophiles are able to extract energy and reducing power to thrive there from a variety of sources (see Table 2.1 for a non-exhaustive list of the inorganic electron donors and acceptors used by acidophiles). However, the standard midpoint potential (E_m) of some couples is pH dependent. In particular, the $E_{m,7}$ for the O_2/H_2O couple is +820 mV while its $E_{m,2}$ is +1120 mV. Please note that standard conditions are defined at 1 M concentrations of reactants or at 1 atmosphere in the case of gases. Thus, the only electron acceptor for exergonic ferrous iron (Fe^{2+}) oxidation is O_2, since $E_{m,2}$ for Fe^{2+}/Fe^{3+} is +680 to +770 mV in acidic sulfate (SO_4^{2-}) solutions. This oxidative reaction will bring little energy since the difference between the $E_{m,2}$ is low. Nevertheless, bacteria need to reconstitute their reducing pool (NADH) for anabolic processes such as CO_2 or N_2 fixation. The $E_{m,7}$ of the NAD^+/NADH couple at the pH of the cytoplasm is −320 mV. Therefore, the electrons have to be 'pushed uphill' from Fe^{2+} to NAD^+ against the redox potential. This 'reverse' electron transfer needs energy that is provided by the proton motive force.

These aspects of the bioenergetics, specific to the extreme acidophiles, will be analysed and discussed in this chapter. Also, the strategies to gain energy for growth in this harsh environment and the known respiratory pathways employed by these particular microorganisms will be described and compared.

General concepts of bioenergetics

All life depends on the maintenance of two basic disequilibria to drive metabolic reactions: (a) a high degree of reduction of the essential cofactors (mainly NADH) and (b) a very strong displacement of the ATP/ADP couple towards high ATP levels. NADH is necessary to provide reducing equivalents for a plethora of metabolic reactions whereas high ATP levels are necessary for powering metabolism. Maintenance of high ATP/ADP ratios is generally considered to define the realm of bioenergetics although in many cases keeping the pool of NADH reduced is intimately coupled to the ATP-synthesizing processes. Although pointed out repeatedly in the bioenergetic literature, we remind the reader that no free energy is contained in an individual molecule of NADH or ATP, but that it is the displacement of the NADH/NAD^+ and ATP/ADP ratios from their equilibrium values that serves to drive life's chemical reactions.

ATP synthesis via substrate-level phosphorylation

ATP synthesis via substrate level phosphorylation occurs in the cytoplasm through chemical reactions with a very low energy yield compared to the chemiosmotic systems discussed below. Organism fully relying on substrate level phosphorylation are

Table 2.1 Electron donors and acceptors used by the extreme acidophilic prokaryotes

	Electron donors						Electron acceptors		
Prokaryotes	H_2	H_2S	$S_4O_6^{2-}$	$S_2O_3^{2-}$	S^0	Fe^{2+}	O_2	Fe^{3+}	S^0
Proteobacteria									
Acidicaldus organivorans					+	−	+	+	−
Acidiphilium acidophilum	−		+	+	+	−	+	+	−
‘*Ferrovum myxofaciens*’	−		−		−	+	+		
Acidihalobacter prosperus		+			+	+	+		
Acidiferrobacter thiooxydans	−	+	+		+	+	+	+	−
Acidithiobacillus ferrooxidans	+	+	+	+	+	+	+	+	−[a]
Acidithiobacillus ferridurans	+		+		+	+	+	+	
Acidithiobacillus ferrivorans	−[b]	+	+	+	+	+	+	+	−
Acidithiobacillus thiooxidans	−	+	+	+	+	−	+	−	−
Acidithiobacillus albertensis					+		+		−
Acidithiobacillus caldus	+[c]	+	+	+	+	−	+	−	−
Nitrospira									
Leptospirillum ferrooxidans	−			−	−	+	+	−	−
Leptospirillum ferriphilum	−			−	−	+	+	−	−
‘*Leptospirillum ferrodiazotrophum*’	−			−	−	+	+		
Firmicute									
Sulfobacillus spp.	+		+		+	+	+	+	−
Alicyclobacillus spp.					+[c]	+[c]	+	+[c]	−
Aquificae									
Hydrogenobaculum acidophilum	+				+	−	+	−	−
Actinobacteria									
Acidimicrobium ferrooxidans	+				−	+	+	+	−
Ferrimicrobium acidiphilum	−				−	+	+	+	−
‘*Acidithiomicrobium*’ sp.					+	+	+		
Ferrithrix thermotolerans	−				−	+	+	+	
Euryarchaeota									
Ferroplasma acidiphilum			−	−	−	+	+	+	−
‘*Ferroplasma acidarmanus*’					−	+	+	+	
Acidiplasma cupricumulans			−	−	−	+	+	+	
Crenarchaeota									
Sulfolobus metallicus		+			+	+	+	−	−
Sulfolobus tokodaii					+	+	+	−	−
Acidianus spp.	+	+			+	+	+	+	+
Metallosphaera spp.					+[c]	+	+	−	−
Sulfurococcus spp.					+	+	+	−	−
Stygiolobus azoricus									+
Sulfurisphaera ohwakuensis					+		+		+

[a]One strain has been shown to grow with H_2 as electron donor and S^0 as electron acceptor (Ohmura *et al*., 2002).
[b]One strain has been shown to grow on H_2 (D. Barrie Johnson, College of Natural Sciences, Bangor University, Bangor, UK, personal communication).
[c]Some species.

rare and the molecular bases of the process seem to be evolutionarily younger than those of chemiosmosis (Lane *et al.*, 2010). Fermentative metabolism is apparently rarer among extreme acidophiles than it is in prokaryotes in general. The main reason for this is that low molecular weight organic acids (such as acetic, glycolic and pyruvic acids) are uncharged/undissociated and lipophilic at the low pH of the medium. When they migrate through the cell membrane to the cytoplasm, which has a pH close to neutral, they dissociate. This results in severe acidification of the cytoplasm and cell death. Chemoautotrophs are generally more sensitive to these small molecular weight organic compounds than the mixotrophs and the heterotrophs. In fact, many acidophilic mixotrophs and heterotrophs can use di- and tricarboxylic organic acids (such as citric acid), glutamic acid and complex organic substrates (e.g. yeast extract). So far, the only acidophiles known to ferment an organic polymer are the thermoacidophilic *Crenarchaeota* belonging to the *Acidilobus*, *Caldisphaera* and *Vulcanisaeta* genera which have been shown to ferment carbohydrate substrates (Prokofeva *et al.*, 2005, 2009).

Chemiosmosis

In contrast to the 'direct' generation of ATP achieved by substrate level phosphorylation, chemiosmosis couples electron transfer reactions between environmental electron donors and acceptors to the enzymatic phosphorylation of ADP by the ATP synthase. The coupling agent is an electrical and osmotic disequilibrium over the bioenergetic membrane (i.e. the cytoplasmic membrane in prokaryotes and the organellar one in eukaryotes). The sum of the electrical ($\Delta\psi$) and the osmotic (ΔpH, ΔpNa) contributions is called the proton (or sodium, see below) motive force (pmf), and is in general expressed in millivolts (mV) as $pmf = \Delta\psi - 2.3(RT/F)\Delta pH$. The proton is in the vast majority of cases the ion used to build up the chemiosmotic potential, but in a few species sodium can fulfil a similar role. However, sodium-based bioenergetics has to the best of our knowledge not been found in acidophiles as of yet and will therefore not be dealt with in this contribution. The chemiosmotic membrane potential drives the molecular-motor type ATP synthase located in the bioenergetic membrane and phosphorylating ADP to ATP on the inner, cytoplasmic side of cells (or organelles). As we will see below, it is precisely the bipartite nature of the pmf, *i.e.* being composed of two physico-chemically differing contributions, which renders bioenergetics in acidic environments possible.

The overwhelming majority of species uses electrochemical disequilibria between reducing and oxidizing substrates to drive more or less extended electron transfer chains and thus to generate chemiosmotic membrane potential. These chains are associated to the bioenergetic membrane and coupled to proton translocation, thus generating chemiosmotic potential. With the exception of methanogenesis and acetogenesis, these redox-couple-based mechanisms could be considered 'respiratory' since they 'respire' oxidizing substrates in the environment. All acidophiles follow this scheme of redox-gradient driven bioenergetic mechanisms.

Bioenergetic particularities of the acidophilic lifestyle

Transporting 'standard' chemiosmosis into acidic environments

As outlined above, pmf, the driver of ATP synthesis, has two distinct components, the electrostatic membrane potential resulting from the imbalance of positive and negative charges between both sides of the cytoplasmic membrane, and an osmotic contribution corresponding to a concentration gradient of H^+ (or Na^+) ions. In all neutrophilic species for which pmf has been determined so far, these contributions sum up to roughly 180 mV (negative inside) of pmf over energized membranes of an active microorganism. However, the distribution of these 180 mV on the electrostatic and the osmotic terms, varies between species. In many *Proteobacteria*, it has been found that pmf is essentially contained in the electrostatic membrane potential, meaning that nearly no pH gradient is present. By contrast, in algal or plant chloroplasts and in cyanobacterial thylakoids, up to 2 units of pH difference, between the inside and the outside of the membrane have been reported, together with a correspondingly lower electric field. Acidophiles take this flexibility of dividing total pmf into variable shares of $\Delta\psi$ and ΔpH to the extreme in order to survive in low-pH environments, as will be detailed in what follows.

In extreme acidophiles the extracellular pH imposed by the environment varies typically between 0.5 and 3. With only one currently known exception (the archaeon *Picrophilus torridus*) acidophiles keep their cytoplasmic pH in the neutral range (typically above pH 5.5). The fact that acidophiles maintain near-neutral internal pH values is frequently interpreted, as due to acid lability of many biomolecules. However, it should be noted that acidophiles synthesize many biomolecules that are located in or exposed to the periplasm, which essentially is at the pH of the surrounding environment. A more likely explanation is that acidophiles simply derive from neutrophilic ancestors and have adapted only their extracellular proteome to the hard conditions imposed by their extreme acidic habitat. This scenario is in line with the fact that bioenergetic adaptation to acidic environments appears to have occurred independently several times in the evolution of life.

There have been many reports, some speculative others more factual, on how acidophiles maintain near-neutral cytoplasmic pH values. Most of these focus on ways to keep protons from diffusing across the cytoplasmic membrane. For example, it has been shown that vesicles reconstituted from lipids derived from acidophiles feature comparatively lower proton permeability than vesicles reconstituted from neutrophile-derived lipids (Slonczewski *et al.*, 2009). However, these findings cannot rationalize the maintenance of near-neutral cytoplasmic pH due to a very simple but fundamental chemiosmotic fact: a bioenergetic membrane is made to conduct protons. In every bioenergetic membrane known to date there are built-in proton channels called rotor/stator-type ATP synthases. No cellular entity relying on a chemiosmotic potential for active transport and/or energy conversion can dispose of the rotor/stator-type ATP synthase. While it has been shown that these ATP synthases get blocked in the absence of a pmf, preventing dissipation of the ATP/ADP disequilibrium through proton efflux, in acidophiles, the protons following the gradient from high concentrations outside to lower ones inside would flow exactly as those driving the ATP synthase in the direction of ATP synthesis. It therefore is difficult to imagine how ATP synthases could be regulated as to impede pH equilibration while still doing what they have been selected for.

The most likely explanation is the fundamentally chemiosmotic one proposed by Ingledew and co-workers for *Acidithiobacillus ferrooxidans* (Alexander *et al.*, 1987; Cox *et al.*, 1979). The key to the solution of the problem lies in the bipartite nature of the pmf, which needs to exceed 150 mV (negative inside) for efficient ATP synthesis to occur. As mentioned above, this magnitude of pmf can be attained by variable (but mutually compensating) contributions of $\Delta\psi$ and ΔpH. Acidophiles ingeniously take advantage of this principle. Fig. 2.1 depicts this mechanism in an idealized system. In a starved cell with a non-energized chemiosmotic membrane, the pmf needs to be close to zero so that there is no tendency for protons to enter the cytoplasm. Assuming a ΔpH between the cytoplasm and the environment of about 5 units, the osmotic contribution to pmf will amount to some −300 mV. Attaining zero pmf therefore requires the presence of a counterbalancing electric field over the membrane of roughly the same magnitude but opposite polarity. This can be achieved by the presence of the required concentrations of membrane-impermeant cations in the cytoplasm. Under these conditions, protons will not enter the cytoplasm even in the presence of proton channels since the repulsive force of the positive charges counterbalances the tendency to achieve osmotic equilibrium. If such a bioenergetic membrane then starts to respire, it will generate a pmf which, in order to maintain inside and outside pH unperturbed, should optimally mainly consist of an electrostatic component $\Delta\psi_{et}$ (negative inside), i.e. similar to what happens in neutrophilic *Proteobacteria*. As shown in Fig. 2.1b, this bioenergetic (electron transfer dependent) $\Delta\psi_{et}$ will partially counteract the static $\Delta\psi_{st}$ resulting in an effective $\Delta\psi_{eff}$ of $\Delta\psi_{st} - \Delta\psi_{et}$. Assuming a typical bioenergetic $\Delta\psi_{et}$ of about −180 mV (see above) yields a $\Delta\psi_{eff}$ of about +120 mV and the resulting total pmf will be at around −180 mV, perfectly in line with what is observed in standard neutrophilic systems. The mentioned values are only chosen to illustrate the principle and certainly will vary between organisms. The values determined experimentally for *At. ferrooxidans* (Cox *et al.*, 1979), however, are in the range of those assumed in our idealized system and are indicated in parenthesis in Fig. 2.1b. The validity of the model proposed

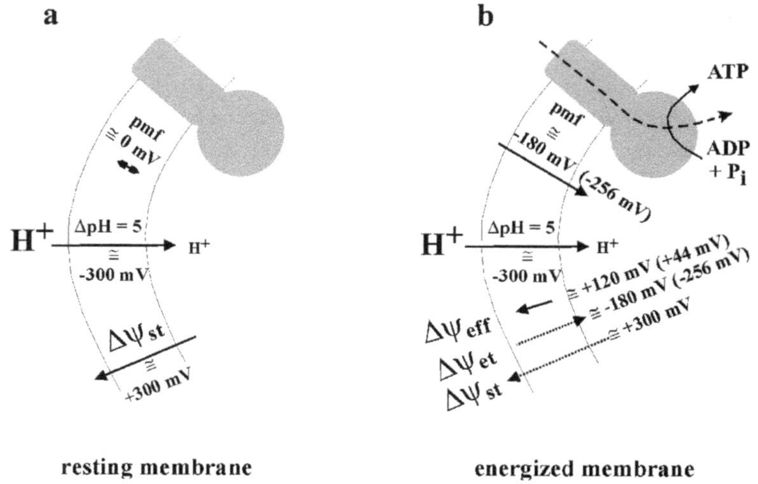

Figure 2.1 Schematic representation of the distribution of proton-motive-force (pmf) onto osmotic (ΔpH) and electrostatic (Δψ) terms yielding (a) a vanishing proton translocation potential (and hence maintenance of a pH gradient) in a non-energized membrane and (b) an ATP synthase-driving pmf under conditions of bioenergetic electron flow. The indicated magnitudes are idealized values considering a typical electron-flow-induced pmf of −180 mV and a pH-difference of 5 units. The numbers in brackets correspond to the pmf value determined experimentally in the Fe^{2+}-oxidizing acidophile *At. ferrooxidans* and the resulting magnitudes of $\Delta\psi_{et}$ (field built up by electron transfer) and $\Delta\psi_{eff}$ (effective field contribution). The grey shape stands for ATP synthase.

by Ingledew and coworkers is in fact strongly corroborated by several observations, including (i) that acidophiles maintain near-neutral cytoplasmic pH values even under conditions of substrate starvation, i.e. when the membrane cannot be kept energized and (ii) that the static pH gradient is readily collapsed by membrane-permeant anions in the medium, such as nitrate or nitrite. Despite the global similarity of the potentials assumed for our idealized system and those actually measured in *At. ferrooxidans*, it should be noted that the measured pmf of −256 mV substantially exceeds the values typically observed in neutrophiles. As discussed below, it appears highly likely that such a high pmf is indeed indispensable for proper functioning of the *At. ferrooxidans* bioenergetic chain and in particular for efficient reverse electron transport towards NAD^+ reduction.

The free energies of inorganic redox couples used by acidophiles are pH dependent

As described above, acidophiles use the same chemiosmotic principle as neutrophiles. The genuinely free energy conserving part of pmf in acidophiles, i.e. Δψ, is generated in the standard way by coupling proton translocation to electron transport and collapsing redox disequilibria between reducing donor and oxidizing acceptor substrates. The standard redox potentials of these substrates, however, may feature dissimilar pH dependences depending on how many protons per electron are involved in the respective redox transition. In a redox event formally described as ($Ox + ne^- + mH^+ \leftrightarrow Red$), the redox potential will decrease by $(m/n) \times 60\,mV$ per pH unit when the pH is lowered. Oxygen reduction to water requires four electrons and four protons, thus yielding a redox potential dependence of −60 mV/pH while NAD^+ reduction to NADH needs two electrons but only one proton and therefore results in a dependence of −30 mV/pH. Therefore, substrate couples viable for energy harvesting at pH 7 are not necessarily so at pH 2 and vice versa. The Fe^{2+}/O_2 couple dealt with in what follows is probably one of the most biologically relevant examples.

Although strongly dependent on its counterions (Bird *et al.*, 2011), iron compounds show a relatively high standard redox midpoint potential which theoretically makes them attractive electron acceptors but very poor electron donors. Most iron compounds have pH-independent standard redox midpoint potentials down to the region of pH 2 and become more oxidizing below their p*K* values. Iron's redox chemistry, however, is complicated by the fact that the solubility of the oxidized form, Fe^{3+}, usually decreases with increasing pH. Therefore, iron's effective redox potential at concentrations above the solubility limit of the reduced form become more reducing at higher pH due to the law of mass-action and the constant removal of the oxidized compound from the reaction mixture through precipitation. The combination of the weak pH dependence of its redox midpoint potential and the strong pH dependence of its effective potential at high concentrations make iron's usefulness as a bioenergetic substrate a complicated issue.

Under a wide range of conditions (type of counterions and concentrations as discussed in the preceding paragraph), the redox span between Fe^{2+} and O_2 at pH 7 will remain inferior to the typical pmf of a growing cell. At low pH values, however, the redox span between the Fe^{3+}/Fe^{2+} and the O_2/H_2O couples exceeds 200 mV and therefore allows proton pumping by an O_2 reductase (cytochrome oxidase). However, as argued by Ferguson and Ingledew (2008), the energy span may still be insufficient for translocation of two protons per transferred electron, which would require redox energies exceeding two times the typical pmf value of 180 mV (~360 mV). Considering the actually measured pmf in *At. ferrooxidans* (Cox *et al.*, 1979), the necessary redox energy would even exceed 512 mV. The O_2 reductases so far found to operate, for example in *At. ferrooxidans* and *Acidithiobacillus ferridurans*, under such growth conditions (Appia-Ayme *et al.*, 1999; Quatrini *et al.*, 2009) belong to the subgroup of haem-copper O_2-reductases. These are called SoxM- (or A-) type and include, for example, the mitochondrial Complex IV. These enzymes typically translocate two protons per electron (Ducluzeau *et al.*, 2014; Rauhamäki and Wikström, 2014; Han *et al.*, 2011). Therefore, the electrochemical driving force of the Fe/O_2 couple appears to be thermodynamically insufficient to drive such an O_2 reductase. As a possible rationalization of this apparent contradiction, Ferguson and Ingledew (2008) noted that key residues involved in proton translocation through one of the two proton channels, the so-called K-channel in O_2 reductases from neutrophiles, are absent in several acidophilic species. It is therefore possible that, in line with thermodynamic constraints, the O_2 reductases in acidophiles pump only one proton per electron through the only functional (so-called D-) channel remaining. This raises the question why the respective iron-oxidizing acidophiles do not simply use other types of O_2 reductases, which may be able to handle lower proton-to-electron stoichiometries than typical SoxM (A)-type enzymes. The so-called SoxB (B)-, SoxC (cbb_3)- or *bd*-type, as well as alternative oxidases (AOX), would all be able to operate at such lower proton-to-electron stoichiometries. It seems likely that the presence of a modified SoxM (A)-type O_2-reductase in the respective species reflects the gene heritage of these organisms' neutrophilic evolutionary ancestors.

As an alternative to Fe^{2+}, both reduced inorganic sulfur compounds (RISCs) and molecular hydrogen (H_2) are frequently found to serve as electron donors for bioenergetic chains in acidophiles. In contrast to iron compounds for which the dependence of redox potential on pH mainly results from differential solubilities and hence is concentration-dependent, all RISCs show redox potential dependences on pH in the range of −60 mV/pH or even exceeding this value. Whereas the redox span between Fe^{2+} and O_2 becomes more favourable at low pH, as outlined above, the span between RISC donors and O_2 remains relatively constant. In turn, Fe^{3+} reduction by RISCs and H_2 becomes less favourable.

NAD^+ reduction at low pH values necessitates reverse electron transfer

At low pH, a distinctive problem arises with respect to reduction of NAD^+. Since this redox cofactor is confined to the circumneutral cytoplasm, its redox potential will correspond to the standard $E_{m,7}$ of −320 mV only somewhat modified by the actual ratio of $NAD^+/NADH$ through mass action. If present at sufficiently high concentrations, several of the RISC compounds and certainly also H_2, are capable of directly reducing NAD^+ at neutral pH values and a number of enzymes catalysing such reduction reactions are known (e.g. NAD^+-reducing [Ni–Fe]-hydrogenases (Lauterbach *et al.*, 2013)). However, at pH 2, even H_2 with its redox midpoint potential of −120 mV, is likely not enough reducing to perform this reaction. NAD^+ reduction, therefore, has to be achieved via the so-called 'reverse electron transfer'. This is, the redox energy of an electron resulting from substrate oxidation must be sufficiently enhanced by the pmf so that its transfer onto NAD^+ becomes exergonic. NAD^+ reduction in reverse electron transfer is ultimately achieved by Complex I operating 'backwards' (if 'forward' is defined in terms of the mitochondrio-centric world of bioenergetics). Considering standard midpoint potentials, the electron transfer from the ubiquinone (UQ) pool (i.e. the ensemble of membrane-diffusing UQ; $E_m = +90$ mV) to NAD^+/NADH ($E_m = -320$ mV) would appear strongly 'uphill'. Whereas 'uphill' electron transfers are indeed transiently feasible if the electron proceeds on to an acceptor with an even more positive redox potential than that of the ultimate donor (e.g. Alric *et al.*, 2006), an uphill electron transfer generating non-equilibrium reduction states of the lower potential acceptor is thermodynamically simply out of the question. This apparent paradox is rationalized by the fact that electron transport and proton translocation in Complex I (as in other chemiosmotically active enzymes) are strictly coupled. Whereas electron transfer from NADH (−320 mV) to UQ (+90 mV) is strongly exergonic if the pmf over the membrane is absent or small, under conditions of a fully energized membrane (assuming a pmf of −180 mV), the driving force becomes small. Since Complex I pumps 2 protons per transferred electron, it has to overcome 2×180 mV of pmf and the redox driving force swivels to a mere 50 mV (320 − 360 + 90) for half-reduced NADH and UQ-pool (at pH 7) and even vanishes when the reduction state of the UQ-pool approaches 90%. Complex I, very much like Complex III (in the following called 'Rieske/cytochrome *b* complex' for the sake of generality), therefore operates close to thermodynamic equilibrium in an energized membrane. This translates into facile reversibility. An at least 90% reduced pool of UQ in such energized conditions can theoretically drive NAD^+ reduction up to 50%. The presence of a full-scale pmf, of course, requires that at least part of the enzymes in the electron transfer chain operate in the forward, i.e. proton-translocating, direction with electrons thermodynamically drawn towards a high potential acceptor substrate.

With respect to the split of electron flow into a pmf-generating pathway towards the terminal acceptor and a pmf-driven one towards NAD^+, two distinct situations can (and do) occur.

Substrate-electron input at the level of the quinone pool

H_2 and RISCs are sufficiently reducing, even at low pH, to directly inject their electrons into the quinone pool. This pathway is shown in Fig. 2.2. In real-life bioenergetic chains, the reduction of the

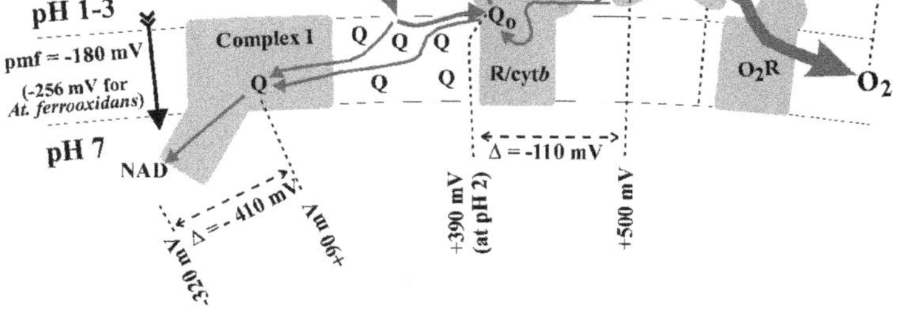

Figure 2.2 Schematic representation of forward and reverse electron flow during aerobic respiration in acidophiles using either (moderately, due to the low pH) reducing (H_2/RISC) or weakly reducing (Fe^{2+}) substrates as electron donors. The thickness of electron flow arrows is meant to indicate the strengths of electron currents in the forward and reverse directions. Q, Rus, c_4, O_2R stand for quinone, rusticyanin, dihaem cytochrome c_4 and oxygen reductase, respectively. Q_o denotes the Q_o-site of the Rieske/cytochrome *b* (marked as R/cyt *b*) complex (the bc_1 complex in the case of *At. ferrooxidans*).

UQ-pool can be performed by very diverse enzyme systems. Group I [Ni–Fe] hydrogenases, for example, contain integral membrane subunits that connect periplasmic oxidation of H_2 to intra-membrane reduction of the UQ-pool (Pandelia *et al.*, 2012). In a very different manner, sulfide-quinone-reductase (SQR) forms a hydrophobic dome on the membrane surface sucking the lipophilic quinone out of the membrane plane and towards the site of H_2S oxidation. In bioenergetic terms, however, the outcome of these strongly differing systems is similar, i.e. reduction of the UQ-pool which can attain substantial levels if high amounts of reducing substrates are present and if the substrate-oxidizing enzymes can turn over at sufficient rates of catalysis. Under these conditions, two proton-pumping enzymes, the Rieske/cyt *b* complex and the terminal reductase (most frequently O_2 reductases) cooperate to generate pmf which then can be used to both drive ATP synthesis and to render NAD^+ reduction by UQH_2 exergonic.

Substrate–electron input at the level of the soluble periplasmic carrier proteins

The vast majority of bioavailable Fe^{2+} compounds feature effective midpoint potentials too high to allow direct injection of electrons into the UQ-pool. *At. ferrooxidans* is arguably the best-studied organism to date, which faces this thermodynamic challenge. As shown in Fig. 2.2, the only remaining option under such circumstances is to inject the substrate-derived reducing equivalents at the level of the periplasmic electron carrier protein usually connecting the Rieske/cyt *b* complex to the terminal reductases, that is, soluble cytochromes, cupredoxins or high-potential iron–sulfur proteins (HiPIPs). This leaves the terminal reductase as the only pmf-generating element of the chain, whereas the Rieske/cyt *b* complex transforms from a proton-translocating into a pmf-consuming enzyme. This permits reverse electron flow from the soluble carrier protein to the UQ-pool to become exergonic. The further increase of pmf-consuming electron transfer steps at the expense of pmf-generating ones obviously has to be counterbalanced by a much higher proportion of electrons flowing in the forward direction, as compared to those going towards NAD^+ reduction (schematically indicated by the thickness of electron flow arrows in Fig. 2.2). The wealth of data gathered on the Fe^{2+}-oxidizing bioenergetic chain of *At. ferrooxidans* allows first glimpses into the more detailed evolutionary adaptations permitting this organism to cope with the thermodynamic challenges of aerobic oxidation of Fe^{2+} at low pH, which is outlined in the following sections.

Optimizing the efficiency of forward electron transfer in the face of weak driving forces

Although Fe^{2+}/Fe^{3+} occurs in a wide range of redox midpoint potentials depending on its chelating counterions (Bird *et al.*, 2011), the bioenergetic electron transport chain of *At. ferrooxidans* seems to be well-adapted to deal with even very weakly reducing kinds of Fe^{2+} compounds. The redox midpoint potential of the key electron shuttle protein, rusticyanin (Rus), mediating between the iron-oxidizing system and the O_2-reductase, has been determined at almost +600 mV at low pH (Ingledew and Cobley, 1980). Rus appears to be the branching-point of electron flow from iron oxidation preceding the split into the (dominant) forward-oriented 'roaring stream' towards molecular oxygen and the reverse 'creek' *via* the Rieske/cyt *b* complex and Complex I towards NAD^+ reduction (Fig. 2.2). These two distinct pathways of electron flow are proposed to each use a dedicated dihaem c_4-type cytochrome for directing electrons towards the O_2 reductase (Cyc1) and the Rieske/cyt *b* complex (CycA1), respectively (Brasseur *et al.*, 2004; Bruscella *et al.*, 2007). The haems in these two c_4-type cytochromes have been determined to titrate in the region of about +400 to +500 mV (Giudici-Orticoni *et al.*, 2000; Cavazza *et al.*, 1996), that is, more than 100 mV less positive than their physiological electron donor rusticyanin. Such a comparatively low redox potential makes obvious sense for facilitating reverse electron transfer through the Rieske/cytochrome *b* complex, since it is this pool of cytochromes which has to provide the driving force (together with the pmf) for the reverse electron transfer reaction. However, a more than 100 mV uphill redox step from Rus to cyt c_4 would certainly represent a major barrier to efficient electron flow. *At. ferrooxidans* has solved this problem by implementing a quite dramatic complex-induced lowering of Rus's midpoint potential. In the Rus:c_4 complex (more specifically the Rus:Cyc1 complex), Rus E_m drops by roughly

100 mV (Giudici-Orticoni *et al.*, 1999) thereby removing the uphill barrier and ensuring efficient electron flow from Fe^{2+} into the c_4-pool. At first glance, this may appear to contradict basic thermodynamic principles and indeed be tantamount to postulating the existence of a *perpetuum mobile.* However, it must be kept in mind that a 100 mV decrease in redox potential upon complex formation, is thermodynamically equivalent to an almost 100 fold stabilization of the oxidized form of Rus in its complex with cyt c_4. Therefore, the cytochrome binds preferentially the oxidized Rus, which of course strongly hampers electron transfer from Rus to the cytochrome. Therefore, the indispensable second half of this intricate redox strategy has to involve inflating the amount of Rus with respect to the remainder of the electron transfer chain. This is precisely what is found in *At. ferrooxidans*; Rus makes up for roughly 5% of total soluble protein in this organism. A high reduction state of this enormous Rus pool therefore counterbalances cyt c_4's preference for oxidized Rus, and the majority of cytochromes c_4 will correspondingly be forced to bind a reduced Rus. Since differences in redox potentials can be seen as microscopic reaction free energies (referring to a single donor–acceptor pair), the energy contained in Rus's drop in E_m upon complex formation is in fact drawn from the macroscopic free energy of an abundant and highly reduced Fe^{2+} reservoir, and a resulting highly reduced bulk of Rus. It is apparent, however, that the above detailed considerations agree with a dynamic complex and do not fit with the proposed model of a stoichiometrically fixed supercomplex involving the Fe^{2+} oxidase, Rus, cyt c_4 and the O_2 reductase (Castelle *et al.*, 2008). Also, as discussed below, a midpoint potential of +600 mV (and higher) for the pool of c_4-cytochromes would make efficient reverse electron flow through the Rieske/cyt *b* complex virtually impossible and the step lowering E_m-values of the chain by 100 mV between Rus and cyt c_4 thus appears crucial for the viability of the Fe^{2+} oxidizing metabolism in *At. ferrooxidans*.

Reverse electron transfer through the Rieske/*cyt* b complex

At. ferrooxidans contains two Rieske/cyt *b* gene cluster copies and one of them has been proposed to be optimized for reverse electron transfer (Brasseur *et al.*, 2002). Although electrochemical data on the redox cofactors of at least one of these complexes have been determined (Elbehti *et al.*, 1999), functional characterization of these enzymes is still rudimentary and we therefore refrain from interpreting the determined E_m values with respect to potential influence on the forward/reverse bias. However, a few more general thermodynamic considerations concerning the Rieske/cyt *b* complex working in reverse are required.

The redox species that need to be reduced by reverse electron transfer through this complex are the UQ and, more specifically, UQ binding at the Qo-site. This binding site is in proton equilibrium with the periplasmic space and the relevant pH to consider for estimating the quinones' E_m value therefore is in the range of pH 2. Thus, in the case of UQ, this E_m is about +390 mV. As mentioned above, the immediate electron donor for reverse transfer, certainly one of the cytochromes c_4, has an E_m in the range of +500 mV. Since the Rieske/cytochrome *b* complex in its forward mode translocates one proton per electron, the pmf has to be taken into account for the difference in redox potentials between the electron acceptor (UQ, +390 mV) and the electron donor (cyt c_4, +500 mV), yielding $390 - 500 + 256 = +146$ mV. In the fully energized membrane of *At. ferrooxidans* the reduction of UQ from cyt c_4 is therefore sufficiently exergonic to drive the reduced form ubiquinol (UQH_2) to appreciably high levels. If *At. ferrooxidans* would not resort to the above described redox-lowering step involving Rus, the nominal driving force for reverse transport would shrink to a mere 46 mV. If UQ and cyt c_4 were present at equal amounts, this would allow more than 50% reduction state of the UQ-pool. In *Proteobacteria* that have UQ, however, the UQ-pool in general is 5 to 10 times in excess of the soluble electron carriers (cytochromes, cupredoxins, etc.). The UQ to cyt c_4 stoichiometry in *At. ferrooxidans* has to the best of our knowledge not yet been determined but we find it unlikely that it would deviate significantly from that of other *Proteobacteria*. This implies that the reduction state of the UQ-pool attainable in redox equilibrium in an energized membrane will be substantially lower than the first order estimates based on redox equilibration between stoichiometric redox couples as derived above. This suggests that the lowering of rusticyanin's redox potential from +600 mV to +500 mV and the unusually high pmf of 256 mV in *At. ferrooxidans* work together to achieve high reduction states of the UQ-pool in pmf-assisted reverse electron transfer through the Rieske/cytochrome *b* complex.

Reverse electron flow through Complex I

As discussed above, high levels of UQH_2 are indispensable for turning Complex I into an NAD^+-reducing enzyme. The site of UQ redox conversion in Complex I is located on the cytoplasmic side of the membrane. The UQ is thus in proton-equilibrium with the circumneutral cytoplasm and its midpoint potential is in the vicinity of +90 mV. Electron transfer from UQ to the NAD^+ pool, therefore, has to climb a nominal redox barrier of 410 mV. With a typical pmf of 180 mV, the $2H^+/e^-$ coupling-ratio of Complex I would afford $2 \times 180 = 360$ mV, reducing the uphill barrier to about 50 mV. Very high reduction levels of the UQ-pool (exceeding 90%) would have to be permanently sustained to ensure persistent replenishing of the NADH pool. Constantly maintaining such high reduction levels in species such as *At. ferrooxidans* appears challenging. Once again, the higher-than-usual pmf in *At. ferrooxidans* may be the key to understanding the bioenergetic processes in this species. With a pmf-contribution of 2×256 mV = 512 mV, reduction of NAD^+ by UQH_2 indeed becomes exergonic and exceedingly high UQH_2 levels are not needed any more. Since the energy required for pmf build-up is drawn from the forward electron transfer from Fe^{2+} to O_2, macroscopic ΔG (difference in Gibbs free energy) from the bulk of the substrate redox couples is converted into an increase of the 'microscopic' redox energy of a specific couple of redox centres in the chain, very much like in the case of the Rus:cyt c_4 redox step.

As discussed above, the midpoint potentials of the quinones were taken as +390 mV at the Qo-site (in contact with periplasmic pH 2) and +90 mV in Complex I (in contact with the neutral cytoplasm), respectively. Since there is only one UQ-pool, this may at first glance appear thermodynamically untenable. The corresponding energy has, by necessity, to be afforded by the 'permanent' electrostatic field (and resulting pH gradient) discussed above. Obviously, this field is not really permanent but only long-lived on the timescale of turnover of a bioenergetic chain. It is likely generated by membrane-impermeant intracellular organic cations, which have to be constantly (although on the timescale of hours rather than milliseconds) replenished in energy consuming syntheses or transport. The free energy contained in the persistent pH gradient established between the cytoplasm and the extracellular space therefore plays an integral part in the total energy balance of both forward and reverse electron transfer in *At. ferrooxidans* and similar bacteria.

Respiration in acidophiles

Electron donors

One of the main characteristics of extreme acidophiles is their ability to use inorganic compounds as electron donors (chemolithotrophy) (Table 2.1). Fe^{2+}, elemental sulfur (S^0) and RISCs represent the major available energy sources. Under aerobic conditions, Fe^{2+}, S^0, RISCs, as well as H_2, can be used as electron donors. In the absence of O_2, these same electron donors can be used for Fe^{3+} reduction while H_2 is used for S^0 reduction (see below). In the case of acidophilic heterotrophic iron reducing bacteria, glycerol or glucose serve as electron donor. While soluble iron exists mainly in two oxidation states, +2 (ferrous) and +3 (ferric), sulfur is found in various oxidation states from −2 in sulfide (H_2S) to +6 in SO_4^{2-}. It should be noticed that in the compounds containing more than one sulfur atom, such as thiosulfate ($S_2O_3^{2-}$) and tetrathionate ($S_4O_6^{2-}$), the sulfur atoms can have different oxidation states. For example, thiosulfate has one sulfone (+5) and one sulfane (−1).

Electron acceptors

Acidophiles can use various electron acceptors, in addition to O_2 (Table 2.1). The high $E_{m,2}$ of Fe^{2+}/Fe^{3+} at pH 2 makes of Fe^{3+} a thermodynamically attractive electron acceptor. In addition, Fe^{3+} is soluble at pH ≤3.5 and therefore more available. A number of acidophilic prokaryotes have been shown to catalyse the dissimilatory reduction of Fe^{3+} (e.g. Johnson *et al.*, 2012, and references therein). In contrast, the capacity for dissimilatory reduction of SO_4^{2-} and/or S^0 is rather uncommon. S^0 reduction has been reported in the thermoacidophilic *Crenarchaeota* (*Acidianus, Stygiolobus, Sulfurisphaera*), in the thermophilic *Euryarchaeota* (*Thermoplasma*) and also in one strain of *At. ferrooxidans* (Dopson and Johnson, 2012; Liu *et al.*, 2012; Jameson *et al.*, 2010 and references therein). So far, no extreme acidophilic (pH optimum ≤3) sulfate reducing bacteria have been identified. At first sight, this is rather surprising since (i) SO_4^{2-} is usually the major anion in acidic environments and (ii) the $E_{m,2}$ value of the SO_4^{2-}/H_2S couple is higher at acidic than at neutral pH, making it a more thermodynamically feasible electron acceptor. However, hydrogen sulfide is a weak acid with two pK_a values (6.9 for H_2S/HS^- and 14 for HS^-/S^{2-}) and its speciation depends on the pH as illustrated in the following equation:

$$H_2S \rightarrow HS^- + H^+ \rightarrow S^{2-} + 2H^+$$

Below pH 5, only the undissociated free H_2S is found and this is the most toxic form of sulfide. It reacts with metallic cofactors, amino acids and metabolic enzymes, and therefore inhibits microbial growth. In addition, the organic acids produced during anaerobiosis, such as lactate or pyruvate, are also poisonous at low pH (see the 'ATP synthesis via substrate level phosphorylation' section). Consequently, most sulfate reducing bacteria are highly sensitive to even mild acidity (pH < 5; Koschorreck, 2008).

Respiratory pathways

Knowledge of the respiratory pathways in the extreme acidophiles remains rudimentary. This is mainly due to the problems described above impeding both biochemistry and construction of mutants by marker exchange mutagenesis, which while feasible, remain fastidious. Some molecular mechanisms involved in the dissimilatory oxidation of the electron donors described in Table 2.1 have, however, been deciphered mainly through 'global' analysis, i.e. bioinformatics coupled with transcriptomics and/or proteomics (Ilbert and Bonnefoy, 2013; Bonnefoy and Holmes, 2012). So far, the aerobic Fe^{2+} and RISC oxidation pathways were described in a number of prokaryotes, while S^0 and Fe^{3+} respiration models were proposed only in *Acidianus ambivalens* and *At. ferrooxidans*, respectively.

Ferrous iron oxidation

Rather than describe the Fe^{2+} oxidation pathways present in acidophiles, which have been extensively reviewed elsewhere (Ilbert and Bonnefoy, 2013; Bonnefoy and Holmes, 2012; Bird *et al.*, 2011), in this chapter the proposed models will be compared (Table 2.2). The reader is referred to these reviews and to the references given in Table 2.2.

The first Fe^{2+} oxidation pathways to be deciphered were those of the type strains of *At. ferridurans* and *At. ferrooxidans*. They appear to be identical and to agree with the pioneering hypothesis put forward by Ingledew in 1982 (Ingledew, 1982) and involving a 'downhill' pathway to oxygen with rusticyanin (RusA), cytochromes *c* and aa_3 oxidase (CoxBACD) and an 'uphill' pathway to NAD^+ through a bc_1 complex functioning in reverse (Fig. 2.2 and Table 2.2). The bifurcation between these two pathways is likely to occur at the level of rusticyanin that gives electrons to two different c_4 cytochromes, Cycl (downhill) or CycA1 (uphill) (Table 2.2). The initial Fe^{2+} electron acceptor is the cytochrome *c* Cyc2 located in the outer membrane with some domains exposed to the external environment (Table 2.2). The four components Cyc2, Cycl, rusticyanin and the aa_3-type cytochrome *c* oxidase were required to reconstitute *in vitro* the 'downhill' pathway with Fe^{2+} oxidation activity (Taha *et al.*, 2008). A computational model of the structural organization of this 'downhill' pathway was recently constructed (Patra *et al.*, 2013). The molecular interactions of charged and hydrophobic amino acids were shown to allow a hypothetical electron wire involving Cyc2, rusticyanin, Cycl and CoxAB in the form of a transient supercomplex (see section on 'Optimizing the efficiency of forward electron transfer in the face of weak driving forces'), thus supporting the model proposed by Bonnefoy's lab.

These same pathways have been proposed in *Acidithiobacillus ferrivorans*. However, an alternative pathway through the iron oxidase Iro and/or the iso-rusticyanin B (RusB) might also have evolved in this species (Table 2.2). Iro, a periplasmic high potential iron–sulfur protein (HiPIP), was described as the first electron acceptor from Fe^{2+} in strains belonging to Group IV '*Acidithiobacillus ferriphilus*' of the Fe^{2+}- and S^0-oxidizing *Acidithiobacillus*. Concerning RusB, its exact function in aerobic Fe^{2+} oxidation is uncertain, all the more

Table 2.2 Electron carriers involved in the Fe^{2+} pathways in acidophiles. Copper proteins are highlighted in dark grey, **cytochromes** are in bold, haem copper oxidase are underlined, Rieske/cytochrome *b* (*bc*$_1$ or *ba*) complexes are underlined in dotted line and iron–sulfur proteins are double-underlined (adapted from Ilbert and Bonnefoy, 2013; Bonnefoy and Holmes, 2012). References not cited in these papers are given in the last line of the table

	Bacteria											Archaea		
Subcellular localization	*Acidithiobacillus ferrooxidans*		*Acidithiobacillus ferridurans*		*Acidithiobacillus ferrivorans*			*Acidihalobacter prosperus*		*Leptospirillum* spp.		*Ferroplasma* spp.	*Sulfolobus* spp., *Metallosphaera* spp.	
Outer membrane	**Cyc2**		**Cyc2**		**Cyc2**			**Cyc2-like**		**Cyt$_{572}$**				
Periplasm	RusA		RusA		RusA		RusB? Iro?	Rus-like		**Cyt$_{579}$**				
Membrane bound	**Cyc1**	**CycA1**	**Cyc1**	**CycA1**	**Cyc1**	**CycA1**		?	?	**Cytc?**	**Cytc?**	Sulfocyanin		
Inner membrane	*aa*$_3$-oxidase	*bc*$_1$ complex; Q; NDH1	*aa*$_3$-oxidase	*bc*$_1$ complex; Q; NDH1	*aa*$_3$-oxidase	*bc*$_1$ complex; Q; NDH1		*aa*$_3$ oxidase	?	*cbb*$_3$-oxidase	*bc*$_1$ complex; Q; NDH1	*aa*$_3$ oxidase	**Cyt*b*$_{558/566}$**; Mco; Haem copper oxidase	**Cyt*b*$_{558/566}$**; [Fe-S]; Mco; *ba* complex; Q; NDH1
References					Liljeqvist *et al.* (2013), Talla *et al.* (2014)			Ossandon *et al.* (2014, Genome announcements 2)				Castelle *et al.* (2015)	Auernik and Kelly (2010)	

NDH1, NADH dehydrogenase (complex I).

so, as its kinetic rate constant for electron transfer from Fe^{2+} was approximately one-half that of RusA. In *Acidihalobacter prosperus*, the electron transfer chain from Fe^{2+} to O_2 was proposed to strongly resemble that described in *At. ferridurans*, *At. ferrooxidans* and *At. ferrivorans*, with the exception of the membrane-bound cytochrome c_1 that seems to be absent (Table 2.2). The 'uphill' and 'downhill' iron oxidation pathways of *Leptospirillum* spp. have features similar to those described above: outer membrane and membrane-bound cytochromes *c*, a bc_1 complex functioning in reverse and a terminal cytochrome oxidase (Table 2.2). The main difference is that no blue copper protein equivalent to rusticyanin is involved.

In *Archaea*, two types of Fe^{2+} oxidation pathways have been described: one in the *Euryarchaeota Ferroplasma* spp. and the other one in the *Crenarchaeota Sulfolobus* spp. and *Metallosphaera* spp. In *Ferroplasma* spp., as in the acidithiobacilli, a blue copper protein (sulfocyanin) and an aa_3 cytochrome oxidase are involved (Table 2.2). Similarly, in *Sulfolobus* spp. and *Metallosphaera* spp., the electrons from Fe^{2+} have been proposed to be extracted by a cytochrome $b_{558/566}$ (FoxCD) and transferred through a multicopper oxidase (Mco) to either oxygen by a haem copper oxidase (FoxAB) or to NAD^+ through a cytochrome *ba* complex, a member of the Rieske/cytochrome *b* superfamily (Table 2.2).

Although the Fe^{2+} oxidation pathways in acidophiles are diverse, nevertheless common themes can be derived from current knowledge:

- The vertical spatial arrangement of the electron carriers in Gram-negative bacteria allows a physical separation of Fe^{2+} oxidation and O_2 reduction. This organization avoids (i) potential oxidative stress due to reactive oxygen species formation by the Fenton reaction, (ii) Fe^{3+} precipitation at the circumneutral pH of the cytoplasm and (iii) acidification of the cytoplasm by consuming and extruding the protons entering the cell through ATP synthase, by the terminal cytochrome oxidase.
- The electrons are generally extracted from Fe^{2+} by a cytochrome *c* (or by a cytochrome *b* or a copper protein in the case of the thermoacidophilic archaea in which there is no current evidence of cytochromes *c*). In Gram-negative bacteria, this cytochrome is localized in the outer membrane.
- With the exception of *Leptospirillum* spp., a blue copper protein is generally involved in electron transfer: for example rusticyanin in *Acidithiobacillus* spp., sulfocyanin in *Ferroplasma* spp. and multicopper oxidase Mco in *Sulfolobus* spp. and *Metallosphaera* spp..
- A cytochrome *c* transfers the electrons to the inner membrane complex (the cytochrome oxidase in the 'downhill' pathway and the Rieske/cytochrome *b* complex in the 'uphill' pathway).
- In the 'downhill' pathway, the electrons from the Fe^{2+} oxidation are used to reduce O_2 by a cytochrome oxidase, generally of the aa_3-type.
- A Rieske/cytochrome *b* complex functioning in reverse is involved in the 'uphill' pathway.

Taken together, these data indicate that the Fe^{2+} oxidation pathways arose independently more than once in evolution and evolved convergently (Ilbert and Bonnefoy, 2013).

Reduced inorganic sulfur compound oxidation

Deciphering RISC oxidation pathways in acidophiles has been difficult for a number of reasons, the main one being that enzymatic reactions are difficult to distinguish from purely chemical transformations given the chemical reactivity of several sulfur compounds (Suzuki, 1999). Abiotic reactions may occur (i) because of the instability of some compounds at low pH (e.g. disproportionation of thiosulfate into sulfite (SO_3^{2-}), SO_4^{2-} and S^0 and disproportionation of polysulfides into S^0 and H_2S); (ii) due to the presence of reductants such as H_2S, especially at high temperature (e.g. tetrathionate reduction to thiosulfate), and (iii) because of the reactivity of some sulfur compounds (e.g. SO_3^{2-} reaction with S^0 to form thiosulfate). Another complexity is that sulfur exists in a variety of oxidation states and a number of intermediates and enzymes are involved in sulfur redox reactions (Fig. 2.3 and Appendix). In all acidophiles, the main inorganic sulfur oxidation pathway is:

$$H_2S \rightarrow S^0 \rightarrow SO_3^{2-} \rightarrow SO_4^{2-}$$

In addition, these prokaryotes are also able to oxidize polythionates, e.g. thiosulfate ($S_2O_3^{2-}$)

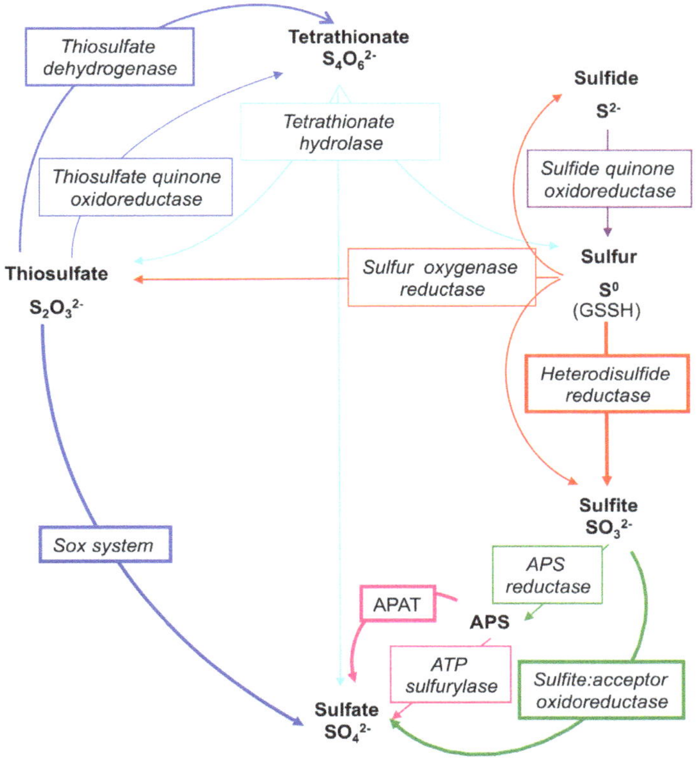

Figure 2.3 Intermediate sulfur compounds (in bold) and relevant enzymes (in italics and boxed) involved in RISCs oxidation in acidophilic prokaryotes. The same motif was used to depict the enzymes using the same substrate. Different thickness was used to discriminate the reactions catalysed by different enzymes using the same substrate.

and tetrathionate ($S_4O_6^{2-}$). To make things even more complex, the same reaction can be catalysed by different enzymes, as shown in Fig. 2.3, which in some cases are present in the same microorganism (see Appendix). For example, (i) thiosulfate can be oxidized directly to SO_4^{2-} through the Sox (sulfur-oxidizing) enzyme system or to tetrathionate through the thiosulfate:quinone oxidoreductase (TQO) or the thiosulfate dehydrogenase (TSD); (ii) S^0 is disproportionated to SO_3^{2-}, H_2S and thiosulfate using the sulfur oxygenase reductase (SOR) or it may be oxidized to SO_3^{2-} by the heterodisulfide reductase complex; (iii) SO_3^{2-} is oxidized directly to SO_4^{2-} by the sulfite:acceptor oxidoreductase (SAOR) or indirectly *via* the formation of the intermediate adenosine-phosphosulfate (APS) by the APS reductase followed by SO_4^{2-} release by adenylylsulfate:phosphate adenylyltransferase (APAT) or ATP sulfurylase (Sat).

In this section, rather than describing the different pathways involved in each acidophilic sulfur-oxidizers for which a model has been proposed, the enzymes involved will be briefly presented (see Appendix) and an overall view of the respiratory chains will be given to highlight the differences and the similarities in the diverse models proposed to date (Fig. 2.4). The reader is referred to a number of recent reviews on the aerobic dissimilatory RISC oxidation pathways in acidophilic prokaryotes (Dopson and Johnson, 2012; Liu *et al.*, 2012; Quatrini *et al.*, 2009) and to the references given in Fig. 2.4 and the Appendix.

Sulfide:quinone reductase

Sulfide:quinone reductases (SQR) are ubiquitous membrane-bound flavoproteins essentially involved in H_2S detoxification and respiration, as well as in heavy metal tolerance. They have been

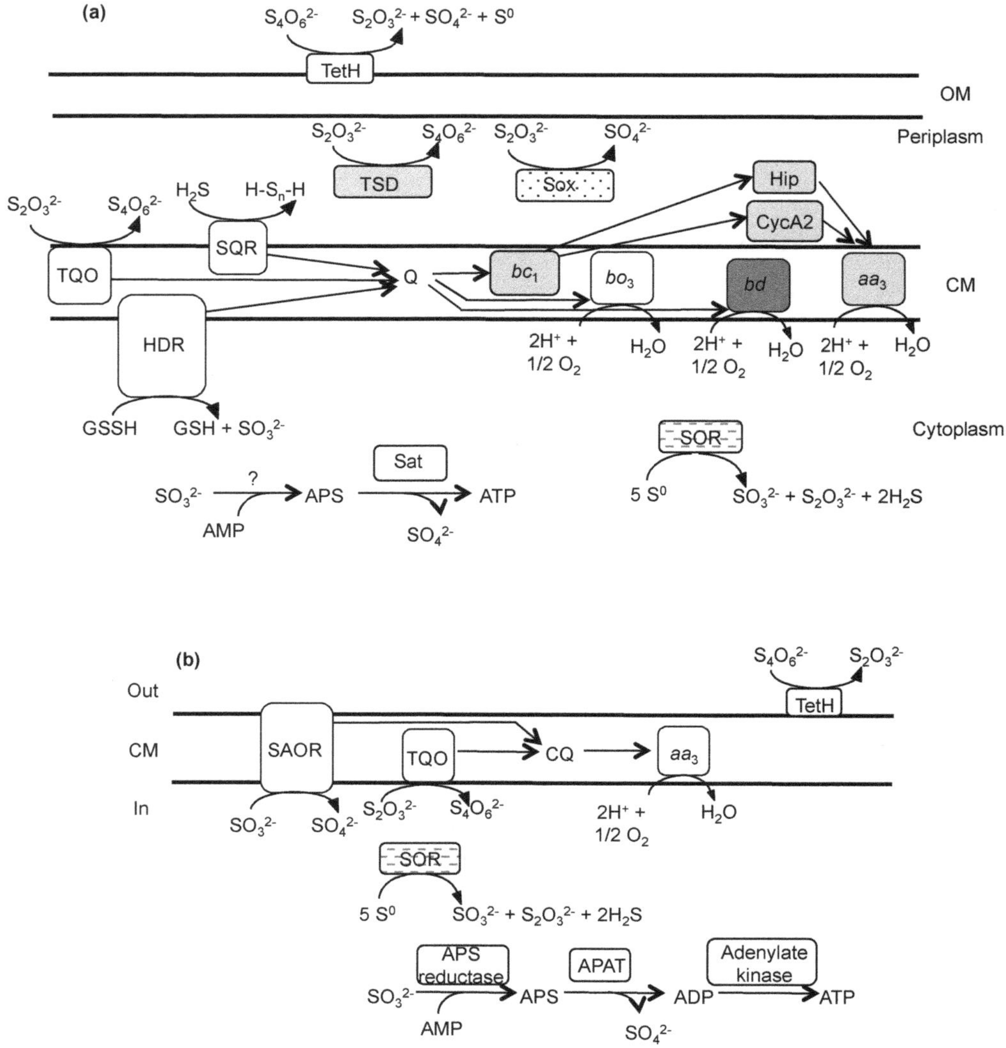

Figure 2.4 Models of the RISCs oxidation pathways in acidithiobacilli (a) and in *Ad. ambivalens* (b). (a) In white, the 'core set' of enzymes/complexes shared between the acidithiobacilli; in light grey, the enzymes/complexes only found in the iron-oxidizers (*At. ferrooxidans*, *At. ferridurans and At. ferrivorans*); in dark grey, the enzymes/complexes only found in *At. caldus*, *At. thiooxidans* and *At. ferrooxidans*; in dotted background, the enzymes/complexes only found in *At. ferrivorans*, *At. thiooxidans* and *At. caldus*; in striped background, the enzymes/complexes only found in *At. ferrivorans* and *At. caldus*. OM, outer membrane; CM, cytoplasmic membrane; Q, quinone pool; CQ, caldariella quinone. Other abbreviations can be found in the text. These models are adapted from Bruscella *et al.* (2007) for *At. ferridurans*, Quatrini *et al.* (2006, 2009) for *At. ferrooxidans*, Liljeqvist *et al.* (2013) and Talla *et al.* (2014) for *At. ferrivorans*, Chen *et al.* (2012) for *At. caldus*, Bobadilla Fazzini *et al.* (2013) for *At. thiooxidans* and Protze *et al.* (2011) for *Ad. ambivalens*.

classified in six groups according to their structure (Gregersen *et al.*, 2011). In the seven acidophilic sulfur-oxidizing prokaryotes analysed, the gene encoding the respiratory-type SQR has been detected (see Appendix). This SQR catalyses the oxidation of H_2S, HS^- and S^{2-} to soluble polysulfide chains (H-S_n-S) and transfers the electrons to the quinone pool in the membrane. The crystal structure of SQR from *Ad. ambivalens* has been determined. Two potential redox active sites are seen in the structure: a covalently bound FAD cofactor and an adjacent pair of cysteine residues. In *At. ferrooxidans*, higher SQR activity and more *sqr* transcripts were observed in S^0-grown than in Fe^{2+}-grown cells. The enzyme has been purified and its three-dimensional structure has been determined. In contrast to the other members of the disulfide oxidoreductase family, the FAD cofactor is non-covalently bound to the N-terminal domain of the *At. ferrooxidans* SQR. It has been proposed that a conserved cysteine residue either acts as a nucleophile, attacking the FAD cofactor, or directly transfers the electrons to this cofactor, which in turn passes them onto the quinone pool. Surprisingly, two putative *sqr* genes were detected in *Leptospirillum* spp. considered as strict Fe^{2+} oxidizers, one of which is clustered with a cytochrome *bd* oxidase operon suggesting that dissimilatory sulfide oxidation could possibly occur in these bacteria.

Sox system

The periplasmic sulfur-oxidizing system (Sox), also called the TOMES system (thiosulfate-oxidizing multienzyme system) or PSO (*Paracoccus* sulfur oxidation) pathway, has mainly been studied in the photo- and chemolithotrophic *Alphaproteobacteria* (reviewed in Frigaard and Dahl, 2009; Ghosh and Dam, 2009; Gregersen *et al.*, 2011). The Sox system catalyses the direct oxidation of thiosulfate, SO_3^{2-}, H_2S and S^0 to SO_4^{2-}, without the formation of free intermediates. The electron acceptor is a soluble small cytochrome *c*. This complex is composed of the periplasmic proteins SoxB, SoxYZ, SoxXA and SoxCD that are by themselves inactive. Thiosulfate, linked covalently to a cysteine residue of SoxY by SoxXA (both being *c* type cytochromes), is proposed to be the substrate for SoxB, a sulfate thiohydrolase, that cleaves the SO_4^{2-} from the bound substrate. SoxCD, a sulfur dehydrogenase, then oxidizes the remaining sulfane sulfur. This latter enzyme, however, is dispensable since a SoxCD truncated system has been characterized in purple and green sulfur bacteria, leading to intracellular sulfur globule formation. Two gene clusters harbouring *soxYZB-soxAX* and *soxYZA-soxB*, i.e. encoding a truncated Sox system, have been detected in the genome of *Acidithiobacillus* (*At.*) *caldus* and *Acidithiobacillus* (*At.*) *thiooxidans*. These genes have been shown to be up-regulated in the presence of tetrathionate in agreement with proteomic data. In the two sequenced strains of *At. ferrivorans*, the *soxYZB–soxAX* cluster has been detected. However, it does not seem to be functional since *soxX* (SS3 and CF27 strains) and *soxA* (SS3 strain) are pseudogenes. No *sox* genes have been identified in *At. ferrooxidans*, *Metallosphaera* (*M.*) *sedula*, *Sulfolobus* (*S.*) *tokodaii* and *Ad. ambivalens* (see Appendix).

Thiosulfate:quinone oxidoreductase

The thiosulfate:quinone oxidoreductase (TQO) is a membrane-bound enzyme that couples thiosulfate oxidation (yielding tetrathionate) to the reduction of quinone (see Appendix). The TQO of *Ad. ambivalens* consists of two subunits, DoxA and DoxD. It binds quinone non-covalently. No other redox-active cofactors were detected. It has been proposed that quinone shuttles the electrons between TQO and the terminal quinol:oxygen reductase leading to transmembrane proton gradient enhancement, and thus to ATP formation. The *doxDA* genes have been detected in *M. sedula* and in *S. tokodaii* (see Appendix). In *At. ferrooxidans*, *At. ferrivorans*, *At. caldus* and *At. thiooxidans*, *doxDA* genes are fused (see Appendix). A duplication of this gene was found divergently oriented in close vicinity in *At. ferrooxidans* (strains ATCC 23270^T and ATCC 53993). Interestingly, only one *doxDA* gene was shown to be transcribed more in S^0-grown than in Fe^{2+}-grown *At. ferrooxidans* ATCC 23270^T cells.

Thiosulfate dehydrogenase

Another tetrathionate-forming thiosulfate dehydrogenase (TSD) was identified in *At. ferrooxidans*. Unlike the membrane-bound TQO described above, TSD is periplasmic and is detected in S^0- or tetrathionate- but not in Fe^{2+}-grown cells. No cofactor has been identified. The *tsd* gene is located in the same cluster as the *doxDA* gene and, like that gene, is also duplicated. One orthologue of the *tsd* gene was identified in *At. ferrivorans* but not in the sulfur oxidizing acidithiobacilli *At. caldus* and *At. thiooxidans* (see Appendix).

Tetrathionate hydrolase

Tetrathionate is hydrolysed to thiosulfate, S^0 and SO_4^{2-} by the tetrathionate hydrolase TetH (see Appendix). The gene encoding this enzyme has only been detected, so far, in acidophilic prokaryotes. Surprisingly, this homodimeric enzyme has been localized in differing subcellular compartments, being periplasmic in *At. caldus* and *At.*

thiooxidans and extracellular in *Ad. ambivalens* and *Ad. hospitalis.* This disagreement on TetH localization does not seem to be species-dependent, since in *At. ferrooxidans* TetH was found periplasmic, membrane-associated, in the outer membrane or extracellular. Interestingly, in *Ad. ambivalens* the TetH activity is enhanced significantly after exposure of cells to increasing pH, suggesting that TetH is released by pH shock and that this enzyme therefore is attached extracellularly to the cell S-layer. This could explain the discrepancies observed and described above. TetH is a homodimer and belongs to the pyrroloquinoline quinone (PQQ) dehydrogenase superfamily and displays a β-propeller structure. However, the PQQ binding residues are missing and no PQQ was detected in the purified active enzyme. The *tetH* gene has been shown to be more strongly expressed in S^0- than in Fe^{2+}-grown cells in *At. ferrooxidans,* and in the presence of tetrathionate compared with S^0 in *At. caldus* and *Ad. ambivalens,* in agreement with its role in tetrathionate hydrolysis. In addition, *tetH* is more strongly expressed under aerobic than under anaerobic growth conditions suggesting that the hydrolysis of tetrathionate mediated by TetH occurs preferentially in aerobic conditions. Unexpectedly, some *Leptospirillum* spp. have a gene encoding a putative tetrathionate hydrolase, while these bacteria have been shown to use Fe^{2+} as sole electron donor.

Sulfur oxygenase reductase

The sulfur oxygenase reductase (SOR) was first identified and then extensively studied in *Ad. ambivalens* (see Appendix). This cytoplasmic enzyme catalyses the oxygen-dependent disproportionation of S^0 to H_2S and SO_3^{2-}, and SO_3^{2-} chemically reacts with S^0 to give thiosulfate. SOR is composed of 24 monomers that form a large hollow sphere enclosing a positively charged nanocompartment. The only cofactor detected is a low-potential mononuclear non-haem iron centre ligated to histidine and glutamic acid residues. A cysteine was proposed to be involved in sulfur binding as a polypeptide chain ($R\text{-}S_n\text{-}SH$). Being cytoplasmic, SOR is not coupled to the respiratory chain and its role in energy conservation is unclear. SOR homologues have been detected not only in archaea, such as *Acidianus* spp., *Sulfolobus* (*S.*) *tokodaii* and *S. metallicus,* but also in *At. caldus* and *At. ferrivorans* (see Appendix). Oddly, the *sor* gene was detected in *Ferroplasma* (*Fp.*) *acidarmanus* which is not described to be a sulfur-oxidizer. The *sor* gene was shown to be more transcribed in aerobic versus anaerobic growth conditions in *Ad. ambivalens,* in S^0- with respect to tetrathionate-grown cells in *At. caldus,* in agreement with its involvement in aerobic S^0 oxidation. An *At. caldus sor* mutant is still able to grow on S^0, indicating the existence of further S^0 oxidizing pathways in this bacterium.

Heterodisulfide reductase

A seven-gene cluster *rdh-tusA-dsrE-hdrC1-hdrB1-hdrA-orf-hdrC2-hdrB2* and a gene *hdrB* potentially encoding a heterodisulfide reductase (HDR) were shown to be highly up-regulated in S^0- compared to Fe^{2+}-grown *At. ferrooxidans* cells. HDR has been extensively studied in methanogenic archaea and sulfate-reducing archaea and bacteria (reviewed in Grein *et al.*, 2013) where it has been shown to catalyse the reversible reduction of the disulfide bond X-S-S-X coupled to energy conservation. Being sulfane sulfur the actual substrate of the sulfur oxidizing enzymes in *Acidithiobacillus* spp., the HdrABC complex was proposed to function in reverse in *At. ferrooxidans* and to oxidize the sulfane sulfur of GSSH into sulfite and to deliver the electrons to the quinone pool. This complex is composed of the flavoprotein HdrA, which anchors the complex to the cytoplasmic membrane, the HdrB subunit with the cysteine-rich domain which binds an unusual [4Fe–4S] cluster involved in disulfide oxidoreduction, and the ferredoxin-like HdrC. HdrC1 and HdrC2 have been shown to bind only one [4Fe–4S] cluster while having two iron–sulfur cluster binding motifs, which could explain why there are two *hdrC* copies. The *rdh, tusA* and *dsrE* genes present in the *hdr* locus from *At. ferrooxidans,* and also up-regulated in S^0-conditions, encode sulfurtransferases involved in a sulfur-relay system proposed to transfer the sulfane sulfur to the heterodisulfide reductase. It is noteworthy that similar *hdr* clusters have been detected in other sulfur oxidizing prokaryotes, including *At. ferrivorans, At. caldus, At. thiooxidans, S. tokodaii* and *M. sedula* (see Appendix). In *M. sedula,* as in *At. ferrooxidans,* these genes have been shown to be stimulated by S^0 or tetrathionate. Proteomics and transcriptomics performed on *At. caldus* also suggests the involvement of HDR and the sulfur relay system in sulfur oxidation. The comparison of the proteome and transcriptome of

At. ferrooxidans cultures grown with S^0 as electron donor, and O_2 or Fe^{3+} as electron acceptor, indicated that the levels of the proteins encoded by the *hdr* genes were greater in aerobic than in anaerobic conditions, supporting the role of heterodisulfide reductase in aerobic oxidation of S^0.

APS reductase, adenylylsulfate:phosphate adenyltransferase, adenylate kinase and ATP sulfurylase

Adenylylsulfate (APS) reductase activity was detected in the cytoplasm of *Ad. ambivalens*. This enzyme catalyses the formation of APS from SO_3^{2-} and AMP. It is a cytoplasmic heterodimer with a FAD-containing α-subunit and a two [4Fe–4S] clusters-containing β-subunit. The genes *aprBA* encoding this heterodimer have not been detected in *At. ferrooxidans*, *At. ferrivorans*, *At. caldus*, *At. thiooxidans*, *S. tokodaii* nor *M. sedula* (see Appendix). In *Ad. ambivalens*, the release of SO_4^{2-} from APS is catalysed by a cytoplasmic adenylylsulfate:phosphate adenyltransferase (APAT) and the ADP formed is subsequently converted to ATP and AMP by the adenylate kinase.

In sulfur-oxidizers, a reverse cytoplasmic ATP sulfurylase (Sat) has been shown to release SO_4^{2-} and ATP from APS (Kappler and Dahl, 2001). In *At. ferrooxidans*, a gene encoding Sat that exhibits 60% similarity with the bifunctional ATP sulfurylase-APS kinase Sat-CytC from the thermophilic sulfur-oxidizer *Aquifex aeolicus* has been detected. Sat from *At. ferrooxidans* has recently been shown to have at least the ATP-sulfurylase activity. The *sat* gene encoding ATP sulfurylase is also present in *At. ferrivorans*, *At. caldus*, *At. thiooxidans*, *S. tokodaii* and *M. sedula* (see Appendix). Therefore, in these sulfur-oxidizers indirect SO_3^{2-} oxidation occurs in the cytoplasm, allowing substrate-level phosphorylation.

Sulfite:acceptor oxidoreductase

In addition to the cytoplasmic APS reductase, APAT and the adenylate kinase system allowing SO_3^{2-} oxidation to SO_4^{2-} via APS formation (see above), a sulfite:acceptor oxidoreductase (SAOR) activity was detected in the membrane of *Ad. ambivalens* (see Appendix). SAOR probably belongs to the molybdopterin-containing enzymes of the sulfite oxidase family. It has been proposed to oxidize SO_3^{2-} to SO_4^{2-} directly, with the electrons being transferred to the quinone pool. Such enzymatic activity was not detected in the other acidophiles.

Other sulfite oxidases

Two distinct sulfite oxidases have been purified from different *At. ferrooxidans* strains. One was membrane-bound, composed of two subunits (61 and 59 kDa) and did not transfer electrons to cytochrome *c* or ferricyanide but to Fe^{3+}, while the other one of 41.5 kDa was soluble and could use cytochrome *c* and ferricyanide as electron acceptors. In addition, the aa_3-type cytochrome oxidase was shown to catalyse SO_3^{2-} oxidation (Sugio *et al.*, 2010). In *At. caldus*, an oxidoreductase molybdopterin-binding protein was proposed to be the sulfite oxidase based on weak similarities (26%) with the sulfite oxidase of *Deinococcus radiodurans*. The corresponding gene is not present in other known acidithiobacilli. In addition, *sorAB* and *soxCD* genes known to encode the periplasmic enzymes involved in direct oxidation of SO_3^{2-} have not been detected in the acidithiobacilli. Recently, a new sulfite oxidase, SoeABC, has been characterized in *Allochromatium vinosum* (Dahl *et al.*, 2013). This membrane-bound iron–sulfur molybdoprotein catalyses direct oxidation of SO_3^{2-} to SO_4^{2-} in the cytoplasm. The *soeABC* genes have been detected in a large number of chemotrophic bacteria, including *Acidithiobacillus* spp. according to the authors. However, a Blastp analysis has shown that the closest match in *At. ferrooxidans* (25/41% identity/similarity for SoeA, 34/49% identity/similarity for SoeB and no obvious similarity with SoeC, unpublished data) is annotated as the sulfur reductase SreABC.

Rieske/cytochrome *b* complex

In *At. ferridurans* and in *At. ferrooxidans* a second Rieske/cytochrome *b* complex (bc_1) functioning in the forward direction in S^0-growth conditions has been found. The $petA_2B_2C_2$ operon encoding this complex was shown to be slightly more expressed in aerobic conditions with S^0- than Fe^{2+} as electron donor (Bruscella *et al.*, 2007; Quatrini *et al.*, 2009) and with Fe^{3+} rather than O_2 as electron acceptor in the presence of S^0 (Osorio *et al.*, 2013), suggesting that this complex transfers electrons injected into the quinone pool *via* RISC oxidation towards the terminal oxidase or the ferric reductase. In the same transcriptional unit as *petA₂B₂C₂* the genes encoding a cytochrome c_4 (*cycA₂*) and a high-potential iron–sulfur protein (*hip*) are present, and have consequently been proposed to serve as physiological electron acceptors of this Rieske/cytochrome *b* complex. Interestingly, no *petABC* operon was detected in *At. caldus* and in *At. thiooxidans* while the same two operons than in *At. ferrooxidans* and *At. ferridurans* are present in *At. ferrivorans* (Talla *et al.*, 2014). It is noteworthy that *At. caldus* and *At. thiooxidans* are unable to oxidize or reduce iron. On the other hand, two operons encoding Rieske/cytochrome *b* complexes have been detected in the obligate Fe^{2+} oxidizer *Leptospirillum* spp. (ten Brink *et al.*, 2013; Levican *et al.*, 2012).

In archaea (e.g. *Ad. ambivalens*), the cytochrome *ba* complex formed by the SoxN (cytochrome *b*), CbsA (cytochrome *b*) and SoxL2 (Rieske protein) proteins, homologous to a Rieske/cytochrome *b* complex, has been characterized (Bandeiras *et al.*, 2009). The corresponding genes have been detected in *S. tokodaii* and *M. sedula* (Bandeiras *et al.*, 2009). In *M. sedula*, these genes are more strongly expressed in Fe^{2+}- than in S^0- or tetrathionate-growth conditions (Auernik and Kelly, 2008; Auernik *et al.*, 2008). *M. sedula* and *S. tokodaii* have two other Rieske/cytochrome *b* complexes (SoxF/SoxG and SoxL1/SoxC), the first one being up-regulated in heterotrophic versus autotrophic conditions with Fe^{2+}, while the second is more expressed in S^0- or tetrathionate- than Fe^{2+}-grown cells. Therefore, SoxL1/SoxC is likely involved in the dissimilatory oxidation of RISCs. It has been recently proposed that these microorganisms have adapted their three Rieske/cytochrome *b* complexes to interact with different oxidases (ten Brink *et al.*, 2013).

Terminal oxidases

Prokaryotes usually have several terminal oxidases reflecting metabolic versatility and specialization. Those involved in Fe^{2+} oxidation are listed in Table 2.2. In *At. ferrooxidans*, the *cydAB* and *cyoABCD* operons have been shown to be more highly expressed in S^0- than in Fe^{2+}-growth conditions, in agreement with optical spectroscopy and activity measurements (Quatrini *et al.*, 2009). Six *cydAB* operons are present in *At. caldus*, five in *At. thiooxidans* but none in *At. ferrivorans*, while three *cyoABCD* operons have been detected in *At. caldus* and *At. ferrivorans*, as well as two in *At. thiooxidans* (Talla *et al.*, 2014). In *At. ferrivorans*, one *cyoABCD* operon is located in the same locus as *doxDA*, *tetH*, *sor* and *sqr* suggesting that this bo_3 quinol oxidase is involved in RISC oxidation (Talla *et al.*, 2014). In addition, expression levels of another *cyoB* have been shown to increase in the presence of tetrathionate as compared with Fe^{2+} (Liljeqvist *et al.*, 2013), likely implying that it is this terminal oxidase that operates when tetrathionate is the electron donor. In *M. sedula*, oxidases are expressed as a function of growth conditions: one (SoxMH) in heterotrophic condition, another one (FoxAA'B) in Fe^{2+}-grown cells and two in S^0 or tetrathionate (DoxBC and SoxBA) (Auernik and Kelly, 2008; Auernik *et al.*, 2008). These last two oxidases are therefore likely involved in RISC dissimilatory oxidation.

Conclusions

Several conclusions can be drawn with respect to RISC oxidation pathways in acidophiles:

- Although the same intermediaries and final products are involved (Fig. 2.3), the enzymes used by different acidophiles differ widely (see Appendix).
- The RISC oxidation pathways are redundant since a substrate can be oxidized by different enzymes in the same microorganism (Fig. 2.3 and Appendix).
- A 'core' set of enzymes/complexes is shared by all acidophilic sulfur oxidizers, including the sulfide:quinone reductase, the thiosulfate:quinone oxidoreductase, the tetrathionate hydrolase and the heterodisulfide reductase (see Appendix).
- Previously, sulfur-oxidizing chemolithotrophic bacteria have been divided into two groups according to their capacity to utilize polythionates (e.g. tetrathionate) and two major exclusive pathways have been proposed: (i) the sulfur-oxidizing pathway (Sox) and (ii) the S_4-intermediary pathway (S_4I) involving polythionates (Frigaard and Dahl, 2009; Ghosh and Dam, 2009). Yet some acidophilic sulfur-oxidizers, namely *At. caldus* and *At. thiooxidans* have the Sox system, and at the same time the tetrathionate hydrolase (see Appendix), indicating that they can use both pathways.

- In contrast to the Fe^{2+} oxidation pathways, the RISC respiratory pathways are organized horizontally along the cytoplasmic membrane with a specific dehydrogenase injecting electrons resulting from the oxidation of RISCs into the quinone pool which in turn transfers them (via a Rieske/cytochrome *b* complex or not) to a terminal oxidase, where O_2 reduction takes place.
- The enzymes are located in different subcellular compartments: the tetrathionate hydrolase likely is attached extracellularly to the outer membrane; the Sox system and the thiosulfate dehydrogenase are in the periplasm; the sulfide:quinone reductase, the thiosulfate:quinone oxidoreductase and the sulfite:acceptor oxidoreductase are membrane-attached on the periplasmic face; the heterodisulfide reductase is anchored to the cytoplasmic membrane on its cytoplasmic side; the sulfur oxygenase reductase, the APS reductase, the adenylylsulfate:phosphate adenylyltransferase, the adenylate kinase and the ATP sulfurylase are cytoplasmic.
- Energy harvested from RISC oxidation may occur by substrate level phosphorylation as well as by oxidative phosphorylation. In the first case, SO_3^{2-} is oxidized to APS that is converted to ATP in the cytoplasm. In the second case, the sulfide:quinone reductase, the thiosulfate:quinone oxidoreductase, the sulfite:acceptor oxidoreductase (when present) and the heterodisulfide reductase transfer the electrons to the respiratory chain at the level of the quinone pool.
- In the iron- and sulfur-oxidizing bacteria, only two cytochromes *c* are involved in the RISC respiratory chains (namely $PetC_2$ of the Rieske/cytochrome *b* complex and the cytochrome $CycA_2$) in contrast to the Fe^{2+} oxidation pathways in which four *c*-type cytochromes have been identified (Cyc2, Cyc1, $CycA_1$, and $PetC_1$). This is in agreement with previous data reporting lower levels of total cytochromes *c* in S^0- than in Fe^{2+}-grown cells of *At. ferridurans* and *At. ferrooxidans* (Yarzabal *et al.*, 2002).
- No genes known to encode sulfite oxidases have been detected in the acidithiobacilli. Whether this reaction occurs (i) abiotically, that is, through SO_3^{2-} reacting with S° to yield thiosulfate or by a free radical mechanism initiated by Fe^{3+} trapped on the cell surface (Harahuc and Suzuki, 2001); (ii) aspecifically, as demonstrated for the aa_3- type cytochrome *c* oxidase (Sugio *et al.*, 2010); or whether (iii) these bacteria possess an, as yet, unidentified sulfite oxidase remains unclear.
- In *Leptospirillum* spp., reported to use only Fe^{2+} as electron donor and O_2 as electron acceptor, the genes encoding sulfide:quinone reductase, tetrathionate hydrolase as well as a bc_1 and a b_6f complexes, have been detected. The presence of these genes is rather puzzling and could suggest alternative metabolisms than aerobic Fe^{2+} oxidation. However, whether these genes are functional is currently not known. These genes could have been acquired recently via horizontal gene transfer (HGT). Another possibility is that the ancestor of the leptospirilli was a RISC oxidizer and has lost most of the genes involved in RISC oxidation, except these ones. The same holds true for '*Fp. acidarmanus*' in which the *sor* gene encoding the sulfur oxygenase reductase was predicted. However, in that case, S° oxidation will not allow the cells to grow and this is probably why this feature was not detected.

Ferric iron reduction

The reduction of Fe^{3+} in acidic conditions features two important advantages: (i) Fe^{3+} is generally soluble at pH <2.5, (ii) the E_m value for Fe^{2+}/Fe^{3+} is dependent on associated ligands and is more electropositive when soluble at acidic pH in SO_4^{2-} solutions ($E_{m,2}$ = 0.68 to 0.77 V) than when complexed at neutral pH ($E_{m,7} \approx 0.3$V; Bird *et al.*, 2011). Therefore, iron is more energetically favourable as electron acceptor in acidic conditions than as electron donor. Dissimilatory Fe^{3+} reduction is widespread among acidophiles (Johnson *et al.*, 2012). The majority of Fe^{2+} oxidizers are also Fe^{3+} reducers. Thus, *At. ferrooxidans* and *At. ferrivorans* are able to reduce Fe^{3+} while *At. caldus* and *At. thiooxidans* are not. *At. ferrooxidans* is able to reduce Fe^{3+} with S^0, H_2 or formate as electron donor and *At. ferrivorans* with S^0 but not H_2 (Hedrich and Johnson, 2013; Ohmura *et al.*, 2002; Pronk *et al.*, 1991).

Little is known about the Fe^{3+} reduction pathways. Two contradictory mechanisms for the oxidation of S^0 by Fe^{3+} in *At. ferrooxidans* were proposed in the 80s (Fig. 2.5 and references

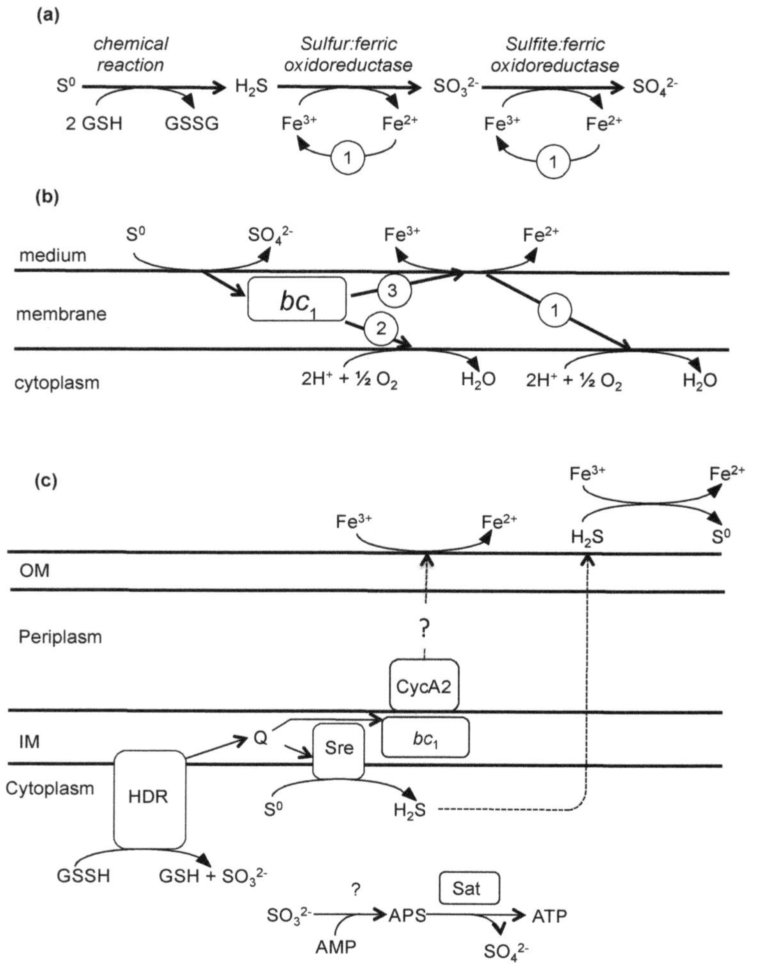

Figure 2.5 Models of ferric iron reduction in *At. ferrooxidans*. (a) Sugio's model according to Sugio *et al.* (1985, 1989); (b) Corbett and Ingledew's model according to Corbett and Ingledew (1987) and Pronk *et al.* (1991); (c) Osorio's model after Osorio *et al.* (2013). (1) Aerobic Fe^{2+} oxidation pathway; (2) aerobic S^0 oxidation pathway; (3) Fe^{3+} reduction with S^0 as electron donor. Abbreviations can be found in the text.

therein). In the first one (Fig. 2.5a), S^0 oxidation to SO_3^{2-} is mediated by a periplasmic sulfur:ferric oxidoreductase, the actual substrate of which is H_2S, and SO_3^{2-} oxidation to SO_4^{2-} is mediated by a membrane-bound sulfite:ferric oxidoreductase. These two enzymes are not coupled to the respiratory chain and therefore, this process does not provide energy for cell growth. In spite of this, several studies indicate that cells do grow by respiring Fe^{3+} with S^0 as electron donor (Ohmura *et al.*, 2002; Pronk *et al.*, 1991). The second model (Fig. 2.5b), based on kinetics and inhibition reaction studies, indicated that the Rieske/cytochrome *b* (bc_1) complex is likely implicated and that components that are soluble in the cytoplasm or attached to the cytoplasmic side of the membrane are also involved. Furthermore, it was proposed that the anaerobic Fe^{3+} reduction and the aerobic Fe^{2+} oxidation pathways might share some common components since the majority of the Fe^{2+} oxidizers can also reduce Fe^{3+}. Nevertheless, the role of rusticyanin in Fe^{3+} reduction in the Group IV iron-oxidizing '*At. ferriphilus*' JCM 7811 strain was called into question by Ohmura *et al.* (2002). Ohmura and colleagues detected this protein in cells grown with H_2 and Fe^{3+}, but devoid of copper and therefore inactive in electron transfer. In addition, a lag phase of several weeks was observed when these cells were switched to aerobic growth with Fe^{2+}, suggesting that the Fe^{3+} reduction system is distinct from the Fe^{2+} oxidation system. According to these authors, a 28 kDa periplasmic cytochrome *c* was highly expressed when Fe^{3+} was the final electron acceptor irrespective of the donor (H_2 or S^0). Furthermore, this cytochrome reduced Fe^{3+} in the presence of H_2 and is therefore likely involved in Fe^{3+} reduction. Its N-terminal sequence appeared to be different from that of Cyc1, Cyc2 and CycA1, the three cytochromes demonstrated to be involved in the Fe^{2+} respiratory pathways, again arguing against a role of the Fe^{2+} oxidizing system in Fe^{3+} reduction. Unfortunately, the sequence was not provided in the respective publication and the corresponding gene cannot be searched in the available genome sequences. Recently, the pathway between S^0 and Fe^{3+} was analysed in *At. ferrooxidans* by transcriptomics and/or proteomics (Osorio *et al.*, 2013; Kucera *et al.*, 2012). However, the data reported in these two papers diverge on several points and a consensus model cannot be reached at present. The most puzzling disagreement is about the redox proteins/genes involved in the 'downhill' pathway between Fe^{2+} and O_2: while Kucera *et al.* (2012) detected a significant increase of rusticyanin and cytochrome c_{552} (Cyc1) under anoxic conditions, Osorio *et al.* (2013) reported the opposite: higher transcription of several genes of the *rus* operon and higher level of Cyc2 in aerobic conditions. Another contradictory result concerns the short-chain dehydrogenase/reductase SdrA2 that was less abundant under anaerobic conditions according to Kucera *et al.* (2012). This protein is encoded by the *petII* operon together with the Rieske/cytochrome *b* complex that functions in the forward direction during S^0 oxidation. According to Osorio *et al.* (2013), several genes of this operon are up-regulated in anaerobic conditions. These discrepancies may reflect differences in the strain utilized and/or in the experimental design. In one case, the type strain (ATCC 23270^T) was grown in a bioreactor after adaption to aerobic or anaerobic growth with S^0 (Osorio *et al.*, 2013), while in the other case, the CCM 4253 strain was grown in Erlenmeyer flasks and the inocula resulted from cultures grown aerobically on Fe^{2+} (Kucera *et al.*, 2012). Osorio *et al.* (2013) proposed a model presented in Fig. 2.5c in which the electrons coming from S^0 oxidation are transferred to the quinone pool and then to the Rieske/cytochrome *b* complex and to the cytochrome CycA2. How the electrons are then channelled on to Fe^{3+} is unknown. However, evidences are provided that S^0 is reduced to H_2S, likely by the sulfur reductase SreABC and that H_2S reduced Fe^{3+} to Fe^{2+} abiotically (Osorio *et al.*, 2013).

In conclusion, Fe^{3+} reduction has so far been shown to occur abiotically by metabolic intermediates, e.g. H_2S, or by an enzyme having a non-specific Fe^{3+} reductase activity, such as the tetrathionate hydrolase (Sugio *et al.*, 2009) and ArsH (Mo *et al.*, 2011), although no specific iron reductase has yet been identified.

Sulfur and sulfate reduction

As mentioned previously, the dissimilatory reduction of S^0 or SO_4^{2-} is rather uncommon among the acidophiles. Only scarce data are available for *At. ferrooxidans* and *Ad. ambivalens*.

In *At. ferrooxidans* ATCC 23270^T cells grown anaerobically in the presence of S^0 and Fe^{3+}, H_2S production was detected as well as up-regulation of the *sreABCD* genes predicted to encode sulfur reductase (see above) (Osorio *et al.*, 2013). In addition, Group IV iron-oxidizing '*At. ferriphilus*' JCM 7811 was shown to grow anaerobically with H_2 as electron donor and S^0 as electron acceptor (Ohmura *et al.*, 2002). Finally, a sulfur reductase was purified from the *At. ferrooxidans* NASF-1 strain (Ng *et al.*, 2000). This enzyme, composed of three different subunits (54, 36 and 35 kDa), appeared to be quite unstable under aerobic conditions and to be absolutely dependent on NADH. It has both sulfur reductase and hydrogenase activities and must therefore be considered a sulfhydrogenase

Concerning *Ad. ambivalens*, the only way it can grow anaerobically is by reducing S^0 with H_2 as electron donor via two membrane-bound enzymes, a hydrogenase and a sulfur reductase. The electron transfer between these two enzymes is mediated by quinones. Both enzymes have been purified (Laska *et al.*, 2003). They are composed of at least three main subunits each, one of which carries the catalytic site (NiFe for hydrogenase and molybdopterin for sulfur reductase), in addition to electron transfer subunits harbouring iron–sulfur clusters, and membrane anchor proteins. The catalytic site

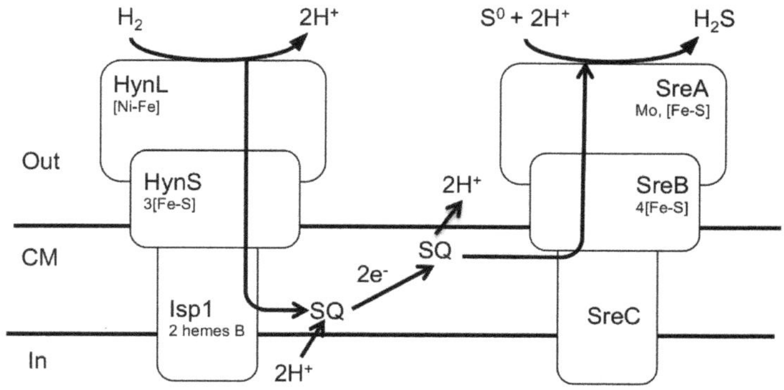

Figure 2.6 Model of sulfur respiration in *Ad. ambivalens*. Subunits of hydrogenase (left) and sulfur reductase (right) complexes with unknown function are omitted for clarity (drawn after Laska *et al.*, 2003) and (Liu *et al.*, 2012). CM, cytoplasmic membrane' SQ, Sulfolobus quinone; e⁻, electron; [Ni–Fe], nickel-fer; Mo, molybdopterin; [Fe–S], iron–sulfur cluster.

of the hydrogenase and that of sulfur reductase are predicted to be located on the periplasmic face of the membrane. With such an arrangement, the generation of an electrochemical gradient requires a redox loop mechanism in which protons are taken up by quinone on the cytoplasmic side of the membrane and released to the outside upon reoxidation (Laska *et al.*, 2003) (Fig. 2.6).

Concluding remarks

Most of the extreme acidophiles feature a versatile energetic metabolism, being able to utilize more than one electron donor and electron acceptor. From what is known today, it is obvious that for the same electron donor and the same electron acceptor, diverse respiratory pathways have evolved suggesting that they arose independently more than once in evolution. As for the case of the acidophiles, various strategies have been developed to oxidize Fe^{2+} in neutrophiles (Bird *et al.*, 2011; Ilbert and Bonnefoy, 2013). Concerning the Fe^{3+} respiratory chain, and based on what we currently know in acidophiles (which is admittedly still rudimentary) it seems that these pathways are also completely different from those present in the neutrophiles, in which essentially cytochromes *c* are involved (Bird *et al.*, 2011). In the case of the aerobic RISC oxidation pathways, apart from Sat and SQR, and the Sox system (which was likely transferred into *At. caldus*, *At. thiooxidans* and *At. ferrivorans* by HGT), most of the other enzymes seem to be specific to the acidophiles. In addition, several steps in the pathways are redundant and possibly one route is preferred over the other depending on growth conditions. This provides flexibility to these microorganisms, which may account for their efficient adaptation to environmental changes.

References

Albers, S.V., Van de Vossenberg, J.L., Driessen, A.J., and Konings, W.N. (2001). Bioenergetics and solute uptake under extreme conditions. Extremophiles 5, 285–294.

Alexander, B., Leach, S., and Ingledew, W.J. (1987). The relationship between chemiosmotic parameters and sensitivity to anions and organic acids in the acidophile *Thiobacillus ferrooxidans*. J. Gen. Microbiol. *133*, 1171–1179.

Alric, J., Lavergne, J., Rappaport, F., Vermeglio, A., Matsuura, K., Shimada, K., and Nagashima, K.V. (2006). Kinetic performance and energy profile in a roller coaster electron transfer chain: a study of modified tetraheme-reaction center constructs. J. Am. Chem. Soc. *128*, 4136–4145.

Appia-Ayme, C., Guiliani, N., Ratouchniak, J., and Bonnefoy, V. (1999). Characterization of an operon encoding two *c*-type cytochromes, an *aa*(3)-type cytochrome oxidase, and rusticyanin in *Thiobacillus ferrooxidans* ATCC 33020. Appl. Environ. Microbiol. *65*, 4781–4787.

Auernik, K.S., and Kelly, R.M. (2008). Identification of components of electron transport chains in the extremely thermoacidophilic crenarchaeon *Metallosphaera sedula* through iron and sulfur compound oxidation transcriptomes. Appl. Environ. Microbiol. *74*, 7723–7732.

Auernik, K.S., and Kelly, R.M. (2010). Physiological versatility of the extremely thermoacidophilic archaeon *Metallosphaera sedula* supported by transcriptomic analysis of heterotrophic, autotrophic, and mixotrophic growth. Appl. Environ. Microbiol. *76*, 931–935.

Auernik, K.S., Maezato, Y., Blum, P.H., and Kelly, R.M. (2008). The genome sequence of the metal-mobilizing, extremely thermoacidophilic archaeon *Metallosphaera sedula* provides insights into bioleaching-associated metabolism. Appl. Environ. Microbiol. *74*, 682–692.

Bandeiras, T.M., Refojo, P.N., Todorovic, S., Murgida, D.H., Hildebrandt, P., Bauer, C., Pereira, M.M., Kletzin, A., and Teixeira, M. (2009). The cytochrome *ba* complex from the thermoacidophilic crenarchaeote *Acidianus ambivalens* is an analog of *bc*(1) complexes. Biochim. Biophys. Acta *1787*, 37–45.

Bird, L.J., Bonnefoy, V., and Newman, D.K. (2011). Bioenergetic challenges of microbial iron metabolisms. Trends Microbiol. *19*, 330–340.

Bobadilla Fazzini, R.A., Cortes, M.P., Padilla, L., Maturana, D., Budinich, M., Maass, A., and Parada, P. (2013). Stoichiometric modeling of oxidation of reduced inorganic sulfur compounds (Riscs) in *Acidithiobacillus thiooxidans*. Biotechnol. Bioeng. *110*, 2242–2251.

Bonnefoy, V., and Holmes, D.S. (2012). Genomic insights into microbial iron oxidation and iron homeostasis in extremely acidic environments. Environ. Microbiol. *14*, 1597–1611.

Brasseur, G., Bruscella, P., Bonnefoy, V., and Lemesle-Meunier, D. (2002). The *bc*(1) complex of the iron-grown acidophilic chemolithotrophic bacterium *Acidithiobacillus ferrooxidans* functions in the reverse but not in the forward direction. Is there a second *bc*(1) complex? Biochim. Biophys. Acta *1555*, 37–43.

Brasseur, G., Levican, G., Bonnefoy, V., Holmes, D., Jedlicki, E., and Lemesle-Meunier, D. (2004). Apparent redundancy of electron transfer pathways via *bc*(1) complexes and terminal oxidases in the extremophilic chemolithoautotrophic *Acidithiobacillus ferrooxidans*. Biochim. Biophys. Acta *1656*, 114–126.

Bruscella, P., Appia-Ayme, C., Levican, G., Ratouchniak, J., Jedlicki, E., Holmes, D.S., and Bonnefoy, V. (2007). Differential expression of two *bc*(1) complexes in the strict acidophilic chemolithoautotrophic bacterium *Acidithiobacillus ferrooxidans* suggests a model for their respective roles in iron or sulfur oxidation. Microbiology *153*, 102–110.

Castelle, C., Guiral, M., Malarte, G., Ledgham, F., Leroy, G., Brugna, M., and Giudici-Orticoni, M.T. (2008). A new iron-oxidizing/O_2-reducing supercomplex spanning both inner and outer membranes, isolated from the extreme acidophile *Acidithiobacillus ferrooxidans*. J. Biol. Chem. *283*, 25803–25811.

Castelle, C.J., Roger, M., Bauzan, M., Brugna, M., Lignon, S., Nimtz, M., Golyshina, O.V., Giudici-Orticoni, M.T., and Guiral, M. (2015). The aerobic respiratory chain of the acidophilic archaeon *Ferroplasma acidiphilum*: a membrane-bound complex oxidizing ferrous iron. Biochim. Biophys. Acta *1847*, 717–728.

Cavazza, C., Giudici-Orticoni, M.T., Nitschke, W., Appia, C., Bonnefoy, V., and Bruschi, M. (1996). Characterisation of a soluble cytochrome c_4 isolated from *Thiobacillus ferrooxidans*. Eur. J. Biochem. *242*, 308–314.

Chen, L., Ren, Y., Lin, J., Liu, X., Pang, X., and Lin, J. (2012). *Acidithiobacillus caldus* sulfur oxidation model based on transcriptome analysis between the wild type and sulfur oxygenase reductase defective mutant. PLoS One *7*, e39470.

Corbett, C.M., and Ingledew, W.J. (1987). Is $Fe^{3+/2+}$ cycling an intermediate in sulphur oxidation by Fe^{2+}-grown *Thiobacillus ferrooxidans*. FEMS Microbiol. Lett. *41*, 1–6.

Cox, J.C., Nicholls, D.G., and Ingledew, W.J. (1979). Transmembrane electrical potential and transmembrane pH gradient in the acidophile *Thiobacillus ferro-oxidans*. Biochem. J. *178*, 195–200.

Dahl, C., Franz, B., Hensen, D., Kesselheim, A., and Zigann, R. (2013). Sulfite oxidation in the purple sulfur bacterium *Allochromatium vinosum*: identification of SoeABC as a major player and relevance of SoxYZ in the process. Microbiology *159*, 2626–2638.

Dopson, M., and Johnson, D.B. (2012). Biodiversity, metabolism and applications of acidophilic sulfur-metabolizing microorganisms. Environ. Microbiol. *14*, 2620–2631.

Ducluzeau, A.L., Schoepp-Cothenet, B., van Lis, R., Baymann, F., Russell, M.J., and Nitschke, W. (2014). The evolution of respiratory O_2/NO reductases: an out-of-the-phylogenetic-box perspective. J. R. Soc. Interface *11*, 20140196.

Elbehti, A., Nitschke, W., Tron, P., Michel, C., and Lemesle-Meunier, D. (1999). Redox components of cytochrome *bc*-type enzymes in acidophilic prokaryotes. I. Characterization of the cytochrome *bc*1-type complex of the acidophilic ferrous ion-oxidizing bacterium *Thiobacillus ferrooxidans*. J. Biol. Chem. *274*, 16760–16765.

Ferguson, S.J., and Ingledew, W.J. (2008). Energetic problems faced by micro-organisms growing or surviving on parsimonious energy sources and at acidic pH: I. *Acidithiobacillus ferrooxidans* as a paradigm. Biochim. Biophys. Acta *1777*, 1471–1479.

Frigaard, N.U., and Dahl, C. (2009). Sulfur metabolism in phototrophic sulfur bacteria. Adv. Microb. Physiol. *54*, 103–200.

Ghosh, W., and Dam, B. (2009). Biochemistry and molecular biology of lithotrophic sulfur oxidation by taxonomically and ecologically diverse bacteria and archaea. FEMS Microbiol. Rev. 33, 999–1043.

Giudici-Orticoni, M.T., Guerlesquin, F., Bruschi, M., and Nitschke, W. (1999). Interaction-induced redox switch in the electron transfer complex rusticyanin-cytochrome *c*(4). J. Biol. Chem. *274*, 30365–30369.

Giudici-Orticoni, M.T., Leroy, G., Nitschke, W., and Bruschi, M. (2000). Characterization of a new dihemic *c*(4)-type cytochrome isolated from *Thiobacillus ferrooxidans*. Biochemistry *39*, 7205–7211.

Gregersen, L.H., Bryant, D.A., and Frigaard, N.U. (2011). Mechanisms and evolution of oxidative sulfur metabolism in green sulfur bacteria. Front. Microbiol. *2*, 116.

Grein, F., Ramos, A.R., Venceslau, S.S., and Pereira, I.A. (2013). Unifying concepts in anaerobic respiration: insights from dissimilatory sulfur metabolism. Biochim. Biophys. Acta *1827*, 145–160.

Han, H., Hemp, J., Pace, L.A., Ouyang, H., Ganesan, K., Roh, J.H., Daldal, F., Blanke, S.R., and Gennis, R.B. (2011). Adaptation of aerobic respiration to low O_2 environments. Proc. Natl. Acad. Sci. U.S.A. *108*, 14109–14114.

Harahuc, L., and Suzuki, I. (2001). Sulfite oxidation by iron-grown cells of *Thiobacillus ferrooxidans* at pH 3 possibly involves free radicals, iron, and cytochrome oxidase. Can. J. Microbiol. *47*, 424–430.

Hedrich, S., and Johnson, D.B. (2013). Aerobic and anaerobic oxidation of hydrogen by acidophilic bacteria. FEMS Microbiol. Lett. *349*, 40–45.

Ilbert, M., and Bonnefoy, V. (2013). Insight into the evolution of the iron oxidation pathways. Biochim. Biophys. Acta *1827*, 161–175.

Ingledew, W.J. (1982). *Thiobacillus ferrooxidans*. The bioenergetics of an acidophilic chemolithotroph. Biochim. Biophys. Acta *683*, 89–117.

Ingledew, W.J., and Cobley, J.G. (1980). A potentiometric and kinetic study on the respiratory chain of ferrous-iron-grown *Thiobacillus ferrooxidans*. Biochim. Biophys. Acta *590*, 141–158.

Jameson, E., Rowe, O.F., Hallberg, K.B., and Johnson, D.B. (2010). Sulfidogenesis and selective precipitation of metals at low pH mediated by *Acidithiobacillus* spp. and acidophilic sulfate reducing bacteria. Hydrometallurgy *104*, 488–493.

Johnson, D.B., Kanao, T., and Hedrich, S. (2012). Redox transformations of iron at extremely low pH: fundamental and applied aspects. Front. Microbiol. *3*, 96.

Kappler, U., and Dahl, C. (2001). Enzymology and molecular biology of prokaryotic sulfite oxidation. FEMS Microbiol. Lett. *203*, 1–9.

Koschorreck, M. (2008). Microbial sulphate reduction at a low pH. FEMS Microbiol. Ecol. *64*, 329–342.

Kucera, J., Bouchal, P., Cerna, H., Potesil, D., Janiczek, O., Zdrahal, Z., and Mandl, M. (2012). Kinetics of anaerobic elemental sulfur oxidation by ferric iron in *Acidithiobacillus ferrooxidans* and protein identification by comparative 2-DE-MS/MS. Antonie van Leeuwenhoek *101*, 561–573.

Lane, N., Allen, J.F., and Martin, W. (2010). How did LUCA make a living? Chemiosmosis in the origin of life. Bioessays *32*, 271–280.

Laska, S., Lottspeich, F., and Kletzin, A. (2003). Membrane-bound hydrogenase and sulfur reductase of the hyperthermophilic and acidophilic archaeon *Acidianus ambivalens*. Microbiology *149*, 2357–2371.

Lauterbach, L., Lenz, O., and Vincent, K.A. (2013). H(2)-driven cofactor regeneration with NAD(P)(+)-reducing hydrogenases. FEBS J. *280*, 3058–3068.

Levican, G., Gomez, M.J., Chavez, R., Orellana, O., Moreno-Paz, M., and Parro, V. (2012). Comparative genomic analysis reveals novel facts about *Leptospirillum* spp. cytochromes. J. Mol. Microbiol. Biotechnol. *22*, 94–104.

Liljeqvist, M., Rzhepishevska, O.I., and Dopson, M. (2013). Gene identification and substrate regulation provide insights into sulfur accumulation during bioleaching with the psychrotolerant acidophile *Acidithiobacillus ferrivorans*. Appl. Environ. Microbiol. *79*, 951–957.

Liu, Y., Beer, L.L., and Whitman, W.B. (2012). Sulfur metabolism in archaea reveals novel processes. Environ. Microbiol. *14*, 2632–2644.

Mo, H., Chen, Q., Du, J., Tang, L., Qin, F., Miao, B., Wu, X., and Zeng, J. (2011). Ferric reductase activity of the ArsH protein from *Acidithiobacillus ferrooxidans*. J. Microbiol. Biotechnol. *21*, 464–469.

Ng, K.Y., Sawada, R., Inoue, S., Kamimura, K., and Sugio, T. (2000). Purification and some properties of sulfur reductase from the iron-oxidizing bacterium *Thiobacillus ferrooxidans* NASF-1. J. Biosci. Bioeng. *90*, 199–203.

Ohmura, N., Sasaki, K., Matsumoto, N., and Saiki, H. (2002). Anaerobic respiration using Fe(3+), S(0), and H(2) in the chemolithoautotrophic bacterium *Acidithiobacillus ferrooxidans*. J. Bacteriol. *184*, 2081–2087.

Ossandon, F.J., Cárdenas, J.P., Corbett, M., Quatrini, R., Holmes, D.S., Watkin, E. (2014). Draft genome sequence of the iron-oxidizing, acidophilic, and halotolerant '*Thiobacillus prosperus*' type strain DSM 5130. Genome Announc. *2*, e01042-14.

Osorio, H., Mangold, S., Denis, Y., Nancucheo, I., Esparza, M., Johnson, D.B., Bonnefoy, V., Dopson, M., and Holmes, D.S. (2013). Anaerobic sulfur metabolism coupled to dissimilatory iron reduction in the extremophile *Acidithiobacillus ferrooxidans*. Appl. Environ. Microbiol. *79*, 2172–2181.

Pandelia, M.E., Lubitz, W., and Nitschke, W. (2012). Evolution and diversification of Group 1 [NiFe] hydrogenases. Is there a phylogenetic marker for O(2)-tolerance? Biochim. Biophys. Acta *1817*, 1565–1575.

Patra, M.C., Pradhan, S.K., Rath, S.N., and Maharana, J. (2013). Structural analysis of respirasomes in electron transfer pathway of *Acidithiobacillus ferrooxidans*: a computer-aided molecular designing study. ISRN Biophys. *2013*, 14.

Prokofeva, M.I., Kublanov, I.V., Nercessian, O., Tourova, T.P., Kolganova, T.V., Lebedinsky, A.V., Bonch-Osmolovskaya, E.A., Spring, S., and Jeanthon, C. (2005). Cultivated anaerobic acidophilic/acidotolerant thermophiles from terrestrial and deep-sea hydrothermal habitats. Extremophiles *9*, 437–448.

Prokofeva, M.I., Kostrikina, N.A., Kolganova, T.V., Tourova, T.P., Lysenko, A.M., Lebedinsky, A.V., and Bonch-Osmolovskaya, E.A. (2009). Isolation of the anaerobic thermoacidophilic crenarchaeote *Acidilobus saccharovorans* sp. nov. and proposal of *Acidilobales* ord. nov., including *Acidilobaceae* fam. nov. and *Caldisphaeraceae* fam. nov. Int. J. Syst. Evol. Microbiol. *59*, 3116–3122.

Pronk, J.T., Liem, K., Bos, P., and Kuenen, J.G. (1991). Energy transduction by anaerobic ferric iron respiration in *Thiobacillus ferrooxidans*. Appl. Environ. Microbiol. *57*, 2063–2068.

Protze, J., Muller, F., Lauber, K., Nass, B., Mentele, R., Lottspeich, F., and Kletzin, A. (2011). An extracellular tetrathionate hydrolase from the thermoacidophilic archaeon *Acidianus ambivalens* with an activity optimum at pH 1. Front Microbiol. *2*, 68.

Quatrini, R., Appia-Ayme, C., Denis, Y., Ratouchniak, J., Veloso, F., Valdes, J., Lefimil, C., Silver, S., Roberto, F., Orellana, O., *et al.* (2006). Insights into the iron and sulfur energetic metabolism of *Acidithiobacillus ferrooxidans* by microarray transcriptome profiling. Hydrometallurgy *83*, 263–272.

Quatrini, R., Appia-Ayme, C., Denis, Y., Jedlicki, E., Holmes, D.S., and Bonnefoy, V. (2009). Extending the models for iron and sulfur oxidation in the extreme acidophile *Acidithiobacillus ferrooxidans*. BMC Genomics *10*, 394.

Rauhamäki, V., and Wikström, M. (2014). The causes of reduced proton-pumping efficiency in type B and C respiratory heme-copper oxidases, and in some mutated variants of type A. Biochim. Biophys. Acta *1837*, 999–1003.

Slonczewski, J.L., Fujisawa, M., Dopson, M., and Krulwich, T.A. (2009). Cytoplasmic pH measurement and homeostasis in bacteria and archaea. Adv. Microb. Physiol. *55*, 1–79, 317.

Sugio, T., Domatsu, C., Munakata, O., Tano, T., and Imai, K. (1985). Role of a ferric ion-reducing system in sulfur oxidation of *Thiobacillus ferrooxidans*. Appl. Environ. Microbiol. *49*, 1401–1406.

Sugio, T., Katagiri, T., Inagaki, K., and Tano, T. (1989). Actual substrate for elemental sulfur oxidation by sulfur: ferric ion oxidoreductase purified from *Thiobacillus ferrooxidans*. Biochim. Biophys. Acta *973*, 250–256.

Sugio, T., Taha, T.M., and Takeuchi, F. (2009). Ferrous iron production mediated by tetrathionate hydrolase in tetrathionate-, sulfur-, and iron-grown *Acidithiobacillus ferrooxidans* ATCC 23270 cells. Biosci. Biotechnol. Biochem. *73*, 1381–1386.

Sugio, T., Ako, A., and Takeuchi, F. (2010). Sulfite oxidation catalyzed by *aa*(3)-type cytochrome *c* oxidase in *Acidithiobacillus ferrooxidans*. Biosci. Biotechnol. Biochem. *74*, 2242–2247.

Suzuki, I. (1999). Oxidation of inorganic sulfur compounds: chemical and enzymatic reactions. Can. J. Microbiol. *45*, 97–105.

Taha, T.M., Kanao, T., Takeuchi, F., and Sugio, T. (2008). Reconstitution of iron oxidase from sulfur-grown *Acidithiobacillus ferrooxidans*. Appl. Environ. Microbiol. *74*, 6808–6810.

Talla, E., Hedrich, S., Mangenot, S., Ji, B., Johnson, D.B., Barbe, V., and Bonnefoy, V. (2014). Insights into the pathways of iron- and sulfur-oxidation, and biofilm formation from the chemolithotrophic acidophile *Acidithiobacillus ferrivorans* CF27. Res. Microbiol. *165*, 753–760.

ten Brink, F., Schoepp-Cothenet, B., van Lis, R., Nitschke, W., and Baymann, F. (2013). Multiple Rieske/cyt*b* complexes in a single organism. Biochim. Biophys. Acta *1827*, 1392–1406.

Yarzabal, A., Brasseur, G., and Bonnefoy, V. (2002). Cytochromes *c* of *Acidithiobacillus ferrooxidans*. FEMS Microbiol. Lett. *209*, 189–195.

Appendix

Enzymes involved in reduced inorganic sulfur compounds oxidation in acidophiles for which a model has been proposed. References not cited in (Dopson and Johnson, 2012; Liu *et al.*, 2012; Quatrini *et al.*, 2009) are given in the last column.

Reactions	*At. ferrooxidans*	*At. ferrivorans*	*At. caldus*	*At. thiooxidans*	*M. sedula*	*S. tokodaii*	*Ad. ambivalens*	References
$H_2S \rightarrow H\text{-}S_n\text{-}H$	SQR	SQR	SQR	SQR	SQR	SQR	SQR	Brito *et al.*, 2009. Biochemistry *48*, 5613–5622 Cherney *et al.*, 2010. J. Mol. Biol. *398*, 292–305 Goltsman *et al.*, 2009. Appl. Environ. Microbiol. *75*, 4599–4615 Mi *et al.*, 2011. J. Microbiol. *49*, 890–901
$S_2O_3^{2-} \rightarrow SO_4^{2-}$	–	nf-Sox	Sox	Sox	–	–	–	Chen *et al.*, 2012. PLoS One *7*, e39470 Talla *et al.*, 2014. Res. Microbiol. *165*, 753–760 Travisany *et al.*, 2014. Res. Microbiol. *165*, 743–752
$S_2O_3^{2-} \rightarrow S_4O_6^{2-}$	TQO	TQO	TQO	TQO	TQO	TQO	TQO	Bobadilla Fazzini *et al.*, 2013. Biotechnol. Bioeng. *110*, 2242–2251 Chen *et al.*, 2012. PLoS One *7*, e39470 Liljeqvist *et al.*, 2013. Appl. Environ. Microbiol. *79*, 951–957 *Talla et al.*, 2014. Res. Microbiol. 165, 753–760 Travisany *et al.*, 2014. Res. Microbiol. *165*, 743–752.
$S_2O_3^{2-} \rightarrow S_4O_6^{2-}$	TSD	TSD	–	–	–	–		Kikumoto *et al.*, 2013. Appl. Environ. Microbiol. *79*, 113–120. Talla *et al.*, 2014. Res. Microbiol. *165*, 753–760.
$S_4O_6^{2-} \rightarrow S_2O_3^{2-} + S^0 + SO_4^{2-}$	TetH	TetH	TetH	TetH	TetH	TetH	TetH	Beard *et al.*, 2011. Front. Microbiol. *2*, 79 Bugaytsova and Lindstrom, 2004. Eur. J. Biochem. *271*, 272–280 Buonfiglio *et al.*, 1999. J. Biotechnol. *72*, 85–93 de Jong *et al.*, 1997. Microbiology *143*, 499–504 Kanao *et al.*, 2010. FEMS Microbiol. Lett. *309*, 43–47 Krupovic *et al.*, 2012. Appl. Environ. Microbiol. *78*, 5463–5465 Osorio *et al.*, 2013. Appl. Environ. Microbiol. *79*, 2172–2181 Protze *et al.*, 2011. Front. Microbiol. *2*, 68 Tano *et al.*, 1996. Biosci. Biotechnol. Biochem. *60*, 224–227

Continued...

Reactions	*At. ferrooxidans*	*At. ferrivorans*	*At. caldus*	*At. thiooxidans*	*M. sedula*	*S. tokodaii*	*Ad. ambivalens*	References
$S^0 \rightarrow SO_3^{2-}$ + H_2S+ $S_2O_3^{2-}$	–	SOR	SOR	–	–	SOR	SOR	Chen *et al.*, 2012. PLoS One 7, e39470 He *et al.*, 2000. FEMS Microbiol. Lett. *193*, 217–221 Janosch *et al.*, 2009. Adv. Mat. Res. *71–73*, 239–242 Liljeqvist *et al.*, 2013. Appl. Environ. Microbiol. *79*, 951–957 Sun *et al.*, 2003. Extremophiles *7*, 131–134 Talla *et al.*, 2014. Res. Microbiol. *165*, 753–760
$S^0 \rightarrow SO_3^{2-}$	HDR	HDR	HDR	HDR	HDR	HDR		Chen *et al.*, 2012. PLoS One *7*, e39470 Liljeqvist *et al.*, 2013. Appl. Environ. Microbiol. *79*, 951–957 Liu *et al.*, 2013. Curr. Microbiol. *66*, 88–95 Liu *et al.*, 2012. Curr. Microbiol. *65*, 416–423 Osorio *et al.*, 2013. Appl. Environ. Microbiol. *79*, 2172–2181 Talla *et al.*, 2014. Res. Microbiol. *165*, 753–760
$SO_3^{2-} \rightarrow$ APS	–	–	–	–	–	–	APS reductase	
APS $\rightarrow SO_4^{2-}$	Sat	Sat	Sat	Sat	Sat	Sat	APAT	Chen *et al.*, 2012. PLoS One *7*, e39470 Jaramillo *et al.*, 2012. Bioinformation *8*, 695–704 Liljeqvist *et al.*, 2013. Appl. Environ Microbiol. *79*, 951–957 Talla *et al.*, 2014. Res Microbiol *165*, 753–760
$SO_3^{2-} \rightarrow SO_4^{2-}$	–	–	–	–	–	–	SAOR	

ND, not determined; nf, non-functional (pseudogene); –, not detected.

Adaptation to Extreme Acidity and Osmotic Stress

3

Carla M. Zammit and Elizabeth L.J. Watkin

Abstract

Environments that are either acidic or have high osmotic potentials are found across the globe in a range of natural and anthropogenic systems. The organisms capable of inhabiting these systems are diverse, including archaea, bacteria and eukaryotes. However, environments where extreme acidity is combined with osmotic stress deriving from elevated concentrations of sodium chloride are seemingly rare. Subsequently, there is a relatively small number of species which have been identified and shown to tolerate both of these stresses simultaneously, and as a result the mechanisms that permit life in these harsh conditions has not been extensively studied. Recent genomic and proteomic studies indicate that several strategies may be employed by acidophilic microorganisms to combat the combined effects of low pH and high osmotic stress, most notably the production of osmo-protectants and the maintenance of membrane integrity. This chapter focuses on iron- and sulfur-oxidizing microorganisms, which are able to tolerate acidic conditions, the effect of osmotic stress induced by salinity on their survival, and mechanisms used to survive these stresses both independently and in combination.

Introduction

Osmotic stress is caused when external concentrations of soluble salts and/or non-ionic solutes differ markedly from intracellular (homeostatic) conditions. When the external environment contains higher concentrations of salts, solutes and/or soluble substrates, water is drawn out across the cell membrane, eventually leading to cell death via metabolic dysregulation. Alternatively, when the external environment contains less salt, solute and/or substrate, water is drawn into the cell leading to cell lysis. This chapter will focus mainly on osmotic stress induced by elevated salinity.

Halophiles are organisms that are 'salt loving'. They require elevated concentrations of salt (NaCl) for growth and are able to tolerate high levels of dissolved salts, while halotolerant microorganisms can grow under saline conditions, but do not require NaCl. While these classifications are a useful frame of reference for discussing these organisms, naturally occurring systems vary greatly in NaCl concentrations. While halophilic microorganisms are restricted to environments of high salt concentrations, halotolerant organisms may encounter transient changes in NaCl concentration outside their ideal growth conditions and, as a result, must adapt to fluctuations in environmental salt conditions.

There are a number of adaptions that halophiles and halotolerant microorganisms employ in order to survive in environments with high osmotic pressure. These microorganisms have adapted their intracellular macromolecules to saline conditions, export salts once they have entered the cytoplasm, and/or prevent salts from entering the cytoplasm. Similarly, acidophilic microorganisms expend energy to maintain a circumneutral intracellular pH through adaptions to the cell membrane, the use of cellular pumps and the production of cytoplasmic buffers. The majority of research into halophiles has been undertaken under neutral or alkaline conditions (de la Haba *et al.*, 2011). The mechanisms that microorganisms may employ to deal with both osmotic stress and acidic conditions simultaneously, and how these differ to the mechanisms employed to deal with both stresses individually, is not well understood. In this chapter the adaptations microorganisms exhibit to survive in acidic conditions and osmotic stress (specifically resulting from high levels of NaCl), the potential environments where microorganisms capable of dealing with both acidic and osmotic stresses may be isolated and the tolerances of known acidophilic microorganisms to osmotic stress induced by salinity, are described.

Adaptation to extreme acidity

Acidophiles maintain a circumneutral intracellular pH, and they can tolerate pH gradients across the cytoplasmic membrane several orders of magnitude greater than neutrophiles (Baker-Austin and Dopson, 2007). Maintaining a circumneutral intracellular pH is important as an influx of protons into the cytoplasm will result in acidification which can adversely affect a number of cellular functions as most proteins function across a narrow pH range (Krulwich *et al.*, 2011). One of the most severe consequences of cytoplasmic acidification is the disruption to pH gradients across the cytoplasmic membrane resulting in changes to the electrochemical gradient (as this is the major contribution to the proton motive force (PMF); Baker-Austin and Dopson, 2007). Many of the mechanisms employed by neutrophiles are conserved in acidophiles, but the methods by which these mechanisms are altered to provide the protective function in the acidophilic environment are poorly understood (Krulwich *et al.*, 2011). The current published literature suggests that acidophiles employ a number of mechanisms to deal with pH stress. Of these mechanisms, the adaptations thought to be key are (i) changes to the cell membrane; (ii) use and structure of cellular pumps; (iii) production of cytoplasmic buffers; (iv) the modification and repair of DNA; (v) the generation of a negative transmembrane potential (reviewed in Baker-Austin and Dopson, 2007).

Acidophilic archaea have rigid and impermeable cytoplasmic membranes that limit the influx of protons into the cytoplasm (Jones *et al.*, 2012; Macalady *et al.*, 2004). Acidophilic archaea have membranes composed of glycerol ethers, which are less susceptible to acid hydrolysis than bacterial glycerol esters (Baker-Austin and Dopson, 2007). The cytoplasmic membrane of thermoacidophilic archaea is composed of a monolayer of tetraether lipids which are physico-chemically more stable, allowing these microorganisms to deal with extremes in pH and temperature (Macalady *et al.*, 2004). Increasing the number of sugar moieties and cyclopentane rings in these lipids reduces proton permeability of the membranes enabling the archaea to adapt to pH stress (Chong *et al.*, 2010). Additionally, some acidophilic archaea have high glycolipid content; under low-pH conditions there is an increased attachment of these glycolipids to the outer surface of the cell, which is thought to protect the cell from the influx of protons (Chong, 2010).

The role and structure of acidophilic bacteria cytoplasmic membranes under pH stress is not as well understood. In response to a decrease in pH from 2.3 to 1.3, the type strain of acidophilic bacterium, *Acidithiobacillus ferrooxidans* has been shown to modulate multiple cell envelope components to decrease proton permeability (Chao *et al.*, 2008). With decreasing external pH, both *At. ferrooxidans* and *Acidithiobacillus caldus* have been shown to have a reduction in cell membrane fluidity, limiting proton influx (Mangold *et al.*, 2013; Mykytczuk *et al.*, 2010). This rigidity can be partly attributed to the presence of hopanoid lipids that have been detected as a component of the cytoplasmic membrane of acidophiles (Jones *et al.*, 2012). The cell surfaces of acidophilic bacteria have high isoelectric (pI) points in comparison to neutrophilic bacteria (Chi *et al.*, 2007). The high pI assists in repulsion of protons from the membrane surface impeding the inflow of protons into the cell thus protecting the cell from decreasing pH.

In addition to changes to cell membrane lipid composition, acidophiles have specialized membrane proteins to deal with pH stress. The Omp40 protein from *At. ferrooxidans* is another example of structural adaptations used to impede proton influx. Omp40 is a trimeric porin which forms an ion channel. In *Escherichia coli* the L3 loop (a region forming part of the internal surface of the channel) of Omp40 is small and negatively charged however, in *At. ferrooxidans* this loop is both larger in size and positively charged at acidic pH (Guiliani and Jerez, 2000). This change assists in passive exclusion of protons by electrostatic repulsion. The membrane proteins, PspA has been shown to bind to membrane phospholipids in a way that may prevent proton leakage assisting in the maintenance of intracellular pH (Kobayashi *et al.*, 2007). This assertion is supported by evidence that shows PspA was found to be more abundant in *At. caldus* cells grown at pH 1.1 than in cells grown at pH 2.5 or 4.0 (Mangold *et al.*, 2013).

Unlike neutrophiles, acidophilic bacteria are able to maintain an internal positive membrane potential under acidic conditions (Matin, 1990). Putative cation transporters, abundant in many acidophilic genomes, pump cations (e.g. potassium or sodium) into the cytoplasm in order reduce the influx of protons by electrostatic repulsion (Jones *et al.*, 2012). In addition putative proton efflux systems directly remove protons from the cytoplasm in an energy dependent manner and have been found in a range of acidophilic microorganisms (Futterer *et al.*, 2004; González *et al.*, 2014). Greater activity of the cation transporters than the proton efflux systems is thought to generate the internal positive membrane potential (Baker-Austin and Dopson, 2007). However, the role of putative proton efflux and cation transporters in the maintenance of acidophiles intracellular pH is unclear and warrants further research (Baker-Austin and Dopson, 2007).

Microorganisms have been shown to buffer cytoplasmic pH by sequestering cations or degrading acid generating compounds. Cytoplasm buffering molecules have an abundance of alkaline amino acids (e.g. lysine, histidine and arginine) that help to stabilize the cytoplasmic pH under acidic conditions. The synthesis of these alkaline amino acids has been shown to increase with decrease in environmental pH in the acidophilic bacterium *Leptospirillum ferrooxidans* (Parro *et al.*, 2007). Under low-pH conditions, protonated organic acids (such as acetic, lactic, formic or propionic) can pass into the cytoplasm where, due to the circumneutral nature of the internal cellular environment, the dissociation to a proton and conjugate base can lead to rapid cytoplasmic acidification (Ciaramella *et al.*, 2005). To counteract this, the genomes of acidophilic microorganisms contain genes coding for enzymes that are responsible for organic acid degradation (Futterer *et al.*, 2004). Despite the suggestive evidence that these enzymes are used for pH homeostasis, it is unclear if acidophilic microorganisms use organic acid degradation specifically to stabilize intracellular pH or purely for metabolic activity (Baker-Austin and Dopson, 2007).

Finally, acidophiles have been shown to have an abundance of DNA modifications and repair mechanisms. While low pH has been shown to be a source of significant DNA damage, it should be noted that these adaptations assist microorganisms in dealing with a range of stresses rather than as a specific response to pH stress alone (Parro *et al.*, 2007).

There are a diverse number of modifications which acidophiles may employ to deal with acidic conditions, however many of these systems remain poorly understood. To date the majority of research into pH homeostasis in extreme acidic conditions has been guided by mechanisms elucidated from transient acidic stress in *Helicobacter pylori* and *E. coli*. Obligate acidophiles, however, are likely to display novel modifications that this strategy will not reveal and, as such, future research needs to be guided by genomic and proteomic studies of obligate acidophiles.

Adaption to osmotic stress

Turgor pressure is the driving force for cell expansion, growth and division therefore must be maintained by microorganisms to survive variations in osmotic pressure (Csonka and Hanson, 1991). In general, microorganisms have developed two distinct mechanisms to deal with high osmotic pressure. The first mechanism, which is mostly found in halophilic archaea, involves the intracellular accumulation of chloride and potassium, which leads to sodium being extruded from the cell. The second mechanism is generally present in halotolerant bacteria and eukaryotes that have to deal with fluctuating levels of salt and involves the accumulation of osmo-protectants in the cytoplasm (Empadinhas and da Costa, 2008).

Osmo-adaption in halotolerant bacteria is biphasic. In both Gram-positive and Gram-negative bacteria an initial immediate and rapid stimulation of potassium uptake is seen (Empadinhas and da Costa, 2008). In *E. coli* the loss of turgor pressure due to osmotic stress activates major potassium transport systems, particularly the high-affinity potassium uptake system, Kdp, which serves as an osmotically inducible potassium scavenger when potassium is present in low concentrations (Sleator and Hill, 2002). At low levels of osmotic stress the increasing charge of potassium inside the cell is compensated for by an accumulation of amino acids and derivatives such as α-glutamate. However at higher levels of osmotic challenge compatible solutes are required (Empadinhas and da Costa, 2008).

Compatible solutes (osmo-protectants) are necessarily highly soluble, low molecular weight organic molecules that carry no net charge at physiological pH, and do not interact with proteins. Sequestration of these compounds in the cytoplasm aids in reducing osmotic stress by balancing internal solute levels with those in the immediate environment. Compatible solutes can accumulate either through *de novo* synthesis or uptake from the environment (Csonka and Hanson, 1991). There are a number of classes of compatible solutes produced by microorganisms. Many bacteria and archaea synthesize and accumulate zwitterionic molecules; derived from amino acids, examples being ectoine and its derivative hydroxyectoine which are the most common osmotic solutes in bacteria (Empadinhas and da Costa, 2008). Betaine and β-glutamine are other examples of this class of osmolytes. Trehalose, a non-reducing glucose disaccharide that occurs in a wide variety of bacteria, both halotolerant and halophilic, is an example of a non-charged solute (Empadinhas and da Costa, 2008).

In addition to maintaining turgor, halotolerant bacteria will alter the lipid composition of their membranes in response to osmotic stress. By increasing the proportion of anionic phospholipids and/or glycolipids in the outermost membrane the surface charge of the cell will change. This is thought to help maintain the hydration of the interface between cell and environment enabling the maintenance of the phospholipid bilayer (Sleator and Hill, 2002). Additionally, membrane porins are also modified. Under increased external osmolality *E. coli* reduces the expression of genes coding for outer membrane channel proteins OmpC and OmpF, reducing the non-specific diffusion of small hydrophilic molecules (Csonka and Hanson, 1991).

Modifications employed by halotolerant microorganisms under osmotic stress are utilized to maintain turgor pressure and membrane integrity. As a number of these have implications on the maintenance of the transmembrane potential, the question arises as to the impact of these on pH homeostasis in acidophilic bacteria. In addition, some of the adaptations halophiles and acidophiles possess may inhibit them from persisting in acidic saline conditions. For example, the cell surface of acidophilic microorganisms has a high pI, whereas the cell surface of halophilic microorganisms has a low pI (Chi *et al.*, 2007; Stan-Lotter *et al.*, 1988.

Adaptation of acidophilic microorganisms to osmotic stress

Acidophiles possess adaptations to their cytoplasmic membranes which help to lower permeability to water and/or ions, these changes help to regulate the osmotic pressure within cells and prevent intracellular acidification (Macalady *et al.*, 2004). However, membrane-permeable anions (such as chloride and nitrate) are able to enter the cell membrane, reducing the transmembrane potential, resulting in the uptake of ions and subsequently the acidification of the cytoplasm (Suzuki *et al.*, 1999).

The impact of chloride on transmembrane potential is possibly the most detrimental effect to the growth of acidophilic microorganisms. Chloride has been demonstrated to disrupt the near neutral internal pH of acidophilic microorganisms, as the difference in ionic strength between the intracellular and extracellular environment causes the cell to uptake more ions, including protons, leading the acidification of the cytoplasm (Matin, 1991). This hypothesis is supported by various studies such as the investigation by Suzuki *et al.* (1999), which proposed that membrane-permeable anions (chloride included) were able to enter the cell membrane reducing the transmembrane potential. Other salts, such as Na_2SO_4 and K_2SO_4, do not inhibit acidophiles beyond the impact they have on osmotic potential (Blight and Ralph, 2004). Using *L. ferriphilum*, Ojumu *et al.* (2008) showed that an increase in ionic strength decreased iron oxidation rates and the amount of biomass generated. They suggested that the osmotic gradient formed between the cellular interior and the external environment placed a great energetic stress on the bacterium, and that increases in ionic stress potentially limit microbial growth but to a lesser extent than chloride induced intracellular acidification.

Shiers *et al.* (2005) investigated the impact of chloride on the growth of an undefined mixed culture of acidophiles, demonstrating that there was a 50% reduction in growth at 0.11 M NaCl or 0.28 M Na_2SO_4. Deveci *et al.* (2008) observed a significant inhibition of the bioleaching of a sulfide ore by a mixed consortium of mesophilic bacteria in the presence of 0.17 M NaCl, whereas Bevilaqua *et al.* (2013) observed enhanced leaching of chalcopyrite in the presence of *At. ferrooxidans* and *At. thiooxidans* with 0.10 M NaCl, although it is unclear whether this was a result of the microorganisms or due to the inhibition of secondary phases which form during chalcopyrite bioleaching. Additionally, it has been shown that many species of acidophiles cannot be adapted to increasing chloride concentrations (Shiers *et al.*, 2005). These studies demonstrate that many acidophilic microorganisms have little tolerance to chloride (Shiers *et al.*, 2005; Blight and Ralph, 2004). Adaptations that acidophilic microorganisms employ to deal with osmotic stress are similar to those that have been identified as neutrophilic microorganisms, utilizing osmo-protectants and through the accumulation of intracellular potassium.

Osmo-protectants

Compatible solutes, as described in halotolerant microorganisms, have been identified in acidophiles exposed to osmotic stress; trehalose, taurine, ectoine and hydroxyectoine are examples of such compounds. By maintaining osmotic balance, acidophiles under osmotic stress are able to prevent the leakage of undesirable free anions into the cell and the downstream effects this has on the maintenance of the positive transmembrane potential. There is some evidence that osmo-protectants are utilized by acidophilic microorganisms under osmotic stress. Kieft and Spence (1988) conducted a study on the effect of osmoprotectants on the ability of *At. ferrooxidans*, to tolerate various salts. The addition of 10 mM betaine to the media acted as an osmo-protectant, although this was demonstrated only at the highest NaCl level tested (0.4 M) whereas 10 mM proline was shown to increase the tolerance of *At. ferrooxidans* to NaCl at all salt concentrations tested (0.2 and 0.4 M). It remains unknown if *At. ferrooxidans* manufactures these osmo-protectants, though *At. ferrooxidans* ATCC 23270^{T} has gene pathways for other known osmo-protectants, trehalose and taurine biosynthesis (Valdés *et al.*, 2008).

A study of the metabolites from biofilms at the Richmond Mine, Iron Mountain (pH 1.0), found that taurine, ectoine and hydroxyectoine were used as osmo-protectants by resident microorganisms in order to cope with osmotic stress (Mosier *et al.*, 2013). It was shown that the bacteria and archaea present in the biofilm were unable to synthesis taurine; however, the low-abundance community member, *Sulfobacillus*, was found to possess genes encoding two taurine transport proteins. The most likely source of taurine was thought to be originating from the fungi present in the biofilm; it was proposed that the bacteria transport the compound into the cell where it then acts as an osmo-protectant.

A number of compatible solutes have been identified in *Leptospirillum* spp. Production of trehalose has been demonstrated in all *Leptospirillum* groups (Goltsman *et al.*, 2009; Parro *et al.*, 2007). In a community genomic and proteomic analysis of biofilms from the Richmond Mine it was shown that both *Leptospirillum* Group II (*L. ferriphilum*, '*L. rubarum*') and Group III ('*L. ferrodiazotrophum*') had three of four known pathways for the synthesis of trehalose (Goltsman *et al.*, 2009). Parro *et al.* (2007) demonstrated that the transcriptome of an environmental strain of *L. ferrooxidans* (Group I) was able to biosynthesise and transport ectoine and hydroxyectoine, while the study by (Goltsman *et al.*, 2009) identified these pathways in both Group I and Group II *Leptospirillum*. This study also demonstrated that the proteins involved in these pathways were present in greater levels at the early stages of biofilm formation where extracellular polymeric substance (EPS) levels were low and microorganisms are presumably more likely to be exposed to greater osmotic stress without the protection of the biofilm. On the other hand, *Leptospirillum* Group III ('*L. ferrodiazotrophum*') does not contain the genes for these pathways and the authors propose that the Group II *Leptospirillum* are better equipped to cope with osmotic challenge than Group III (Goltsman *et al.*, 2009). Genes encoding enzymes for both ectoine and trehalose biosynthetic pathways were identified in the recently published draft genome of the *L. ferriphilum* DSM 14647 (Cárdenas *et al.*, 2014). However, this appears not to have been demonstrated experimentally.

There have been a number of other studies that demonstrate the use of osmo-protectants in acidophilic microorganisms. The acidophilic heterotroph *Acidiphilium cryptum* produced increasing amounts of trehalose and hydroxyectoine in

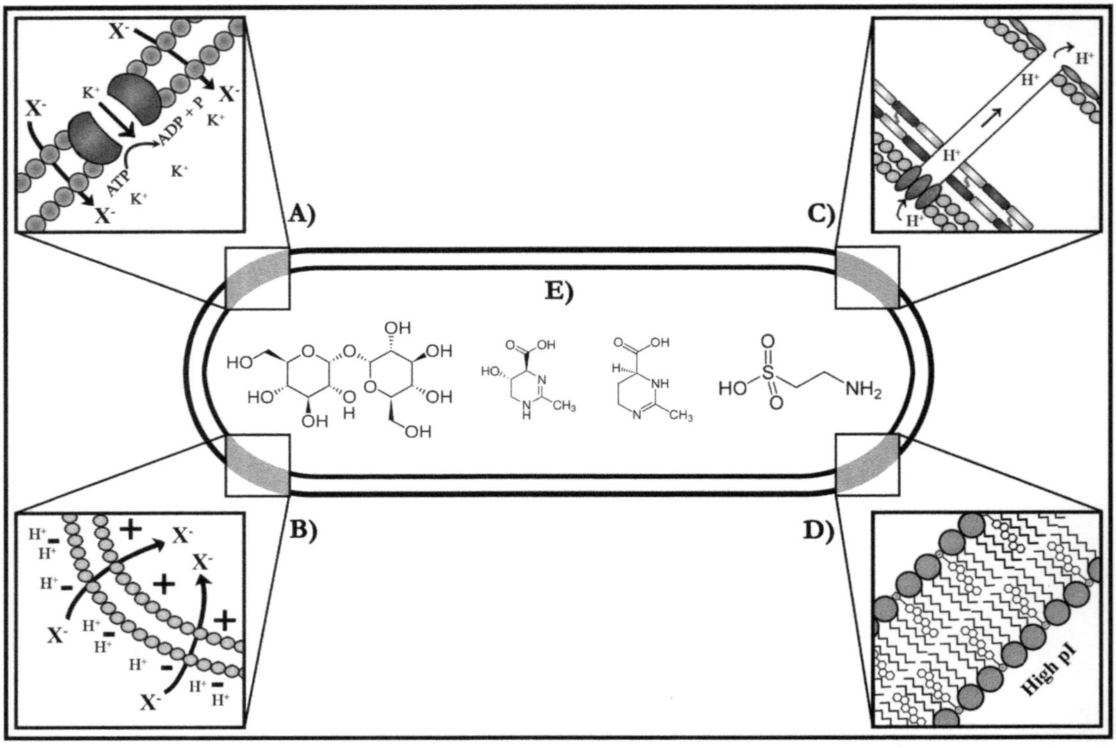

Figure 3.1 Strategies employed by acidophilic microorganisms to combat the effects of osmotic stress have implications on pH homeostasis. (A) Soluble anions mitigate the effects of cation sequestration. (B) The influx of soluble anions reduces the positive membrane potential. (C) Active proton export is up-regulated to counter-act the influx of protons due to changes in osmotic balance. (D) Densely packed, minimally fluid membranes rich in hopanoid lipids maintain a high external pI excluding cations from the cell. (E) Trehalose, ectoine, hydroxyectoine and taurine are commonly produced compatible solutes. Balancing internal solute levels prevents leakage of free anions into the cell.

response to increasing osmotic stress caused by NaCl or $Al_2(SO_4)_3$ with hydroxyectoine produced at a ratio of 5:1 with trehalose (Moritz *et al.*, 2015). In many bacterial species, proline has been shown to function as a compatible solute that accumulates within cells that are challenged by osmotic stress (Sleator and Hill, 2002). In a proteomic study of *At. caldus* SM-1 the production of enzymes involved in proline synthesis was only observed when cells were grown in the presence of 0.5 M NaCl (Guo *et al.*, 2014). Genome analysis of *Thiobacillus prosperus* revealed genes involved in the biosynthesis of the compatible solutes ectoine, sucrose and glycine betaine (Ossandon *et al.*, 2014).

Other systems

A number of studies have identified other means of halotolerance in acidophilic microorganisms. Parro *et al.* (2007) demonstrated the up-regulation of genes for a two-component osmo-sensitive potassium channel system in isolates of *L. ferrooxidans* isolated from environmental samples that had been subjected to an increase in salinity levels. This study also demonstrated the up-regulation of an ABC high-affinity potassium transport system. The authors proposed that potassium may play an important role as an osmo-regulator in natural environments but, due to the known role of potassium sequestration in pH homeostasis, elucidating the extent to which potassium serves as an osmoregulator has proven experimentally challenging.

Dew and du Plessis (2002) observed a negative effect of high NaCl levels on cell membrane integrity, implying the need for increased biosynthesis of cell membrane proteins. A proteomic study of *At. caldus* (DSM8584^T) and *Acidimicrobium ferrooxidans* (DSM10331^T) exposed to 0.10 and 0.17 M NaCl (respectively) found that proteins involved in membrane biosynthesis were differentially abundant in both microorganisms upon exposure to NaCl (Zammit *et al.*, 2012). Guo *et al.* (2014) also demonstrated an increased abundance of proteins involved in membrane biosynthesis when *At. caldus* was grown in the presence of NaCl. Both groups also identified other proteins differentially abundant upon exposure to NaCl, including those involved in energy production, central carbon metabolism and carbon dioxide fixation, implying a metabolic shift of carbon flux in bacteria under NaCl stress. A greater expression of genes involved in the key metabolic reactions of carbon metabolism and energy production in *L. ferrooxidans* under salinity stress was also demonstrated by Parro *et al.* (2007). These results suggested that several strategies might be employed by acidophilic microorganisms to combat the effects of chloride on osmotic stress and pH homeostasis. These involve (i) offsetting the increase in intracellular pH; (ii) stabilizing the outer membrane; (iii) the production of amino acids, and other compatible solutes possibly to serve as osmo-protectants (Fig. 3.1).

The adaptations employed to mitigate osmotic stress have been demonstrated across a range of acidophilic microorganisms. However, most of this research has focused on microorganisms that do not tolerate high levels of osmotic stress and on the mechanisms of the effect of osmotic stress in general. As chloride salts has a far greater inhibitory effect on acidophiles than sulfate salts, the specific effect of chloride on acidophiles and the adaptations that these microorganisms have to deal with increased levels of this anion is an area that requires more research.

Acidic saline environments

As highlighted in Chapter 1, acid can be generated biologically or abiotically and acidic environments vary vastly in size and nature. Environments high in osmotic stress range from saltpans to the surface of leaves. Both acidic and high osmotic stress environments can be found globally, in a diverse range of locations. However, acidic–saline environments are found in only a few known locations on earth, such as a shallow geothermally heated seafloor at the beach of Porto di Levante (Italy) and the acid saline lakes of Australia, Chile and Bolivia. Hence, there may have been little evolutionary pressure for acidophilic microorganisms to develop systems to deal with saline conditions. Acidic saline environments on Earth can develop naturally while others have been created as a result of anthropogenic activities. Acidic saline environments result when acidic systems mix with water carrying salts. The acid can be generated by acid rain or proximity to iron or sulfide deposits, which oxidize upon exposure to oxygen to create acid. Salt can enter these systems from the influx of seawater or from marine evaporite deposits (McArthur *et al.*, 1991). Naturally occurring environments, which have these properties, include acidic saline lakes, volcanoes that occur in coastal areas and evaporite deposits with high iron and/or sulfide content.

Acidic saline environments that have developed due to anthropogenic activities mainly stem from the mining industry, for example heaps in biomining operations, tailings dams and AMD. Heaps in biomining operations are acidic, as a consequence of the oxidative dissolution of metal sulfides, As a consequence of either the presence of salts within the ore or the recycling of leachate solutions (or both), these systems can also develop saline conditions. Tailings dams are used to contain the wastes from mining operations. Again the dissolution of iron and/or sulfide minerals can lead to acidification, while saline conditions can develop through the concentration of salts due to evaporation. In AMD, sulfide minerals from an exposed ore bodies and mineral wastes (tailings etc.) oxidize and produce acid, in some cases this can occur in environments that already have high salt concentrations or become saline due to evaporation.

Acidic saline lakes, drains and deposits

Salt lakes develop in arid or semi-arid regions, where their waters are enriched with dissolved solids. These lakes can be acidic, neutral or alkaline, and multiple lakes often form within close proximity to one another (Benison and LaClair, 2003). The source of this acidity is contested amongst scientists, though it is commonly attributed to the oxidation of sulfidic deposits which occur within close proximity to the lake (McArthur *et al.*, 1991). The salt from these lakes originates from seawater, in some cases in the form of ancient evaporite deposits. Interestingly, acidic saline environments have been proposed to serve as an analogue to early

Martian environments. The acidic saline lakes of southern Western Australia and north-west Victoria (Australia) have high amounts of iron oxides and sulfates with either very low concentrations, or the complete absence of carbonates, which are also features that have been identified in Martian environments (Benison and LaClair, 2003). Therefore investigating microorganisms inhabiting acidic saline lakes and the mechanisms they employ to live in such harsh environments may provide us with an insight into life on early Mars (Chapter 17).

Salt lakes are a prominent feature across the Yilgarn (southern Western Australia and South Australia), as the geological conditions in this region are conductive to their formation (Dickson and Giblin, 2006). Due to the mineralogy, climatology and terrestrial flora (dominated by *Eucalyptus* sp.) of the region, the groundwater pH can range from acidic to alkaline (Peiffer *et al.*, 2009). In many agricultural areas saline ground waters are rising due to the clearing of deep-rooted vegetation (Dickson and Giblin, 2006). When these waters approach the surface a condition called 'dry land salinity' occurs, leading to the salinization of the soil and above-ground water supplies. One solution used to address dry land salinity is to drain the rising groundwater. Bores are drilled to bring groundwater to the surface, where the water is then drained into rivers or lakes or evaporated off in man-made channels.

Northern Chile is also home to acidic saline lakes; however, the Chilean lakes are found at high altitudes (above 3000 m) in intra-volcanic basins with the acidity attributed to the oxidation of native sulfur. The mineralogy of these lake regions is significantly different from that seen in the Australian systems (Escudero *et al.*, 2013). An interesting site described by Korehi *et al.* (2013) is located at Chañaral Bay, in the Atacama Desert, northern Chile, where a 1-km tailings dump was created from copper mining activities during 1938–1975 with close proximity to the ocean and an evaporite deposit. Parts of the deposit have oxidized, forming a highly saline (1.0 M NaCl), low-pH (2–4) crust (10–15 m depth).

These locations provide the conditions for acidophilic halophiles to evolve; however, very few microbiological studies have been carried out in these systems. Both acidophiles (two *Acidiphilium*-like strains and *T. prosperus*) and the neutrophile *Geobacter* have been identified in the Western Australian acidic saline drains (Peiffer *et al.*, 2009; Zammit *et al.*, 2012), while at Chañaral Bay microorganisms belonging to the genera *Acidithiobacillus*, *Alicyclobacillus*, *Sulfobacillus*, *Halothiobacillus*, *Thiohalomonas* and *Thiomicrospira* were identified (Korehi *et al.*, 2013). Many opportunities exist for comprehensive microbiological studies on acidic saline environments that may provide the opportunity to discover novel microorganisms and functional capacities.

Volcanoes near seawater

Magma ascends from deep within the earth to the crust, bringing it in contact with silicate materials rich in iron and magnesium. When these mix with seawater and are left to form lakes or shallow pools acidic saline conditions can occur (Huber and Stetter, 1989). These conditions occur wherever tectonic plates cross the ocean floor and create volcanic ridges or islands; alternatively, they can form along coastal regions. From shallow water neighbouring hydrothermal vents near a Volcano in Italy and Milos, Greece, *T. prosperus* and several *T. prosperus*-like strains have been isolated (Simmons and Norris, 2002).

Bioleaching environments

Biomining utilizes acidophilic microorganisms to break down ores (bioleaching) and facilitate metal recovery (Chapter 16). Biomining is carried out under acidic conditions; however, in some operations, especially in Australia and Chile, high levels of salt can also be encountered either due to lack of fresh water or the presence of salts within the ore. Although the combination of these conditions is rare they occur in regions that are high in low-grade copper deposits, areas that would otherwise be ideal candidates for biomining operations. As the life of the mine progresses, salts become increasingly concentrated through the recycling of leaching liquor. This can lead to significant impacts on the viability of current biomining operations and limits the use of biomining technologies in areas high in salt.

Acidophilic halophiles

Only a limited number of acidophilic halophiles have been isolated and characterized. While many species of extreme acidophiles (pH growth optima

< 3.0) and halophiles have been characterized, only a few species of microorganisms that have the ability to tolerate both extreme acidity and elevated concentrations of NaCl have been described. Bacteria which have been isolated or detected in acidic saline environments include members of the genera *Acidiphilium* and *Thiobacillus*; *T. prosperus* is the only acidophile described which requires chloride for growth (Nicolle *et al.*, 2009). Some archaea can also tolerate acidic saline environments, such as *Acidianus* and *Sulfolobus*.

Bacteria

Acidiphilium spp.

Acidiphilium spp. are strictly aerobic, chemo-organotrophic bacteria (Hiraishi and Imhoff, 2005) with the exception of *A. acidophilum* which is also able to oxidize elemental sulfur and reduced inorganic sulfur compounds (RISCs) autotrophically. All known *Acidiphilium* spp. can catalyse the dissimilatory reduction of ferric iron (Coupland and Johnson, 2007). *Acidiphilium* spp. vary in their ability to withstand pH, but typically grow between pH 2.0 and 5.9, with some species reported to grow at pH 1.5–2.0 (Hiraishi and Imhoff, 2005). *Acidiphilium* spp. appear to be more tolerant to chloride than other acidophiles, and have been identified in acidic saline groundwater (Peiffer *et al.*, 2009) and acidic saline drains (Zammit *et al.*, 2012) in Western Australia. Whether or not these isolates are capable of autotrophic metabolism is yet to be determined. However, the clones from the acidic saline drain only shared a 96% identity of their 16S rRNA genes to other *Acidiphilium* spp. and therefore appear to be distinct species (Zammit *et al.*, 2012). Some known species of *Acidiphilium* have been tested for their ability to withstand NaCl. *A. cryptum* was shown to be able to survive up to 1.2 M NaCl, although cell numbers were ~50% when grown with 0.5 M NaCl compared to cultures grown in chloride-free medium (Hiraishi and Imhoff, 2005). The same strain of *A. cryptum* has also been shown to tolerate up to 0.3 M $Al_2(SO_4)_3$ with a 50% reduction in growth rate (Moritz *et al.*, 2015).

Thiobacillus prosperus and *T. prosperus*-like bacteria

T. prosperus and related microorganisms have been identified in sulfur-rich shallow water hydrothermal pools at a Volcano in Italy and a similar environment in Milos, Greece (Simmons and Norris, 2002), an acidic saline drain in Western Australia (Zammit *et al.*, 2012) and identified from the stem-post of a 500-year old warship, the *Mary Rose* (Preston *et al.*, 2012). The type strain (DSM 5130) was described as a mesophilic (optimal growth at 37°C), halophilic, acidophilic (pH 1.0–4.5), Gram-negative bacterium that oxidizes sulfur and iron. One strain of *T. prosperus* (VM17) can grow at 1 M NaCl, while most other strains, including the type strain (DSM 5130) tolerate up to 0.6 M (Huber and Stetter, 1989). At first it was reported that *T. prosperus* was able to grow without NaCl, but later this was retracted and it is now known that *T. prosperus* requires a minimum of 40 mM chloride (Davis-Belmar *et al.*, 2008) and RISCs for growth (Nicolle *et al.*, 2009). Two '*T. prosperus*'-like strains were isolated from an acidic saline drain in the southwest of Western Australia (Zammit *et al.*, 2012) and were shown to grow at up to 0.5 M NaCl and to dominate a mixed culture of iron- and sulfur-oxidizing micro-organisms containing up to 0.35 M NaCl (Zammit *et al.*, 2012). Unfortunately, the '*T. prosperus*'-like strain from the *Mary Rose* has not been cultivated *in vitro* (Preston *et al.*, 2012).

Initially ferrous iron oxidation by '*T. prosperus*' was reported to be very slow in comparison with that catalysed by *At. ferrooxidans* (Huber and Stetter, 1989). However, it was later demonstrated to have similar iron oxidation rates to *At. ferrooxidans* when the medium was supplemented with RISCs (Nicolle *et al.*, 2009). '*T. prosperus*' has also been shown to have similar genetic elements for iron oxidation as *At. ferrooxidans* (Nicolle *et al.*, 2009; Davis-Belmar *et al.*, 2008).

Acidithiobacillus spp.

The genus *Acidithiobacillus* includes several species that have been isolated from a range of environments and display great phylogenetic heterogeneity and distinct physiological characteristics (Hallberg *et al.*, 2009). *Acidithiobacillus* spp. typically demonstrate optimal growth below pH 4.0 and are mesophilic chemolithotrophs. *Acidithiobacillus* spp. have been shown to have relatively limited tolerance to chloride; however, some isolates from marine sediments have a slightly higher tolerance to chloride. *Acidithiobacillus* spp. were identified from Chañaral Bay and were subsequently shown to be a part of a mixture which oxidized pyrite a 1.0 M NaCl (Korehi *et al.*, 2013). Harahuc *et al.* (2000) noted that while iron-grown cells of *At. ferrooxidans* were more sensitive to chloride (0.21 M NaCl) than sulfur-grown cells (0.39 M NaCl) sulfur oxidation is more sensitive to osmotic pressure than iron oxidation. The authors purport this to be due to the differing water solubilities of iron and sulfur. A study on the toxic effect of chloride on the pyrite leaching efficiency by *At. caldus* demonstrated a tenfold decrease in cell numbers after seven days exposure to just 85 mM NaCl indicating significant sensitivity to the presence of chloride (Gahan *et al.*, 2009). In a later study, however, the type strain of *At. caldus* was shown to be able to tolerate up to 0.45 M NaCl when cells were grown with tetrathionate (Zammit *et al.*, 2012). A marine strain of *At. thiooxidans* (SH) requires NaCl for growth; optimal growth was recorded when media was supplemented with 0.35 M NaCl, although the optimum pH for growth for this strain was 4.0 (Kamimura *et al.*, 2003). The addition of 0.2 M of various anions (sulfate, phosphate, chloride, nitrate and fluoride) and cations (potassium, sodium, lithium, rubidium and caesium) to growth media at pH 2.3 and 4.5 were shown to inhibit sulfur oxidation by *At. thiooxidans*, possibly due to high osmotic pressures (Suzuki *et al.*, 1999). A similar effect was seen at 0.5 M sucrose (Suzuki *et al.*, 1999).

Leptospirillum spp.

Leptospirillum spp. are extremely acidophilic iron-oxidizing chemolithotrophs that have been reported to have limited tolerance of chloride (Gahan *et al.*, 2010; Romero *et al.*, 2003). A *L. ferrooxidans*-like bacterium (strain LA), isolated from a copper heap bioleaching operation in Chile, could oxidize iron when grown in media containing either 0.3 M NaCl and 0.4 M Na_2SO_4, or 0.05 M NaCl and 0.08 M Na_2SO_4 (Romero *et al.*, 2003). Batch experiments of *L. ferriphilum* showed that growth was completely inhibited by 0.35 M chloride and in the presence of pyrite the cell numbers of *L. ferriphilum* in a mixed culture, initially became undetectable after the addition of just 50 mM chloride but begun to re-emerge in low numbers after 7 days' exposure (Gahan *et al.*, 2009, 2010). Other studies have demonstrated that *L. ferriphilum* (DSM 14647) can grow at up to 0.3 M NaCl, but in mixed cultures containing NaCl at concentrations as low as 20 mM it is outcompeted by other microorganisms (Wang *et al.*, 2012; Zammit *et al.*, 2012). Despite this, *L. ferriphilum* (DSM 22399) was shown to dominate the column effluent in leaching chalcocite rich ore in the presence of 0.17 M NaCl (Davis-Belmar *et al.*, 2014).

Sulfobacillus spp.

Sulfobacillus spp. are iron- and sulfur-oxidizing acidophilic bacteria that are moderately thermophilic and can grow either autotrophically or heterotrophically (Golovacheva and Karavaiko, 1978). The *Sb. thermosulfidooxidans* type strain (DSM 9293) has been shown to tolerate up to 0.3 M NaCl in pure and mixed culture (Zammit *et al.*, 2012). Elsewhere, a microorganism closely related to *S. thermosulfidooxidans* (99.5% identity over the 5′-terminal of 400 bp of the 16S rRNA gene) isolated from saline waters, was reported to be partially inhibited by just 30 mM NaCl (Simmons and Norris, 2002). Australian-native isolates of *Sulfobacillus* spp. have been shown to be inhibited by 90 mM NaCl (Keeling *et al.*, 2006). *Sulfobacillus* spp. were also identified from Chañaral Bay and shown to grow at 1 M NaCl (Korehi *et al.*, 2013). *Sulfobacillus* sp. TPY, originally isolated from hydrothermal vents in the Pacific Ocean, was shown to tolerate 300 mM NaCl during the bioleaching of chalcopyrite (at pH 1.8) and during a mixed culture experiment, was shown to outcompete other microorganisms at pH 2.0, 45°C and 0.02 M NaCl (Wang *et al.*, 2012).

Acidiferrobacter thiooxydans

Acidiferrobacter thiooxydans is an acidophilic thermo-tolerant, iron- and sulfur-oxidizing bacterium, which is more closely related to the alkalophile *Ectothiorhodospira* spp. than other acidophilic bacteria (Hallberg *et al.*, 2011). Growth of *Af. thiooxydans* is enhanced under elevated osmotic stress – an effect observed with a range of salts regardless of the cation and with either chloride or sulfate as the anion. *Af. thiooxydans* is therefore reported to be a moderate osmophile, rather than a halophile, which requires a minimum external osmotic potential of 2 bar for growth with optimal growth observed at 5 bar (Hallberg *et al.*, 2011).

Archaea

Crenarchaeotes

In general, it appears that acidophilic archaea are slightly more tolerant to salts than acidophilic bacteria. *Acidianus* spp. are reported to grow at up to 0.7 M NaCl (Segerer *et al.*, 1986), though there is large species and strain variations, e.g. *Ad. infernus* grows optimally at 0.03 M NaCl (Segerer *et al.*, 1986) and *Ad. sulfidivorans* at 0.09 M NaCl (Plumb *et al.*, 2007). *Acidianus* spp. are facultative aerobes that both oxidize and reduce sulfur, dependent on the availability of oxygen, and can grow autotrophically. During aerobic growth S°, mineral sulfide or Fe^{2+} oxidation occurs and anaerobically Fe^{3+} or S° are reduced (Plumb *et al.*, 2007). *Acidianus* spp. have a relatively large pH and temperature growth range, pH 0.35 to 6 and 45 to 90°C, respectively (Chapter 6).

Sulfolobus spp. also grow in hot environments with their optimum growth at ~75°C, can oxidize sulfur and have been isolated from a variety of geothermal sites (see Chapter 6). *Sulfolobus* spp. have a range of tolerance to chloride, strains of *S. solfataricus* and *S. acidocaldarius* were inhibited by 0.32 M NaCl (Grogan, 1989), whereas two different species of *S. metallicus* were shown to grow optimally at 0.13 M NaCl but were inhibited by 0.51 M NaCl (Salo-Zieman *et al.*, 2006).

Metallosphaera cuprina has been reported to grow on up to 0.17 M NaCl (Liu *et al.*, 2011); however, no data on other *Metallosphaera* species are available.

Euryarchaeotes

Thermoplasma acidophilum and *Thermoplasma volcanium* are thermophilic (optimum temperature 55 and 60°C, respectively), acidophilic (optimum pH 1.0 and 2.0, respectively), sulfur-oxidizing euryarchaeotes that have been isolated from solfatara fields in the Azores, Iceland, Indonesia, Italy and the USA (Segerer *et al.*, 1988). In a mixed culture of microorganisms related to *Sulfolobus metallicus* and *T. acidophilum*, optimal sulfur oxidation was recorded at 50°C with 0.13 M chloride; however, oxidation was completely inhibited at 0.51 M chloride (Salo-Zieman *et al.*, 2006).

Ferroplasma spp. have been shown to be important members of AMD and bioleaching environments, especially at the extremes of pH (pH 1.4 or less), and are mesophilic or moderately thermophilic, heterotrophic iron-oxidizing acidophiles (Dopson *et al.*, 2004). The type strain, *F. acidiphilum* DSM 12658, has been shown to grow at up to 0.34 M NaCl at pH 1.8 (Zammit *et al.*, 2012).

Conclusions

Microorganisms have developed extensive mechanisms for dealing with environmental extremes. Survival strategies employed by halophilic and acidophilic microorganisms involve adaptations to cellular membranes, the use of cellular pumps or the production of cytoplasmic buffers. Environments in which both acidic and saline conditions persist are rare, and therefore there have only been a limited number of microorganisms that have been shown to simultaneously deal with acidic conditions and osmotic stress. This makes the study of the microbial communities of these systems of particular interest as little is known about the microbial diversity and of the mechanisms employed by microorganisms to deal with both acidity and salinity simultaneously.

References

Baker-Austin, C., and Dopson, M. (2007). Life in acid: pH homeostasis in acidophiles. Trends Microbiol. *15*, 165–171.

Benison, K.C., and LaClair, D.A. (2003). Modern and ancient extremely acid saline deposits: terrestrial analogs for martian environments? Astrobiology *3*, 609–618.

Bevilaqua, D., Lahti, H., Suegama, P.H., Garcia, O., Benedetti, A.V., Puhakka, J.A., and Tuovinen, O.H. (2013). Effect of Na chloride on the bioleaching of a chalcopyrite concentrate in shake flasks and stirred tank bioreactors. Hydrometallurgy *138*, 1–13.

Blight, K.R., and Ralph, D.E. (2004). Effect of ionic strength on iron oxidation with batch cultures of chemolithotrophic bacteria. Hydrometallurgy *73*, 325–334.

Cárdenas, J.P., Lazcano, M., Ossandon, F.J., Corbett, M., Holmes, D.S., and Watkin, E. (2014). Draft genome sequence of the iron-oxidizing acidophile *Leptospirillum ferriphilum* type strain DSM 14647. Genome Announc. *2*, e01153-14.

Chao, J., Wang, W., Xiao, S., and Liu, X. (2008). Response of *Acidithiobacillus ferrooxidans* ATCC 23270 gene expression to acid stress. World J. Microbiol. Biotechnol. *24*, 2103–2109.

Chi, A., Valenzuela, L., Beard, S., Mackey, A.J., Shabanowitz, J., Hunt, D.F., and Jerez, C.A. (2007). Periplasmic proteins of the extremophile *Acidithiobacillus ferrooxidans:* A high throughput proteomics analysis. Mol. Cell. Proteomics *6*, 2239–2251.

Chong, P.L.-G. (2010). Archaebacterial bipolar tetraether lipids: physico-chemical and membrane properties. Chem. Phys. Lipids *163*, 253–265.

Ciaramella, M., Napoli, A., and Rossi, M. (2005). Another extreme genome: how to live at pH 0. Trends Microbiol. *13*, 49–51.

Coupland, K., and Johnson, D.B. (2007). Evidence that the potential for dissimilatory ferric iron reduction is widespread among acidophilic heterotrophic bacteria. FEMS Microbiol. Lett. *279*, 30–35.

Csonka, L.N., and Hanson, A.D. (1991). Prokaryotic osmoregulation: genetics and physiology. Annu. Rev. Microbiol. *45*, 569–606.

Davis-Belmar, C.S., Nicolle, J.L.C., and Norris, P.R. (2008). Ferrous iron oxidation and leaching of copper ore with halotolerant bacteria in ore columns. Hydrometallurgy *94*, 144–147.

Davis-Belmar, C.S., Cautivo, D., Demergasso, C., and Rautenbach, G. (2014). Bioleaching of copper secondary sulfide ore in the presence of chloride by means of inoculation with chloride-tolerant microbial culture. Hydrometallurgy *150*, 308–312.

Devici, H., Jordan, M.A., Powell, N., and Alp, I. (2008). Effect of salinity and acidity on bioleaching activity of mesophilic and extremely thermophilic bacteria. Trans. Nonferrous Met. Soc. China *18*, 714–721.

Dew, D.W., and du Plessis, C. (2002). Bioleaching of a sulphide concentrate in a saline solution. Patent number WO2002081761 A3.

Dickson, B., and Giblin, A. (2006). Features and effects of some acid-saline groundwaters of southern Australia. Chinese J. Geochem *25*, 227–227.

Dopson, M., Baker-Austin, C., Hind, A., Bowman, J.P., and Bond, P.L. (2004). Characterization of Ferroplasma isolates and *Ferroplasma acidarmanus* sp. nov., extreme acidophiles from acid mine drainage and industrial bioleaching environments. Appl. Environ. Microbiol. *70*, 2079–2088.

Empadinhas, N., and da Costa, M. (2008). Osmoadaptation mechanisms in prokaryotes: distribution of compatible solutes. Int. Microbiol. *11*, 151–161.

Escudero, L.V., Bijman, J., Chong, G., Pueyo, J.J., and Demergasso, C. (2013). Geochemistry and microbiology in an acidic, high altitude (4,000 m) salt flat. High Andes, Northern Chile. Adv. Mat. Res. *825*, 28–32.

Futterer, O., Angelov, A., Liesegang, H., Gottschalk, G., Schleper, C., Schepers, B., Dock, C., Antranikian, G., and Liebl, W. (2004). Genome sequence of *Picrophilus torridus* and its implications for life around pH 0. Proc. Natl. Acad. Sci. U.S.A. *101*, 9091–9096.

Gahan, C.S., Sundkvist, J.-E., and Sandström, Å. (2009). A study on the toxic effects of chloride on the biooxidation efficiency of pyrite. J. Hazard. Mater. *172*, 1273–1281.

Gahan, C.S., Sundkvist, J.E., Dopson, M., and Sandström, Å. (2010). Effect of chloride on ferrous iron oxidation by a *Leptospirillum ferriphilum*-dominated chemostat culture. Biotechnol. Bioeng. *106*, 422–431.

Golovacheva, R.S., and Karavaiko, G.I. (1978). *Sulfobacillus*, a genus of spore forming thermophilic bacteria. Mikrobiologiya *47*, 815–822.

Goltsman, D.S.A., Denef, V.J., Singer, S.W., VerBerkmoes, N.C., Lefsrud, M., Mueller, R.S., Dick, G.J., Sun, C.L., Wheeler, K.E., Zemla, A., *et al.* (2009). Community genomic and proteomic analyses of chemoautotrophic iron-oxidizing *"Leptospirillum rubarum"* (group II) and *"Leptospirillum ferrodiazotrophum"* (group III) bacteria in acid mine drainage biofilms. Appl. Environ. Microbiol. *75*, 4599–4615.

González, C., Yanquepe, M., Cardenas, J.P., Valdes, J., Quatrini, R., Holmes, D.S., and Dopson, M. (2014). Genetic variability of psychrotolerant *Acidithiobacillus ferrivorans* revealed by (meta)genomic analysis. Res. Microbiol. *165*, 726–734.

Grogan, D.W. (1989). Phenotypic characterization of the archaebacterial genus *Sulfolobus*: comparison of five wild-type strains. J. Bacteriol. *171*, 6710–6719.

Guiliani, N., and Jerez, C.A. (2000). Molecular cloning, sequencing, and expression of omp-40, the gene coding for the major outer membrane protein from the acidophilic bacterium *Thiobacillus ferrooxidans*. Appl. Environ. Microbiol. *66*, 2318 – 2324.

Guo, X., Jiang, C., Luo, Y., Zhang, M., Poetsch, A., and Liu, S. (2014). Proteomic and molecular investigations revealed that *Acidithiobacillus caldus* adopts multiple strategies for adaptation to NaCl stress. Chin. Sci. Bull. *55*, 301–309.

de la Haba, R.R., Sánchez-Porro, C., Marquez, M.C., and Ventosa, A. (2011). Taxonomy of halophiles. In Extremophiles Handbook, Horikoshi, K. ed. (Springer, Tokyo, Japan), pp. 255–308.

Hallberg, K.B., Amouric, A., Brochier-Armanet, C., Bonnefoy, V., and Johnson, B.D. (2009). Physiological and phylogenetic heterogeneity among iron-oxidizing *Acidithiobacillus* spp., and characteristics of the novel species *Acidithiobacillus ferrivorans*. Adv. Mat. Res. *71–73*, 167–170.

Hallberg, K.B., Hedrich, S., and Johnson, D. (2011). *Acidiferrobacter thiooxydans*, gen. nov. sp. nov.; an acidophilic, thermo-tolerant, facultatively anaerobic iron- and sulfur-oxidizer of the family Ectothiorhodospiraceae. Extremophiles *15*, 271–279.

Harahuc, L., Lizama, H.M., and Suzuki, I. (2000). Selective inhibition of the oxidation of ferrous iron or sulfur in *Thiobacillus ferrooxidans*. Appl. Environ. Microbiol. *66*, 1031–1037.

Hiraishi, A., and Imhoff, J.F. (2005). Acidiphilium Harrison 1981, 331 VP emend. Kishimoto, Kosako, Wakao, Tano and Hiraishi 1995b, 90. In Bergey's Manual of Systematic Bacteriology Volume Two: The Proteobacteria, Part C: The Alpha, Beta-, Delta-, and Epsilonbacteria, Garrity, G.M., Brenner, D.J., Krieg, N.R., and Staley, J.T., eds (Springer, New York, NY, USA), pp. 54–62.

Huber, H., and Stetter, K.O. (1989). *Thiobacillus prosperus* sp. nov., represents a new group of halotolerant metal-mobilizing bacteria isolated from a marine geothermal field. Arch. Microbiol. *151*, 479–485.

Jones, D.S., Albrecht, H.L., Dawson, K.S., Schaperdoth, I., Freeman, K.H., Pi, Y., Pearson, A., and Macalady, J.L. (2012). Community genomic analysis of an extremely acidophilic sulfur-oxidizing biofilm. ISME J. *6*, 158–170.

Kamimura, K., Higashino, E., Moriya, S., and Sugio, T. (2003). Marine acidophilic sulfur-oxidizing bacterium requiring salts for the oxidation of reduced inorganic sulfur compounds. Extremophiles *7*, 95–99.

Keeling, S.E., Davies, K.L., Palmer, M.L., Townsend, D.E., Watkin, E., Johnson, J.A., and Watling, H.R. (2006). Utilisation of native microbes from a spent chalcocite test heap. Hydrometallurgy *83*, 124–131.

Kieft, T.L., and Spence, S.D. (1988). Osmoregulation in *Thiobacillus ferrooxidans:* Stimulation of iron oxidation by proline and betaine under salt stress. Curr. Microbiol. *17*, 255–258.

Kobayashi, R., Suzuki, T., and Yoshida, M. (2007). *Escherichia coli* phage-shock protein A (PspA) binds to membrane phospholipids and repairs proton leakage of the damaged membranes. Mol. Microbiol. *66*, 100–109.

Korehi, H., Blöthe, M., Sitnikova, M.A., Dold, B., and Schippers, A. (2013). Metal mobilization by iron- and sulfur-oxidizing bacteria in a multiple extreme mine tailings in the Atacama Desert, Chile. Environ. Sci. Technol. *47*, 2189–2196.

Krulwich, T.A., Sachs, G., and Padan, E. (2011). Molecular aspects of bacterial pH sensing and homeostasis. Nat. Rev. Microbiol. *9*, 330–343.

Liu, L.-J., You, X.-Y., Guo, X., Liu, S.-J., and Jiang, C.-Y. (2011). *Metallosphaera cuprina* sp. nov., an acidothermophilic, metal-mobilizing archaeon. Int. J. Syst. Evol. Micr. *61*, 2395–2400.

Macalady, J., Vestling, M., Baumler, D., Boekelheide, N., Kaspar, C., and Banfield, J. (2004). Tetraether-linked membrane monolayers in *Ferroplasma* spp: a key to survival in acid. Extremophiles *8*, 411–419.

McArthur, J.M., Turner, J.V., Lyons, W.B., Osborn, A.O., and Thirlwall, M.F. (1991). Hydrochemistry on the yilgarn block, western Australia: Ferrolysis and mineralisation in acidic brines. Geochim. Cosmochim. Ac. *55*, 1273–1288.

Mangold, S., Jonna, V., Dopson, M., and Rao Jonna, V. (2013). Response of *Acidithiobacillus caldus* toward suboptimal pH conditions. Extremophiles *17*, 689–696.

Matin, A. (1990). Keeping a neutral cytoplasm; the bioenergetics of obligate acidophiles. FEMS Microbiol. Lett. *75*, 307–318.

Matin, A. (1991). pH Homeostasis in Acidophiles. In Novartis Foundation Symposium 221 – Bacterial Responses to pH, Chadwick, D., and Cardew, G., eds (John Wiley & Sons, Ltd., Chichester, UK), pp. 152–166.

Moritz, K.D., Amendt, B., Witt, E.M., and Galinski, E.A. (2015). The hydroxyectoine gene cluster of the non-halophilic acidophile *Acidiphilium cryptum*. Extremophiles *19*, 87–99.

Mosier, A.C., Justice, N.B., Bowen, B.P., Baran, R., Thomas, B.C., Northen, T.R., and Banfield, J.F. (2013). Metabolites associated with adaptation of microorganisms to an acidophilic, metal-rich environment identified by stable-isotope-enabled metabolomics. MBio *4*, e00484-12.

Mykytczuk, N.C.S., Trevors, J.T., Ferroni, G.D., and Leduc, L.G. (2010). Cytoplasmic membrane fluidity and fatty acid composition of *Acidithiobacillus ferrooxidans* in response to pH stress. Extremophiles *14*, 427–441.

Nicolle, J.L.C., Simmons, S., Bathe, S., and Norris, P.R. (2009). Ferrous iron oxidation and rusticyanin in halotolerant, acidophilic '*Thiobacillus prosperus*'. Microbiology *155*, 1302–1309.

Ojumu, T.V., Petersen, J., and Hansford, G.S. (2008). The effect of dissolved cations on microbial ferrous-iron oxidation by *Leptospirillum ferriphilum* in continuous culture. Hydrometallurgy *94*, 69–76.

Ossandon, F.J., Cárdenas, J.P., Corbett, M., Quatrini, R., Holmes, D.S., and Watkin, E. (2014). Draft genome sequence of the iron-oxidizing, acidophilic, and halotolerant "*Thiobacillus prosperus*" type strain DSM 5130. Genome Announc. *2*, e01042-14.

Parro, V., Moreno Paz, M., and Gonzalez Toril, E. (2007). Analysis of environmental transcriptomes by DNA microarrays. Environ. Microbiol. *9*, 453–464.

Peiffer, S., Oldham, C., Salmon, U., Lillicrap, A., and KuÃàsel, K. (2009). Does iron cycling trigger generation of acidity in groundwaters of Western Australia? Environ. Sci. Tech. *43*, 6548–6552.

Plumb, J.J., Haddad, C.M., Gibson, J.A.E., and Franzmann, P.D. (2007). *Acidianus sulfidivorans* sp. nov., an extremely acidophilic, thermophilic archaeon isolated from a solfatara on Lihir Island, Papua New Guinea, and emendation of the genus description. Int. J. Syst. Evol. Microbiol.*57*, 1418–1423.

Preston, J., Watts, J.E., and Jones, M. (2012). Novel bacterial community associated with 500-year-old unpreserved archaeological wood from King Henry VIII's Tudor warship the Mary Rose. Appl. Environ. Microbiol. *78*, 8822–8828.

Romero, J., Yañez, C., Vásquez, M., Moore, E.R.B., and Espejo, R.T. (2003). Characterization and identification of an iron-oxidizing, *Leptospirillum*-like bacterium, present in the high sulfate leaching solution of a commercial bioleaching plant. Res. Microbiol. *154*, 353–359.

Salo-Zieman, V.A., Sivonen, T., Plumb, J., Haddad, C., Laukkanen, K., Kinnunen, P.M., Kaksonen, A., Franzmann, P., and Puhakka, J. (2006). Characterization of a thermophilic sulfur oxidizing enrichment culture dominated by a *Sulfolobus* sp. obtained from an underground hot spring for use in extreme bioleaching conditions. J. Ind. Microbiol. Biotechnol. 33, 984–994.

Segerer, A., Neuner, A., Kristjansson, J.K., and Stetter, K.O. (1986). *Acidianus infernus* gen. nov., sp. nov. and *Acidianus brierleyi* comb. nov.: Facultatively aerobic, extremely acidophilic thermophilic sulfur-metabolizing archaebacteria. Int. J. Syst. Bacteriol. *36*, 559–564.

Segerer, A., Langworthy, T.A., and Stetter, K.O. (1988). *Thermoplasma acidophilum* and *Thermoplasma volcanium* sp. nov. from Solfatara Fields. Syst. Appl. Microbiol. *10*, 161–171.

Shiers, D.W., Blight, K.R., and Ralph, D.E. (2005). Sodium sulfate and sodium chloride effects on batch culture of iron oxidising bacteria. Hydrometallurgy *80*, 75–82.

Simmons, S., and Norris, P. (2002). Acidophiles of saline water at thermal vents of Vulcano, Italy. Extremophiles *6*, 201–207.

Sleator, R.D., and Hill, C. (2002). Bacterial osmoadaptation: the role of osmolytes in bacterial stress and virulence. FEMS Microbiol. Rev. *26*, 49–71.

Stan-Lotter, H., Lang Jr, F.J., and Hochstein, L.I. (1988). Electrophoresis and isoelectric focusing of whole cell and membrane proteins from the extremely halophilic archaebacteria. Appl. Theor. Electrophor. *1*, 147–153.

Suzuki, I., Lee, D., Mackay, B., Harahuc, L., and Oh, J.K. (1999). Effect of various ions, pH, and osmotic pressure on oxidation of elemental sulfur by *Thiobacillus thiooxidans*. Appl. Environ. Microbiol. *65*, 5163–5168.

Valdés, J., Pedroso, I., Quatrini, R., Dodson, R., Tettelin, H., Blake, R., Eisen, J., and Holmes, D. (2008). *Acidithiobacillus ferrooxidans* metabolism: from genome sequence to industrial applications. BMC Genomics 9, 597.

Wang, Y., Su, L., Zhang, L., Zeng, W., Wu, J., Wan, L., Qiu, G., Chen, X., and Zhou, H. (2012). Bioleaching of chalcopyrite by defined mixed moderately thermophilic consortium including a marine acidophilic halotolerant bacterium. Bioresource Technol. *121*, 348–354.

Zammit, C., Mangold, S., rao Jonna, V., Mutch, L., Watling, H., Dopson, M., and Watkin, E. (2012). Bioleaching in brackish waters—effect of chloride ions on the acidophile population and proteomes of model species. Appl. Microbiol. Biotechnol. 93, 319–329.

Oxidative Stress and Metal Tolerance in Extreme Acidophiles

4

Alonso Ferrer, Omar Orellana and Gloria Levicán

Abstract

In the extremely acidic conditions found in natural acid rock environments and bioleaching operations, microorganisms have to deal with an abundant supply of (heavy) metals. Iron is required as micronutrient by all acidophiles, and as a primary energy source or alternative electron acceptor by some of them. However iron and other metals can also induce oxidative damage to biomolecules by generating reactive oxygen species (ROS). Acidophilic microorganisms, therefore, are often exposed to highly oxidizing conditions and face the problem of maintaining intracellular metal and redox homeostasis. Although acidophiles fight oxidative stress via a number of ROS scavenging enzymes, they seem to depend mainly on biomolecule repair using a plethora of protein and lipid repair systems. Evidence collected to date indicates that the molecular determinants of this response also serve to protect the cells from other extreme environmental factors such as temperature, pH and metal(loid)s. Acidophiles are well equipped with mechanisms to export, sequester and/or store metals. Reducing the bioavailability of metals also seems to play an important role in their tolerance, via complex formation (with sulfate anions) and binding to extracellular polymeric substances. This chapter summarizes the mechanisms of protection to oxidative stress and metal toxicity in acidophiles, how these are configured to build the robust protection system required to thrive under extreme stress conditions, and how the ability of acidophiles to respond to those conditions may be critical in determining their growth and activity.

Introduction

Oxidative stress is a phenomenon experienced by both aerobic and anaerobic organisms from all three domains of life. It occurs when cellular antioxidant mechanisms are overwhelmed by the excessive generation of oxidant free radicals (Imlay, 2008). Thus, oxidative stress is characterized by a major increase in the amount of oxidized cellular components. Given the link between chemical oxidation of cellular molecules and disease, and its impact on the activity of microorganisms and the organization and dynamics of microbial communities, understanding oxidative stress and associated cellular responses is of great interest.

Aerobic organisms use molecular oxygen (O_2) as electron acceptor for energy conservation and ATP synthesis. However, most organisms possess limited tolerance to this molecule, primarily due to the oxidative damage it produces (Cabiscol *et al*, 2000). Its toxicity is mainly attributable to the partially reduced reactive oxygen species (ROS) that are inevitably produced during aerobic respiration (Imlay, 2008). ROS react and damage all major cellular components, such as DNA, RNA, proteins, lipids and cofactors, thereby having a substantial impact on cell physiology. ROS include the superoxide anion ($O_2{\cdot}^-$), the hydroxyl radical ($OH{\cdot}$) and the non-radical oxidant hydrogen peroxide (H_2O_2). Aerobic organisms experience the continuous production of $O_2{\cdot}^-$ and H_2O_2 derived from respiratory processes, when freely diffusing O_2 takes exposed electrons from the enzymes involved in electron transfer. Flavo-enzymes are particularly relevant to this phenomenon being so ubiquitous and abundant.

ROS are also generated in a number of different ways, including exposure to environmental factors

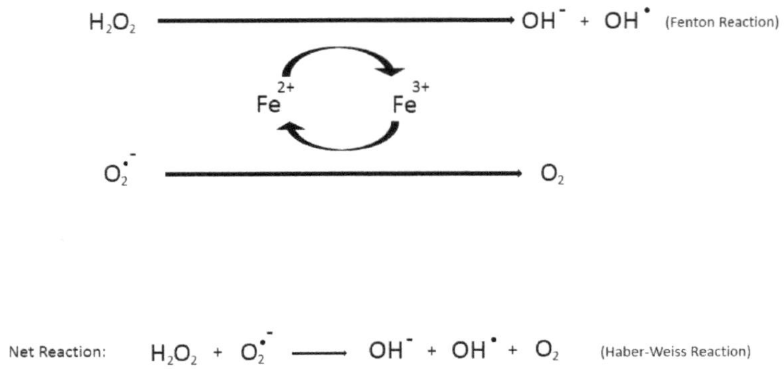

Figure 4.1 ROS formation by Fenton and Haber–Weiss reactions.

such as light, oxidative chemical agents, increased partial pressure of O_2, and metals. It is well documented that the presence of reduced transition metals, such as iron and copper, contributes to ROS generation. In the presence of these metals, $O_2^{•-}$ and $OH^•$ can be generated through the Fenton or the Haber–Weiss reactions (Fig. 4.1).

Many acidophiles are chemolitho-autotrophic bacteria or archaea able to grow aerobically via the oxidation of ferrous to ferric iron or reduced inorganic sulfur compounds to sulfuric acid. Low-pH environments ($pH < 3$) are often metal-rich, due to increased solubility of metal(loid)s in acid, including copper (Cu), iron (Fe), zinc (Zn), cobalt (Co), cadmium (Cd), and arsenic (As) among others. Heavy metal contamination of these environments is further enhanced by biological oxidation of metal-containing sulfide ores. Therefore, microorganisms inhabiting natural and man-made acidic niches need to cope with the threats imposed by unusually high metal concentrations and metal induced ROS species. It has thus been assumed that acidophilic microorganisms possess efficient molecular mechanisms for both oxidative stress response and heavy metal tolerance. This chapter provides an overview of recent progress in the understanding of the different systems evolved by acidophiles to face both challenges, while preserving the integrity and functionality of their cellular components.

ROS generation in acidic bioleaching microorganisms

It is well documented that redox active metals such as iron, copper, cobalt and chromium (Cr), undergo redox cycling reactions and may induce generation of reactive radicals such as $O_2·^-$ and OH· in biological systems. Copper, for example, which is the main metal mobilized during mineral ore leaching processes can reach concentrations as high as 19 g/l (300 mM) in certain industrial settings (Watkin, 2009). This element is also required as micronutrient, since it is part of the redox centres of several essential proteins (e.g. haem-copper oxidases). However, because of its ability to induce ROS formation it is considered a strong inducer of oxidative stress. A second group of metals, including cadmium, nickel (Ni) and mercury (Hg), induce the depletion of glutathione and bond to sulfhydryl groups of proteins. Arsenic, as trivalent [As(III)] or reduced arsenic, has a high affinity for sulfhydryl groups of proteins and also may react with the thiol groups present in the active sites of many enzymes and on glutathione, glutaredoxin and thioredoxin, thereby affecting intracellular redox homeostasis, protein folding and xenobiotic detoxification. It has been also proposed that arsenic is able to induce generation of hydrogen peroxide.

Therefore, acidophiles thriving in extreme acidic and metal loaded environments are exposed to severely oxidizing conditions that are bound to exert negative consequences for the cell. Mechanisms involved in protection against oxidative stress are thus deemed essential for adaptation and survival in these environments (Table 4.1).

Spontaneous environmental ROS production derived from mineral surfaces colonized by acidophiles also impose oxidizing conditions. ROS such as H_2O_2, $O_2·^-$ and OH· can be formed on the surface of pyrite particles as consequence of the direct electron transfer established from pyrite to O_2 via the Fenton and Haber–Weiss reactions (Schoonen

Table 4.1 Antioxidant mechanisms described to be involved in stress response in acidophiles

Microorganism	Environmental factor/ growth condition	Enzyme/gene	Reference
At. ferrooxidans	Energy source (iron vs. elemental sulfur)	Superoxide dismutase, thioredoxin peroxidase	Ramírez *et al.* (2004)
	Copper	DsbG domain protein (DsbG-L), glutaredoxin, glutathione S-transferase, glutathione synthetase, Glutathione reductase, superoxide dismutase, alkyl hydroperoxidase AhpC, GrosES chaperones, AhpC/TSA family/glutaredoxin domain protein. polyphosphate metabolism	Almárcegui *et al.* (2013, 2014), Álvarez and Jerez (2004), Orell *et al.* (2010)
	Cadmium	Cysteine and GSH intracellular levels	Zheng *et al.* (2014)
	Chalcopyrite	Glutathione reductase, superoxide dismutase, thioredoxin reductase, glutathione	Rodrigues *et al.* (2010), Martínez *et al.* (2013)
	Bornite	Alkyl hydroperoxidase, chaperone GroEL	Felício *et al.* (2010)
At. caldus	Zinc	Alkyl hydroperoxidase reductase	Mangold (2013)
	NaCl	Amino acid biosynthesis (glutamine, lysine. arginine), S-adenosylmethionine synthetase. Superoxide dismutase MnSOD, thiol peroxidase Tpx, chaperone DnaK	Zammit *et al.* (2012), Guo *et al.* (2014)
L. ferriphilum	Ferric iron, H_2O_2, diamide	Thioredoxin/thioredoxin reductase	Norambuena *et al.* (2012)
L. ferrooxidans	Environmental biofilm sample	Proteases ClpX/ClpP, chaperones ClpB, ClpC, GrpE, DnaKJ, GroESL, carotenoid biosynthesis genes, OxyR regulator, DNA helicase RecG	Moreno-Paz *et al.* (2010), Parro *et al.* (2007)
Leptospirillum group II	Temperature stress	Cysteine and histidine, chaperones DnaK and ClpB	Mosier *et al.* (2015)
	Acid Mine Drainage	Thioredoxin, rubrerythrin, peroxiredoxin	Ram *et al.* (2005), Goltsman *et al.* (2009)
	9K-BR medium	Peroxiredoxin, thiol peroxidase, methionine sulfoxide reductase	Belnap *et al.* (2009)
Leptospirillum group III	Temperature stress	Aminoacid synthesis (alanine, lysine, glutamate, cysteine, isoleucine and tryptophan), chaperones GroEL and HscA	Mosier *et al.* (2015)
Am. ferrooxidans	NaCl	Alkyl hydroxiperoxidase	Zammit *et al.* (2012)
'*Fp. acidarmanus*'	Copper	RecA/RadA recombinase, endonuclease IV, DNA repair protein RadA, thermosome subunit Hsp60 family, chaperone DnaK	Baker-Austin *et al.* (2005)
	Arsenic	Thioredoxin reductase TrxB, peroxiredoxin, alkyl hydroperoxidase	Baker-Austin *et al.* (2007b)
S. solfataricus	H_2O_2	Superoxide dismutase FeSOD, peroxiredoxin, DPS-like protein (ferritin-like protein), ruberythrin, disulfide oxidoreductas, Dps-like protein	Maaty *et al.* (2009), Wiedenheft *et al.* (2005)
	Ni^{2+}	Putative peroxiredoxin, NAD-specific glutamate dehydrogenases	Salzano *et al.* (2007)
	H_2O_2, paraquat, tert-butyl hydroperoxide	Peroxiredoxins Bcp, disulfide oxidoreductase PDO	Limauro *et al.* (2006, 2008), Pedone *et al.* (2006)
	UV irradiation, actinomycin D	NER genes XPF, XPG and XPB	Salerno *et al.* (2003)
	Glucose-rich medium	Superoxide dismutase FeSOD	Cannio *et al.* (2000)

Table 4.1 Continued

Microorganism	Environmental factor/ growth condition	Enzyme/gene	Reference
S. metallicus	Copper/chalcopyrite	Amino acids biosynthesis, chaperone Hsp60, proteasome subunit, metallochaperone (CopM), a Cu-exporting ATPase (CopA), polyphosphate metabolism	Orell *et al.* (2013), Remonsellez *et al.* (2006)
	Energy source (iron vs. elemental sulfur)	Alkyl hydroxiperoxidase	Bathe and Norris (2007), Burton *et al.* (1995)
P. torridus	pH and temperature stress	Thioredoxin/thioredoxin reductase	Thümer *et al.* (2011)

et al., 2010). Pyrite spontaneously forms ROS with a kinetic that is directly proportional to surface area of the particles, explaining why large pyrite surface area loading results in oxidative stress and inhibition of activity of bioleaching microorganisms (Jones *et al.*, 2013). Furthermore, it has been demonstrated that these ROS are capable of degrading RNA and that they negatively affect cell growth and Fe(II) oxidation rates of *Sulfolobus metallicus* cultures (Jones *et al.*, 2013). Yet, impact of the mineral-derived ROS on the dynamics of surface ore colonization, particularly during the early stages of biofilm establishment, remains to be assessed. Thus, the ability of bioleaching organisms, in particular primary colonizers of mineral surfaces, to endure ROS and preserve their physical integrity, growth and colonization skills involves the existence of antioxidants mechanisms that allow them to cope with external ROS that are permanently formed in their surrounding environments.

Besides the external factors that favour ROS generation, iron-oxidizing acidophiles have to deal with additional ROS derived from their energy metabolism. In these bacteria, the respiratory process involves dissimilatory oxidation of ferrous iron (Fe(II)) to ferric (Fe(III)). Because the redox couple Fe(III)/(II) has a very positive redox potential (+0.77 V at pH 2, though it is ~70 mV lower in sulfate-rich liquors), only the O_2 can act as natural electron acceptor (O_2/H_2O +1.12 V at pH 7). Therefore, the use of iron as electron donor is necessarily coupled to the reduction of molecular oxygen. Given the relatively small difference in redox potential between the Fe(III)/Fe(II) couple and O_2/H_2O, a high amount of Fe(II) is required to produce a slight increase in the number of cells (Rawlings, 2005). For this reason, these bacteria oxidize Fe(II) at high rates, resulting in high respiratory rates. To maintain high Fe(II) oxidation rate, iron-oxidizing microorganisms require a high cellular content of redox proteins (Rawlings, 2005). In fact, *Acidithiobacillus ferrooxidans* growing in different substrates is able to modulate the content of cytochromes, reaching up to 10% of the total proteins in cells grown in iron medium (Levicán *et al.*, 2007; Yarzábal *et al.*, 2002). A high number of cytochromes c were predicted to be encoded in the genomes (18–20 genes) of *Leptospirillum* spp., suggesting that elevated intracellular concentration of these proteins could also be attained in these obligate iron-oxidizers (Levicán *et al.*, 2012). In *At. ferrooxidans* it has been demonstrated that the level of cytochromes correlates with the haem content being the concentration 1.8-fold higher in cells grown in Fe(II) than in S° (Levicán *et al.*, 2007). Haem is a prostetic group comprising protopophyrin IX and iron; therefore, haem excess is toxic to cells owing to its ability to catalyse the Fenton reaction. Given this background, iron-oxidizing acidophiles are predicted to have a greater tendency to produce intracellular ROS than other acidophilic microorganisms.

An additional aspect is that acidic pH *per se* might induce damage to proteins and DNA. In some neutrophilic organisms, pH stress generates an increase in the content of thio-protecting molecules like glutathione and induces the biosynthesis of proteins involved in redox protection as peroxiredoxin AhpC, thioredoxin and thiol peroxidase Tpx, among others. In the case of acidophiles, it has been suggested that proton homeostasis occurs by means of highly impermeable cell membranes, by resistance to the inflow of protons through the generation of a Donnan potential, and by pumping the protons outside the cell (Baker-Austin and Dopson, 2007). Moreover, in these organisms, proteins attached to membranes or located in the periplasm have a basic pI that favours their stability in acid (e.g. Ram *et al.*, 2005). However, some acidophilic microorganisms such as *Leptospirillum* spp., *Ferroplasma* spp. and *Picrophilus* spp. are more resistant to acid than others since they can grow at pH < 1. This fact suggests that these organisms have additional mechanisms to tolerate extreme acidity. Such mechanisms are likely connected to the oxidative stress response. Evidence in this direction has been obtained in the case of the extremely acidophilic thermophilic archaeon *Picrophilus torridus*, which lives optimally at pH 0.7 and 65°C, and can even survive negative pH values (Ciaramella *et al.*, 2005). When this microorganism was exposed to high (1.8) or low (0.5) pH at 70°C, a marked up-regulation of thioredoxin/thioredoxin reductase system was detected. In the same way, alkyl hydroxyperoxide reductase seems to represent an important protective mechanism under low pH and high temperature (Thürmer *et al.*, 2011). Clearly, the combination of extreme acidity, high temperature, high concentrations of iron and heavy metals, is highly stressing and defence mechanisms associated to oxidative stress response seem to play a critical role alleviating the individual and combined effects of these stressors. Further research should be done to decipher the interplay of different ROS-generating stressors and to determine whether these antioxidant components have a role increasing tolerance to pH and temperature in other extreme acidophiles.

Oxidative stress response in acidophiles

Numerous studies have investigated single enzymes, pathways, and regulatory factors involved in the detoxification of ROS in neutrophilic microorganisms. However, the underlying mechanisms mediating the oxidative stress response in acidophilic microorganisms are only just beginning to be elucidated. In addition, the importance of this response on the organization and dynamics of acidophilic microorganism communities remains largely unexplored.

General considerations about the oxidative stress response

In general, microorganisms have developed and maintain a range of enzymatic and non-enzymatic mechanisms of defence against ROS and their adverse effects. Commonly, neutrophilic Gram-negative bacteria synthesize cytoplasmic and periplasmic superoxide dismutases (MnSOD, FeSOD and CuSOD) as a first line of defence against $O_2^{\cdot -}$. In most organisms H_2O_2 is removed by catalases, peroxidases and peroxiredoxins. The AhpCF peroxiredoxin is a NADH peroxidase which is the main defence against hydrogen and organic peroxides in bacteria. Beside enzymatic defence against ROS, most cells use a wide number of organic non-enzymatic antioxidants. In this context L-γ-glutamyl-L-cysteinyl-glycine (glutathione, GSH) and thioredoxins are essential to maintain a reducing environment in the cytosol and to reduce disulfide bonds resulting from the oxidation of proximal sulfhydryl groups on proteins. Genes encoding proteins of glutathione-dependent system are present in most Gram-negative bacteria and some *archaea*, but are under-represented in Gram-positive bacteria (Fahey, 2013). In addition, a number of different low molecular weight (LMW) thiols such as bacillithiol, mycothiol, cysteine, ergothioneine and coenzyme A, which complement or replace the function of glutathione, have also been described (Fahey, 2013).

Bacteria have developed complex gene regulatory responses to oxidative stress. The response to superoxide and peroxide is largely mediated through the induction of genes encoding superoxide dismutase and catalases. A number of transcriptional regulators controlling cellular response to ROS have been described. Due to the fact that ROS are produced during the normal course of aerobic metabolism, expression of some ROS scavenging systems is constitutive. However, bacteria can also experience transient exposure to high-levels of ROS. Consequently, transcriptional regulators have evolved to sense the levels of ROS and coordinate the appropriate and prompt response. Thus, while OxyR and PerR are sensors of H_2O_2, OhrR senses organic peroxide (ROOH). On the other hand the SoxRS system affects the overproduction of large number of proteins in response to $O_2^{\cdot -}$ (e.g. more than 40 in *Escherichia coli*). The alternative sigma factor RpoS is a stationary phase regulator found in *Proteobacteria*, coordinating the expression of around 200 genes, several of which respond to oxidative stress. It should be noted that oxidative stress defence and iron homeostasis are intimately linked and cross-talking between regulatory networks that control these responses can be predicted. It has been described that in some bacteria, the iron responsive transcriptional regulator Fur (ferric uptake regulator) plays an important role controlling the expression of genes associated to oxidative stress response. Fur encoding genes have been detected in several acidophiles genomes. Although its functional relevance in maintenance of redox balance had not been aimed, a possible role in control of antioxidant defence can be deduced. This issue will be approached in more detail below.

Beyond their role in the coordination of the cellular response against oxidative insults, these transcriptional regulators influence other complex biological phenomena, such as biofilm formation, pathogenicity and antibiotic resistance via the direct regulation of specific proteins. These data suggest that complex regulatory networks involving regulators other than those associated with redox homeostasis are present in most microorganisms. These findings are concurrent with the idea that upon exposure to diverse environmental stresses, bacteria evoke a common stress response that involves the molecular mechanisms for antioxidant defence. Thus, the ability to maintain intracellular concentrations of toxic ROS within safe limits largely influences the activity and fitness of microorganisms in natural environments.

Antioxidants and biomolecule repairing systems of acidophiles

As in other well-studied microorganisms, the oxidative stress response of acidophiles is thought to involve molecular mechanisms for scavenging ROS as a first line of defence, and enzymes to repair damaged chemical groups in proteins, nucleic acid and lipids, as a second one. Both aspects of the molecular strategies used by acidophiles to confront ROS have been studied using bioinformatic tools and sequenced genomes of acidophilic bacteria and archaea (Cárdenas *et al.*, 2012). With the exception of *Leptospirillum* spp., all genomes analysed contained at least one copy of a superoxide dismutase-encoding gene (Fe/Mn-DOD, Cu/Zn-SOD and/or NiSOD). Conversely, catalase-encoding genes were found only in a reduced number of genomes. To compensate for this lack, and in order to achieve effective detoxification of peroxides, most acidophiles encode several peroxidases and peroxiredoxins that use a number of reducing agents and have different substrate specificity. Accordingly, these data suggest that strategies used by acidophiles to transform $O_2{\cdot}^-$ into H_2O_2 are conserved, but reduction of H_2O_2 proceeds by alternative pathways.

A remarkable diversity of genes encoding different biomolecule-repair systems have been detected in acidophilic microorganisms. Almost all bacterial genomes analysed contained genes *mutM*, *mutY* and *nth* for the detection and removal of modified bases (base excision repair pathway). Endonuclease III- and V-encoding genes were also present in some acidophilic archaea. Genes for Nucleotide excision repair UvrABC and transcription–repair coupling factor Mfd have been found to be present in nearly all acidophilic bacteria. As it was previously described (Maaty *et al.*, 2009), a gene encoding the ferritin-like protein Dps, which is involved in DNA protection, was predicted in the Sulfolobales. This gene is also present in the genomes of members of *Acidithiomicrobium*, *Alicyclobacillus* and *Sulfobacillus*, although its physiological significance has not been addressed yet.

Lipid (L) peroxidation is a free radical chain reaction involving mostly polyunsaturated fatty acids and, at least in *E. coli*, also monounsaturated fatty acids. Intermediates of this chain reaction include lipid hydroperoxide (LOOH), and several radicals (L·, OLOO· and LOO·) considered to be highly toxic. LOOH has an increased polarity, a long lifetime, and promotes the production of harmful radicals such as L·, OLOO· and LOO·. Peroxiredoxin from Bcp and Ahp families and peroxidases from Dyp families are able to reduce LOOH to its corresponding alcohol (Wood *et al.*, 2003). In acidophiles, peroxiredoxin encoding genes are widely represented and they are present in multiple copies per genome (Cárdenas *et al.*, 2012). The C subunit of peroxiredoxin from the AhpAC complex is consistently conserved among bacterial and archaeal acidophiles (Cárdenas *et al.*, 2012); however, NADH transfer subunit was found only in *Alicyclobacillus* spp. On the other hand, the haem-dependent Dyp peroxidase involved in organic peroxide reduction was exclusively found in the leptospirilli, while the lipoyl-dependent organic peroxidase was detected in leptospirilli and *Acidiphilium cryptum*. As mentioned above, a role for these peroxidases in inorganic peroxide reduction is also plausible.

Protein repair systems which restore the reduced status of cysteines and methionines involve thioredoxin/thioredoxin reductase, thiol peroxidase Tpx and thioredoxin-dependent methionine sulfoxide oxidoreductase MsrAB (Cabiscol *et al.*, 2000). As mentioned above, glutathione and other low molecular weight thiols are also involved in re-establishing the redox status of sulfides in proteins (Fahey, 2013). Bioinformatic inspection in database containing 44 acidophiles genomes (Cárdenas *et al.*, 2012) indicated that all of the acidophiles possessed genes associated with the thioredoxin/thioredoxin reductase system. A couple of these systems have been characterized in *At. ferrooxidans* (Wang *et al.*, 2009) and *Leptospirillum ferriphilum* (Norambuena *et al.*, 2012). In these bacteria thioredoxin/thioredoxin reductase system has proved to be active and redundant. Details of these finding will be further described below. Tpx encoding gene was detected in members of the genera *Acidithiobacillus*, *Acidimicrobium* and *Sulfolobus*. *Msr* genes are also well conserved. Remarkably, with the exception of the acidithiobacilli and *A. cryptum*, glutathione-related genes are absent in all analysed acidophilic genomes. Whether acidophiles microorganisms use alternative LMW thiols as redox buffer is an issue that requires further study.

One of the main effects of ROS on proteins is the damage of iron–sulfur (Fe-S) clusters present in a number of metalloproteins. The structural versatility and chemical reactivity of these clusters have facilitated their selection as cofactors for many proteins, and their participation in multiple cellular processes such as respiration, central metabolism, gene regulation, modification of RNA and DNA repair. Acidophiles thrive in environments where both sulfur compounds (predominantly sulfate), and iron are abundant, thus Fe–S cluster biosynthesis should not be limited by availability of the structural elements. However, ROS and cobalt might impair Fe–S cluster-containing proteins function. ROS-induced iron release from these clusters further exacerbates oxidative stress in the affected cells. It has been reported that in *Deinococcus radiodurans*, a neutrophile bacterium that can withstand severe oxidative stress, the content of Fe-S proteins is significantly reduced compared with sensitive bacteria (Ghosal *et al.*, 2005). In the case of acidophiles, reduction of the metalloproteome has been reported for at least a couple of species, but does not seem to occur in all. While the proteome of the iron-oxidizing archaeon *Ferroplasma acidiphilum* is dominated by iron-proteins (Ferrer *et al.*, 2007), only a few metalloproteins are present in the proteomes of its phylogenetic neighbour, *P. torridus*, and its habitat neighbour, *At. ferrooxidans* Thus, more studies are needed to determine the number and abundance of iron-proteins, particularly of Fe–S clusters-containing proteins, in iron-oxidizers and other acidophilic microorganisms.

In prokaryotes, three systems for Fe-S biosynthesis encoded in different operons are described. The three pathways identified to date are the Isc (iron sulfur cluster) system, the Suf (sulfur formation) system and the Nif (nitrogen fixation) system. Isc is considered to be the 'housekeeping' system, while Suf is functional under oxidative stress and iron limitation conditions and Nif is specific for proteins of nitrogen fixation (Ayala-Castro *et al.*, 2008). The occurrence of these pathways has not yet been investigated in acidophiles, however a preliminary bioinformatic inspection of sequenced genomes has indicated that acidophilic bacteria like *Acidimicrobium ferrooxidans*, *A. cryptum* and *Alicyclobacillus hesperidum*, as well as acidophilic crenarchaea, possess the Suf system, but not the *isc* genes (G. Levicán, unpublished). These data suggest that acidophilic microorganisms use Suf system for the Fe-S cofactor assembly, which is lined with higher resistance of Suf proteins to oxidative conditions and ROS in other model bacteria (Dai and Outten, 2012). Exceptions include the *Leptospirillum* spp. and *At. ferrooxidans*. In these bacteria *isc*, but not *suf* genes have been detected, together with a large number of genes (>80 in each bacterium) encoding Fe–S cluster-containing proteins (G. Levicán, unpublished). Some iron–sulfur proteins, such as high-potential iron protein (Hipip) and Rieske subunit of the bc_1 complex from *At. ferrooxidans* (e.g. Bruscella *et al.*, 2007) have been already experimentally characterized as containing Fe–S clusters, suggesting that the Isc system from this bacterium is active under highly oxidizing conditions. At least in *At. ferrooxidans*, activity of Isc system proteins has indeed been demonstrated (Zeng *et al.*, 2007).

According to Cárdenas *et al* (2012), in general, acidophiles lack genes encoding known oxidative stress response regulators. Nevertheless, PerR and Irr transcription regulators for peroxide and iron response, respectively, were found scattered in members of several genera. Additionally, an OxyR-like regulator was described in *Leptospirillum ferrooxidans* as up-regulated in a natural site where high metal concentrations were detected (Parro *et al.*, 2007), although its role in metal tolerance and oxidative stress response has not yet been addressed. *At. ferrooxidans* encodes a gene sharing significant sequence similarity to known Fur orthologues, including conservation of functional motifs. This regulator has been proved functional against *E. coli* Fur-responsive promoters and a number of targets associated with cellular functions including metal trafficking and utilization and redox balance (*grx, trx, gst*) have been predicted to occur in the genome of the type strain (Quatrini *et al.*, 2007). Thus, it could be suggested that Fur regulator from acidithiobacilli might have a role in coordinating metal homeostasis and oxidative stress responses. Yet, little is known about the regulatory networks that control redox homeostasis in extreme acidophiles; thus further studies aimed at deepening our understanding of this issue are required.

Antioxidant defence components have been experimentally tested in a limited number of acidophiles. High-throughput expression experiments carried out to evaluate global response of acidophilic microorganisms to several external factors have shown induction of antioxidant molecular elements, as part of the response to conditions such as metals, shift of energy source and temperature changes. Molecular determinants shown experimentally to be involved in stress protection in bacterial and archaeal acidophiles are shown in Table 4.1. According to proteomic data the SOD level is 2-fold higher in *At. ferrooxidans* grown in iron than in cultures grown on elemental sulfur (Ramírez *et al.*, 2004), suggesting differences in intracellular redox status of cells grown in either substrate. Transcription of *sod* also showed an increase in *At. ferrooxidans* cells upon exposure to 1% w/v pyrite or chalcopyrite (for 10 days) and to copper (Rodrigues *et al.*, 2010). In *Acidithiobacillus caldus sod* mRNA expression was also found to increase upon exposure to NaCl stress (Guo *et al.*, 2014). Moreover, in the archaeon *Sulfolobus solfataricus*, an extracellular Fe-SOD showed to be 10- and 20-fold higher in cultures supplemented with air than in static cultures (Cannio *et al.*, 2000). Production of the protein was also strongly dependent on the carbon source added to the medium, being glucose the most effective inducer. It has been suggested that *S. solfataricus* Fe-SOD along with a peroxiredoxin and a DPLS protein are part of a molecular complex that can efficiently remove ROS by converting $O_2^{\cdot-}$ into water (Maaty *et al.*, 2009). Formation of this stable complex, along with a reported compact tetrameric SOD, are indicative of some adaptations evolved to resist extreme temperature and pH in this thermoacidophilic archaeon. Taken together, these data suggest that superoxide dismutation activity is a minimum basal response required to deal several kinds of stress in a wide range of acidophiles.

Contrary to what has been show to occur in neutrophiles, catalase appears to play no role in H_2O_2 scavenging and stress response in acidophiles. However, peroxidases and peroxiredoxins are well documented as playing a pivotal role in protection against several stress conditions. The up-regulation of alkyl hydroperoxide reductase Ahp has been observed in *S. metallicus* (Bathe and Norris, 2007) after switching from elemental and other reduced forms of sulfur to ferrous iron and pyrite, in '*Ferroplasma acidarmanus*' after exposure to arsenic (Baker-Austin *et al.*, 2007) and in *At. ferrooxidans* after exposure to copper (Almárcegui *et al.*, 2014). In turn, the expression of thiol peroxidase Tpx encoding gene showed to be increased under NaCl stress in *At. caldus* (Guo *et al.*, 2014). As mentioned above these proteins might also have a role in detoxification of organic peroxides derived from lipid oxidation.

In case of the leptospirilli, experimental studies combined with genomic analysis have shown that they lack canonical genes encoding SOD enzyme, catalase and glutathione-related proteins (Norambuena *et al.*, 2012; Cortés *et al*, 2011). In turn, several copies of the *trx* genes ($n = 12$), encoding thioredoxins, have been identified in *L. ferriphilum*. Comparative analysis of oxidative stress response in *L. ferriphilum* and *At. ferrooxidans* has revealed that, after treatment with the oxidizing agents Fe_3SO_4,

H_2O_2 and K_2CrO_4, intracellular levels of ROS were significantly higher in *L. ferriphilum*. The higher ability of *L. ferriphilum* to avoid lipid damage has been attributed, at least in part, to the presence of Dyp peroxidase and lipoyl-dependent organic peroxidase described above. Accordingly, the level of lipoperoxidation measured by thiobarituric acid-reactive substances turned out to be significantly lower in *L. ferriphilum* for all tested compounds (Cortés *et al.*, 2011). These results have suggested that the leptospirilli may favour repair of the cellular damage over degradation of ROS. In addition, exposure to ROS-generating compounds led to up-regulation of several *trx* genes and a concomitant increase of the thioredoxin activity (Norambuena *et al.*, 2012), which proved to be significantly higher than in *E. coli* and *Bacillus subtilis*. These results are consistent with metaproteomic data derived from the Iron Mountain acid mine drainage biofilm community where *Leptospirillum* group II was found to be dominant (48% of metaproteome). In the communal proteome both thioredoxins and disulfide isomerases (Dsb) were highly represented (9% of the metaproteome) (Ram *et al.*, 2005).

The thioredoxin-based system also seems to be physiologically relevant in *At. ferrooxidans*, as deduced from proteomic studies aimed at identifying proteins associated with thiosulfate oxidation (Chi *et al.*, 2007) and biofilm formation on pyrite (Vera *et al.*, 2013). In both studies, high amounts of disulfide isomerases (Dsb) were detected. These are likely involved in catalysing the rearrangement of disulfide bonds, allowing the proper folding of proteins. In addition to these proteins, *At. ferrooxidans* SOD, thioredoxin and glutathione-based system are induced upon exposure to chalcopyrite (Rodrigues *et al.*, 2010) and copper (Orell *et al.*, 2010). Unlike *Leptospirillum*, *At. ferrooxidans* possesses all the genes encoding the proteins involved in glutathione metabolism and function. Up-regulation of these genes upon exposure to copper further support a role of this metabolite in protecting the cells against ROS induced by metal stress (Xia *et al.*, 2011; Orell *et al.*, 2010).

Relevance of the thioredoxin/thioredoxin reductase system have also been reported for the archaeon *S. solfataricus* exposed to H_2O_2 (Maaty *et al.*, 2009) and *P. torridus* exposed to extreme pH and temperature (Thürmer *et al.*, 2011) (Table 4.1). This system showed also to be up-regulated at both gene and protein levels upon exposure of '*F. acidarmanus*' to arsenic (Baker-Austin *et al.*, 2007). Hence, thiol/disulfide restoring systems based on small redox active thioredoxins seem to be required as important stress-evoked determinants in a wide range of microorganisms that inhabit extreme acidic environments.

Finally, a number of elements envisaged to play an important role in cellular defence against oxidative stress are small organic molecules. These have been suggested to play a role in the solubilization of minerals and the mobilization of metals due to their ability to complex such elements. Examples of such compounds are siderophores, which are produced under iron-limiting conditions to enhance the uptake of iron. Moreover, several compounds including sugars, amino acids and derivatives might also have a role as compatible solutes to maintain homeostasis under high salt concentration and dehydration. In acidophiles, metabolites directly linked to their physiology and stress response are less known. Recently, metabolites such as taurine and hydroxyectoine were identified and suggested to provide protection against osmotic stress in natural acidophilic biofilm community (Mosier *et al.*, 2013; Parro *et al.*, 2007). It has also been reported that in this community, *Leptospirillum* group II bacteria decrease production of ectoine and hydroxyectoine as biofilms mature, suggesting that biofilm structure and specially the extracellular polymeric substances (EPS) that are involved in the formation of biofilm provides some resistance to extreme external conditions. Although the role of EPS to complex (heavy) metals has been well documented, the relevance of the metabolites and EPS to prevent oxidative stress in acidophilic microorganisms has not been determined, and this could be of great importance for their applications in industrial and environmental biotechnologies.

Other relevant molecule involved in a number of redox reactions including respiratory chains is haem. As mentioned before, haem is the cofactor of many enzymes and cytochromes. Biosynthesis of haem in *At. ferrooxidans* as in other microorganisms, is dependent of glutamyl-tRNA, which is also a substrate for protein synthesis. In this bacterium both biosynthetic pathways are closely coordinated, particularly under oxidative stress, where a preference for protein synthesis seems to take place (Farah *et al.*, 2014).

Metal stress response in acidophilic microorganisms

In this section, a brief description of the strategies used by acidophiles and a comparison to neutrophiles to cope with metal ions from the environment are described. Molecular mechanisms that account for these strategies are described in detail in specific and comprehensive reviews (Dopson *et al.*, 2014; Navarro *et al.*, 2013; Orell *et al.*, 2013; Slyemi and Bonnefoy, 2012; Bonnefoy and Holmes, 2011).

Metals ions are essential nutrients to all living cells since they play crucial roles in many cellular processes. Bound to macromolecules, mainly to proteins as cofactors, metal ions participate in the catalysis of redox reactions, the transport of electron in respiratory chains, in gene expression as part of transcription factors and bound to nucleic acids or accumulate in storage proteins for later use. Owing to the redox potential of some metal ions a number of redox reactions use metals bound directly or as part of an organic cofactor, serving as intermediates that facilitates the exchange of electrons between molecules. Thus, the acquisition of metal ions is a relevant task for many microorganisms, particularly for all those that live in environments where the concentration of metal ions is very low. This is the case of several relevant pathogens, whose hosts use metal limitation strategies to avoid (or at least decrease) the susceptibility of infection. The opposite happens in acidophilic microorganisms. For instance, some strains of *At. ferrooxidans* can proliferate in concentrations of Fe, Ni, Cu or Zn ions approaching 1 M, approximately 1000-fold the concentration tolerated by *E. coli* (Dopson *et al.*, 2014). Since the natural environments where acidophiles proliferate are metal rich, with ion concentrations exceeding by far those tolerated by neutrophiles, strategies used by these microorganisms to acquire metal ions or to resist their potentially deleterious effects are of great interest from both fundamental and applied perspectives.

While microbial pathogens, and other neutrophiles, need to scavenge poorly bioavailable metals for their function, acidophiles are usually in the presence of environmental concentrations of soluble metal ions that easily exceed their needs. Thus, pathogens have developed mechanisms to import metal ions that allow them to concentrate intracellularly the metal ions required. These mechanisms include among others, siderophores that chelate metals themselves, complexing molecules (e.g. haem) or even storage proteins (e.g. bacterioferritins). In the case of intracellular metal overload, export mechanisms take place to reduce the potentially damaging metal ions. Conversely, acidophiles confronting an excess of external metal ions, require, mechanisms to resist or to dispose of the metals in order to preserve homeostasis. When the external metal ion concentrations that neutrophiles and acidophiles are capable to resist without deleterious effects are compared, questions concerning the metal export mechanisms arise.

The possibility of novel mechanisms of metal tolerance by acidophiles has been the focus of much research (Orell *et al.*, 2010). Both, active and passive mechanisms have been proposed to account for the tolerance in acidophiles. Among the active mechanisms, a finely tuned iron homeostasis response controlled by the Fur regulator in *At. ferrooxidans* has been described (Quatrini *et al.*, 2007). Efflux ATPases of the Cop system, metal chaperones (Navarro *et al.*, 2013) as well as the proton gradient-driven RND system (Mangold *et al.*, 2013) are used by different acidophiles to export metal ions, such as Cu. In certain acidophiles, increased metal tolerance has been correlated to higher numbers of metal exporters, several of which seem to have been acquired by horizontal gene transfer; e.g. the export of Cu in *At. ferrooxidans* strain ATCC 53993 (Orellana *et al.*, 2011). Other alternative mechanism to account for metal tolerance is the existence of intracellular polyphosphates, which have been proposed to act as potential carriers of metals to the extracellular milieu (e.g. Remonsellez *et al.*, 2006). Limited potential for uptake of Fe(II) in *Leptospirillum* or efficient storage of the ion in bacterioferritins have also been described (Osorio *et al.* 2008).

Concerning passive mechanisms, some questions that arise are whether acidophiles are really exposed to high concentrations of metals or rather to low concentrations of bioavailable ions, or alternatively if they face marked gradients in different environmental settings (e.g. planktonic versus biofilms communities). Although the real concentrations of metal ions in acidic environments are in the range of 0.1–1 M, it has been proposed that the free metal concentrations might be very low because of complexation to sulfates and hydroxides, and that these chelates might not be accessible to cells for the use or uptake of metals (Dopson *et al.*,

2014). However, rigorous evaluation of the solubility of these chelates and their equilibrium dynamics are required to further clarify this matter. The fact that acidophiles exist as sessile communities, or biofilms protected by exopolysaccharides (EPS), might also affect bioavailability of metal ions. It has been indicated that EPS might have differential permeability to different metal ions and affect their diffusion into the matrix, thereby contributing to reduce their bioavailability to microorganisms (Harrison *et al.*, 2007). Exposure to gradients of metal concentrations might explain why certain microorganisms from the consortium of bioleaching microorganisms are highly resistant while others from the same population are not. Another relevant issue is whether intrinsic features of acidophiles help them to counteract the effect of high concentrations of metals. A positive membrane potential (instead of a negative one in case of neutrophiles) has been proposed to exist in acidophilic microorganisms and underlie their capacity to keep protons from entering the cell. This potential might contribute to avoid the entry of metal cations to the cytoplasm (Baker-Austin and Dopson, 2007) or to facilitate their exclusion. Among the many factors that might contribute to modify the membrane potential, a novel processes that participates in the modification of the lipids of the cellular membrane has been described. It involves the aminoacylation of diacyl glycerol by alanine or arginine in an aminoacyl-tRNA dependent mechanism, which seems to contribute to increase the positive charge of the membrane (Roy *et al.*, 2009). Whether this mechanism takes place in acidophiles is still to be analysed.

Thus, in either natural or human made environments the exposure of acidophilic microorganisms to high metal ions concentrations might vary considerably compared to laboratory conditions. Both passive and active mechanisms might account for the tolerance of microorganisms to elevated metal exposure. But also the physiological status might contribute substantially to reduce the exposure to metal ions based on the fact that biofilms can alter the permeability of ions and the accessibility to the cells in the matrix. Probably this environment makes each bacterium in the biofilm to be exposed to a completely different concentration and bioavailability of the metal ions compared to planktonic microorganisms (Harrison *et al.*, 2007). Thus a number of considerations from physical, biochemical and genetics have to be taken in account before understanding the real magnitude as acidophilic microorganisms are exposed to metal ions in nature.

Concluding remarks

Acidophilic microorganisms often thrive in high metal-laden and ROS-generating environments, and thus oxidative stress is an important risk factor for them. While these microorganisms can scavenge ROS oxidants such superoxide and peroxide, they primarily rely on its antioxidant system to prevent cellular damage of proteins and lipids. Thioredoxin-based thiol/disulfide exchange systems are abundant and widely distributed among bacterial and archaeal acidophiles, and they seem to play a central role in maintenance of redox status of thiol-containing proteins under oxidative stress induced by metals and other factors. Unlike other highly ROS tolerant microorganisms, in acidophiles Fe-S proteins seem to be abundant and biosynthesis of Fe–S cluster seems to occur by ROS-tolerating Suf system. On the other hand, the presence of the Suf, but not the Isc system in leptospirilli is indicative of functional and structural variations that deserve special attention. Further studies are mandated to confirm these possibilities and to determine whether acidophiles possess mechanisms to compensate for impairment of the Fe–S clusters and other metalloproteins. Alkyl hydroxyperoxidases are also well conserved and widely distributed in sequenced acidophiles, and have proven to be up-regulated under several stressful environmental conditions. Indeed, these peroxidases represent one the most important determinants in acidophiles to scavenge organic and inorganic peroxides, and to protect lipids. Evidence collected so far suggests that acidophiles have developed a strong response to palliate the impact of ROS on biomolecules rather than ROS removal systems under oxidative conditions and other kind of challenges such as metals, temperature, pH, radiation and nutrient source shifts. In addition, absence of classical oxidative stress regulators raises the question about the nature of acidophiles regulators and the possible cross-talk between these regulators and those involved in other stress responses.

Acidophilic microorganisms are remarkably tolerant to many cationic metals and some metalloids, and they possess robust metal resistance mechanisms that involve export, sequestration and storage of metal ions. Whether acidophiles actually confront such elevated metal concentrations or, due to external complexation mediated by sulfate ions or adsorption onto extracellular polymeric substances of biofilm structures, metal bio-availability is much reduced, has been debated. Further efforts are required to determine the magnitude of metal concentrations that acidophiles face in the natural environments when biofilm organized communities rather than isolated planktonic microorganisms are present.

It is apparent from the discussion above that acidophilic microorganisms are heavily dependent on antioxidant defence for survival, thus the ability of these microorganisms to adapt quickly and efficiently to environmental redox variations can explain their abundance and activity in these extreme environments. However, additional studies taking account oxidative stress response and involvement of antioxidant elements in the general stress response and adaptation to natural environment are essential before drawing a final conclusion on the general antioxidant strategies of acidophiles.

Acknowledgements

This work was supported by FONDECYT (Grants Nos. 1120746, 1110203 and 1150834). We also thank to CONICYT for a doctoral fellowship to A. Ferrer.

References

Almárcegui, R.J., Navarro, C.A., Paradela, A., Albar, J.P., Von Bernath, D., and Jerez, C.A. (2014). New copper resistance determinants in the extremophile *Acidithiobacillus ferrooxidans:* a quantitative proteomic analysis. J. Proteome Res. *13*, 946–960.

Ayala-Castro, C., Saini, A., and Outten, F.W. (2008). Fe–S cluster assembly pathways in bacteria. Microbiol. Mol. Biol. Rev. *72*, 110–125.

Baker-Austin, C., and Dopson, M. (2007). Life in acid: pH homeostasis in acidophiles. Trends Microbiol. *15*, 165–171.

Baker-Austin, C., Dopson, M., Wexler, M., Sawers, R.G., and Bond, P.L. (2005). Molecular insight into extreme copper resistance in the extremophilic archaeon *"Ferroplasma acidarmanus"* Fer1. Microbiology *151*, 2637–2646.

Baker-Austin, C., Dopson, M., Wexler, M., Sawers, R.G., Stemmler, A., Rosen, B.P., and Bond, P.L. (2007). Extreme arsenic resistance by the acidophilic archaeon *"Ferroplasma acidarmanus"* Fer1. Extremophiles *11*, 425–434.

Bathe, S., and Norris, P.R. (2007). Ferrons iron- and sulfur-induced genes in *Sulfolobus metallicus*. Appl. Environ. Microbiol. *73*, 2491–2497.

Belnap, C.P., Pan, C., VerBerkmoes, N.C., Power, M.E., Samatova, N.F., Carver, R.L., Hettich, R.L., and Banfield, J.F. (2010). Cultivation and quantitative proteomic analyses of acidophilic microbial communities. ISME J. *4*, 520–530.

Bonnefoy, V., and Holmes, D.S. (2012). Genomic insights into microbial iron oxidation and iron uptake strategies in extremely acidic environments. Environ. Microbiol. *14*, *1597–1611*.

Bruscella, P., Appia-Ayme, C., Levicán, G., Ratouchniak, J., Jedlicki, E., Holmes, D.S., and Bonnefoy, V. (2007). Differential expression of two bc1 complexes in the strict acidophilic chemolithoautotrophic bacterium *Acidithiobacillus ferrooxidans* suggests a model for their respective roles in iron or sulfur oxidation. Microbiology *153*, 102–110.

Burton, N.P., Williams, T.D., and Norris, P.R. (1995). A potential anti-oxidant protein in a ferrous iron-oxidizing *Sulfolobus* species. FEMS Microbiol. Lett. *134*, 91–95.

Cabiscol, E., and Ros, J. (2000). Oxidative stress in bacteria and protein damage by reactive oxygen species. Int. Microbiol. *3*, 3–8.

Cannio, R., D'Angelo, A., Rossi, M., and Bartolucci, S. (2000). A superoxide dismutase from the archaeon *Sulfolobus solfataricus* is an extracellular enzyme and prevents the deactivation by superoxide of cell-bound proteins. Eur. J. Biochem. *267*, 235–243.

Cárdenas, J.P., Moya, F., Covarrubias, P., Shmaryahu, A., Levicán, G., Holmes, D.S., and Quatrini, R. (2012). Comparative genomics of the oxidative stress response in bioleaching microorganisms. Hydrometallurgy *127–128*, 162–167.

Chi, A., Valenzuela, L., Beard, S., Mackey, A.J., Shabanowitz, J., Hunt, D.F., and Jerez, C.A. (2007). Periplasmic proteins of the extremophile *Acidithiobacillus ferrooxidans*: a high throughput proteomics analysis. Mol. Cell. Proteomics *6*, 2239–2251.

Ciaramella, M., Napoli, A., and Rossi, M. (2005). Another extreme genome: how to live at pH 0. Trends. Microbiol. *13*, 49–51.

Cortés, A., Flores, R., Norambuena, J., Cárdenas, J.P., Quatrini, R., Chávez, R., Orellana, O., and Levicán, G. (2011). Comparative study of redox stress response in the acidophilic bacteria *Leptospirillum ferriphilum* and *Acidithiobacillus ferrooxidans*. In 19th International Biohydrometallurgy Symposium, Qiu, G., Jiang, T., Qin, W., Liu, X., Yang, Y., and Wang, H., eds. (Central South University Press, Changsha, China) pp. 354–357.

Dai, Y., and Outten, F.W. (2012). The *E. coli* SufS–SufE sulfur transfer system is more resistant to oxidative stress than IscS–IscU. FEBS Lett. *586*, 4016–4022.

Dopson, M., Ossandon, F.J., Lövgren, L., and Holmes, D.S. (2014). Metal resistance or tolerance? Acidophiles confront high metal loads via both abiotic and biotic mechanisms. Front. Microbiol. *5*, 10–13.

Fahey, R.C. (2013). Glutathione analogs in prokaryotes. Biochim. Biophys. Acta *1830*, 3182–3198.

Farah, C., Levicán, G., Ibba, M., and Orellana, O. (2014). Effect of hydrogen peroxide on the biosynthesis of heme and proteins: potential implications for the partitioning of Glu-tRNA(Glu) between these pathways. Int. J. Mol. Sci. *15*, 23011–23023.

Felício, A.P., De Oliveira, E., Odena, M.A., Garcia, O., Bertolini, M.C., Ferraz, L.F.C., Ottoboni, L.M.M., and Novo, M.T.M. (2011). Differential proteomic analysis of *Acidithiobacillus ferrooxidans* cells maintained in contact with bornite or chalcopyrite: Proteins involved with the early bacterial response. Process Biochem. *46*, 770–776.

Ferrer, M., Golyshina, O.V., Beloqui, A., Golyshin, P.N., and Timmis K.N. The cellular machinery of *Ferroplasma acidiphilum* is iron-protein-dominated. Nature *445*, 91–94.

Ghosal, D., Omelchenko, M.V., Gaidamakova, E.K., Matrosova, V.Y., Vasilenko, A., Venkateswaran, A., Zhai, M., Kostandarithes, H.M., Brim, H., Makarova, K.S., *et al.* (2005). How radiation kills cells: Survival of *Deinococcus radiodurans* and *Shewanella oneidensis* under oxidative stress. FEMS Microbiol. Rev. *29*, 361–375.

Goltsman, D.S.A., Denef, V.J., Singer, S.W., VerBerkmoes, N.C., Lefsrud, M., Mueller, R.S., Dick, G.J., Sun, C.L., Wheeler, K.E., Zemla, A., *et al.* (2009). Community genomic and proteomic analyses of chemoautotrophic iron-oxidizing *"Leptospirillum rubarum"* (Group II) and *"Leptospirillum ferrodiazotrophum"* (Group III) bacteria in acid mine drainage biofilms. Appl. Environ. Microbiol. *75*, 4599–4615.

Guo, X., Jiang, C., Luo, Y., Zhang, M., Poetsch, A., and Liu, S. (2014). Proteomic and molecular investigations revealed that *Acidithiobacillus caldus* adopts multiple strategies for adaptation to NaCl stress. Chinese Sci. Bull. *59*, 301–309.

Harrison, J.J., Ceri, H., and Turner, R.J. (2007). Multimetal resistance and tolerance in microbial biofilms. Nat. Rev. Microbiol. *5*, 928–938.

Imlay, J.A. (2008). Cellular defences against superoxide and hydrogen peroxide. Annu. Rev. Biochem. *77*, 755–776.

Jones, G.C., Van Hille, R.P., and Harrison, S.T.L. (2013). Reactive oxygen species generated in the presence of fine pyrite particles and its implication in thermophilic mineral bioleaching. Appl. Microbiol. Biotechnol. *97*, 2735–2742.

Levicán, G., Katz, A., de Armas, M., Núñez, H., and Orellana, O. (2007). Regulation of a glutamyl-tRNA synthetase by the heme status. Proc. Natl. Acad. Sci. U.S.A. *104*, 3135–3140.

Levicán, G., Gómez, M.J., Chávez, R., Orellana, O., Moreno-Paz, M., and Parro, V. (2012). Comparative genomic analysis reveals novel facts about *Leptospirillum* spp. cytochromes. J. Mol. Microbiol. Biotechnol. *22*, 94–104.

Limauro, D., Pedone, E., Pirone, L., and Bartolucci, S. (2006). Identification and characterization of 1-Cys peroxiredoxin from *Sulfolobus solfataricus* and its involvement in the response to oxidative stress. FEBS J. *273*, 721–731.

Limauro, D., Pedone, E., Galdi, I., and Bartolucci, S. (2008). Peroxiredoxins as cellular guardians in *Sulfolobus solfataricus*: characterization of Bcp1, Bcp3 and Bcp4. FEBS J. *275*, 2067–2077.

Maaty, W.S., Wiedenheft, B., Tarlykov, P., Schaff, N., Heinemann, J., Robison-Cox, J., Valenzuela, J., Dougherty, A., Blum, P., Lawrence, C.M., *et al.* (2009). Something old, something new, something borrowed; how the thermoacidophilic archaeon *Sulfolobus solfataricus* responds to oxidative stress. PLoS One 4, e6964.

Mangold, S., Potrykus, J., Björn, E., Lövgren, L., and Dopson, M. (2013). Extreme zinc tolerance in acidophilic microorganisms from the bacterial and archaeal domains. Extremophiles *17*, 75–85.

Martínez, P., Gálvez, S., Ohtsuka, N., Budinich, M., Cortés, M.P., Serpell, C., Nakahigashi, K., Hirayama, A., Tomita, M., Soga, T., *et al.* (2013). Metabolomic study of Chilean biomining bacteria *Acidithiobacillus ferrooxidans* strain Wenelen and *Acidithiobacillus thiooxidans* strain Licanantay. Metabolomics *9*, 247–257.

Moreno-Paz, M., Gómez, M.J., Arcas, A., and Parro, V. (2010). Environmental transcriptome analysis reveals physiological differences between biofilm and planktonic modes of life of the iron oxidizing bacteria *Leptospirillum* spp. in their natural microbial community. BMC Genomics *11*, 404.

Mosier, A.C., Justice, N.B., Bowen, B.P., Baran, R., Thomas, B.C., Northen, T.R., and Banfield, J.F. (2013). Metabolites associated with adaptation of microorganisms to an acidophilic, metal-rich environment identified by stable-isotope-enabled metabolomics. MBio. *4*, e00484-12.

Mosier, A.C., Li, Z., Thomas, B.C., Hettich, R.L., Pan, C., and Banfield, J.F. (2015). Elevated temperature alters proteomic responses of individual organisms within a biofilm community. ISME J. *9*, 180–194.

Navarro, C.A., von Bernath, D., and Jerez, C.A. (2013). Heavy metal resistance strategies of acidophilic bacteria and their acquisition: Importance for biomining and bioremediation. Biol. Res. *46*, 363–371.

Norambuena, J., Flores, R., Cárdenas, J.P., Quatrini, R., Chávez, R., and Levicán, G. (2012). Thiol/Disulfide system plays a crucial role in redox protection in the acidophilic iron-oxidizing bacterium *Leptospirillum ferriphilum*. PLoS One 7, e44576.

Orell, A., Navarro, C.A., Arancibia, R., Mobarec, J.C., and Jerez, C.A. (2010). Life in blue: copper resistance mechanisms of bacteria and archaea used in industrial biomining of minerals. Biotechnol. Adv. *28*, 839–848.

Orell, A., Remonsellez, F., Arancibia, R., and Jerez, C.A. (2013). Molecular characterization of copper and cadmium resistance determinants in the biomining thermoacidophilic archaeon *Sulfolobus metallicus*. Archaea *2013*, 289236.

Orellana, L.H., and Jerez, C.A. (2011). A genomic island provides *Acidithiobacillus ferrooxidans* ATCC 53993 additional copper resistance: a possible competitive advantage. Appl. Microbiol. Biotechnol. *92*, 761–767.

Osorio, H., Martínez, V., Nieto, P.A., Holmes, D.S., and Quatrini, R. (2008). Microbial iron management mechanisms in extremely acidic environments: comparative genomics evidence for diversity and versatility. BMC Microbiol. *8*, 203–220.

Parro, V., Moreno-Paz, M., and González-Toril, E. (2007). Analysis of environmental transcriptomes by DNA microarrays. Environ. Microbiol. *9*, 453–464.

Pedone, E., Limauro, D., D'Alterio, R., Rossi, M., and Bartolucci, S. (2006). Characterization of a multifunctional protein disulfide oxidoreductase from *Sulfolobus solfataricus*. FEBS J. *273*, 5407–5420.

Quatrini, R., Lefimil, C., Veloso, F.A., Pedroso, I., Holmes, D.S., and Jedlicki, E. (2007). Bioinformatic prediction and experimental verification of Fur-regulated genes in the extreme acidophile *Acidithiobacillus ferrooxidans*. Nucleic Acids Res. *35*, 2153–2166.

Ram, R.J., VerBerkmoes, N.C., Thelen, M.P., Tyson, G.W., Baker, B.J., Blake, R.C., Shah, M., Hettich, R.L., and Banfield, J.F. (2005). Community proteomics of a natural microbial biofilm. Science *308*, 1915–1920.

Ramírez, P., Guiliani, N., Valenzuela, L., Beard, S., and Jerez, C.A. (2004). Differential protein expression during growth of *Acidithiobacillus ferrooxidans* on ferrous iron, sulfur compounds, or metal sulfides. Appl. Environ. Microbiol. *70*, 4491–4498.

Rawlings, D.E. (2005). Characteristics and adaptability of iron- and sulfur-oxidizing microorganisms used for the recovery of metals from minerals and their concentrates. Microb. Cell Fact. *4*, 13.

Remonsellez, F., Orell, A., and Jerez, C.A. (2006). Copper tolerance of the thermoacidophilic archaeon *Sulfolobus metallicus*: possible role of polyphosphate metabolism. Microbiology *152*, 59–66.

Rodrigues, V.D., Martins, P.F., Gaziola, S.A., Azevedo, R.A., and Ottoboni, L.M.M. (2010). Antioxidant enzyme activity in *Acidithiobacillus ferrooxidans* LR maintained in contact with chalcopyrite. Process Biochem. *45*, 914–918.

Roy, H., Dare, K., and Ibba, M. (2009). Adaptation of the bacterial membrane to changing environments using aminoacylated phospholipids. Mol. Microbiol. *71*, 547–550.

Salerno, V., Napoli, A., White, M.F., Rossi, M., and Ciaramella, M. (2003). Transcriptional response to DNA damage in the archaeon *Sulfolobus solfataricus*. Nucleic Acids Res. *31*, 6127–6138.

Salzano, A.M., Febbraio, F., Farias, T., Cetrangolo, G.P., Nucci, R., Scaloni, A., and Manco, G. (2007). Redox stress proteins are involved in adaptation response of the hyperthermoacidophilic archaeon *Sulfolobus solfataricus* to nickel challenge. Microb. Cell Fact. *6*, 25.

Schoonen, M.A.A., Harrington, A.D., Laffers, R., and Strongin, D.R. (2010). Role of hydrogen peroxide and hydroxyl radical in pyrite oxidation by molecular oxygen. Geochim. Cosmochim. *74*, 4971–4987.

Slyemi, D., and Bonnefoy, V. (2012). How prokaryotes deal with arsenic. Environ. Microbiol. Rep. *4*, 571–586.

Thürmer, A., Voigt, B., Angelov, A., Albrecht, D., Hecker, M., and Liebl, W. (2011). Proteomic analysis of the extremely thermoacidophilic archaeon *Picrophilus torridus* at pH and temperature values close to its growth limit. Proteomics *11*, 4559–4568.

Vera, M., Krok, B., Bellenberg, S., Sand, W., and Poetsch, A. (2013). Shotgun proteomics study of early biofilm formation process of *Acidithiobacillus ferrooxidans* ATCC 23270 on pyrite. Proteomics *13*, 1133–1144.

Wang, Y., Zhang, X., Liu, Q., Ai, C., Mo, H., and Zeng, J. (2009). Expression, purification and molecular structure modeling of thioredoxin (Trx) and thioredoxin reductase (TrxR) from *Acidithiobacillus ferrooxidans*. Curr. Microbiol. *59*, 35–41.

Watkin, E.L.J., Keeling, S.E., Perrot, F.A., Shiers, D.W., Palmer, M.L., and Watling, H.R. (2009). Metals tolerance in moderately thermophilic isolates from a spent copper sulfide heap, closely related to *Acidithiobacillus caldus, Acidimicrobium ferrooxidans* and *Sulfobacillus thermosulfidooxidans*. J. Ind. Microbiol. Biotechnol. *36*, 461–465.

Wiedenheft, B., Mosolf, J., Willits, D., Yeager, M., Dryden, K.A., Young, M., and Douglas, T. (2005). An archaeal antioxidant: characterization of a Dps-like protein from *Sulfolobus solfataricus*. Proc. Natl. Acad. Sci. U.S.A. *102*, 10551–10556.

Wood, Z.A., Schröder, E., Robin Harris J., and Poole, L.B. (2003). Structure, mechanism and regulation of peroxiredoxins. Trends Biochem. Sci. *28*, 32–40.

Xia, J., Wu, S., Zhang, R., Zhang, C., He, H., Jiang, H., Nie, Z., and Qiu, G. (2011). Effects of copper exposure on expression of glutathione-related genes in *Acidithiobacillus ferrooxidans*. Curr. Microbiol. *62*, 1460–1466.

Yarzábal, A., Brasseur, G., and Bonnefoy, V. (2002). Cytochromes *c* of *Acidithiobacillus ferrooxidans*. FEMS Microbiol. Lett. *209*, 189–195.

Zammit, C.M., Mangold, S., Jonna, Vr., Mutch, L.A., Watling, H.R., Dopson, M., and Watkin, E.L. (2012). Bioleaching in brackish waters-effect of chloride ions on the acidophile population and proteomes of model species. Appl. Microbiol. Biotechnol. *93*, 319–329.

Zeng, J., Geng, M., Jiang, H., Liu, Y., Liu, J., and Qiu, G. (2007). The IscA from *Acidithiobacillus ferrooxidans* is an iron–sulfur protein which assemble the [Fe4S4] cluster with intracellular iron and sulfur. Arch. Biochem. Biophys. *463*, 237–244.

Zheng, C., Chen, M., Tao, Z., Zhang, L., Zhang, X.F., Wang, J.Y., and Liu, J. (2015). Differential expression of sulfur assimilation pathway genes in *Acidithiobacillus ferrooxidans* under Cd2+ stress: evidence from transcriptional, enzymatic, and metabolic profiles. Extremophiles *19*, 429–436.

Part III

Diversity of Acidophilic Microorganisms

Physiological and Phylogenetic Diversity of Acidophilic Bacteria

5

Mark Dopson

Abstract

Acidophilic bacteria can be found in natural and anthropogenic acidic environments such as acid sulfate soils and biomining operations. These environments range in temperatures from below zero where low temperature adapted, acidophilic bacteria accelerate metal and acid release from sulfide minerals, through mesophilic environments, to hot solfataric fields containing *Hydrogenobaculum acidophilum* with a temperature optimum of 65°C. Acidophilic bacteria have been isolated from the *Actinobacteria, Aquificae, Firmicutes, Nitrospira, Proteobacteria,* and *Verrucomicrobia* phyla, and are capable of oxidizing both inorganic and organic electron donors coupled to the reduction of oxygen or ferric iron, though no extremely acidophilic bacteria are known to ferment organic substrates. Acidophilic bacteria also exhibit a range of carbon metabolisms, from obligate autotrophs such as *Leptospirillum* spp., facultative autotrophs such as *Sulfobacillus* spp. that can both fix carbon dioxide (CO_2) or assimilate organic carbon, to obligate heterotrophs such as *Alicyclobacillus tolerans*. This chapter summarizes present knowledge of the physiological and phylogenetic diversity of acidophilic bacteria and highlights differences in growth characteristics between the various species.

Natural and anthropogenic environments

Acidophilic bacteria (pH optimum <5) exist in natural acidic environments such as solfatara and hydrogen sulfide fed cave systems, as well as anthropogenic environments such as bioreactors for treatment of wastewaters containing reduced inorganic sulfur compounds (RISCs). Many of these milieus have small concentrations of organic carbon resulting in the dominant bacteria being autotrophs, such as *Acidithiobacillus ferrooxidans* and *Leptospirillum ferriphilum*, which fix inorganic carbon for growth via the Calvin–Benson–Bassham (CBB) cycle or reductive citric acid cycle, respectively. However, other acidophilic bacteria are able to utilize organic carbon either as mixotrophs (e.g. *Alicyclobacillus tolerans*; Karavaiko *et al.*, 2005) or heterotrophs (e.g. *Acidiphilium cryptum*; Harrison, 1981). Mixotrophic and/or heterotrophic acidophiles are often able to grow on organic carbon released by autotrophic acidophiles (Nancucheo and Johnson, 2010).

Acidic environments encompass a wide range of temperatures from cold milieus at latitudes close to the poles and high altitudes to hot environments in bioheaps for extraction of metals from sulfide minerals and geothermal sites. Acidophilic bacteria have adapted to grow in temperatures from below 0°C (Hallberg *et al.*, 2010; Johnson *et al.*, 2014) to a known maximum of 65°C (Shima and Suzuki, 1993) (Table 5.1). In general, the low temperature acidophiles are predominantly bacteria; mesophilic and moderately thermophilic species include both acidophilic bacteria and archaea, while extremely thermophilic acidophiles are mostly archaea. The general division between the optimal growth temperatures is due to the greater thermal stability at high temperature of isoprenoid alcohols ether-linked to glycerol in archaeal membranes compared to ester lipids in bacteria and eukaryotes.

Phylogeny of acidophilic bacteria

The first acidophile to be isolated was *Acidithiobacillus thiooxidans* from the *Proteobacteria* (see Chapter 1) and since then acidophilic bacteria have been

Table 5.1 Properties of type strains of selected acidophilic bacteria

Species	Electron donors[a]	Electron acceptors	Carbon metabolism
Low temperature adapted mesophiles (able to grow at 5°C)			
Acidithiobacillus ferrivorans^T NO-37	Fe^{2+}/S^0/RISC	O_2/Fe^{3+}	Autotroph
'*Ferrovum myxofaciens*' P3G	Fe^{2+}	O_2	Autotroph
Mesophilic bacteria (optimum between 15 to 39°C)			
Acidithiobacillus albertensis^T ATCC 35403	S^0/RISC	O_2	Autotroph
Acidithiobacillus ferridurans^T ATCC 33020	Fe^{2+}/S^0/RISC/H_2	O_2/Fe^{3+}	Autotroph
Acidithiobacillus ferrooxidans^T ATCC 23270	Fe^{2+}/S^0/RISC/H_2	O_2/Fe^{3+}	Autotroph
Acidithiobacillus thiooxidans^T ATCC 19377	S^0/RISC	O_2	Autotroph
Acidiferrobacter thiooxydans^T m-1	Fe^{2+}/S^0/RISC	O_2/Fe^{3+}	Autotroph
Leptospirillum ferriphilum^T ATCC 49881	Fe^{2+}	O_2	Autotroph
'*Leptospirillum ferrodiazotrophum*' UBA1	Fe^{2+}	O_2	Autotroph
Leptospirillum ferrooxidans^T L15	Fe^{2+}	O_2	Autotroph
Acidiphilium acidophilum^T ATCC 27807	S^0/RISC/C	O_2/Fe^{3+}	Facultative autotroph
Sulfobacillus benefaciens^T BRGM2	Fe^{2+}/S^0/RISC/C/H_2	O_2/Fe^{3+}	Facultative autotroph
Alicyclobacillus aeris^T ZJ-6	Fe^{2+}/S^0/RISC/C	O_2	Facultative autotroph
Acidiphilium cryptum^T Lhet2	C	O_2/Fe^{3+}	Heterotroph
Acidiphilium rubrum^T ATCC 35905	RISC/C	O_2/Fe^{3+}	Heterotroph
'*Acidithrix ferrooxidans*' Py-F3	Fe^{2+}/C	O_2	Heterotroph
Acidocella facilis^T PW2	C	O_2	Heterotroph
Acidocella aromatica^T PFBC	C	O_2/Fe^{3+} mineral	Heterotroph
Alicyclobacillus disulfidooxidans^T SD-11	Fe^{2+}/S^0/RISC/C	O_2	Heterotroph
Alicyclobacillus ferrooxydans^T TC-34	Fe^{2+}/RISC/C	O_2	Heterotroph
Alicyclobacillus tolerans^T K1	Fe^{2+}/S^0/C	O_2/Fe^{3+}	Heterotroph
Ferrimicrobium acidiphilum^T T23	Fe^{2+}/C	O_2/Fe^{3+}	Heterotroph
Acidomonas methanolica^T TK 0705	C	O_2	Facultative methylotroph
Moderately thermophilic and thermo-tolerant acidophiles (optimum between 40 and 59°C)			
Acidithiobacillus caldus^T KU	S^0/RISC/H_2	O_2	Autotroph
'*Acidithiomicrobium*' P2	Fe^{2+}/S^0	O_2	Autotroph
Acidimicrobium ferrooxidans^T ICP	Fe^{2+}/C/H_2	O_2/Fe^{3+}	Facultative autotroph
Aciditerrimonas ferrireducens^T IC-180	C/H_2	O_2/Fe^{3+}	Facultative autotroph
Sulfobacillus acidophilus^T NAL	Fe^{2+}/S^0/RISC/C/H_2	O_2/Fe^{3+}	Facultative autotroph
Sulfobacillus sibiricus^T N1	Fe^{2+}/S^0/RISC/C	O_2	Facultative autotroph
Sulfobacillus thermosulfidooxidans^T AT-1	Fe^{2+}/S^0/RISC/C/H_2	O_2/Fe^{3+}	Facultative autotroph[b]
Sulfobacillus thermotolerans^T Kr1	Fe^{2+}/S^0/RISC/C	O_2	Facultative autotroph
Acidicaldus organivorans^T Y008	S^0/C	O_2/Fe^{3+}	Heterotroph
Ferrithrix thermotolerans^T Y005	Fe^{2+}/C	O_2/Fe^{3+}	Heterotroph
Thermophilic bacteria (temperature optimum ^{3}60°C)			
Hydrogenobaculum acidophilum^T 3H-I	S^0/RISC/H_2	O_2	Autotroph
Methylacidiphilum infernorum V4	CH_4	O_2	Heterotroph

[a]S^0, elemental sulfur; RISC, reduced inorganic sulfur compounds; C, organic carbon.
[b]Autotrophic growth with sulfur is weak.

isolated from the phyla *Actinobacteria*, *Aquificae*, *Firmicutes*, *Nitropsirae* and *Verrucomicrobia* (Fig. 5.1). These bacteria include both Gram-negative species (e.g. *Leptospirillum* spp.) as well as Gram-positive species (e.g. *Sulfobacillus* spp.). The *Proteobacteria* are a heterogeneous phylum of Gram-negative bacteria of which the *Acidithiobacillia*, α-, β-, and γ-classes contain acidophiles. Species from the *Acidithiobacillus* genus, such as the low temperature-adapted *Acidithiobacillus ferrivorans* and the moderate thermophile *Acidithiobacillus caldus*, were originally classified as *Acidithiobacillales* in the γ-*Proteobacteria* class. However, this has been amended as multi-protein phylogenetic analysis

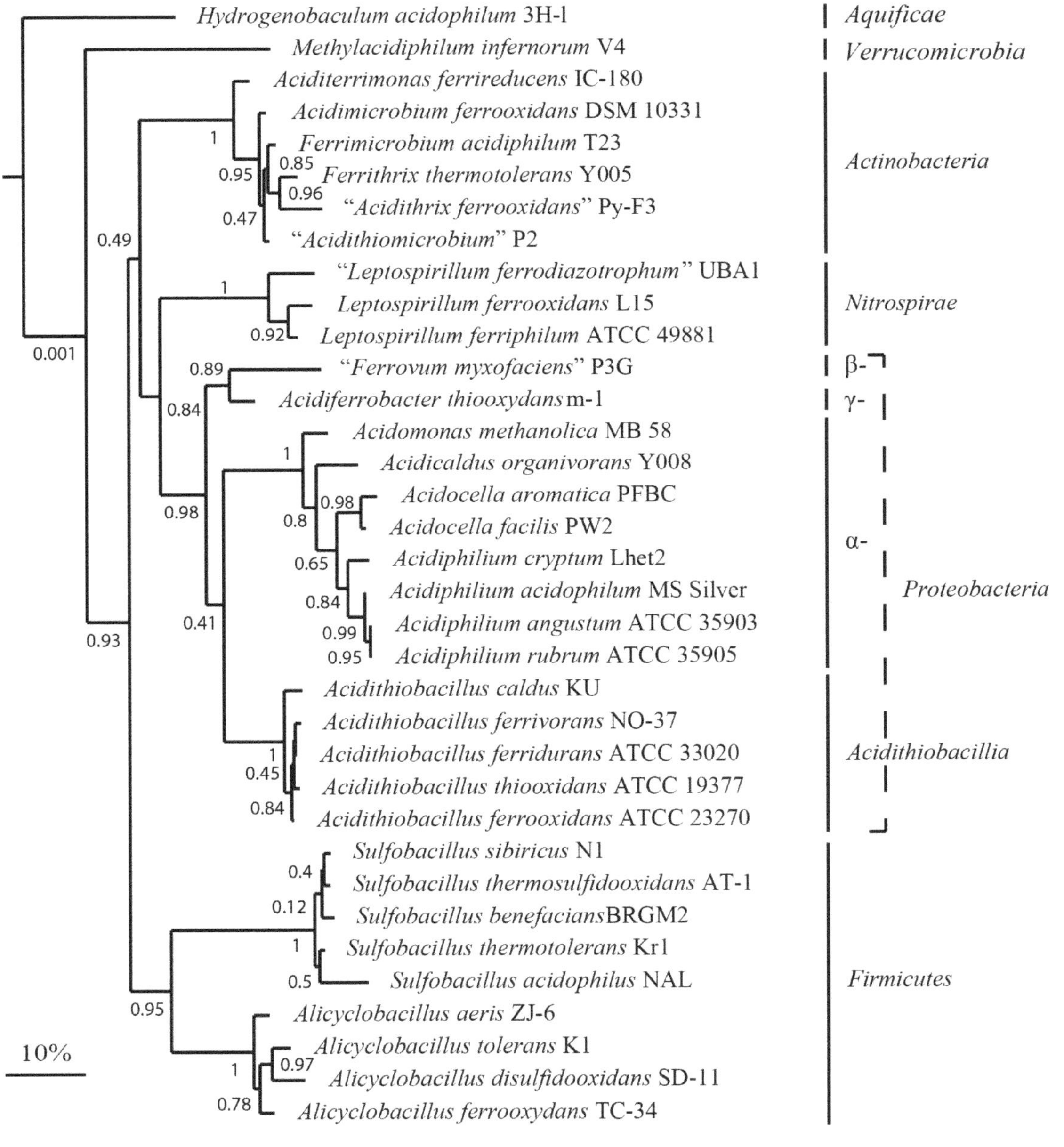

Figure 5.1 Phylogenetic tree of 16S rRNA gene sequences from selected acidophilic bacteria. The tree was constructed using the maximum likelihood method in the PhyML program. The tree was rooted with the acidophilic archaeon *Thermoplasma acidophilum* 122–1B2 (not shown), bootstrap values are given, and the scale bar denotes 10% sequence divergence.

suggests the establishment of a new *Proteobacteria* class, proposed to be called the *Acidithiobacillia* (Williams and Kelly, 2013). The *Actinobacteria* is a phylum of Gram-positive acidophilic and neutrophilic bacteria containing many uncultured soil microorganisms. The phylum has six genera containing acidophiles although '*Acidithiomicrobium*' and '*Acidithrix*' are currently candidate genera. Species from the bacterial genus *Aquificae* grow in harsh environments including acidic milieus such as hot springs and sulfur pools but contain only a single acidophile species, *Hydrogenobaculum acidophilum*. The *Firmicutes* are low GC, primarily spore-forming, and Gram-positive bacteria that includes the two genera, *Sulfobacillus* and *Alicyclobacillus*, that contain species that are extremely acidophilic. The phylum *Nitropsirae* includes *Leptospirillum* spp., which are Gram-negative bacteria with characteristic vibrioid and spirillum-like cell morphologies. The other phylum containing acidophiles is the recently described *Verrucomicrobia* with few cultured species, although molecular data suggests that they are abundant in the environment.

From low temperature adapted bacteria to thermophiles

In this chapter, acidophilic bacteria are categorized according to their optimum growth temperatures. Low temperature-adapted bacteria are defined as being able to grow close to 0°C but having temperature optima above 15°C; mesophiles have optimum growth temperatures between 15 and 39°C; moderately thermophilic and thermo-tolerant acidophiles have an optimum growth temperature from 40 to 59°C but can sometimes out-compete the mesophiles at temperatures around 30°C; and extreme thermophiles are defined as having an optimum growth temperature at or above 60°C.

Low-temperature-adapted bacteria

The occurrence of acidophiles able to grow at low temperatures was first shown in a study of psychrotolerant (but not psychrophilic) strains able to grow heterotrophically (Berthelot *et al.*, 1993). Later studies presented microbial oxidation of ferrous iron (Fe^{2+}) and pyrite (FeS_2) oxidation in Arctic environments. This was supported by a laboratory experiment with sulfidic waste and tailings from northern Canada that showed biotic sulfide oxidation coupled to oxygen consumption as low as −11°C. Recent molecular phylogenetic studies of FeS_2 weathering in the Antarctic (Dold *et al.*, 2013), acid sulfate soils in northern Finland (Wu *et al.*, 2013) and bioreactors for low temperature removal of reduced sulfur compounds (Liljeqvist *et al.*, 2011a) also suggested the presence of low temperature acidophiles. However, none of the above reports carefully characterized the bacterial species present.

Two species of psychrotolerant acidophiles have since been characterized: *At. ferrivorans* (Hallberg *et al.*, 2010) and '*Ferrovum myxofaciens*' (Johnson *et al.*, 2014). *At. ferrivorans*^{T} (strain NO-37) is from the *Acidithiobacillia* class within the *Proteobacteria* and was first isolated from an abandoned copper mine in Norway. It oxidizes Fe^{2+} and reduced sulfur compounds with a temperature optimum of ~28°C and a pH optima of 2.5 (Hallberg *et al.*, 2010). *At. ferrivorans* strain SS3 (formerly designated *At. ferrooxidans* SS3) has a lower temperature optimum at ~22°C and oxidizes Fe^{2+} (Kupka *et al.*, 2007), elemental sulfur (S^0) and reduced sulfur compounds (Kupka, 2009), and sulfide minerals (Liljeqvist *et al.*, 2013; Dopson *et al.*, 2007) at temperatures as low as 6°C. The genomes of *At. ferrivorans* SS3 and another strain (CF27, originally isolated from a cobalt mine in Idaho) have been sequenced (Talla *et al.*, 2013; Liljeqvist *et al.*, 2011b) and genes suggested to code for Fe^{2+} and reduced sulfur compound oxidation identified (Liljeqvist *et al.*, 2013). '*Fv. myxofaciens*' P3G (the proposed type strain) is an autotrophic Fe^{2+} oxidizing β-proteobacterium identified from macroscopic 'streamer' biofilms in a pH 2.5, low-temperature acid mine drainage (AMD) stream (Johnson *et al.*, 2014) though closely related strains have been found in clone libraries worldwide. Along with the iron-oxidizing actinobacterium *Acidithrix ferrooxidans* (Jones and Johnson, 2015), *At. ferrivorans* and '*Fv. myxofaciens*' dominated the streamer biofilm and are likely to be the major providers of organic carbon via CO_2 fixation (Kay *et al.*, 2013).

The existence of low temperature-adapted acidophiles facilitates the oxidation of Fe^{2+} and RISCs in low-temperature environments that cover a large area of the earth's surface (e.g. polar regions and high altitudes). These bacteria have been exploited for biotechnological processes in cold environments, such as removing RISCs from mineral processing wastewaters at low ambient temperatures (Liljeqvist *et al.*, 2011a).

Mesophilic bacteria

Mesophilic acidophilic bacteria (optimum growth temperature of 15–39°C) comprise more classified species than low- or high-temperature-adapted species. These include obligate autotrophs such as *Leptospirillum* spp., facultative autotrophs such as *Sulfobacillus benefaciens* (Johnson *et al.*, 2008), and heterotrophs such as *Alicyclobacillus disulfidooxidans* (Karavaiko *et al.*, 2005).

The most widely studied acidophilic bacteria are from the *Acidithiobacillia* genus *Acidithiobacillus* that includes the first acidophile to be isolated, *At. thiooxidans*, as well as *At. ferrooxidans*, the first species identified to catalyse the oxidation of ferrous iron at low pH and thereby accelerate the genesis of AMD. *At. ferrooxidans* strains grow more rapidly than most other species in media used to enrich iron-oxidizing acidophiles, and also more readily on solid media. This led to their selection in classical plate growth studies, despite often constituting a minor fraction of the microbial community, and resulted in the misconception that *At. ferrooxidans* is the dominant acidophile in many acidic environments. This view persisted until the advent of molecular techniques that identified the full range of acidophilic bacteria described here. All the acidithiobacilli are obligate autotrophs and fix CO_2 by the CBB cycle, and the Fe^{2+} oxidizing species have also been demonstrated to fix nitrogen (N_2) (Hedrich *et al.*, 2011; Kelly and Wood, 2000). *At. ferrooxidans* serves as a model species for the acidophiles and the type strain has a temperature optimum between 30 and 35°C and a pH optimum of 2.5 (Kelly and Wood, 2000). The genome sequences from two strains have been released and pathways/systems have been elucidated for (i) the oxidation of ferrous iron (Fe^{2+}), elemental sulfur (S^0) and RISCs coupled to the reduction of O_2 (Quatrini *et al.*, 2009); (ii) the oxidation of S^0 coupled to ferric iron (Fe^{3+}) reduction in the absence of air (Osorio *et al.*, 2013); (iii) fixation of carbon dioxide by the CBB cycle (Esparza *et al.*, 2010); (iv) central carbon metabolism (Appia-Ayme *et al.*, 2006); (v) iron homeostasis (Osorio *et al.*, 2008); (vi) quorum sensing and biofilm formation (Ruiz *et al.*, 2007) as well as (vii) its bioenergetics strategies to grow in the presence of a proton gradient across the cytoplasmic membrane up to approximately four orders of magnitude (Ferguson and Ingledew, 2008).

At. ferrooxidans was originally characterized as a mesophilic iron- and sulfur-oxidizing acidophile, and many new isolates were wrongly attributed to the species, resulting in strains being highly heterogeneous, both in terms of the phenotypic characteristics and phylogenies. This issue has recently been addressed in a number of studies. For example, the molecular taxonomy of twenty-one iron-oxidizing acidithiobacilli strains was analysed and subsequently split into four species (Amouric *et al.*, 2011). These were the original species (*At. ferrooxidans*), the low temperature adapted *At. ferrivorans* (described above), *Acidithiobacillus ferridurans* (Hedrich and Johnson, 2013a), and an as yet uncharacterized *Acidithiobacillus* clade (Group IV). Traits used to assign the strains to different species are that they form distinct clades based on multilocus gene sequence analysis, differences in Fe^{2+} and S^0 compound oxidation pathways, and various phenotypic factors. The name *At. ferridurans* refers to its ability to tolerate high concentrations of iron and to grow aerobically with Fe^{2+}, S^0, RISCs, and H_2 as electron donor as well as coupling H_2, S^0, or tetrathionate oxidation to Fe^{3+} reduction under anaerobic conditions (Hedrich and Johnson, 2013a). The type strain (ATCC 33020) has temperature and pH optima of 29°C and 2.1, respectively, and most strains are motile. Several *At. ferridurans* isolates grow at a lower pH than either *At. ferrooxidans* or *At. ferrivorans* as some strains were able to grow at pH < 1.3 (Hedrich and Johnson, 2013a). Other mesophilic acidithiobacilli are *At. thiooxidans* and *Acidithiobacillus albertensis* (Kelly and Wood, 2000). The *At. thiooxidans* type species (ATCC 19377) is motile, has temperature and pH optima of 30 to 35°C and 2.0 to 3.0, respectively, and the species is differentiated from the Fe^{2+} oxidizing *Acidithiobacillus* spp. by its obligate growth on S^0 and reduced sulfur compounds. Although it cannot oxidize Fe^{2+}, *At. thiooxidans* grows in biomining environments in combination with iron-oxidizers such as *At. ferrooxidans*. *At. albertensis* has temperature and pH optima of 25–30°C and 3.5–4.0, respectively and has significant phenotypic differences from other *Acidithiobacillus* spp. such as the presence of a glycocalyx and motility by a tuft of flagella (Kelly and Wood, 2000). The *At. albertensis* type strain (ATCC 35403) is an obligate autotroph capable of growth via oxidation of S^0 and reduced sulfur compounds (Bryant *et al.*, 1983). Another *At. albertensis* strain (BY-05) was demonstrated to contribute to copper dissolution from chalcopyrite ($CuFeS_2$) when grown in co-culture with *At. ferrooxidans* presumably due to its ability to oxidize S^0 (Xia *et al.*, 2007).

The *γ-proteobacterium Acidiferrobacter* (*Acf.*) *thiooxydans* (strain m-1^T) was originally isolated from coal waste in the USA by a pioneering scientist in acidophile microbiology, Arthur Harrison Jr. While *Acf. thiooxydans* also oxidizes Fe^{2+}, S^0 and RISCs using O_2 as electron acceptor, reduces Fe^{3+} when cultured with S^0 in the absence of air, and grows optimally at approximately pH 2.0, it is only distantly related to *Acidithiobacillus* spp., and is more thermo-tolerant (growth optimum at 38°C and maximum 47°C (Hallberg, 2011). *Acf. thiooxydans* is an obligate autotroph, a facultative diazotroph, and its growth in liquid culture requires an osmotic potential >2 bar (optimum 5 bar) making it a moderate osmophile. *Acf. thiooxydans* is more closely related to the currently mis-classified, halophilic iron- and sulfur-oxidizer *Thiobacillus prosperus* (Huber and Stetter, 1989) than to *Acidithiobacillus* spp. suggesting it may be involved in sulfide mineral dissolution in areas containing water with a high osmotic strength.

Other mesophilic acidophiles are from the *Nitrospira* phylum and include *Leptospirillum ferrooxidans* (Hippe, 2000), *L. ferriphilum* (Coram and Rawlings, 2002), '*Leptospirillum ferrodiazotrophum*' (Goltsman *et al.*, 2009; Tyson *et al.*, 2005), and two uncultured candidate species identified from community genomic and proteomic analyses, '*Leptospirillum rubarum*' (Group II) and '*Leptospirillum* Group IV' (Goltsman *et al.*, 2013; Goltsman *et al.*, 2009). *Leptospirillum* spp. are often identified in biomining bioreactors due to their greater affinity for Fe^{2+} and ability to grow at lower pH, which allows them to outcompete *At. ferrooxidans* in high redox potentials and in solutions containing small concentrations of Fe^{2+} (Rawlings *et al.*, 1999). Despite their importance in biomining, considerably less is known about their energy and nutrient metabolism compared to the acidithiobacilli. *Leptospirillum* spp. oxidize Fe^{2+} coupled to O_2 reduction as their sole electron donor and acceptor, fix CO_2, and can tolerate lower pH values than iron-oxidizing acidithiobacilli (Hippe, 2000). Genome data analysis suggests *L. ferriphilum* and '*L. ferrodiazotrophum*' iron homeostasis strategies include Fe^{3+}-siderophore outer membrane receptors but no Fe^{2+} uptake systems (Osorio *et al.*, 2008). One *L. ferrooxidans* genome (of an isolate, strain C2–3, obtained from a volcanic ash deposit in Japan) is publicly available. This was noted not to contain genes coding for ribulose 1,5-bisphosphate carboxylase/oxygenase (RubisCO; used for CO_2 fixation via the CBB cycle), though it did contain all *nif* genes required for nitrogen fixation except *nifL* (Fujimura *et al.*, 2012). The type strain of *L. ferriphilum* (ATCC 49881) is described as having pH and temperature optima between 1.4°C to 1.8°C and 30°C to 37°C (Coram *et al.*, 2002). However, many other strains have higher temperature optima including *L. ferriphilum* strain MT6 that has an optimum of 43°C (Okibe *et al.*, 2003) and the genome sequenced strain (ML-04) is described as moderately thermophilic (Mi *et al.*, 2011). The *L. ferriphilum* ML-04 genome analysis suggests it fixes CO_2 via the reductive citric acid cycle and assimilates ammonium. '*L. ferrodiazotrophum*' also contains *nif* genes and this trait was exploited for targeted isolation from a mixed AMD culture from Iron Mountain, California (Tyson *et al.*, 2005). The Iron Mountain AMD environment is limited by nitrogen availability and even though '*L. ferrodiazotrophum*' is only present in a low abundance, its ability to fix nitrogen was considered to be a key component of the mixed community. '*L. rubarum*' was also identified from the highly acidic and metal contaminated Iron Mountain AMD site and community genomic and proteomic analyses predicts it lacks all the components of the CBB cycle and is suggested to fix CO_2 via the reductive citric acid cycle (Goltsman *et al.*, 2009). The uncultured '*Leptospirillum* Group IV' was also identified from Iron Mountain by cultivation independent techniques and is suggested to be able to fix nitrogen and CO_2 with a similar autotrophic metabolism to that of '*L. ferrodiazotrophum*' (Goltsman *et al.*, 2013).

Several mesophilic acidophiles are able to grow using both organic and inorganic energy sources including *Acidiphilium acidophilum*, which was originally isolated from a culture of *At. ferrooxidans* (Hiraishi *et al.*, 1998). It is unique among *Acidiphilium* spp. in being able to grow on S^0 and

RISCs in the absence of an organic carbon source (optimal growth at 25 to 30°C and pH 3.0 to 3.5). *A. acidophilum* is also able to grow aerobically or anaerobically by coupling the oxidation of organic compounds including sugars and amino acids to the reduction of oxygen or ferric iron. Other facultative autotrophs include *Sb. benefaciens* (Johnson *et al.*, 2008) and *Alicyclobacillus aeris* (Guo *et al.*, 2009). *Sb. benefaciens* was isolated from stirred tank and heap biomining environments and can grow by aerobic oxidation of Fe^{2+}, S^0, RISCs, H_2, and organic carbon. It is also capable of anaerobic coupling of glycerol oxidation to Fe^{3+} reduction (Johnson *et al.*, 2008). A notable feature of *Sb. benefaciens* is its greater efficiency to fix CO_2 than other members of the genus *Sulfobacillus*. This trait may permit *Sb. benefaciens* to grow more efficiently than other *Sulfobacillus* spp. in AMD environments that are typically low in dissolved organic carbon.

The type strain of *Alb. aeris* was also isolated from a biomining environment, has temperature and pH optima of 30°C and pH 3.5, and grows on various organic carbon compounds such as glycerol (Guo *et al.*, 2009). However, some strains of iron- and sulfur-oxidizing *Alicyclobacillus* are capable of growth at temperatures up to 45°C (DB Johnson, personal communication).

Obligately heterotrophic mesophilic bacteria include members of the genera *Acidiphilium*, *Acidocella*, and *Acidomonas* (all *Alphaproteobacteria*), *Ferrimicrobium* and *'Acidithrix'* (both *Actinobacteria*), and *Alicyclobacillus* spp. (*Firmicutes*). *Acidiphilium* spp. were originally described as only being capable of growth under strictly aerobic conditions (Harrison, 1981). It is now known that *Acidiphilium* spp. can reduce Fe^{3+} either in the presence of air or under oxygen-limited conditions (Johnson and McGinness, 1991) though they do not grow in strictly anoxic environments. In addition, *Acidiphilium* SJH can carry out reductive dissolution of solid Fe^{3+} minerals such as goethite and jarosite (Bridge and Johnson, 2000). The ability to reduce Fe^{3+} minerals allows these bacteria to grow in oxygen-limited environments such as sediments of AMD streams and to potentially cycle Fe^{2+}/Fe^{3+} between oxidizing and reducing niches within environments (Fig. 5.2). *A. cryptum* was isolated as a contaminant in cultures of autotrophic acidophiles where it is capable of growth on released organic carbon (Nancucheo and Johnson, 2010). Possibly as a consequence of their original isolation as contaminants growing on small concentrations of organic carbon, *A. cryptum* was first thought to be an oligotroph (Harrison, 1981). However, it has since been shown that high

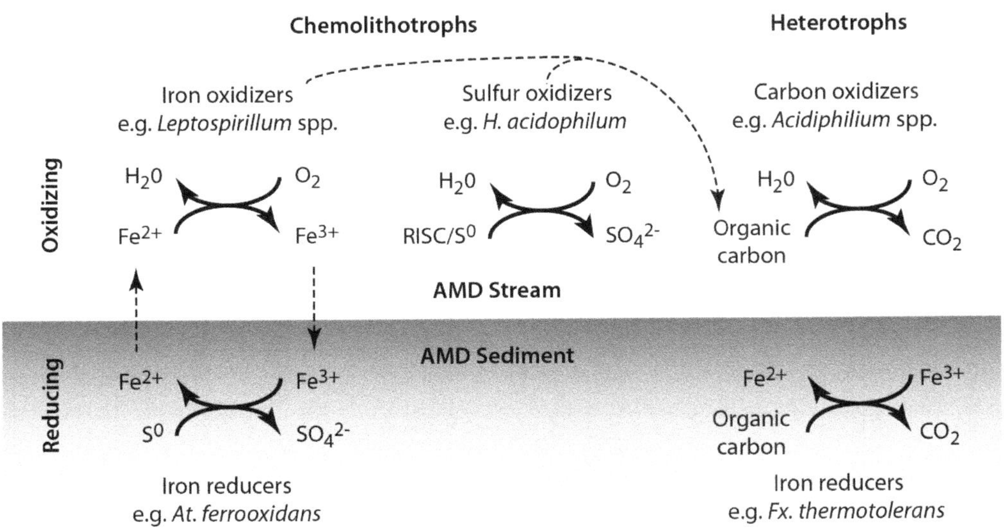

Figure 5.2 Carbon and energy sources exploited by acidophilic bacteria. The relationships between aerobic and anaerobic growth strategies are described as for the water column and sediment of an acid mine drainage (AMD) stream. Dotted lines denote transport processes.

cell densities can be obtained by culture in medium containing glucose at pH 2.5 (Kusel *et al.*, 1999). *Acidiphilium rubrum* was isolated from acidic coal mine stream water and sediment samples and can grow on a range of organic carbon sources (Wichlacz *et al.*, 1986). Being isolated from both stream water (likely oxidizing) and sediment (likely reducing) is reflected in its ability to use both O_2 and Fe^{3+} as terminal electron acceptor. There are several other species of *Acidiphilium*, including *A. organovorum* which was isolated from an *At. ferrooxidans* culture (previously maintained on inorganic carbon and energy sources) by growth with glucose as the electron donor at pH 3.0. The ecological role of *A. organovorum* is the likely removal of organic carbon excreted by *At. ferrooxidans*. *Acidiphilium* spp. also have low concentrations of bacteriochlorophyll *a* chelated with zinc in their cells (Hiraishi *et al.*, 1998).

Acidocella facilis and *Acidocella aminolytica* are also obligately heterotrophic *Alphaproteobacteria*, and were originally described as *Acidiphilium* spp. though have since been reclassified based on phylogenetic analysis of the 16S rRNA genes (Kishimoto *et al.*, 1995). Like *Acidiphilium* spp., *Ac. facilis* is a heterotrophic acidophile that can grow on a range of organic carbon compounds but in contrast to *Acidiphilium* spp., it can grow in the presence of 3.5% (w/v) NaCl (Kishimoto *et al.*, 1995). *Ac. aminolytica* was isolated from acid coal mine drainage and grows at slightly higher pH (3.0 to 6.0) and oxidizes a wide range of organic carbon compounds. In contrast, *Acidocella aromatica* grows on a limited range of organic compounds such as acetate and phenol coupled to reduction of O_2 or ferric minerals (Jones *et al.*, 2013) though not on commonly used substrates such as glucose and yeast extract. Organic acids are typically highly toxic to autotrophic acidophiles as they are protonated in the highly acidic external milieu, but once they enter the near neutral cytoplasm they dissociate to release protons that acidifies the cytoplasm and ultimately, causes cell death. *Ac. aromatica* is unusually tolerant to organic acids and grows in 10 mM, but not 20 mM acetic acid (Jones *et al.*, 2013). The ecological role of *Ac. aromatica* is likely to degrade the organic carbon not utilized by, and toxic to, autotrophic acidophiles. Its unique ability among acidophilic bacteria to oxidize aromatic compounds also provides the potential to treat wastewaters containing organic pollutants (Hallberg *et al.*, 1999). Another classified species, *Acidocella aluminiidurans*, was isolated from an acidic (pH 3) swamp in Vietnam and can tolerate 500 mM aluminium sulfate (Kimoto *et al.*, 2010).

Acidomonas methanolica is the sole species in its genus and is a facultative methylotroph that is differentiated from the other acidophilic *Proteobacteria* spp. based on its DNA–DNA homology (Urakami *et al.*, 1989). It is non-motile; grows at pH 2.0 to 5.5 (optimum between 3.0 and 5.0), has an optimum temperature between 30 to 37°C; and does not reduce nitrogen. *Acm. methanolica* grows on several organic carbon compounds including ethanol and glucose but unlike *Ac. facilis* it does not grow in the presence of salt.

Ferrimicrobium acidiphilum is an obligate heterotroph isolated from an abandoned pyrite mine in north Wales, that can couple glycerol oxidation to dissimilatory Fe^{3+} reduction (Johnson *et al.*, 2009). *Fm. acidiphilum* is also capable of Fe^{2+} oxidation (but not S^0 or RISCs) that increases its growth yield as well as catalysing FeS_2 dissolution in the presence of yeast extract (Johnson *et al.*, 2009). Like *Fm. acidophilum*, the as yet officially unnamed '*Acidithrix ferrooxidans*' is a heterotrophic acidophile that grows in filaments and is capable of oxidizing Fe^{2+} (Kay *et al.*, 2013). Despite being isolated from the same permanently cold (circa 11°C) AMD stream as the low temperature adapted isolates *At ferrivorans* and '*Fv. myxofaciens*', '*Acx. ferrooxidans*' is a mesophile. A combination of a higher optimum growth temperature and the low concentration of organic carbon in the mine water likely explain why it is present as a minor member of the community.

Other classified mesophilic heterotrophic acidophilic bacteria are *Alicyclobacillus* spp., and include *Alb. disulfidooxidans*, *Alicyclobacillus ferrooxydans*, and *Alb. tolerans* (Jiang *et al.*, 2008; Karavaiko *et al.*, 2005; Dufresne *et al.*, 1996). *Alb. disulfidooxidans* and *Alb. tolerans* were originally named as *Sulfobacillus* spp. before being reclassified. *Alb. disulfidooxidans* was isolated after enrichment from wastewater sludge with S^0 as energy source and is able to grow aerobically on organic carbon compounds such as glutamate and glucose; it oxidizes Fe^{2+}, S^0, and RISCs; as well as catalysing FeS_2 dissolution. However, its name refers to the fact that it grows most efficiently on disulfide compounds such as oxidized glutathione (Dufresne *et al.*, 1996). *Alb. ferrooxydans* was isolated from a solfataric soil and like *Alb. disulfidooxidans*, it is able to grow aerobically on a range of organic carbon sources, Fe^{2+}, and RISCs (Jiang *et al.*, 2008). *Alb. tolerans* is named for its ability to tolerate a wide range of hydrogen ion concentrations (pH 1.5 to 5.0) and temperature (< 20°C to 55°C) (Karavaiko *et al.*, 2005). Although it is able to grow at higher temperatures, its optimum of 37 to 42°C places it as a (thermo-tolerant) mesophile. As with the other named *Alicyclobacillus* sp., it is able to grow aerobically on organic carbon sources, Fe^{2+}, and S^0 but is also able to couple substrate oxidation to Fe^{3+} reduction (Karavaiko *et al.*, 2005).

Moderately thermophilic and thermo-tolerant acidophiles

Moderately thermophilic and thermo-tolerant bacteria have an optimum growth temperature from 40°C to 60°C and encompass both Gram-negative and Gram-positive species.

Only two obligately autotrophic, moderate thermophiles have been described, the *Proteobacteria At. caldus* (Hallberg and Lindström, 1994) and the *Actinobacteria 'Acidithiomicrobium'* P2, which has not been formally named (Norris *et al.*, 2011). *At. caldus* is the best characterized moderately thermophilic bacterium and can only grow aerobically on S^0, ISCs, and H_2 at temperature and pH optima of 45°C and pH 2.0 to 2.5, respectively (Hedrich and Johnson, 2013b; Hallberg and Lindström, 1994). The genomes of two strains of *At. caldus* have been sequenced (Acuña *et al.*, 2013; You *et al.*, 2011) and pathways for S^0 and RISC oxidation have been revealed (Chen *et al.*, 2012). Despite not being able to oxidize Fe^{2+}, *At. caldus* has been identified in biomining stirred tanks for treatment of various ores. Its role in biomining has been suggested to be to remove a sulfur/sulfur rich layer from the surface of the mineral and to oxidize soluble reduced sulfur compounds, thereby generating sulfuric acid which assists the process (Dopson and Lindström, 1999). *'Acidithiomicrobium'* P2 was isolated from an enrichment culture previously inoculated with samples from natural geothermal environments, metal sulfide mines, and coal mines (Cleaver *et al.*, 2007) and has a temperature optimum of 55°C. It grows autotrophically using Fe^{2+} and S^0, and has a S^0 oxidizing capacity similar to *At. caldus*, suggesting it has a potential application in biomining (Norris *et al.*, 2011).

Several moderately thermophilic facultative autotrophs are present in the *Sulfobacillus* genus from the *Firmicutes* phylum including *Sb. acidophilus, Sb. sibiricus, Sb. thermosulfidooxidans*, and *Sb. thermotolerans* (Bogdanova *et al.*, 2006; Melamud *et al.*, 2003; Norris *et al.*, 1996). All of these species can grow autotrophically utilizing Fe^{2+}, S^0, and RISCs (though autotrophic growth on S^0 is poor in the case of *Sb. thermosulfidooxidans*), heterotrophically on organic carbon, while *Sb. acidophilus* and *Sb. thermosulfidooxidans* have also been demonstrated to oxidize H_2. *Sb. sibiricus* was isolated from a biomining pilot plant treating a gold/arsenic concentrate and is differentiated from the other *Sulfobacillus* spp. based upon its growth temperature (optimal growth temperature of 55°C) and that it is unable to grow on organic carbon sources such as sucrose without an inorganic energy source (Melamud *et al.*, 2003). *Sb. thermosulfidooxidans* strain Cutipay was isolated from a copper biomining environment; it has an optimum growth temperature of 45°C and its genome has been sequenced (Travisany *et al.*, 2012). Activities of enzymes involved in RISC oxidation (such as thiosulfate-oxidizing enzyme and tetrathionate hydrolase) as well as RubisCO (for CO_2 fixation) were identified in *Sb. thermosulfidooxidans* subsp. *asporogenes* (Egorova *et al.*, 2004) and genes ascribed to these functions have been identified in the *Sb. thermosulfidooxidans* strain Cutipay genome (Travisany *et al.*, 2012). *Sb. thermotolerans* was also isolated from a biomining operation treating a gold containing sulfide mineral and has a pH optimum of 2.0 and a temperature optimum of 40°C (Bogdanova *et al.*, 2006). It primarily grows mixotrophically but is unable to maintain autotrophic or heterotrophic growth after several subcultures, suggesting that it optimally grows as a member of a mixed acidophile community.

Two moderately thermophilic facultative autotrophs have been described, *Acidimicrobium ferrooxidans* and *Aciditerrimonas ferrireducens*, both of which are *Actinobacteria*. *Am. ferrooxidans* grows autotrophically on Fe^{2+} and H_2 as well as heterotrophically on organic carbon such as yeast extract at an optimum pH of 2 and temperature of 45–50°C (Clark and Norris, 1996; Hedrich and Johnson, 2013b). *Am. ferrooxidans* is better able to grow autotrophically than the moderately thermophilic *Sulfobacillus* spp. as it did not require a CO_2 enriched atmosphere (Clark and Norris, 1996). Evidence for a further *Acidimicrobium* species has been presented from a continuous bioreactor treating a nickel containing sulfide mineral at 49°C, but the species was not isolated in pure culture (Cleaver *et al.*, 2007). The *Act. ferrireducens* type species (DSM 45281) was isolated from a solfataric field in Japan and grows heterotrophically via oxidation of organic carbon such as yeast extract coupled to either O_2 or Fe^{3+} at 35 to 58°C (optimal growth at 50°C) and pH 2.0 to 4.5 (optimal growth at pH 3.0) (Itoh *et al.*, 2011). Autotrophic growth only occurred via anaerobic H_2 oxidation coupled to Fe^{3+} reduction.

The moderately thermophilic obligate heterotrophs *Ferrithrix thermotolerans* and *Acidicaldus organivorans* are from the *Actinobacteria* and *Alphaproteobacteria* phyla, respectively. *Fx. thermotolerans* was isolated from the Yellowstone National Park, USA and is named for its ability to grow as filaments up to a maximum of 50°C (optimum growth temperature of 43°C) and it has a pH optimum of 1.8 (Johnson *et al.*, 2009). It is a strict heterotroph capable of oxidizing organic carbon such as glycerol together with yeast extract (suggesting it requires one or more growth factors) coupled to aerobic and anaerobic reduction of O_2 and Fe^{3+} (Johnson *et al.*, 2009). In addition to oxidizing organic carbon, *Fx. thermotolerans* can catalyse the dissimilatory oxidation of Fe^{2+}. *Acd. organivorans* is the most thermophilic acidophilic *Proteobacterium* isolated to date with a temperature optimum of 50–55°C (maximum 65°C) and growing with an optimum pH of 2.5–3.0 (Johnson *et al.*, 2006). *Acd. organivorans* grows heterotrophically on organic carbon coupled to either O_2 or Fe^{3+} but requires growth factor(s) to be able to grow on simple carbon and energy sources. It is also able to oxidize S^0 in the presence of yeast extract, but is not capable of fixing CO_2. Unusually for an acidophile, *Acd. organivorans* oxidizes phenol and higher cell numbers were obtained compared to growth in the presence of yeast extract alone (Johnson *et al.*, 2006).

Thermophilic bacteria

Two thermophilic acidophiles (defined as those having temperature optima at or above 60°C) have been officially named, *H. acidophilum* and *Methylacidiphilum infernorum* from the *Aquificaea* and *Verrucomicrobia* phyla, respectively. *H. acidophilum* (originally described as *Hydrogenobacter acidophilus*) was isolated from a Japanese solfataric area (Stohr *et al.*, 2001; Shima and Suzuki, 1993). It is a Gram-negative, obligate autotroph (via the reductive TCA cycle) that oxidizes H_2, RISCs, and S^0 as electron donors coupled to O_2 reduction at optimum pH and temperature of 3 to 4 and 65°C, respectively. *M. infernorum* was isolated from the 'Hell's Gate' (Tikitere) geothermal area, New Zealand and is the only identified acidophilic (optimum pH 2.0–2.5) methanotroph (Dunfield *et al.*, 2007). The *M. infernorum* genome has been sequenced and contains genes annotated as methane monooxygenases (*pmo*). The ability to oxidize methane suggests that *M. infernorum* inhabits high-temperature niches close to soils and sediments inhabited by methanogenic microorganisms (Dunfield *et al.*, 2007). This metabolic trait may potentially be exploited in biofilters to remove the greenhouse gas methane.

Overview and future prospects

To date, acidophiles have been identified in just 6 out of a total of 29 bacterial phyla containing cultured representatives, making them much less diverse than neutrophilic bacteria. However, they have nearly as diverse a range of electron donors and acceptors, carbon sources, and growth temperature as exhibited by neutrophilic bacteria (Fig. 5.2). Despite this large range of growth strategies, some differences occur such as a lack of acidophilic fermenters. A further difference is a greater propensity to grow utilizing inorganic carbon and energy sources in acidophiles, possibly due to the typically oligotrophic environments in which they live (such as AMD). The exceptions to this rule tend to be heterotrophs that live synergistically oxidizing organic carbon extruded from autotrophs (Nancucheo and Johnson, 2010).

The number of recognized acidophilic bacteria is continuously rising, and the use of deep community DNA sequencing (metagenomics) is revealing the existence of further phylotypes. These novel species may be identified by directed isolation based upon novel traits identified by community 'omics' technologies, as carried out to isolate '*L. ferrodiazotrophum*' (Tyson *et al.*, 2005).

References

Acuña, L.G., Cárdenas, J.P., Covarrubias, P.C., Haristoy, J.J., Flores, R., Nunez, H., Riadi, G., Shmaryahu, A., Valdés, J., Dopson, M., *et al.* (2013). Architecture and gene repertoire of the flexible genome of the extreme acidophile *Acidithiobacillus caldus*. PLoS One *8*, e78237.

Amouric, A., Brochier-Armanet, C., Johnson, D.B., Bonnefoy, V., and Hallberg, K.B. (2011). Phylogenetic and genetic variation among Fe(II)-oxidizing acidithiobacilli supports the view that these comprise multiple species with different ferrous iron oxidation pathways. Microbiology *157*, 111–122.

Appia-Ayme, C., Quatrini, R., Denis, Y., Denizot, F., Silver, S., Roberto, F., Veloso, F., Valdés, J., Cárdenas, J.P., Esparza, M., *et al.* (2006). Microarray and bioinformatic analyses suggest models for carbon metabolism in the autotroph *Acidithiobacillus ferrooxidans*. Hydrometallurgy *83*, 273–280.

Berthelot, D., Leduc, L.G., and Ferroni, G.D. (1993). Temperature studies of iron-oxidizing autotrophs and acidophilic heterotrophs isolated from uranium mines. Can. J. Microbiol. *39*, 384–388.

Bogdanova, T.I., Tsaplina, I.A., Kondrat'eva, T.F., Duda, V.I., Suzina, N.E., Melamud, V.S., Tourova, T.P., and Karavaiko, G.I. (2006). *Sulfobacillus thermotolerans* sp. nov., a thermotolerant, chemolithotrophic bacterium. Int. J. Syst. Evol. Microbiol. *56*, 1039–1042.

Bridge, T.A.M., and Johnson, D.B. (2000). Reductive dissolution of ferric iron minerals by *Acidiphilium* SJH. Geomicrobiol. J. *17*, 193–206.

Bryant, R.D., McGoarty, K.M., Costerton, J.W., and Laishley, E.J. (1983). Isolation and characterisation of a new acidophilic *Thiobacillus* strain (*T. albertis*). Can. J. Microbiol. *29*, 1159–1170.

Chen, L., Ren, Y., Lin, J., Liu, X., and Pang, X. (2012). *Acidithiobacillus caldus* sulfur oxidation model based on transcriptome analysis between the wild type and sulfur oxygenase reductase defective mutant. PLoS One *7*, e39470.

Clark, D.A., and Norris, P.R. (1996). *Acidimicrobium ferrooxidans* gen. nov., sp. nov.: mixed-culture ferrous iron oxidation with *Sulfobacillus* species. Microbiology *142*, 785–790.

Cleaver, A.A., Burton, N.P., and Norris, P.R. (2007). A novel *Acidimicrobium* species in continuous cultures of moderately thermophilic, mineral-sulfide-oxidizing acidophiles. Appl. Environ. Microbiol. *73*, 4294–4299.

Coram, N.J., and Rawlings, D.E. (2002). Molecular relationship between two groups of the genus *Leptospirillum* and the finding that *Leptospirillum ferriphilum* sp. nov. dominates South African commercial biooxidation tanks that operate at 40°C. Appl. Environ. Microbiol. *68*, 838–845.

Dold, B., Gonzalez-Toril, E., Aguilera, A., Lopez-Pamo, E., Cisternas, M.E., Bucchi, F., and Amils, R. (2013). Acid rock drainage and rock weathering in Antarctica: Important sources for iron cycling in the Southern Ocean. Environ. Sci. Technol. *47*, 6129–6136.

Dopson, M., and Lindström, E.B. (1999). Potential role of *Thiobacillus caldus* in arsenopyrite bioleaching. Appl. Environ. Microbiol. *65*, 36–40.

Dopson, M., Halinen, A.K., Rahunen, N., Ozkaya, B., Sahinkaya, E., Kaksonen, A.H., Lindstrom, E.B., and Puhakka, J.A. (2007). Mineral and iron oxidation at low temperatures by pure and mixed cultures of acidophilic microorganisms. Biotechnol. Bioengin. *97*, 1205–1215.

Dufresne, S., Bousquet, J., Boissinot, M., and Guay, R. (1996). *Sulfobacillus disulfidooxidans* sp. nov., a new acidophilic, disulfide-oxidizing, gram-positive, spore-forming bacterium. Int. J. Syst. Bacteriol. *46*, 1056–1064.

Dunfield, P.F., Yuryev, A., Senin, P., Smirnova, A.V., Stott, M.B., Hou, S., Ly, B., Saw, J.H., Zhou, Z., Ren, Y., *et al.* (2007). Methane oxidation by an extremely acidophilic bacterium of the phylum *Verrucomicrobia*. Nature *450*, 879–882.

Egorova, M.A., Tsaplina, I.A., Zakharchuk, L.M., Bogdanova, T.I., and Krasil'nikova, E.N. (2004). Effect of cultivation conditions on the growth and activities of sulfur metabolism enzymes and carboxylases of *Sulfobacillus thermosulfidooxidans* subsp. *asporogenes* strain 41. Appl. Biochem. Microbiol. *40*, 381–387.

Esparza, M., Cardenas, J.P., Bowien, B., Jedlicki, E., and Holmes, D.S. (2010). Genes and pathways for CO_2 fixation in the obligate, chemolithoautotrophic acidophile, *Acidithiobacillus ferrooxidans*. BMC Microbiol. *10*, 229.

Ferguson, S.J., and Ingledew, W.J. (2008). Energetic problems faced by micro-organisms growing or surviving on parsimonious energy sources and at acidic pH: I. *Acidithiobacillus ferrooxidans* as a paradigm. Biochim. Biophys. Acta *1777*, 1471–1479.

Fujimura, R., Sato, Y., Nishizawa, T., Oshima, K., Kim, S.W., Hattori, M., Kamijo, T., and Ohta, H. (2012). Complete genome sequence of *Leptospirillum ferrooxidans* strain C2–3, isolated from a fresh volcanic ash deposit on the island of Miyake, Japan. J. Bacteriol. *194*, 4122–4123.

Goltsman, D.S., Denef, V.J., Singer, S.W., VerBerkmoes, N.C., Lefsrud, M., Mueller, R.S., Dick, G.J., Sun, C.L., Wheeler, K.E., Zemla, A., *et al.* (2009). Community genomic and proteomic analyses of chemoautotrophic iron-oxidizing "*Leptospirillum rubarum*" (Group II) and "*Leptospirillum ferrodiazotrophum*" (Group III) bacteria in acid mine drainage biofilms. Appl. Environ. Microbiol. *75*, 4599–4615.

Goltsman, D.S., Dasari, M., Thomas, B.C., Shah, M.B., VerBerkmoes, N.C., Hettich, R.L., and Banfield, J.F. (2013). New group in the *Leptospirillum* clade: cultivation-independent community genomics, proteomics, and transcriptomics of the new species "*Leptospirillum* group IV UBA BS". Appl. Environ. Microbiol. *79*, 5384–5393.

Guo, X., You, X.Y., Liu, L.J., Zhang, J.Y., Liu, S.J., and Jiang, C.Y. (2009). *Alicyclobacillus aeris* sp. nov., a novel ferrous- and sulfur-oxidizing bacterium isolated from a copper mine. Int. J. Syst. Evol. Microbiol. *59*, 2415–2420.

Hallberg, K.B., and Lindström, E.B. (1994). Characterization of *Thiobacillus caldus*, sp. nov., a moderately thermophilic acidophile. Microbiology *140*, 3451–3456.

Hallberg, K.B., Kolmert, K., Johnson, D.B., and Williams, P.A. (1999). A novel metabolic phenotype among acidophilic bacteria: aromatic degradation and the potential use of these organisms for the treatment of wastewater containing organic and inorganic pollutants. In Process Metallurgy, Amils, R., and Ballester, A., eds (Elsevier, Amsterdam, Netherlands), pp. 719–728.

Hallberg, K.B., Gonzalez-Toril, E., and Johnson, D.B. (2010). *Acidithiobacillus ferrivorans*, sp. nov.; facultatively anaerobic, psychrotolerant iron-, and sulfur-oxidizing acidophiles isolated from metal mine-impacted environments. Extremophiles *14*, 9–19.

Hallberg, K.B., Hedrich, S., and Johnson, D.B. (2011). *Acidiferrobacter thiooxydans*, gen. nov. sp. nov.; an acidophilic, thermo-tolerant, facultatively anaerobic iron- and sulfur-oxidizer of the family Ectothiorhodospiraceae. Extremophiles *15*, 271–279.

Harrison, A.P. (1981). *Acidiphilium cryptum* gen. nov., sp. nov., heterotrophic bacterium from acid mine environments. Int. J. Syst. Bacteriol. *32*, 327–332.

Hedrich, S., and Johnson, D.B. (2013a). *Acidithiobacillus ferridurans* sp. nov., an acidophilic iron-, sulfur- and hydrogen-metabolizing chemolithotrophic gammaproteobacterium. Int. J. Syst. Evol. Microbiol. *63*, 4018–4025.

Hedrich, S., and Johnson, D.B. (2013b). Aerobic and anaerobic oxidation of hydrogen by acidophilic bacteria. FEMS Microbiol. Lett. *349*, 40–45.

Hedrich, S., Schlomann, M., and Johnson, D.B. (2011). The iron-oxidizing proteobacteria. Microbiology *157*, 1551–1564.

Hippe, H. (2000). *Leptospirillium* gen. nov (ex Markoysan 1972), nom. rev., including *Leptospirillium ferrooxidans* sp nov (ex Markoysan 1972), nom. rev. and *Leptospirillium thermoferrooxidans* sp. nov. (Golovacheva *et al.* 1992). Int. J. Syst. Evol. Microbiol. *50*, 501–503.

Hiraishi, A., Nagashima, K.V.P., Matsuura, K., Shimada, K., Takaichi, S., Wakao, N., and Katayama, Y. (1998). Phylogeny and photosynthetic features of *Thiobacillus acidophilus* and related acidophilic bacteria: its transfer to the genus *Acidophilium* as *Acidophilium acidophilum* comb. nov. Int. J. Syst. Bacteriol. *48*, 1389–1398.

Huber, H., and Stetter, K.O. (1989). *Thiobacillus prosperus* sp. nov., represents a new group of halotolerant metal-mobilizing bacteria isolated from a marine geothermal field. Arch. Microbiol. *151*, 479–485.

Itoh, T., Yamanoi, K., Kudo, T., Ohkuma, M., and Takashina, T. (2011). *Aciditerrimonas ferrireducens* gen. nov., sp. nov., an iron-reducing thermoacidophilic actinobacterium isolated from a solfataric field. Int. J. Syst. Evol. Microbiol. *61*, 1281–1285.

Jiang, C.Y., Liu, Y., Liu, Y.Y., You, X.Y., Guo, X., and Liu, S.J. (2008). *Alicyclobacillus ferrooxydans* sp. nov., a ferrous-oxidizing bacterium from solfataric soil. Int. J. Syst. Evol. Microbiol. *58*, 2898–2903.

Johnson, D.B., and McGinness, S. (1991). Ferric iron reduction by acidophilic heterotrophic bacteria. Appl. Environ. Microbiol. *57*, 207–211.

Johnson, D.B., Stallwood, B., Kimura, S., and Hallberg, K.B. (2006). Isolation and characterization of *Acidicaldus organivorus*, gen. nov., sp. nov.: a novel sulfur-oxidizing, ferric iron-reducing thermo-acidophilic heterotrophic Proteobacterium. Arch. Microbiol. *185*, 212–221.

Johnson, D.B., Joulian, C., d'Hugues, P., and Hallberg, K.B. (2008). *Sulfobacillus benefaciens* sp. nov., an acidophilic facultative anaerobic Firmicute isolated from mineral bioleaching operations. Extremophiles *12*, 789–798.

Johnson, D.B., Bacelar-Nicolau, P., Okibe, N., Thomas, A., and Hallberg, K.B. (2009). *Ferrimicrobium acidiphilum* gen. nov., sp. nov. and *Ferrithrix thermotolerans* gen. nov., sp. nov.: heterotrophic, iron-oxidizing, extremely acidophilic actinobacteria. Int. J. Syst. Evol. Microbiol. *59*, 1082–1089.

Johnson, D.B., Hallberg, K.B., and Hedrich, S. (2014). Uncovering a microbial enigma: Isolation and characterization of the streamer-generating, iron-oxidizing, acidophilic bacterium "*Ferrovum myxofaciens*". Appl. Environ. Microbiol. *80*, 672–680.

Jones, R.M., and Johnson, D.B. (2015). *Acidithrix ferrooxidans* gen. nov., sp. nov.; a filamentous and obligately heterotrophic, acidophilic member of the *Actinobacteria* that catalyzes the dissimilatory oxido-reduction of iron. Res. Microbiol. *166*, 111–120.

Jones, R.M., Hedrich, S., and Johnson, D.B. (2013). *Acidocella aromatica* sp. nov.: an acidophilic heterotrophic alphaproteobacterium with unusual phenotypic traits. Extremophiles *17*, 841–850.

Karavaiko, G.I., Bogdanova, T.I., Tourova, T.P., Kondrat'eva, T.F., Tsaplina, I.A., Egorova, M.A., Krasil'nikova, E.N., and Zakharchuk, L.M. (2005). Reclassification of '*Sulfobacillus thermosulfidooxidans* subsp. *thermotolerans*' strain K1 as *Alicyclobacillus tolerans* sp. nov. and *Sulfobacillus disulfidooxidans* Dufresne *et al.* 1996 as *Alicyclobacillus disulfidooxidans* comb. nov., and emended description of the genus *Alicyclobacillus*. Int. J. Syst. Evol. Microbiol. *55*, 941–947.

Kay, C.M., Rowe, O.F., Rocchetti, L., Coupland, K., Hallberg, K.B., and Johnson, D.B. (2013). Evolution of microbial "streamer" growths in an acidic, metal-contaminated stream draining an abandoned underground copper mine. Life 3, 189–210.

Kelly, D.P., and Wood, A.P. (2000). Reclassification of some species of *Thiobacillus* to the newly designated genera *Acidithiobacillus* gen. nov., *Halothiobacillus* gen. nov. and *Thermithiobacillus* gen. nov. Int. J. Syst. Evol. Microbiol. *50 Pt 2*, 511–516.

Kimoto, K., Aizawa, T., Urai, M., Ve, N.B., Suzuki, K., Nakajima, M., and Sunairi, M. (2010). *Acidocella aluminiidurans* sp. nov., an aluminium-tolerant bacterium isolated from *Panicum repens* grown in a highly acidic swamp in actual acid sulfate soil area of Vietnam. Int. J. Syst. Evol. Microbiol. *60*, 764–768.

Kishimoto, N., Kosako, Y., Wakao, N., Tano, T., and Hiraishi, A. (1995). Transfer of *Acidiphilium facilis* and *Acidiphilium aminolytica* to the genus *Acidocella* gen. nov., and emendation of the genus *Acidiphilium*. Syst. Appl. Microbiol. *18*, 85–91.

Kupka, D., Rzhepishevska, O.I., Dopson, M., Lindstrom, E.B., Karnachuk, O.V., and Tuovinen, O.H. (2007). Bacterial oxidation of ferrous iron at low temperatures. Biotechnol. Bioengin. *97*, 1470–1478.

Kupka, D., Liljeqvist, M., Nurmi, P., Puhakka, J.A., Tuovinen, O.H., and Dopson, M. (2009). Oxidation of elemental sulfur, tetrathionate, and ferrous iron by the psychrotolerant *Acidithiobacillus* strain SS3. Res. Microbiol. *160*, 767–774.

Kusel, K., Dorsch, T., Acker, G., and Stackebrandt, E. (1999). Microbial reduction of Fe(III) in acidic sediments: Isolation of *Acidiphilium cryptum* JF-5 capable of coupling the reduction of Fe(III) to the oxidation of glucose. Appl. Environ. Microbiol. *65*, 3633–3640.

Liljeqvist, M., Sundkvist, J.-E., Saleh, A., and Dopson, M. (2011a). Low temperature removal of inorganic sulfur compounds from mining process waters. Biotechnol. Bioengin. *108*, 1251–1259.

Liljeqvist, M., Valdes, J., Holmes, D.S., and Dopson, M. (2011b). Draft genome of the psychrotolerant acidophile *Acidithiobacillus ferrivorans* SS3. J. Bacteriol. *193*, 4304–4305.

Liljeqvist, M., Rzhepishevska, O.I., and Dopson, M. (2013). Gene identification and substrate regulation provides insights into sulfur accumulation during bioleaching with the psychrotolerant acidophile *Acidithiobacillus ferrivorans*. Appl. Environ. Microbiol. *71*, 951–957.

Melamud, V.S., Pivovarova, T.A., Turova, T.P., Kolganova, T.V., Osipov, G.A., Lysenko, A.M., Kondrat'eva, T.F., and Karavaiko, G.I. (2003). *Sulfobacillus sibiricus* sp. nov., a new moderately thermophilic bacterium. Mikrobiologiia *72*, 681–688.

Mi, S., Song, J., Lin, J., Che, Y., Zheng, H., and Lin, J. (2011). Complete genome of *Leptospirillum ferriphilum* ML-04 provides insight into its physiology and environmental adaptation. J. Microbiol. *49*, 890–901.

Nancucheo, I., and Johnson, D.B. (2010). Production of glycolic acid by chemolithotrophic iron- and sulfur-oxidizing bacteria and its role in delineating and sustaining acidophilic sulfide mineral-oxidizing consortia. Appl. Environ. Microbiol. *76*, 461–467.

Norris, P.R., Clark, D.A., Owen, J.P., and Waterhouse, S. (1996). Characteristics of *Sulfobacillus acidophilus* sp. nov. and other moderately thermophilic mineral-sulphide-oxidizing bacteria. Microbiology *142*, 775–783.

Norris, P.R., Davis-Belmar, C.S., Brown, C.F., and Calvo-Bado, L.A. (2011). Autotrophic, sulfur-oxidizing actinobacteria in acidic environments. Extremophiles *15*, 155–163.

Okibe, N., Gericke, M., Hallberg, K.B., and Johnson, D.B. (2003). Enumeration and characterization of acidophilic microorganisms isolated from a pilot plant stirred-tank bioleaching operation. Appl. Environ. Microbiol. *69*, 1936–1943.

Osorio, H., Martinez, V., Veloso, F.A., Pedroso, I., Valdes, J., Jedlicki, E., Holmes, D.S., and Quatrini, R. (2008). Iron homeostasis strategies in acidophilic iron oxidizers: Studies in *Acidithiobacillus* and *Leptospirillum*. Hydrometallurgy *94*, 175–179.

Osorio, H., Mangold, S., Denis, Y., Nancucheo, I., Johnson, D.B., Bonnefoy, V., Dopson, M., and Holmes, D.S. (2013). Anaerobic sulfur metabolism coupled to dissimilatory iron reduction in the extremophile *Acidithiobacillus ferrooxidans*. Appl. Environ. Microbiol. *79*, 2172–2181.

Quatrini, R., Appia-Ayme, C., Denis, Y., Jedlicki, E., Holmes, D., and Bonnefoy, V. (2009). Extending the models for iron and sulfur oxidation in the extreme acidophile *Acidithiobacillus ferrooxidans*. BMC Genomics *10*, 394.

Rawlings, D.E., Tributsch, H., and Hansford, G.S. (1999). Reasons why "*Leptospirillum*"-like species rather than *Thiobacillus ferrooxidans* are the dominant iron-oxidizing bacteria in many commercial processes for the biooxidation of pyrite and related ores. Microbiology *145*, 5–13.

Ruiz, L.M., Gonzalez, A., Frezza, M., Soulere, L., Queneau, Y., Doutheau, A., Rohwerder, T., Sand, W., Jerez, C.A., and Guiliani, N. (2007). Is the quorum sensing type AI-1 system of *Acidithiobacillus ferrooxidans* involved in its attachment to mineral surfaces? In Biohydrometallury: From the Single Cell to the Environment, Schippers, A., Sand, W., Glombitza, F., and Willsher, S., eds (Trans Tech Publications, Pfaffikon, Switzerland), pp. 345–349.

Shima, S., and Suzuki, K.-I. (1993). *Hydrogenobacter acidophilus* sp. nov., a thermoacidophilic, aerobic, hydrogen-oxidizing bacterium requiring elemental sulfur for growth. Int. J. Syst. Evol. Bacteriol. *43*, 703–708.

Stohr, R., Waberski, A., Volker, H., Tindall, B.J., and Thomm, M. (2001). *Hydrogenothermus marinus* gen. nov., sp. nov., a novel thermophilic hydrogen-oxidizing bacterium, recognition of *Calderobacterium hydrogenophilum* as a member of the genus *Hydrogenobacter* and proposal of the reclassification of *Hydrogenobacter acidophilus* as *Hydrogenobaculum acidophilum* gen. nov., comb. nov., in the phylum '*Hydrogenobacter*/*Aquifex*'. Int. J. Syst. Evol. Microbiol. *51*, 1853–1862.

Talla, E., Hedrich, S., Ji, B.Y., Johnson, D.B., and Bonnefoy, V. (2013). Genome analysis of the psychrotolerant acidophile *Acidithiobacillus ferrivorans* CF27. Adv. Mat. Res. *825*, 145–148.

Travisany, D., Di Genova, A., Sepulveda, A., Bobadilla-Fazzini, R.A., Parada, P., and Maass, A. (2012). Draft genome sequence of the *Sulfobacillus thermosulfidooxidans* Cutipay strain, an indigenous bacterium isolated from a naturally extreme mining environment in northern Chile. J. Bacteriol. *194*, 6327–6328.

Tyson, G.W., Lo, I., Baker, B.J., Allen, E.E., Hugenholtz, P., and Banfield, J.F. (2005). Genome-directed isolation of the key nitrogen fixer *Leptospirillum ferrodiazotrophum* sp. nov. from an acidophilic microbial community. Appl. Environ. Microbiol. *71*, 6319–6324.

Urakami, T., Tamaoka, J., Suzuki, K.-I., and Komagata, K. (1989). *Acidomonas* gen. nov., incorporating *Acetobacter methanolicus* as *Acidomonas methanolica* comb. nov. Int. J. Syst. Evol. Bacteriol. 39, 50–55.

Wichlacz, P.L., Unz, R.F., and Langworthy, T.A. (1986). *Acidiphilium angustum* sp. nov., *Acidiphilium facilis* sp. nov., and *Acidiphilium rubrum* sp. nov.: Acidophilic heterotrophic bacteria isolated from acidic coal mine drainage Int. J. Syst. Evol. Bacteriol. 36, 197–201.

Williams, K.P., and Kelly, D.P. (2013). Proposal for a new class within the phylum *Proteobacteria*, *Acidithiobacillia* classis nov., with the type order *Acidithiobacillales*, and emended description of the class *Gammaproteobacteria*. Int. J. Syst. Evol. Microbiol. *63*, 2901–2906.

Wu, X., Wong, Z.L., Sten, P., Engblom, S., Österholm, P., and Dopson, M. (2013). Microbial community potentially responsible for acid and metal release from an Ostrobothnian acid sulfate soil. FEMS Microbiol. Ecol. *84*, 555–563.

Xia, J.-l., Peng, A.-a., He, H., Yang, Y., Liu, X.-d., and Qiu, G.-z. (2007). A new strain *Acidithiobacillus albertensis* BY-05 for bioleaching of metal sulfides ores. Trans. Nonferr. Metals Soc. China *17*, 168–175.

You, X.-Y., Guo, X., Zheng, H.-J., Zhang, M.-J., Liu, L.-J., Zhu, Y.-Q., Zhu, B., Wang, S.-Y., Zhao, G.-P., Poetsch, A., *et al.* (2011). Unraveling the *Acidithiobacillus caldus* complete genome and its central metabolisms for carbon assimilation. J. Genet. Genomics 38, 243–252.

Diversity and Physiologies of Acidophilic *Archaea*

6

Olga V. Golyshina, Manuel Ferrer and Peter N. Golyshin

Abstract

The majority of cultured acidophilic archaea are represented by two phyla, the *Crenarchaeota* and the *Euryarchaeota*, that inhabit sulfur-rich areas in volcanically active regions worldwide and sulfidic deposits rich in metals, acid mine drainage systems, macroscopic growths in forms of streamers, slimes, mats and microbial stalactites. A metabolic network formed by these archaea is determined by niche adaptation and is characterized by sulfur-, hydrogen- and iron-dependency, and by the organotrophy of scavenger species and strains. Acidophily is a major trait in the organisms from the above environments and most characterized species are moderate or extreme thermophiles, though there are some exceptions. Isolation sites of acidophilic archaea are almost exclusively limited to terrestrial habitats, with marine environments being poorly explored as a reservoir for these organisms, with only one such acidophilic euryarchaeote isolated to date. The nature and mechanisms of acidophily are believed to be determined by a number of factors with special archaeal membranes and lipids being of the greatest importance. Genome analysis of a number of species of acidophilic archaea has identified certain metabolic traits that make them distinct from bacterial counterparts and suggests some unusual variants of classical pathways.

Introduction: history, environments, and major physiological characteristics

As described in Chapter 1, two species of archaea (*Sulfolobus* and *Thermoplasma* spp.) were among the first acidophilic prokaryotes to be described, though they were considered, as was the norm at that time, to be bacteria. Further studies of DNA-dependent RNA polymerases in *Sulfolobus* spp. (including '*Caldariella acidophila*', later re-described as *Sulfolobus solfataricus* and the organism isolated by James Brierley and later named as *S. brierleyi*) and in *Tp. acidophilum*, revealed differences with bacterial RNA polymerases and some similarity with eukaryotic RNA polymerases (Zillig *et al.*, 1980). *Sulfolobus* and *Thermoplasma* subsequently became incorporated into the *Archaea*, joining this new domain of life proposed by the landmark publication of Woese and Fox (1977). That coincided with the very beginning of the recognition of *Archaea* as a separate domain of the prokaryotic kingdom, which nowadays embraces several phyla, dozens of classes, orders and families, almost 100 genera and many more species. The best-studied acidophilic archaea are affiliated with the two (firstly proposed and largest) of six currently accepted phyla, namely the *Crenarchaeota* and *Euryarchaeota*, and their physiological and genomic features largely reflect the rather steady physico-chemical conditions in their habitats.

Two principal physical–chemical parameters (low pH and high temperatures) define very often the environments inhabited by acidophilic archaea. These included terrestrial sulfur-containing solfataric fields (Chapter 1) which can vary in temperature from ambient to ~100°C. The main electron donors/acceptors available in these types of environments are sulfur compounds in a different oxidation state S^0, SO_4^{2-}, and SO_3^{2-}; Fe^{2+} and Fe^{3+} are present mostly as ferrous sulfides or ferric iron hydroxides; gases presented can include H_2O (as steam), H_2, CO_2, CO, O_2, CH_4, H_2S and NH_3 (Stetter, 2002). A wide range of temperatures, the large diversity of substrates and availability or lack of oxygen explain the variety of isolated acidophiles from solfatara and suggest some potential for isolating novel prokaryotes. Upper oxygenated layers of solfatara fields with extremely acidic pH and high temperatures are well known habitats of aerobic/facultatively anaerobic thermophilic organisms, whereas deeper zones depleted in oxygen are preferred by anaerobic members of the family *Sulfolobaceae*. Acidophilic archaea have been isolated from solfataric areas in Yellowstone National Park and Lassen National Park (USA), Pozzuoli, (near Naples) and Vulcano Island (Italy), Hokkaido (Japan), Reykjnes, Krisuvik (Iceland), Mutnovskyi, Uzon/Geyzer Valley on Kamchatka (Russia), Sao Miguel, Furnas, Ribeira Quente, Azores (Portugal), Java (Indonesia), Montserrat (Caribbean), Kunashir/Kuril Islands (Russia) and others.

Other geothermal zones with low pH populated by acidophilic archaea include acidic thermal springs and shallow submarine hydrothermal systems, which are typical features of volcanic regions (Fig. 6.1). One example is site Porto di Levante on Vulcano Island (Italy), comprising a zone around the beach and neighbouring areas with a number of shallow submarine and terrestrial fumaroles, gradients of temperatures, pH and electron donors/acceptors pairs. Due to the high buffering capacity of seawater, submarine hydrothermal systems are characterized by steep chemo- and thermoclines, with a range of pH and temperatures within a limited space, which increases the potential diversity of indigenous microorganisms.

Extremely acidic water bodies, such as acid mine drainage, associated with the oxidative dissolution of sulfide mineral-rich deposits, as well as macroscopic growths of streamers, slimes, mats and microbial stalactites, are usually inhabited by acidophilic archaea as well as bacteria (Chapters 8 and 13). Mostly mesophilic and moderately thermophilic, euryarchaeotes of the family *Ferroplasmaceae* (order *Thermoplasmatales*) have been isolated from these biotopes, and less frequently some acidophilic crenarchaeotes.

Acidophilic archaea include aerobic and anaerobic chemolitho-autotrophs and chemo-organotrophs. Electron donors known to fuel energetic metabolism by chemolithotrophic species include sulfur, sulfide, ferrous iron and hydrogen, while electron acceptors include oxygen, nitrate, sulfur, sulfate, thiosulfate and ferric iron. Many chemolithotrophic archaea can switch from autotrophic mode to organic carbon consumption. Heterotrophic lifestyle is more common in acidophilic archaea and is sometimes accompanied with their ability to use inorganic energy sources e.g. reduced sulfur or ferrous iron. Acidophilic archaea are often members of microbial consortia in both natural and man-made acidic environments, and interact with acidophilic bacteria and eukaryotes (Chapter 9).

Figure 6.1 Site exemplifying typical environments for isolation of acidophilic archaea (Sao Miguel Island, Azores, Portugal).

Two major groups of acidophilic archaea can be defined with respect to their pH growth optima and ranges: extreme and moderate acidophiles. Extreme acidophiles are common among the members of the orders *Sulfolobales* (*Crenarchaeota*) and *Thermoplasmatales* (*Euryarchaeota*), whereas moderate acidophiles comprise members of the orders *Acidilobales* and *Thermoproteales* (*Crenarchaeota*).

A characteristic trait for all extremely acidophilic archaea, with a few exceptions for some members of the *Thermoplasmatales*, is their thermophilic or extremely thermophilic nature, which is the reason for the widespread use of the combined term, 'thermoacidophilic archaea'. It is noteworthy that among all acidophilic organisms known so far, acidophilic archaea thrive at the lowest limits of pH and at highest temperature, being the most acidophilic (minimal pH for growth for *Picrophilus* spp. is about 0) and the most thermophilic organisms (the documented maximal temperature for growth revealed as high as 95°C for some moderately and extremely acidophilic *Crenarchaeota*).

The Sulfolobaceae

The most numerous (in terms of known species) acidophilic archaea are crenarchaeotes of the family *Sulfolobaceae* (order *Sulfolobales*, class *Thermoprotei*), and these are exclusively thermoacidophiles. Most of these have been isolated from solfatara fields, and consequently the majority have sulfur-dependent metabolisms. Nevertheless, several distinct physiological groups can be identified within the family *Sulfolobaceae*. Representatives of genera *Sulfolobus*, *Metallosphaera* and *Sulfurococcus* are aerobic sulfur-oxidizers; those of the genus *Stygiolobus* (only current representative *Sg. azoricus*) are anaerobic sulfur reducers; members of genera *Acidianus* and *Sulfurisphaera* are facultatively anaerobic sulfur-respiring prokaryotes. The main characteristics of the type strains (or selected strains, in cases of species with invalidly published names) of the family *Sulfolobaceae* are listed in Table 6.1.

Many members of the genus *Sulfolobus*, e.g. *S. acidocaldarius*, *S. shibatae* and *S. solfataricus*, are characterized by the ability to oxidize molecular hydrogen. *S. metallicus* is able to oxidize ferrous iron and is an obligate chemolithoauthotroph, while all other *Sulfolobus* species are obligate or facultative heterotrophs. The type strains of *S. solfataricus* (DSM 1616) and *S. acidocaldarius* (DSM 639) were initially described as sulfur-oxidizers, but were reported to lose this ability in the laboratory and to become obligate heterotrophs (Huber and Prangishvili, 2006), probably because the originally cultures were mixed. *Sulfolobus* spp. are thermophiles or hyperthermophiles, with temperature optima for growth varying from 65°C for *S. metallicus*, 70–75°C for *S. acidocaldarius* and up to 80–85°C for all other *Sulfolobus* species (Table 6.1). *S. yangmingensis* has been reported to grow at 95°C, though this is a moderate acidophile (pH optimum 4.0, though also reported also to grow at pH 2.0). The pH growth optimum of *Sulfolobus* spp. is generally ~ 2 to 3 (Table 6.1).

Metallosphaera species are sulfur-oxidizing thermoacidophilic archaea, closely related to *Sulfolobus*. A hallmark of this genus is the ability of its members to oxidize metal sulfidic ores, such as pyrite, chalcopyrite and sphalerite; the archaeon *M. prunae*, in contrast to other *Sulfolobaceae*, was isolated from the uranium mine in Thuringia, Germany. *Metallosphaera* spp. share the same general features as other *Sulfolobaceae* – the ability for the oxidation of sulfur and oxidation of molecular hydrogen and organic compounds, and being aerobic facultative chemolithotrophs. *Sulfurococcus* spp. have similar G+C contents and physiologies to *Metallosphaera* spp., though their 16S rRNA gene sequences phylogeny analysis placed strains within the genus *Acidianus*. No strains of *Sulfurococcus* are currently available (T.A. Pivovarova, personal communication).

Two genera of facultatively anaerobic thermoacidophilic archaea are currently recognized: *Acidianus* and *Sulfurisphaera*. *Acidianus* spp. can both oxidize and reduce elemental sulfur ores and some are able to oxidize ferrous or reduce ferric iron. Many strains are obligate chemolithotrophs, and some can oxidize sulfides, respectively. *Acidianus* spp. are the most acidophilic members of the family *Sulfolobaceae*, with optimal pH for *Ad. sulfidivorans* being pH 0.8–1.4 (Table 6.1). *Sulfurisphaera ohwakuensis* share many characteristics with *Acidianus* spp., though its phylogenetic relatedness is quite low (about 90% 16S rRNA sequence identity).

Sg. azoricus is the only strictly anaerobic archaeon of the family *Sulfolobaceae* described to date, and displays sulfur- and hydrogen-dependent lithotrophic metabolism.

Table 6.1 Taxonomy, sources of isolation and main metabolic lifestyles of the isolates of the family *Sulfolobaceae*

Species	Place of isolation	Temperature and pH growth ranges (optima)	G+C (mol%)	Metabolism	Reference
S. acidocaldarius	Terrestrial acidic hot spring (Locomotive), Yellowstone NP (USA)	55–85 (75–80)°C, 1.5–6 (3)	37	Aerobic mixo/heterotroph, oxidation of H_2 and organic C	Brock *et al.* (1972)
S. solfataricus	Terrestrial solfataric spring, Pisciarelli (Italy)	50–87 (85)°C, 2–5.5 (3–4.5)	35	Aerobic mixo/heterotroph, oxidation of H_2 and organic C	Zillig *et al.* (1980)
S. shibatae	Acidic geothermal spring, Kiushu Island (Japan)	n.d.–86 (81)°C, n.d. (3)	35	Aerobic mixo/heterotroph, oxidation of S^0, H_2 and organic C	Grogan *et al.* (1990)
S. metallicus	Solfataric field, (Iceland)	50–75 (65)°C, 1–4.5 (2–3)	38	Aerobic lithoautotroph, oxidation of Fe^{2+}, sulfidic ores and S^0	Huber and Stetter (2006)
S. tokodaii	Acidic geothermal spring, Kiushu Island (Japan)	70–85 (80)°C, 2–5 (2.5–3)	33	Aerobic chemolithotroph/heterotroph, oxidation of S^0 and organic C	Suzuki *et al.* (2002)
'*S. islandicus*'	Lassen volcanic NP (USA)	65–88 (78)°C, 2–5 (2.7)	35	Aerobic chemolithotroph/heterotroph	Whitaker *et al.* (2003)
'*S. technogensis*'	Acidic hot spring, Tengchong, China	65–95 (85)°C, 1.7–6.5 (3.5)	34	Aerobic chemolithotroph/heterotroph	Xiang *et al.* (2003)
M. sedula	Solfataric field, Pisciarelli, Italy	50–80 (75)°C, 1–4.5 (2–3)	45	Aerobic facultative chemolithoautotroph, oxidation of sulfidic ores, S^0, H_2 and organic C	Huber *et al.* (1989)
M. prunae	Uranium mine, Thüringen, Germany	55–80 (75)°C, 1–4.5 (2–3)	46	Aerobic facultative chemolithoautotroph, oxidation of sulfidic ores, H_2, S^0 and organic C	Fuchs *et al.* (1995)
'*M. yellowstonensis*'	Yellowstone NP, USA	n.d. (65°C), n.d. (3)	48	Aerobic facultative chemolithoautotroph, oxidation of sulfidic ores, S^0 and Fe^{2+}	Kozubal *et al.* (2011)
M. hakonensis	Hot spring in Ohwaku-dani, Japan	50–80 (70)°C, 1–4 (3)	46	Aerobic facultative chemolithoautotroph, oxidation of S^0, polythionates, metal sulfides and organic C	Kurosawa *et al.* (2003)
M. cuprina	Sulfuric hot spring in Tengchong, Yunnan, China	55–75 (65)°C, 2.5–5.5 (3.5)	40	Aerobic facultative chemolithoautotroph; oxidation of polythionates, sulfide ores and organic C	Liu, *et al.* (2011)
Sc. mirabilis	Crater of the Uzon, volcanic area, Kamchatka, Russia	50–86 (70–75)°C, 1–5.8 (2–2.6)	44	Aerobic facultative chemolithoautotroph, oxidation of S^0, Fe^{2+}, sulfide minerals and organic C	Golovacheva *et al.* (1987)
Sc. yellowstonii	Hydrotherm of the Yellowstone NP, USA	40–80 (60)°C, 1–5.5 (2–2.6)	44.6	Aerobic facultative chemolithoautotroph, oxidation of S^0, Fe^{2+}, sulfide minerals and organic C	Karavaiko *et al.* (1994)
Ad. brierleyi	Acidic thermal spring, Yellowstone NP, USA	45–75 (70)°C, 1–6 (1.5–2)	31	Facultative anaerobe/chemolithoautotroph; oxidation of H_2, sulfidic ores and organic C; oxidation/reduction of S^0	Segerer *et al.* (1986)
Ad. infernus	Solfatara, Italy	65–95 (85–90)°C, 1–5.5 (2)	31	Facultative anaerobe/chemolithoautotroph; oxidation of H_2, sulfidic ores and organic C; oxidation/reduction of S^0	Segerer *et al.* (1986)

Table 6.1 Continued

Species	Place of isolation	Temperature and pH growth ranges (optima)	G+C (mol%)	Metabolism	Reference
Ad. ambivalens	Solfataric spring, Iceland	n.d.–87 (80)°C, 1–3.5 (2.5)	33	Facultative anaerobic, obligate lithoautotroph, oxidation/reduction of S^0	Fuchs *et al.* (1996)
Ad. sulfidivorans	Solfatara on Lihir Island, Papua New Guinea	45–83 (74)°C, 0.35–3.0 (0.8–1.4)	31	Facultative anaerobe; obligate chemolithotroph, oxidation of sulfidic ores, S^0 and Fe^{2+}	Plumb *et al.* (2007)
'*Ad. manzaensis*'	Hot fumarole in Manza, Japan	60–90 (80)°C, 1.0–5.0 (1.2–1.5)	29.9	Facultative anaerobic/ chemolithoautotroph; oxidation of H_2, S^0, and organic C; reduction of Fe^{3+}	Yoshida *et al.* (2006)
Ss. ohwakuensis	Acidic hot spring, Ohwaku Valley, Hakone, Japan	–92 (84)°C, 1–5 (2.0)	33	Facultative anaerobic, mixo/ heterotroph, reduction of S^0, and oxidation of organic C	Kurosawa *et al.* (1998)
Sg. azoricus	Solfataric field on Sao, Miguel Island, Azores	57–89 (80)°C, 1.5–5 (2–2.6)	38	Obligate anaerobic/chemo-lithoautotroph (S^0/H_2)	Segerer *et al.* (1991)

The *Thermoplasmatales*

Acidophilic archaea of the taxonomic phylum *Euryarchaeota* have been isolated mostly from terrestrial solfataric fields, sulfidic ores and acidic mine waters, and one representative (the moderately acidophile '*Aciduliprofundum boonei*') from the marine environment (a deep-sea vent). Currently, all cultured acidophilic members of the phylum *Euryarchaeota* belong to the single order, the *Thermoplasmatales*, comprising three families (*Thermoplasmataceae*, *Picrophilaceae*, *Ferroplasmaceae*) and five genera (*Thermoplasma*, *Picrophilus*, *Ferroplasma*, *Thermogymnomonas* and *Acidiplasma*), and include the most acidophilic of all known life forms. Characterized species of the *Thermoplasmatales* are mesophilic, thermo-tolerant or moderate thermophiles, in contrast to the predominantly extremely thermophilic *Sufolobales*. They are aerobic or facultatively anaerobic organisms with absolute requirement for complex organic carbon (e.g. yeast, bacterial or beef extracts), and their nutrition in natural biotopes is supposed to be of a scavenger type, i.e. is based on the products of decomposing cells. The archaea of the family *Ferroplasmaceae* are able to metabolize iron, and *Thermoplasma* spp. are characterized by sulfur respiration. Noteworthy, *Thermoplasma* spp., *Ferroplasma* spp., *Acidiplasma* spp. and *Thermogymnomonas acidicola* are cell wall-less archaea, possessing only a monolayer cytoplasmic membrane (Fig. 6.2). *Picrophilus* spp. are the only exception from that general pattern in *Thermoplasmatales*, as they do possess S-layers. Also noteworthy is that the genomes of *Thermoplasma* spp. and *P. torridus* were found to be the smallest for free-living organisms, with sizes of about 1.54–1.58 Mbp.

Tp. acidophilum and *Tp. volcanium* are distributed widely in aerobic and anaerobic solfataric volcanic

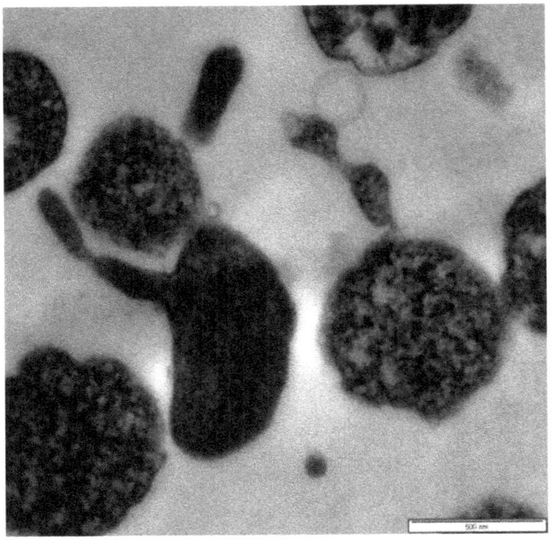

Figure 6.2 Electron micrograph of a thin section of *Acidiplasma aeolicum*. Bar 500 nm. The image was kindly provided by Dr Heinrich Lünsdorf (HZI, Braunschweig, Germany).

zones across the globe and were also found in self-heated coal refuse piles and warm acidic swamps in Indonesia (Huber and Stetter, 2006). *Picrophilus* spp. were isolated from solfataric fields in Japan, and at the moment no other isolates or clones with similar sequences have been documented, suggesting that these organisms could be endemic species. In contrast, archaea of the family *Ferroplasmaceae* are ubiquitous in sulfide ore-based biotopes, solfataric fields and volcanic environments, and 16S rRNA gene sequence-identical cultured and uncultured organisms have been found on five continents (Golyshina, 2011). The main features of acidophilic *Euryarchaeota* are listed in Table 6.2.

Adaptation strategies in extremely acidophilic Archaea

The S-layer, the common component of the cell envelope in acidophilic and many other archaea, is known to possess a paracrystalline structure. The S-layer is also present in some acidophilic Gram-positive bacteria, such as *Sulfobacillus* spp., and could supposedly be a factor in their adaptation to extreme environmental conditions.

The cell wall in the *Sulfolobales* represents a regular quasi-crystalline S-layer with a unique three-fold symmetry consisting of glycoprotein subunits located at a distance of about 20 nm one from another and at approximately 18 nm between an S-layer and the cytoplasmic membrane (Baumeister and Lembcke, 1992). Veith *et al.* (2009) reported that the S-layers of *Acidianus*, *Sulfolobus* and *Metallosphaera* spp. were composed by polypeptides SlaA and SlaB. SlaA homologues are distributed only among other members of the order *Sulfolobales* and SlaB showed distant relation to S-layer proteins from other *Crenarchaeota*. The S-layer consists of three SlaB situated inside and linking the complex with the membrane and six outside-located SlaA, comprising a fairly robust, detergent-resistant sacculus (Veith *et al.*, 2009).

Cytoplasmic membranes and lipids of acidophilic archaea are of a great interest and are

Table 6.2 Taxonomy, sources of isolation and main metabolic lifestyles of the isolates of the order *Thermoplasmatales*

Species	Site of isolation	Temperature and pH growth ranges (optima)	G+C (mol%)	Metabolism	Reference
Tp. acidophilum	Coal refuse pile, USA	45–63 (59)°C, 0.5–4 (1–2)	46	Facultative anaerobic, chemoheterotroph; reduction of S^0	Darland *et al.* (1970)
Tp. volcanium	Submarine solfatara field, Vulcano Island, Italy	33–67 (60)°C, 1–4 (2.0)	38 – 40	Facultative anaerobic chemo-heterotroph, reduction of S^0	Segerer *et al.* (1988)
P. torridus	Volcanic solfataric field, Hokkaido, Japan	47–65 (60)°C, 0–3.5 (0.7)	36	Obligate aerobe; chemoheterotroph	Schleper *et al.* (1995)
P. oshimae	Volcanic solfataric field, Hokkaido, Japan	47–65 (60)°C, 0–3.5 (0.7)	36	Obligate aerobe; chemoheterotroph	Schleper *et al.* (1995)
Fp. acidiphilum	Gold-containing arsenopyritic ore concentrate, Kazakhstan	15–45 (35)°C, 1.3–2.2 (1.7)	36.5	Facultative anaerobe; chemo-lithotroph; oxidation of sulfidic ores and Fe^{2+}	Golyshina *et al.* (2000)
'Fp. acidarmanus'	AMD, Iron Mountain, USA	23–46 (42)°C, 0–1.5 (1.2)	37	Facultative anaerobe; chemo-mixo/heterotroph; oxidation of sulfidic ores and Fe^{2+}	Edwards *et al.* (2000)
Ap. cupricumulans	Chalcocite-containing sulfide ore, Myanmar	22–63 (54)°C, 0.4–1.8 (1.1)	34	Facultative anaerobe; chemo-mixotroph; oxidation of sulfidic ores and Fe^{2+}	Hawkes *et al.* (2006)
Ap. aeolicum	Hydrothermal pool, Vulcano Island, Italy	15–65 (~43)°C, 0–4 (1.4–1.6)	36	Facultative anaerobe; chemomixo/heterotroph; oxidation of Fe^{2+} and organic C	Golyshina *et al.* (2009)
Tg. acidicola	Solfataric soil, Ohwaku-dani, Hakone, Japan	38–68 (60)°C, 1.8–4.0 (3.0)	56	Aerobic; chemoheterotroph	Itoh *et al.* (2007)

considered to play a pivotal role in survival at low pH. One important barrier to withstand the transmembrane proton gradient is a unique cell membrane with a very low proton permeability.

The low proton permeability of membranes to maintain the pH gradient and inverted electrical potential were proposed as main mechanisms to facilitate the survival of acidophiles at low pH (van de Vossenberg *et al.*, 1998). Other important factors that have been identified in acidophilic archaea are the prevalence of secondary over primary transporters, a high ratio of chaperones and other mechanisms of adaptation on all levels from structural to functional alterations in biomolecules, metabolic traits and regulation of gene expression (Fütterer *et al.*, 2004). The membrane is organized as a monolayer consisting of ether lipids; hydrophilic units (glycerols) and hydrophobic parts represented by isoprenoids connected in such a manner that provides a very stable and less fluid compartment. Tetraether core lipids together with small fractions of diethers are components characteristic for acidophilic archaea of both the *Crenarchaeota* and *Euryarchaeota*. Another characteristic allowing membrane lipids of acidophilic archaea to enforce protection from proton ingress into the cell was reported to feature a high number of sugar moieties in glycolipids on the cell surface, as it was shown in experiments with *Tp. acidophilum*. The data of Wang *et al.* (2012) demonstrated that OH-rich sugar coating layer was acting as a proton shelter in the above microorganisms.

The cell membrane in the *Sulfolobales* is composed of isoprenyl ether lipids, including a diglycerol tetraether with C_{40}/C_{20} biphytanyl chains and a glycerol-alkylnonitoltetraether. The main constituents of membrane lipids are tetraethers with up to four pentacyclic rings. The proportion of tetra- to diethers varies in different species and depends also on growth conditions (Huber and Prangishvili, 2006). Other lipid elements unique for the *Sulfolobales* were found to be quinones: caldariellaquinone (benzo-[*b*]-thiophen-4,7-quonone), sulfolobus-quinone [6-(3,7,11,15,19,23-hexamethyltetracosyl)-5-methyl-benzo[*b*]thiophen-4,7-quinone] and tricyclic quinone [1,2-b; 4,5-b′]dithiophen-4,8-quinone) (Huber and Prangishvili, 2006).

Representatives of the families *Thermoplasmataceae*, *Ferroplasmaceae*, and *Tg. acidicola*, lack cell walls, and their cells are surrounded by a rigid cytoplasmic membrane containing unique chemical constituents. The cell membrane of *Tp. acidophilum* is a monolayer and has in its content a mannose-rich glycoprotein forming a hydroskeleton facilitating significant membrane rigidity. Lipids are represented mostly by tetraethers (i.e. acyclic, mono- and bi- pentacyclic C_{40} biphytanyl diglycerol), with minor amounts of diethers (C_{20} phytanyl glycerol), structurally a polar lipid was shown to be a lipoglycan. Lipids of *Tp. acidophilum* were determined as novel glycolipids, represented by dibiphytanyl-diglycerol tetraether (caldarcheol) and sugars gulose and glucose (Huber and Stetter, 2006). *P. torridus* polar lipids were identified as a β-glucosyl derivative of caldarchaeol (Schleper *et al.*, 1995) and membrane lipids of the *Ferroplasmaceae* have been found to be mostly dibiphytanyl tetraether lipids and sugar residues reported as β-D-glucose and β-galactopyranose (Golyshina, 2011). Principal lipids of *Tg. acidicola* were defined as tetraethers with at least three acyclic and cyclic moieties (Itoh *et al.*, 2007). The S-layer of *P. torridus* (the only archaeon from the order *Thermoplasmatales* shown to date to possess an S-layer) was found to have a tetragonal symmetry and a centre-to-centre distance of about 20 nm (Schleper *et al.*, 1995).

The thermosome (chaperonin archaeal complex, type II chaperonin, Hsp60) of *Tp. acidophilum* has been studied in considerable detail and shown to comprise two alternating subunits with rare allosteric features. *Tp. acidophilum* also harbours various further chaperones of different families: three components of small heat shock protein (sHsp) family chaperones, two of GimC, several members of AAA+ ATPase family (Lon and Lon2; VAT and VAT2), PPIase (peptidyl prolyl cis–trans isomerase) and two representatives of CsaA family (Ruepp *et al.*, 2001). Analysis of the genome of *P. torridus* revealed a whole array of genes coding for chaperones Hsp60, Hsp70, Hsp20, VAT and Lon2 (Fütterer *et al.*, 2004).

Trehalose, an osmotically active compound, commonly accumulates in archaea (e.g. in the *Sulfolobales*, *S. solfataricus*, *Ad. ambivalens* and in *M. sedula*) and is considered to be involved into the stress response due to its high hydrophilicity, pH- and thermo-stability, though the physiological role of this disaccharide in *Sulfolobales* remains still unknown. Martins *et al.* (1997) suggested that, since the *Sulfolobales* normally grow in media with very low salt concentrations, trehalose is unlikely to serve as an osmolyte. The production of trehalose is supposed to occur from malto-oligosaccharides via TreY–TreZ pathway, which also exists in *S. islandicus*, *S. acidocaldarius* and *S. tokodaii*. In addition to that, trehalose synthase (TreT) orthologues have been found in the *Sulfolobales* genomes. A glycosyl-transferring TreT is characterized as a unidirectional enzyme, catalysing the formation of trehalose only (Zaparty and Siebers, 2011).

Carbon metabolism

Generally, central metabolism in archaea is characterized by many non-canonical pathways facilitated by unusual enzymes; this is also the case in acidophilic archaea. The modified Entner–Doudoroff (ED) pathway is the known route for sugar metabolism in thermoacidophilic archaea, while the Embden–Meyerhof–Parnas (EMP) pathway is used for gluconeogenesis only (Zaparty and Siebers, 2011).

Acidophilic Crenarchaeota

Central carbon metabolic network of the *Sulfolobaceae* has been studied both biochemically and *in silico*, based on annotation of genomes. The branched ED pathway (used for sugar degradation in *Archaea*) has been shown for galactose and glucose degradation in *S. solfataricus*, which is one of the most-studied organisms for sugar transformation in archaea. The branched ED pathway is characterized by the lack of the initial phosphorylation, which is typical of the classical ED pathway. Key enzymes were shown to be the bifunctional 2-keto-3-deoxy-6-phosphogluconate aldolase, 2-keto-3-deoxygluconate kinase and glycerate kinase, where both kinases catalyse the phosphorylation steps in a semi-phosphorylative and in non-phosphorylative branches. The candidate genes for the same enzymes and homologues of glyceraldehyde dehydrogenases have been identified in all acidophilic archaea of the phylum *Crenarchaeota* (Zaparty and Siebers, 2011). In the EMP pathways of acidophilic crenarchaeotes, the phosphofructokinase was demonstrated to be the only missing enzyme, which was revealed by the lack of corresponding activities in crude extracts and in isolated enzymes. However, the hexokinase activity was detected in crude extracts of *S. solfataricus*, and hexokinase homologues were identified in all sequenced thermo-acidophiles (Zaparty and Siebers, 2011). In addition, a later study indicated that in *S. solfataricus* P2 the missing semi-phosphorylated ED branch in glucose catabolism could be compensated by non-phosphorylated ED branch; semi-phosphorylated branch has an additional role in anabolism/gluconeogenesis by providing sugar phosphates and glycerate kinase may be a major player in the control of transfer between the glycolytic non-phosphorylated and the glycolytic/gluconeogenetic semi-phosphorylated ED pathway (Kouril *et al.*, 2013).

Pentose phosphate metabolism in *Sulfolobaceae* may be carried out using a reversed ribulose monophosphate pathway, as genes encoding two key enzymes (3-hexulose-6-phosphate isomerase and 3-hexulose-6-phosphate synthase) for this process have been identified (Zaparty and Siebers, 2011).

Pyruvate is converted via acetyl-CoA to CO_2 by pyruvate–ferredoxin oxidoreductase, and organic compounds produced in the ED pathway are completely converted to CO_2 via the TCA cycle, with the generation of additional physiological electron carriers, such as NAD(P)H, ferredoxin (red), $FADH_2$ and energy. All necessary TCA cycle candidate genes have been determined in all sequenced *Sulfolobaceae*, and genomes of *Sulfolobaceae* have revealed the presence of genes encoding for isocitrate lyase and malate synthase (though not in *S. tokodai*) and of the glyoxylate bypass (Zaparty and Siebers, 2011).

Fixation of CO_2 by autotrophic *Sulfolobales* spp. has been studied in detail using the obligate aerobe *M. sedula*, and was proposed to occur via the modified 3-hydroxypropionate/4-hydroxybutyrate cycle, known as the 'fifth autotrophic pathway'. The same pathway was considered to operate in the strict anaerobe *Sg. azoricus* (Berg *et al.*, 2010). Data of measured enzyme activities in both strains were reported to be similar and attributed to acetyl-CoA/propionyl-CoA carboxylase, malonyl-CoA reductase, malonate semialdehyde reductase as well as enzymes for the transformation of 3-hydropropionate to propionyl-CoA (Berg *et al.*, 2010). The same pathway has been suggested to function under oxic and anoxic conditions in the facultative anaerobe *Acidianus* spp. (*Ad. infernus, Ad. brierley* and *Ad. ambivalens*; Berg *et al.*, 2010).

Acidophilic Euryarchaeota

The modified ED pathway has been found in the *Thermoplasmatales* and shows stark differences to central metabolism of the crenarchaeotes. Biochemical studies revealed the key enzyme, a novel 2-keto-3-deoxygluconate kinase, was not related to a ribokinase superfamily but represented a member of BadF/BadG/BcrA/BcrD ATPase family and was distributed in all *Thermoplasmatales* studied at the time (Zaparty and Siebers, 2011). Reher *et al.* (2010) concluded that, in *P. torridus,* the strictly non-phosphorylative ED pathway is functional and involves a novel keto-3-deoxygluconate-specific aldolase. These authors found homologues of the ferredoxin-dependent aldehyde oxidoreductase in *P. torridus* but not found in either *Thermoplasma* spp. examined. In addition, no candidate fructose-1,6-bisphosphatase aldolase was found in the genomes of *Tp. acidophilum, Tp. volcanium* and *P. torridus* revealed, supporting the existence of an alternative aldolase involved in the gluconeogenesis in these organisms (Zaparty and Siebers, 2011). Furthermore, inspection of genome for TCA cycle enzymes in *Thermoplasmatales* has identified some differences. While *P. torridus* harbours all required genes for a complete TCA cycle, both *Thermoplasma* genomes lacked *sdhC* and *sdhD* genes for subunits of succinate dehydrogenase (Zaparty and Siebers, 2011), suggesting an apparent bypass of the corresponding reaction.

Further interesting metabolic peculiarities in *Thermoplasmatales* are (i) the lack of genes for pyruvate carboxylase candidates in all genomes analysed; (ii) absence of phosphoenolpyruvate carboxylase in the two *Thermoplasma* spp., of phosphoenolpyruvate carboxykinase in *P. torridus,* which necessary for catalysis of some anaplerotic reactions (Zaparty and Siebers, 2011) while the latter was found to be merely expressed in the proteome of *Fp. acidiphilum* (Golyshina *et al.*, 2006). In contrast to that, all above enzymes were found in all analysed genomes of acidophilic archaea of the order *Sulfolobales* (Zaparty and Siebers, 2011).

Details of pentose phosphate metabolism in the *Thermoplasmatales* are currently unresolved. The two key enzymes operating in the reversed ribulose monophosphate pathway, 3-hexulose-6-phosphate isomerase and 3-hexulose-6-phosphate synthase, are lacking in all three *Thermoplasmatales* genomes that have been analysed (Zaparty and Siebers, 2011).

The polypeptide degradation pathway in *Tp. acidophilum* was proposed to operate with α- and β-proteasomes rather than with tricorn peptidases, and also with the tricorn-interacting factors F1–F3. However, Ruepp *et al.* (2000) found that ubiquitin and proteasomal AAA ATPase were not present in the genome, though the latter enzyme might be substituted by other proteins with a similar activity. Furthermore, numerous putative proteases, some of tricorn proteins, extracellular acid proteases and hydrolases of unknown specificity, have been identified in the genome of this euryarchaeote (Ruepp *et al.*, 2000).

Candidate genes (*otsA* and *otsB*) for the best-studied trehalose biosynthesis pathway TPS/TPP or OtsA/OtsB were found in the genomes of three *Thermoplasmatales,* together with the two trehalose synthase genes *treS* for conversion of the α-1,4-linkage maltose into α-1,1-linkage trehalose present (Zaparty and Siebers, 2011).

Sulfur metabolism

A detailed study of sulfur metabolism by acidophilic archaea has been carried out with *Ad. ambivalens* (Kletzin *et al.*, 2004). It was demonstrated that sulfur oxygenase reductase (SOR) catalysed elemental sulfur oxidation to the reaction products sulfite, thiosulfate and hydrogen sulfide, coupled to oxygenase reaction. The same enzyme also catalysed disproportionation of S^0. Generation of sulfide as a major intermediate was suggested to be characteristic of archaeal sulfur oxidation (Rohwerder and Sand, 2007). The structure and mechanism of the action of the sulfur oxygenase reductase have been resolved (Veith *et al.*, 2011; Kletzin *et al.*, 2004). The enzyme that oxidizes thiosulfate to tetrathionate (thiosulfate:quinone oxidoreductase) in *Ad. ambivalens* has been studied and found to possess subunits similar to DoxA and DoxD, described as a part of terminal quinol:oxygen oxidoreductase, a cytochrome aa_3 (Kletzin *et al.*, 2004). Also, the tetrathionate hydrolase from *Ad. ambivalens,* grown with tetrathionate as a substrate, has been investigated in detail. The protein, which has a very low-pH optimum for activity (pH 1), was found to be S-layer-connected and glycosylated, as is common for extracellular and membrane-bound proteins from *Sulfolobales* (Protze *et al.*, 2011). Auernik and Kelly (2008) identified that '*sox*AB-CDD' cluster was induced by both elemental sulfur and tetrathionate in *M. sedula*. Synthesis of other components of that pathway, e.g. heterodisulfide reductase-like complex, a predicted tetrathionate hydrolase and the putative polysulfide/sulfur/dimethyl sulfoxide reductase, were all stimulated by reduced inorganic sulfur species.

Anaerobic hydrogen oxidation coupled to reduction of elemental sulfur has been reported for *Ad. ambivalens*. The hydrogenase was suggested to be of the Ni-Fe type, with a *Sulfolobus*-type quinone being the electron acceptor, a proximal cluster (3Fe–4S) in place of a (4Fe–4S) and a unique Rieske ferredoxin (Kletzin *et al.*, 2004; Laska *et al.*, 2003). Sulfur reductase was found to be encoded by the gene cluster *sre*ABCDE, consisting of five ORFs and featuring some special biochemical properties (Laska *et al.*, 2003).

Respiratory chains in acidophilic *Archaea*

The components of aerobic electron transfer chains of members of *Sulfolobaceae* contain a number of unique features, which have been studied in detail and confirmed in many cases by genomic annotation. For example, caldariellaquinone is present in all *Sulfolobaceae*/*Sulfolobales* as a main component, accounting for about 60% of the total quinones, the ratio varying between the family members, growth temperature and other conditions (Schäfer *et al.*, 1999).

An abundance of ferredoxins has been found in archaea that metabolize sulfur. These iron–sulfur proteins displayed a high thermostability, an ability to functionally replace pyridine nucleotides in glyceraldehyde dehydrogenase, aldehyde dehydrogenase or 2-oxoacid dehydrogenase and also to play a role in the transfer of hydrogen to NADH (Schäfer *et al.*, 1999). *Sulfolobus* and *Thermoplasma* ferredoxins have been purified and the three-dimensional structures displayed specific zinc-binding site in supplement to the (3Fe–4S) and (4Fe–4S) clusters (Iwasaki and Oshima, 2001).

Novel distinctive terminal oxidase complexes have been found in *S. acidocaldarius*, namely SoxABCD, that combined the features of respiratory complexes III and IV, which commonly occur separately in bacteria (Schäfer *et al.*, 1999) and the supercomplex SoxM with 10 metal-binding sites, unifying SoxE, F, G, H and SoxM – oxygen reductase of A-type (Komorowski *et al.*, 2002). No *c*-type cytochromes have been detected either in the *Sulfolobaceae* or in other acidophilic archaea so far. However, the associations of Rieske iron–sulfur proteins (2Fe–2S) with cytochrome *b* or its homologues suggest they may function in a similar way. A small molecular weight copper protein, sulfocyanin, which acts as an electron carrier to the oxygen reductase from complex III, firstly found to be a part of the supercomplex SoxM, is usually present in respiratory chains of *Sulfolobaceae* and in some acidophilic euryarchaeotes.

The cytochrome complex *ba* was isolated from *Ad. ambivalens* and biochemically characterized; its function is analogous to *bc1* complex (Bandeiras *et al.*, 2009). A gene cluster designated as *fox* encode membrane proteins such as putative cytochromes *b*, a haem copper oxidase (subunits I and II), ferredoxins and alternative proteins with Fe-S domains, this Fox cluster was shown to be spread among ferrous-iron-oxidizing *Sulfolobaceae* (Ilbert and Bonnefoy, 2013). Hypothetical model of ferrous iron oxidation has recently been suggested for '*Metallosphaera yellowstonensis*' (Kozubal *et al.*, 2011). This model, based on the analysis of the draft genome, assumes that electrons are extracted from ferrous iron by a cytochrome *b* (FoxC) and that a multicopper oxidase with two plastocyanin type I copper domains (Mco) is involved in transporting electrons 'uphill' or 'downhill' to the cytochrome *ba* or to the protein complex FoxAB (a haem copper oxidase), respectively (Kozubal *et al.*, 2011).

Prenylated haems of A_s and B types have been found in terminal oxidases in *Sulfolobus* and *Acidianus* spp. (Schäfer *et al.*, 1999). Aerobic electron transfer components of the *Sulfolobaceae* identified in sets of complex II with SDH (succinate dehydrogenase) activity were found in all studied *Sulfolobales*. The terminal oxidases equivalent to complex IV (of aa_3 or ba_3 types) were shown to be fairly ubiquitous too, though complex III, being usually the bc_1 component, was detected only in some isolates (Ilbert and Bonnefoy, 2013; Schäfer *et al.*, 1999).

The unique quinone found in *Thermoplasma*, thermoplasmaquinone, is represented by 2,[5 or 8]-dimethyl-3-heptaprenyl-1,4-naphthoquinone (Collins and Langworthy, 1983). Also menaquinone and methionaquinone were detected in *Thermoplasma*. The relative amount of quinones was shown to be oxygen dependent: an increase in thermoplasmaquinone correlated with conditions of oxygen limitation, while an increase in menaquinone and methionaquinone with elevated oxygen levels (Shimada *et al.*, 2001). The major respiratory quinones of some of *Ferroplasmaceae* members were found to be naphthoquinone derivatives (Golyshina, 2011).

Tp. acidophilum genome annotation revealed the presence of flavoproteins, homologues of cytochrome *b* clustered together with Rieske iron–sulfur protein, and two cytochrome *bd* quinol oxidases. Homologues of *asrA* and *asrB* genes, encoding proteins involved in dissimilatory sulfur reduction (e.g. in *Salmonella*) and a sulfide-quinone reductase, were proposed to be essential for sulfur respiration in this archaeon (Ruepp *et al.*, 2000).

Inspection of the genome of *P. torridus* suggested a complex consisting of the quinone oxidase, sulfocyanin and the terminal oxidase. Sulfocyanin sequence showed homology to *Sulfolobus* spp. sulfocyanin genes and the same chain was also found in the '*Ferroplasma acidarmanus*' genome (Fütterer *et al.*, 2004). Furthermore, a complex I (putative proton pumping NADH dehydrogenase) and genes for an A_0A_1-type ATPase, as suggested by authors, produce a competent respiratory system managing proton removal and restricting over-acidification of cytoplasm (Fütterer *et al.*, 2004). It is worth noting that *P. torridus*, together with '*Fp. acidarmanus*' (but not *Tp. acidophilum*) and acidophiles of the phylum *Crenarchaeota*, harbours a non-classical type of succinate dehydrogenase. Further information on this subject can be found in Chapter 2.

Uncultured acidophilic Archaea

Extensive metagenomic studies conducted with environmental samples from a number of acidic locations have revealed a huge pool of uncultured and unclassified potentially acidophilic archaeal community members, phylogenetically affiliated to both the *Crenarchaeota* and *Euryarchaeota*. Two groups first discovered in Iron Mountain ore body (Chapter 13) and in other AMD systems, acidic springs, aerobic and anaerobic continental volcanic areas, marine hydrothermal systems, sediments and probably many other acidic locations worldwide, and extensively investigated in recent years, are described in brief.

Uncultured *Thermoplasmatales* inhabiting the above biotopes represent phylogenetically diverse clusters with metabolisms (based on the metagenomic data analysis) thought to be principally those typical for acidophilic *Thermoplasmata*. However, major efforts are needed to isolate these unknown archaea, in order to study them *in vitro*. Emerging accumulation of data on diversity of *Thermoplasmata* supports important biogeochemical and ecological roles for this group.

The second group of great interest is represented by 'ultra-micro archaea' or 'nano-sized archaea' (the 'archaeal Richmond Mine acidophilic nano-organisms', ARMAN) found in filtrates (liquids that pass through 0.45 μm pore size membrane filters) which were first detected in the Iron Mountain site (Chapter 13). These archaeal lineages were later confirmed to be quite ubiquitous in AMD (Mendez-Garcia *et al.*, 2014; Baker *et al.*, 2010; Comolli *et al.*, 2009). The reconstruction of metagenomic data suggested their phylogenetic position to be deeply branching within the phylum *Euryarchaeota* with a few proposed putative types of some of those groups of uncultured archaea granted a *Candidatus* status (Mendez-Garcia *et al.*, 2014; Baker *et al.*, 2010). It is interesting to note that 16S rRNA genes of some of the sequences have been found to incorporate introns (Baker *et al.*, 2010). The presence of introns within 16S rRNA genes has earlier been reported for archaea of the family *Thermoproteaceae* (Burggraf *et al.*, 1993). The proposed average genome sizes of ARMAN types were approximately 1 Mbp, and electron microscope images documented some cells with dimensions < 500 nm and interspecies penetration (probably with archaea of the order *Thermoplasmatales*), which brought into question the nature of interactions involving these archaea. Microscopic analysis also provided evidence for possible free-living lifestyle of 'ultra-micro' archaea, the existence of tubular structures of unresolved function and of viral particles associated with cells (Comolli *et al.*, 2009).

Outlook

Metagenomic, metaproteomic and metabolomics studies, and single-cell genomics conducted with samples from acidic environments, have uncovered a great diversity of yet uncultured archaea in these environments, in particular (and also from cold wetland ecosystems, putative acidophilic methanogens and filterable fraction of archaea affiliated with *Euryarchaeota* (Mendez-Garcia *et al.*, 2014). Microbial communities inhabiting sediments in extremely acidic and metal-containing pit lakes of the Iberian Pyrite Belt were found to include uncultivated *Thaumarchaeota* (Falagan *et al.*, 2014). Representatives of *Thaumarchaeota* (members of which phylum are apparently very abundant in the biosphere) were also found to populate sediments of hydrothermal springs in Yellowstone National Park, Wyoming), that contained iron oxide and elemental sulfur. It is noteworthy that at least some of these organisms do not contain ammonium monooxygenase genes commonly ubiquitous in *Thaumarchaeota* that generally rely on ammonium oxidation (Beam *et al.*, 2014). A new candidate phylum, '*Geoarchaeota*', has very recently been suggested with, as yet, uncultured representatives that colonize acidic high-temperature iron oxide mats in Yellowstone National Park (Kozubal *et al.*, 2013). Furthermore, metagenomic analysis has provided evidence for the phylogenetic placement of this lineage being distinct from *Crenarchaeota*, and evidence of a unique combination in the metabolic pathways of both archaeal and bacterial types (Kozubal *et al.*, 2013).

It is apparent that acidophilic archaea could potentially be found across many lineages, some already known and others yet to be described. A considerable effort is needed to gain a novel insight into the metabolic mechanisms, evolution and roles in element cycling of this exciting group of archaea, which should serve to promote the development of new techniques and approaches for isolating these novel prokaryotes and for cultivating them *in vitro*. Future studies combining high resolution OMICS approaches with physiological and biochemical investigations will allow filling numerous gaps in our knowledge on this unique group of archaea.

Acknowledgements

The authors thank EU FP7 Grant No. 287589 'MicroB3' and EU Horizon 2020 Grant No. 634486 'INMARE' for financial support.

References

Auernik, K.S., and Kelly, R.M. (2008). Identification of components of electron transport chains in the extremely thermoacidophilic crenarchaeon *Metallospaera sedula* through iron and sulfur compound oxidation transcriptomes. App. Env. Microbiol. *74*, 7723–7732.

Baker, B.J., Comolli, L.R., Dick, G.J., Hauser, L.J., Hyatt, D., Dill, B.D., Land, M.L., Verberkmoes, N.C., Hettich, R.L., and Banfield, J.F. (2010). Enigmatic, ultrasmall, uncultivated Archaea. Proc. Natl. Acad. Sci. U.S.A. *107*, 8806–8811.

Bandeiras, T.M., Refojo, P.N., Todorovic, S., Murgida, D.H., Hildebrandt, P., Bauer, C., Pereira, M.M., Kletzin, A., and Teixeira, M. (2009).The cytochrome ba complex from the thermoacidophilic crenarchaeote *Acidianus ambivalens* is an analog of bc(1) complexes. Biochim. Biophys. Acta *1787*, 37–45.

Baumeister, W., and Lembcke, G.J. (1992). Structural features of archaebacterial cell envelopes. Bioenerg. Biomembr. *24*, 567–575.

Beam, J.P., Zackary, J., Jay, Z.J., Kozubal, M.A., and Inskeep, W.P. (2014). Niche specialization of novel *Thaumarchaeota* to oxic and hypoxic acidic geothermal springs of Yellowstone National Park. ISME J. *8*, 938–951.

Berg, I.A., Kockelkorn, D., Ramos-Vera, W.H., Say, R.F., Zarzycki, J., Hügler, M., Alber, B.E., and Fuchs, G. (2010). Autotrophic carbon fixation in archaea. Nat. Rev. Microbiol. *8*, 447–460.

Brock, T.D., Brock, K.M., Belly, R.T., and Weiss, R.L. (1972). *Sulfolobus*: a new genus of sulfur-oxidising bacteria living at low pH and high temperature. Arch. Microbiol. *84*, 54–68.

Burggraf, S., Larsen, N., Woese, C.R., and Stetter, K.O. (1993). An intron within the 16S ribosomal RNA gene of the archaeon *Pyrobaculum aerophilum*. Proc. Natl. Acad. Sci. U.S.A. *90*, 2547–2550.

Collins, M.D., and Langworthy, T.A. (1983). Respiratory quinone composition of some acidophilic bacteria. Syst. Appl. Microbiol. *4*, 295–304.

Comolli, L.R., Baker,.J., Downing, K.H., Siegerist, C.E., and Banfield, J.F. (2009). Three-dimensional analysis of the structure and ecology of a novel, ultra-small archaeon. ISME J. 3, 159–167.

Darland, G., Brock, T.D., Samsonoff, W., and Conti, S.F. (1970). A thermophilic acidophilic Mycoplasma isolated from a coal refuse pile. Science *170*, 1416–1418.

Edwards, K.J., Bond, P.L., Gihring, T.M., and Banfield, J.F. (2000). An archaeal iron-oxidising extreme acidophile important in acid mine drainage. Science *287*, 1796–1799.

Falagán, C., Sánchez-España, J., and Johnson, D.B. (2014). New insights into the biogeochemistry of extremely acidic environments revealed by a combined cultivation-based and culture-independent study of two stratified pit lakes. FEMS Microbiol. Ecol. *87*, 231–243.

Fuchs, T., Huber, H., Teiner, K., Burggraf, S., and Stetter, K.O. (1995). *Metallosphaera prunea*, sp. nov. a novel, metal-mobilizing, thermoacidophilic Archaeum, isolated from a uranium mine in Germany. Syst. Appl. Microbiol. *18*, 560–566.

Fuchs, T., Huber, H., Burggraf, S., and Stetter, K.O. (1996). 16S rRNA –based phylogeny of the archaeal order

Sulfolobales and reclassification of *Desulfurococcus ambivalents* as *Acidianus ambivalens* comb. nov. Syst. Appl. Microbiol. *19*, 56–60.

Fütterer, O., Angelov, A., Liesegang, H., Gottschalk, G., Schleper, C., Schepers, B., Dock, C., Antranikian, G., and Liebl, W. (2004). Genome sequence of *Picrophilus torridus* and its implications for life around pH 0. Proc. Natl. Acad. Sci. U.S.A. *101*, 9091–9096.

Golovacheva, R.S., Valieho-Roman, K.M., and Troitskii, A.V. (1987). *Sulfurococcus mirabilis* gen. nov., sp. nov., a new thermophilic archaebacterium with the ability to oxidise sulfur. Mikrobiologiya *56*, 100–107.

Golyshina, O.V., Pivovarova, T.A., Karavaiko, G.I., Kondratéva, T.F., Moore, E.R., Abraham, W.R., Lünsdorf, H., Timmis, K.N., Yakimov, M.M., and Golyshin, P.N. (2000). *Ferroplasma acidiphilum* gen. nov., sp. nov., an acidophilic, autotrophic, ferrous-iron-oxidising, cell-wall-lacking, mesophilic member of the *Ferroplasmaceae* fam. nov., comprising a distinct lineage of the Archaea. Int. J. Syst. Evol. Microbiol. *50*, 997–1006.

Golyshina, O.V., Golyshin, P.N., Timmis, K.N., and Ferrer, M. (2006). The 'pH optimum anomaly" of intracellular enzymes of *Ferroplasma acidiphilum*. Environ. Microbiol. *8*, 416–425.

Golyshina, O.V., Yakimov, M.M., Lünsdorf, H., Ferrer, M., Nimtz, M., Timmis, K.N., Wray, V., Tindall, B.J., and Golyshin, P.N. (2009). *Acidiplasma aeolicum* gen. nov., sp. nov., a euryarchaeon of the family *Ferroplasmaceae* isolated from a hydrothermal pool, and transfer of *Ferroplasma cupricumulans* to *Acidiplasma cupricumulans* comb. nov. Int. J. Syst. Evol. Microbiol. *59*, 2815–2823.

Golyshina, O.V. (2011). Environmental, biogeographic, and biochemical patterns of archaea of the family *Ferroplasmaceae*. Appl. Environ. Microbiol. *77*, 5071–5078.

Grogan, D., Palm, P., and Zillig, W.J. (1990). Isolate B12, which harbors a virus- like element, represents a new species of the archaebacterial genus *Sulfolobus*, *Sulfolobus shibatae*, sp. nov. Arch. Microbiol. *154*, 594–599.

Hawkes, R.B., Franzmann, P.D., O'Hara, G., and Plumb, J.J. (2006). *Ferroplasma cupricumulans* sp. nov., a novel moderately thermophilic, acidophilic archaeon isolated from an industrial-scale chalcocite bioleach heap. Extremophiles *10*, 525–530.

Huber, H., and Prangishvili, D. (2006). *Sulfolobales*. In The Prokaryotes, Falkow, S., Rosenberg, E., Dworkin, M., and Stackenbrandt, E., eds (Springer, New York, NY, USA), pp. 23–51.

Huber, H., and Stetter, K.O. (2006). *Thermoplasmatales*. In The Prokaryotes, Falkow, S., Rosenberg, E., Dworkin, M., and Stackebrandt, E., eds (Springer, New York, NY, USA), pp. 101–112.

Huber, G., Spinnler, C., Gambacorta, A., and Stetter, K.O. (1989). *Metallosphaera sedula* gen. and sp. nov. represents a new genus of aerobic, metal-mobilizing, thermoacifdophilic archaebacteria. Syst. Appl. Microbiol. *12*, 538–547.

Ilbert, M., and Bonnefoy, V. (2013). Insight into the evolution of the iron oxidation pathways. Biochim. Biophys. Acta *1827*, 161–175.

Itoh, T., Yoshikawa, N., and Takashina, T. (2007). *Thermogymnomonas acidicola* gen. nov., sp. nov., a novel thermoacidophilic, cell wall-less archaeon in the order *Thermoplasmatales*, isolated from a solfataric soil in Hakone, Japan. Int. J. Syst. Evol. Microbiol. *57*, 2557–2561.

Iwasaki, T., and Oshima, T. (2001). Ferredoxin and related enzymes from *Sulfolobus*. In Methods in Enzymology, Vol. 334, Hyperthermophilic Enzymes Part C, Adams, M.W.W., and Kelly, R.M., eds (Academic Press, San Diego, CA, USA), pp. 3–22.

Karavaĭko, G.I., Golyshina, O.V., Troitskiĭ, A.V., Val'ekho-Roman, K.M., Golovacheva, R.S., and Pivovarova, T.A. (1994). *Sulfurococcus yellowstonii* sp. nov. – a new species of iron- and sulfur-oxidising thermoacidophilic Archaeobacterium. Mikrobiologiia *63*, 668–682.

Kletzin, A., Urich, T., Müller, F., Bandeiras, T.M., and Gomes, C.M. (2004). Dissimalotory oxidation and reduction of elemental sulfur in thermophilic archaea. J. Bioenerg. Biomemr. *36*, 77–91.

Komorowski, L., Verheyen, W., and Schäfer, G. (2002). The archaeal respiratory supercomplex SoxM from *S. acidocaldarius* combines features of quinole and cytochrome c oxidases. Biol. Chem. *383*, 1791–1799.

Kouril, T., Wieloch, P., Reimann, J., Wagner, M., Zaparty, M., Albers, S.V., Schomburg, D., Ruoff, P., and Siebers, B. (2013). Unraveling the function of the two Entner-Doudoroff branches in the thermoacidophilic Crenarchaeon *Sulfolobus solfataricus* P2. FEBS J. *280*, 1126–1138.

Kozubal, M.A., Dlakic, M., Macur, R.E., and Inskeep, W.P. (2011). Terminal oxidase diversity and function in "*Metallosphaera yellowstonensis*": gene expression and protein modeling suggest mechanisms of Fe(II) oxidation in the *sulfolobales*. Appl. Environ. Microbiol. *77*, 1844–1853.

Kozubal, M.A., Romine, M., Jennings, Rd., Jay, Z.J., Tringe, S.G., Rusch, D.B., Beam, J.P., McCue, L.A., and Inskeep, W.P. (2013). Geoarchaeota: a new candidate phylum in the Archaea from high-temperature acidic iron mats in Yellowstone National Park. ISME J. *7*, 622–634.

Kurosawa, N., Itoh, Y.H., and Itoh, T. (2003). Reclassification of *Sulfolobus hakonensis* Takayanagi *et al.* 1996 as *Metallosphaera hakonensis* comb. nov. based on phylogenetic evidence and DNA G+C content. Int. J. Syst. Evol. Microbiol. *53*, 1607–1608.

Laska, S., Lottspeich, F., and Kletzin, A. (2003). Membrane-bound hydrogenase and sulfur reductase of the hyperthermophilic and acidophilic archaeon *Acidianus ambivalens*. Microbiology *149*, 2357–2371.

Liu, L.J., You, X.Y., Zheng, H., Wang, S., Jiang, C.Y., and Liu, S.J. (2011). Complete genome sequence of *Metallosphaera cuprina*, a metal sulfide-oxidising archaeon from a hot spring. J. Bacteriol. *193*, 3387–3388.

Kurosawa, N., Itoh, Y.H., Iwai, T., Sugai, A., Uda, I., Kimura, N., Horiuchi,T., and Itoh, T. (1998). *Sulfurisphaera ohwakuensis* gen. nov., sp. nov., a novel extremely thermophilic acidophile of the order *Sulfolobales*. Int. J. Syst. Bacteriol. *48*, 451–456.

Martins, L.O., Huber, R., Huber, H., Stetter, K.O., Da Costa, M.S., and Santos, H. (1997). Organic solutes in hyperthermophilic archaea. Appl. Environ. Microbiol. *63*, 896–902.

Méndez-García, C., Mesa, V., Sprenger, R.R., Richter, M., Diez, M.S., Solano, J., Bargiela, R., Golyshina, O.V., Manteca, A., Ramos, J.L., *et al.* (2014). Microbial stratification in low pH oxic and suboxic macroscopic growths along an acid mine drainage. ISME J. *8*, 1259–1274.

Plumb, J.J., Haddad, C.M., Gibson, J.A.E., and Franzmann, P.D. (2007). *Acidianus sulfidivorans* sp. nov., an extremely acidophilic, thermophilic archaeon isolated from a solfatara on Lihir Island, Papua New Guinea, and emendation of the genus description. Int. J. Syst. Evol. Microbiol. *57*, 1418–1423.

Protze, J., Müller, F., Lauber, K., Naß, B., Mentele, R., Lottspeich, F., and Kletzin, A. (2011). An extracellular tetrathionate hydrolase from the thermoacidophilic archaeon *Acidianus ambivalens* with an activity optimum at pH 1. Front. Microbiol. *2*, 68.

Reher, M., Fuhrer, T., Bott, M., and Schönheit, P. (2010). The nonphosphorylative Entner-Doudoroff pathway in the thermoacidophilic euryarchaeon *Picrophilus torridus* involves a novel 2-keto-3-deoxygluconate- specific aldolase. J. Bacteriol. *192*, 964–974.

Rohwerder, T., and Sand, W. (2007). Oxidation of inorganic sulfur compounds in acidophilic prokaryotes. Eng. Life Sci. *7*, 301–309.

Ruepp, A., Graml, W., Santos-Martinez, M.L., Koretke, K.K., Volker, C., Mewes, H.W., Frishman, D., Stocker, S., Lupas, A.N., and Baumeister, W. (2000). The genome sequence of the thermoacidophilic scavenger *Thermoplasma acidophilum*. Nature *407*, 508–513.

Ruepp, A., Rockel, B., Gutsche, I., Baumeister, W., and Lupas, A.N. (2001). The Chaperones of the archaeon *Thermoplasma acidophilum*. J. Struct. Biol. *135*,126–138.

Schäfer, G., Engelhard, M., and Müller, V. (1999). Bioenergetics of Archaea. Microbiol. Mol. Biol. Rev. *63*, 570–620.

Schleper, C., Puehler, G., Holz, I., Gambacorta, A., Janekovic, D., Santarius, U., Klenk, H.P., and Zillig, W. (1995). *Picrophilus* gen. nov., fam. nov.: a novel aerobic, heterotrophic, thermoacidophilic genus and family comprising archaea capable of growth around pH 0. J. Bacteriol. *177*, 7050–7059.

Segerer, A., Neuner, A., Kristjansson, J.K., and Stetter, K.O. (1986a). *Acidianus infernus* gen. nov., sp. nov., and *Acidianus brierleyi* comb. nov.: facultatively aerobic, extremely acidophilic thermophilic sulfur-metabolizing archaeabacteria. Int. J. Syst. Bacteriol. *36*, 559–564.

Segerer, A., Langworthy, T.A., and Stetter, K.O. (1986). *Thermoplasma acidophilum* and *Thermoplasma volcanium* sp. nov. from solfatara fields. Syst. Appl. Microbiol. *10*, 161–171.

Segerer, A.H., Trincone, A., Gahrtz, M., and Stetter, K.O. (1991). *Stygiolobus azoricus* gen. nov., sp. nov. represents a novel genus of anaerobic, extremely thermoacidophilic archaebacteria of the order *Sulfolobales*. Int. J. Syst. Bacteriol. *41*, 495–501.

Shimada, H., Shida, Y., Nemoto, N., Oshima, T., and Yamagishi, A. (2001). Quinone profiles of *Thermoplasma acidophilum* HO-62. J. Bacteriol. *183*,1462–1465.

Stetter, K.O. (2002). Hyperthermophilic microorganisms. In Astrobiology: The Quest for the Conditions of Life, Horneck, G., and Maumstark-Khan, C., eds (Springer Verlag, New York, NY, USA), pp. 169–184.

Suzuki, T., Iwasaki, T., Uzawa, T., Hara, K., Nemoto, N., Kon, T., Ueki, T., Yamagishi, A., and Oshima, T. (2002). *Sulfolobus tokodaii* sp. nov. (f. *Sulfolobus* sp. strain 7), a new member of the genus *Sulfolobus* isolated from Beppu Hot Springs, Japan. Extremophiles *6*, 39–44.

Thurl, S., Witke, W., Buhrow, I., and Schäfer, W. (1986). Quinones from archaebacteria, II. Different types of quinones from sulfur-dependent archaebacteria. Biol. Chem. Hoppe Seyler. *367*, 191–197.

Trincone, A., Lanzotti, V., Nicolaus, B., Zillig, W., DeRosa, M., and Gambacorta, A. (1989). Comparative lipid composition of aerobically and anaerobically grown *Desulfwolobus ambivalens*, an autotrophic thermophilic archaeobacterium. Microbiology *135*, 2751–2757.

van de Vossenberg, J.L., Driessen, A.J., and Konings, W.N. (1998). The essence of being extremophilic: the role of the unique archaeal membrane lipids. Extremophiles *2*, 163–170.

Veith, A., Kling, A., Zolghadr, B., Lauber, K., Mentele, R., Lottspeich, F., Rachel, R., Albers, S.V., and Kletzin, A. (2009). *Acidianus*, *Sulfolobus* and *Metallosphaera* surface layers: structure, composition and gene expression. Mol. Microbiol. *73*, 58–72.

Veith, A., Urich, T., Seyfarth, K., Protze, J., Frazão, C., and Kletzin, A. (2011). Substrate pathways and mechanisms of inhibition in the sulfur oxygenase reductase of *Acidianus ambivalens*. Front. Microbiol. *2*, 37.

Wang, X., Lv, B., Cai, G., Wu, Y., Wang, X., Ren, B., and Ma, H. (2012). A proton shelter inspired by the sugar coating of acidophilic archaea. Sci. Rep. *2*, 892.

Whitaker, R.J., Grogan, D.W., and Taylor, J.W. (2003). Geographic barriers isolate endemic populations of hyperthermophilic archaea. Science *301*, 976–978.

Woese, C.R., and Fox, G.E. (1977). Phylogenetic structure of the prokaryotic domain: the primary kingdoms. Proc. Natl Acad. Sci. U.S.A. *74*, 5088–5090.

Xiang, X., Dong, X., and Huang, L. (2003). *Sulfolobus tengchongensis* sp. nov., a novel thermoacidophilic archaeon isolated from a hot spring in Tengchong, China. Extremophiles *7*, 493–498.

Yoshida, N., Nakasato, M., Ohmura, N., Ando, A., Saiki, H., Ishii, M., and Igarashi, Y. (2006). *Acidianus manzaensis* sp. nov., a novel thermoacidophilic archaeon growing autotrophically by the oxidation of H_2 with the reduction of Fe^{3+}. Curr. Microbiol. *53*, 406–411.

Zaparty, M., and Siebers, B. (2011). Physiology, metabolism and enzymology of thermoacidophiles. Reconstruction of the central carbon metabolic network of thermoacidophilic archaea. In Extremophiles Handbook, Horikoshi, K., ed. (Springer, Japan), pp. 602–639.

Zillig, W., Stetter, K.O., Wunderl, S., Schulz, W., Priess, H., and Scholz, J. (1980). The *Sulfolobus-'Caldariella'* group: taxonomy on the basis of the structure of DNA-dependent RNA polymerases. Arch. Microbiol. *125*, 259–269.

Physiological and Phylogenetic Diversity of Acidophilic Eukaryotes

7

Angeles Aguilera, Sanna Olsson and Fernando Puente-Sánchez

Abstract

The study of extremely acidic environments has grown significantly over the past 50 years. Highly acidic environments are generally unfavourable for the growth of photosynthetic organisms. However, these ecosystems are usually inhabited by acidophilic and acid-tolerant eukaryotic microorganisms such as algae, amoeba, ciliates, heliozoa and rotifers, as well as fungi and yeasts. This chapter provides an overview of our knowledge of the diversity and ecophysiology of eukaryotic acidophilic microorganisms, as well as summarizing recent data from one of the most extensive extreme acidic environments, the Río Tinto, Spain.

Diversity of eukaryotic acidophiles

Exploration of the biosphere has led to continued discoveries of life in environments that were previously considered uninhabitable. Life is now known to survive in and sometimes thrive under what seem to be harsh environmental conditions, and the number of different organisms known to populate these environmentally extreme conditions has grown rapidly in recent years. When of the most familiar extremophiles are prokaryotes, eukaryotic microbial life may be found actively growing in almost any extreme condition where there is a source of energy to sustain it, with the only exception of high temperature (> 70°C; Roberts, 1999) and the deep subsurface biosphere. In this regard, the development of molecular technologies and their application to microbial ecology has increased our knowledge of eukaryotic diversity in many different environments (Caron *et al.*, 2004). This is particularly relevant in extreme environments, that are sometimes more difficult to replicate in the laboratory. Studies based on molecular ecology have demonstrated that eukaryotic organisms are exceedingly adaptable and not notably less so than prokaryotes, although most habitats have not been sufficiently well explored for sound generalizations to be made. Molecular analysis has also revealed novel protist genetic diversity in different extreme environments (Amaral-Zettler *et al.*, 2011; Roberts, 1999).

One group of extremophiles that is becoming increasingly important, both ecologically and economically, are acidophilic organisms that live in low-pH environments (pH < 3). These are in general highly toxic environments because of the high concentrations of protons and heavy metals. Highly acidic environments are relatively scarce worldwide and are generally associated with volcanic activity and mining operations (Baffico *et al.*, 2004). The natural oxidation and dissolution of the sulfidic minerals exposed to oxygen and water results in acid production, and the process can be greatly enhanced by microbial metabolism (Nordstrom and Southman, 1997). At the same time, low pH facilitates metal solubility in water, particularly cationic metals (such as aluminium and many heavy metals), and therefore acidic water tends to have high concentrations of heavy metals (Johnson, 1998).

Although it is usually assumed that these environmental conditions limit eukaryotic growth and diversity due to their toxicity, most of these acidic extreme environments contain an unexpectedly high degree of eukaryotic diversity. Microorganisms found in extremely acidic environments include single-celled and filamentous eukaryotes (microalgae, protozoa, yeast and fungi) and multicellular rotifera [e.g. *Rotaria* sp., *Cephalodella* sp. and *Elosa* sp. (Amaral-Zettler *et al.*, 2011; Bell and Weithoff, 2003; Deneke, 2000); Fig. 7.1]. This fact demonstrates that complex eukaryotic systems can thrive and dominate extremely acidic, heavy, metal-laden environments. Since some of these acidophiles are closely related to cultured neutrophiles, it may be surmised that eukaryotes must have the ability to adapt from neutral to acidic environments over relatively short geological time-scales. Eukaryotic extremophiles are now known to be more widely distributed and phylogenetically diverse than previously thought (Amaral-Zettler *et al.*, 2011).

Acidophilic microalgae

Microalgae are usually the dominant eukaryotic microorganisms in extremely acidic environments. Within them, species belonging to chlorophytes, chrysophytes and euglenophytes are the main components of the eukaryotic communities. The reasons for the dominance of these particular algal groups in these ecosystems are presumed to lie within their adaptive capabilities, their ecophysiological specialization and the ecological advantages that may be offered. In this regard, one metabolic strategy to overcome limiting carbon concentration for photosynthesis in these environments is mixotrophy, as well as the ability of migration to different water layers in order to find the optimal environmental conditions (Moser and Weisse, 2011a).

Chlamydomonas acidophila is one of the most abundant microalgal species in extremely acidic environments (Nishikawa and Tominaga, 2001). Besides its adaptation to low pH, tolerance to high concentrations of dissolved metals has also been reported for this species (Aguilera and Amils, 2005). These extreme chemical conditions induce cellular stress responses resulting in a decreased photosynthetic quantum yield, increased small heat-shock protein levels, changes in total cell protein profiles and changes in cell metal content, but not in growth rates (Cid *et al.*, 2010; Spijkerman

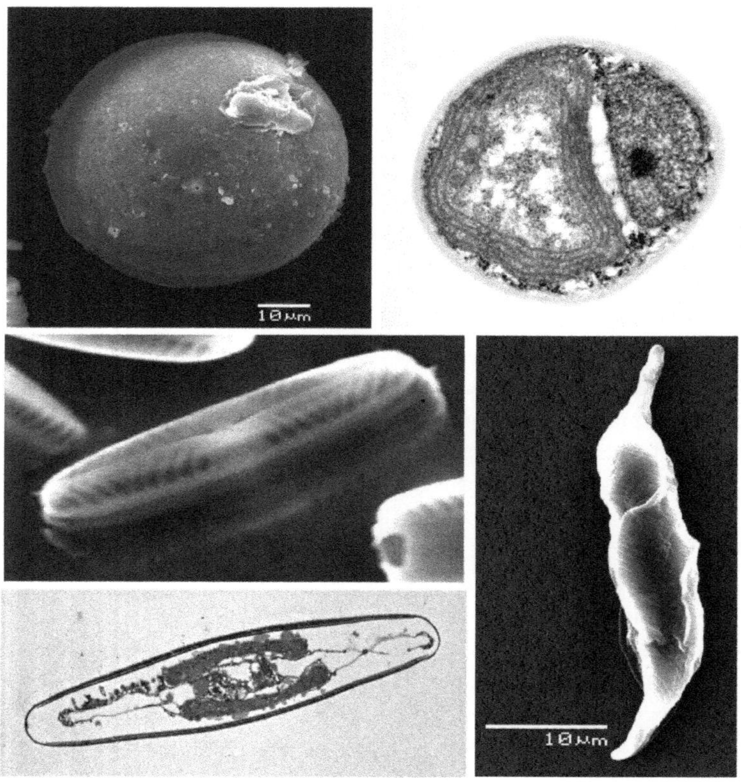

Figure 7.1 Examples of eukaryotic acidophiles: (a) SEM-BSE image of the microalgae *Chlorella* sp.; (b) TEM image of the red microalgae *Cyanidium* sp.; (c) SEM-BSE image of the diatom *Pinnularia* sp.; (d) SEM-BSE image of the protist *Euglena mutabilis*; (e) TEM image of the diatom *Pinnularia* sp.

et al., 2007). These multiple adaptations of *C. acidophila* increase its tolerance to metals and permit survival under such adverse environmental conditions.

The microalgae, *Dunaliella acidophila*, is one of the most extreme of all known acidophile, thriving optimally at a narrow pH range between 0 and 3 while maintaining an intracellular pH of 7 (Gimmler *et al.*, 1989). However, this species deals with low pH in a very different manner. In this microalga, a combination of factors contribute to reduce the influx of protons and to actively extrude protons out of cell. They include a positive inside transmembrane potential, a positive surface charge and the existence of a potent plasma membrane H^+-ATPase. *D. acidophila* overproduces a potent plasma membrane H^+-ATPase that resembles in its catalytic and regulatory properties the higher plant H^+-ATPase (Sekler *et al.*, 1991). However, *D. acidophila* enzymes differ from typical plant H^+-ATPases in several kinetic and structural properties: (i) ATPase and proton transport is stimulated 2–3-fold by K^+ ions, in comparison to 20–30% stimulation in plants; (ii) the K_m for ATP of about 50 _M is considerably lower than for most known plant H^+-ATPases; (iii) the pH optimum of 6.0 is slightly lower than for plant enzymes; and (iv) the apparent molecular weight of the ATPase polypeptide is higher than the plant H^+-ATPase (Sekler *et al.*, 1991). Another factor which has been suggested to decrease the permeability of *D. acidophila* cells to protons is their positive surface charge, which would also explain the high resistance to polyvalent cations and heavy metals (Gimmler *et al.*, 1991).

The only other microalgal species reported to grow at pH 0 is the unicellular red microalgae *Galdieria sulfuraria*. This microalgae is, in this regard, a unique member of the order Cyanidiales, displaying high salt and metal tolerance (Reeb and Bhattacharya, 2010). *G. sulfuraria* naturally inhabits volcanic hot sulfur springs, solfatara soils, and anthropogenic hostile environments. In habitats with high concentrations of arsenic, aluminium, cadmium, mercury, and other toxic metals, *G. sulfuraria* frequently represents up to 90% of total biomass and almost all the eukaryotic biomass. This algae shows an enormous metabolic flexibility, growing either photoautotrophically or heterotrophically on more than 50 different carbon sources. Environmental adaptation seems to have been facilitated by horizontal gene transfer from various bacteria and archaea, often followed by gene family expansion. At least 5% of protein-coding genes of *G. sulfuraria* were probably acquired horizontally. These proteins are involved in ecologically important processes ranging from heavy metal detoxification to glycerol uptake and metabolism (Schönknecht, 2013).

Acidophilic protists

Euglena mutabilis is one of the most globally widespread acidophilic protists in acidic environments and has been considered as a bioindicator of these ecosystems (Aguilera *et al.*, 2006a; Sittenfeld *et al.*, 2002). Recent studies have revealed that *E. mutabilis* secretes metabolites into the media which include a large number of amino acids, polyamine compounds, urea and some sugars but no fatty acids, suggesting a selective organic matter contribution in acidic ecosystems (Halter *et al.*, 2012). Such a production may have a crucial impact on the bacterial community, as it has been suggested previously that prokaryotes transport and recycle *in situ* most of the metabolites secreted by *E. mutabilis*. Consequently, this protist may have an indirect but important role in mining related areas but also in other ecological niches described as nitrogen limited (Halter *et al.*, 2012).

Besides euglenoids, other groups of protozoa are also frequently found in acidic environments, playing a key role in the pelagic food web of acidic mining lakes. Several species of hypotrich ciliates have been described, *Oxytricha* sp. being one of the most frequent species as well such as heliozoa such as *Actinophrys* sp. (Packroff, 2000). However, in general, these species are acid-tolerant rather than acidophilic, and are able to grow under a broad pH range (Weisse *et al.*, 2013). Rotifers and amoeba are also present in these habitats. In acidic lakes, the phytoplankton is dominated by the mixotrophic flagellate species *Ochromonas* spp. (Nixdorf *et al.*, 1998). Ciliates seem to be of minor importance, relative to mesotrophic circumneutral lakes (Packroff, 2000). Since, in acidic waters, algae have to cope with a limited supply of CO_2 for photosynthesis because of the absence of a bicarbonate pool, the relative contribution of mixotrophic flagellates to the algal community tends to increase with decreasing pH and may reach 90% of the pigmented biomass in acidic lakes with pH <3 (Moser and Weisse, 2011b).

Acidophilic fungi and yeast

Among eukaryotic organisms, acidophilic algae and protozoa have received more attention than fungi and yeasts, although fungi have long been recognized as active participants in acidification of sulfide-rich environments (Hujslova *et al.*, 2014; Gross and Robbins, 2000; Armstrong, 1981; Joseph, 1953). Additionally, most studies related to these organisms have been focussed on biodiversity, and little is known about their physiology or their interactions with other species. Most fungi living in acidic habitats should be regarded as acid-tolerant rather than strictly acidophilic because they are also able to grow under neutral or even alkaline pH (Gross and Robbins, 2000). Fungi seem to play an important role in these environments because, together with other microorganisms, they form biofilms on the surface of rocks. These biofilms are the site of metal and mineral precipitation and provide a substrate for other microbial populations. Fungi can display metal resistance and can sequester specific metals allowing less tolerant species to exist (Duran *et al.*, 1999).

Studies on yeasts in acidic environments are even less common and also mostly related to descriptions of their biodiversity. Usually, those that form pink-coloured colonies (*Rhodotorula* sp.) and arthroconidia (*Trichosporon* sp.) have been found to be the most commonly encountered. Among the ascomycetes, species belonging to the Hemiascomycetes and Euascomycetes (*Glomerella* sp. and *Lecytophora* sp.) have been described in the Iberian Pyrite Belt (IPB) as well as basidiomycetous yeasts distributed in the classes Hymenomycetes and Urediniomycetes (Gadanho *et al.*, 2006). These species have also been described in natural acidic geothermal areas (Russo *et al.*, 2008). Three novel asexual basidiomycetous yeast species have been recently described in the IPB (*Cryptococcus aciditolerans* sp. nov., *Cryptococcus ibericus* sp. nov. and *Cryptococcus metallitolerans* sp. nov.) belonging to the order Filobasidiales and form a well-separated clade (Gadanho and Sampaio, 2009). These *Cryptococcus* species are apparently specialists of acidic aquatic environments since they require low pH for growth, a property that has not been observed before in yeasts.

Acidophilic higher plants

Imperata cylindrical is probably the most common higher plant found in acidic environments, at least in the IPB. This species grows in riverine soils forming tall cane-like grass communities in springs and sand-banks and is frequently found growing on the banks of the Tinto River. The study of the elemental distribution of metals in the plant body confirmed that *I. cylindrical* is an iron hyper-accumulator, with Fe accounting for >1% of its dry mass (Rodriguez *et al.*, 2005). Recent analysis has indicated that iron accumulates in this plant mainly as jarosite and ferritin, and that the metal content is much higher in rhizomes than in leaves (Amils *et al.*, 2007). Jarosite accumulates mainly in roots and rhizomes, while ferritin has been detected in all the structures. Iron bioformations are clearly visible on the surface of the rhizome, forming extensive deposits on the abaxial and adaxial epidermis in the leaves. The unequivocal identification of jarosite, a mineral that can be formed only in acidic conditions, in the tissues of *I. cylindrica*, strongly suggests that the metabolism of iron in plants is much more complex and versatile than has been reported so far. It has been suggested that accumulation of iron at structural and cellular level not only has interest as storage for a critical element but also as a means to prevent cell damage produced by oxidative stress. Ferritins protect cells from oxidative stress at the same time that they store iron for use in the cellular functions that require it (Fuente *et al.*, 2012; Amils *et al.*, 2007).

Heather (*Erica* spp.) can also often be found in the proximity of highly acidic and metal-contaminated areas. Besides acidity, soils associated with *Erica* populations are very low in nutrient availability, which means that these plants have to deal with an even more extreme root environment. *Erica* species are able to tolerate the harsh environmental conditions of the mining areas with no apparent symptoms of toxicity even in individuals growing in the river banks with their roots partially submerged into the contaminated water. The ability to survive in such soils depends on different strategies to cope with high metal concentrations, such as decreased element uptake or transport, detoxification by complexation with organic ligands and intracellular compartmentation (Oliva *et al.*, 2012). Thus, *Erica* plants do not hyper-accumulate elements like copper and lead, shows root metal fixation processes and slow growth rates, which provides protection against toxicity by excessive metal presence.

Structure and development of acidophilic eukaryotic biofilms as a mechanism for surviving extreme conditions

Natural biofilms are organized multicellular systems with a structural and functional architecture which influences metabolic processes, response to nutrients, predation and other factors of the ecosystem (Costerton *et al.*, 1987). Consequently, structural studies of microbial biofilms and their formation play a critical role in understanding the ecophysiological processes in natural habitats. This is even more important in extreme environments where forming a structured biofilm might protect the organisms from external stress conditions and allow them to resist more extreme conditions (Seckbach, 1999). Therefore it is important to study how biofilms in highly acidic conditions affect geochemical processes such as metal immobilization (van Hullebusch *et al.*, 2003) and influence the ecophysiological rates of the microorganisms (Messerli *et al.*, 2005) when compared to microorganisms growing in a planktonic form (Macfie *et al.*, 1994). Knowledge of biofilm structure and function has grown significantly in the last few years due to advances in microscopic techniques such as laser scanning confocal (Lawrence *et al.*, 1991) and cryoscanning (Stal and Défarge, 2005) microscopy. However, most of these studies have been carried out using bacterial biofilms and very few studies have explored the structural features of eukaryotic biofilms.

Eukaryotic microbial communities found in acidic environments are usually distributed in extensive biofilms, and their structure and composition have been mainly studied in Río Tinto (SW, Spain). Although the development of biofilms in aquatic and terrestrial extreme environments has been documented (García-Meza *et al.*, 2005), the distribution pattern and factors controlling their development in acidic extreme environments are not yet well understood (Ferris *et al.*, 1989). In the case of the Río Tinto, the distribution of these photosynthetic communities seems to be more influenced by the presence of heavy metals than low pH (Aguilera *et al.*, 2006a). Metal toxicity depends on their bioavailability, and this accessibility depends mainly on physicochemical factors, on the presence of chelators and on the concentration of free ions in the water (Sunda and Huntsman, 1998). In general, at acidic pH most heavy metals are in a soluble cationic form, thus more easily available to microorganisms (Ernst, 1998).

The role of EPS in heavy metal absorption in acidic environments

The biofilm matrix is a dynamic environment that organizes microbial cells (Sutherland, 2001). Their major components, besides water, are extracellular polymeric substances (EPS). The EPS consists of a complex mixture of proteins, carbohydrates, acid polysaccharides, lipids, DNA and humic substances, although the wide range of environments in which biofilms are found, makes it extremely difficult to generalize about their structure and physiological activities (Sutherland, 2001). The adsorption of cationic transition metals by EPS is attributed to their large number of negatively charged functional groups such as carboxyl, phosphate and sulfate at neutral pH (Kaplan *et al.*, 1987). However, the heavy metal complexing properties of EPS may be altered by pH, since pH determines metal solubility and also the organic ligand adsorption capacities (Stone, 1997). In this regard, low pH modifies the ionic status of the different EPS functional groups, changing their electrochemical properties (Ferris *et al.*, 1989). Thus, at low pH the availability of negatively charged sites such as carboxylates and phosphates is greatly reduced so fewer metal cations are absorbed. The reduction of surface charge density with decreasing pH levels also serves to weaken electrostatic free energy contributions to metallic ion adsorption (Jefree and Read, 1991).

Studies carried out with eukaryotic biofilms in the Río Tinto have shown high concentrations of EPS in all samples examined, always reaching concentration of milligrams per gram of biofilm dry weight (Aguilera *et al.*, 2008), which is three orders of magnitude greater than most of the data reported in neutral microphytobenthic environments (Hirst and Jordan, 2003) (Fig. 7.2). Extensive production of EPS has also been described for other extreme environments, such as Arctic ocean ice or deep-sea hydrothermal vents (Mancuso-Nichols *et al.*, 2005; Krembs *et al.*, 2002).

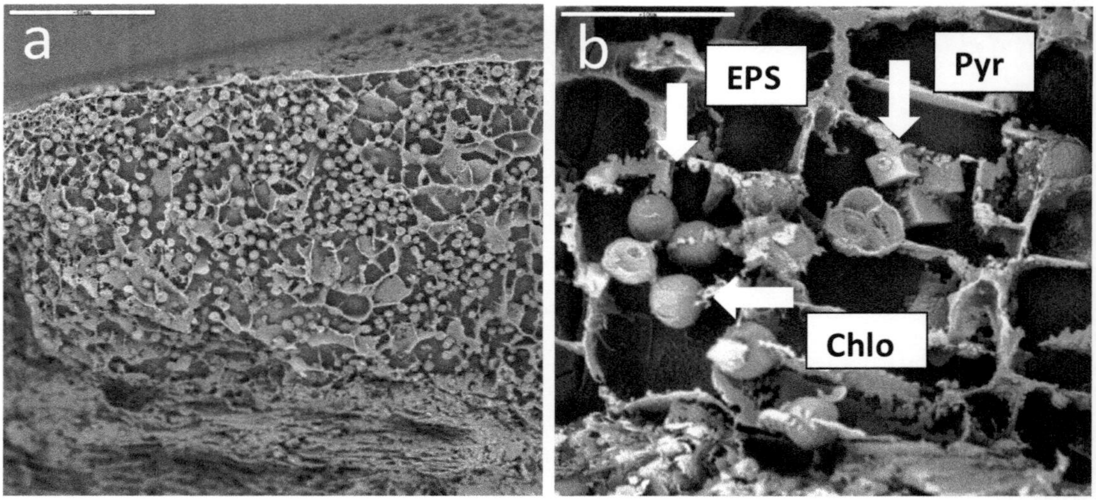

Figure 7.2 Low-temperature scanning electron microscopy images (LT-SEM) of a *Chlorella* sp. photosynthetic biofilm. This technique allows fresh frozen biofilms to be visualized; (a) general view of the biofilm (the bar = 50 µm); (b) detailed image (the bar = 10 µm). The mineralized nature of the EPS around the cells is arrowed in (b), and (arrows) and iron sulfide minerals embedded in the matrix are indicated (Pyr).

In the Río Tinto eukaryotic biofilms, total amounts of EPS are positively correlated with the total concentration of heavy metals present in the water (Aguilera *et al.*, 2008). These results are comparable to those measured in other environments (García-Meza *et al.*, 2005). In addition, the negative correlation between EPS and pH (total amount of EPS decreases as pH increases) is also significant. These results are in agreement with the well documented fact that, at low pH, the availability of negatively charged sites such as carboxylates and phosphates is greatly reduced (Flemming *et al.*, 1999). Since fewer metal cations can be adsorbed as pH decreases, more EPS is required to chelate the same amount of metals.

Ultrastructure of eukaryotic acidophilic biofilms

Many studies have investigated the physiology and structure of bacterial biofilms in order to understand the underlying processes of their formation. Analysis of the internal structure of biofilms by confocal laser scanning microscopy have shown that biofilm communities form highly structured microbial assemblies (Norton *et al.*, 1998). For acidophilic eukaryotic biofilms, the internal structure has been less studied; however scanning electron microscopy in backscattered mode has permitted the observation of transversal sections of the biofilms attached to the substrate in the Río Tinto (Aguilera *et al.*, 2006b). As mentioned previously, most of the eukaryotic microbial communities found in the Río Tinto are distributed in biofilms, and the macroscopic shape and species composition of the biofilms vary greatly throughout the river (Aguilera *et al.*, 2006a). Some of them adopt filamentous morphologies while others form thick colourful patches firmly attached to the mineral substrates. Río Tinto microbial biofilms are tridimensional structures that show a spectrum of structurally heterogeneous forms determined by the dominating photosynthetic organisms in which fungal hyphae and prokaryotic microorganisms are also found (Aguilera *et al.*, 2006a).

Microscale structural differences among naturally grown biofilms have been observed in different localities of the Río Tinto. Although some of the biofilms form a well-defined layered structure, others show several layers of cells loosely packed between minerals or decaying organic matter. Many of these biofilms are dominated by a specific photosynthetic microorganism, while most of the bacteria species were common to all of them, showing no preference for association with a specific eukaryotic species. Some biofilms, such as the ones formed by *Dunaliella* (Chlorophyta), only appear in the most acidic localities, and have a clear association with fungal communities. In addition to the vertical distribution of the same organisms, different biofilms are distributed horizontally under the same environmental conditions. In general, the surface is colonized non-uniformly, and while the factors responsible for this distributional pattern are not well known they may be related to specific growth of different microorganisms and surface characteristics (Hutchinson *et al.*, 2006). The structural study has shown the importance of metabolic interactions within the microbial communities themselves, and also between different biofilms.

These observations on biofilm ultrastructure indicate a close link between the photosynthetic communities and other components of the trophic web. Some studies support a possible role for the large biomass of photosynthetic organisms as primary producers for the prokaryotic-fungal consuming communities (Sabater *et al.*, 2003; Gross and Robbins, 2000). While the diversity of fungi found in harsh water conditions has been described (Gadanho *et al.*, 2006; Baker *et al.*, 2004), little is known about the fungal communities associated with photosynthetic biofilms. Preliminary results using fluorescence *in situ* hybridization (FISH and CARD-FISH) have shown that in some areas of the Río Tinto, the chemolithotrophic prokaryotic communities are important structural components of photosynthetic biofilms (Souza-Egipsy *et al.*, 2008).

In addition, heterotrophic prokaryotes and fungi have been found associated with photosynthetic biofilms (Aguilera *et al.*, 2006b). It has also been shown that the interaction between yeasts and chemolithotrophic bacteria stimulate ferrous iron oxidation (Wong and Gu, 2004). These interactions may be of crucial importance in extreme oxidative environments where Fe (III) reduction to Fe(II) is rapid and strongly coupled to oxidation of organic compounds in water and sediments (Lovley *et al.*, 1991). In the case of the acidic photosynthetic biofilms, synergistic associations between prokaryotic microorganisms and fungi may stimulate photosynthetic biofilm formation through mechanisms that could include improved nutrient acquisition. Future mesocosm experiments will seek to improve understanding of the functional mechanisms behind such microbial interactions.

Heavy metal resistance: functional transcriptomic approaches

In spite of their morphological simplicity, microalgae are very diverse both genetically and in their lifestyles. When exposed to environments modified by anthropological activities, organisms are forced to adapt much more rapidly than their natural rate of adaptation to changes, but the toxic effects of heavy metals on acidophilic organisms that have been exposed during long time periods might be completely different from the effects on non-exposed populations. Therefore, studies using acidophilic organisms in their natural environment provide essential information on metal toxicity and the evolution of eukaryotic resistance mechanisms to heavy metal stress.

Recent progress in genomic research on acidophilic microalgae has given new insights on the versatile metabolism of these extremophiles. In addition to being good subjects for evolutionary and ecological studies, they are potential sources for novel enzymes which might be useful for biotechnological applications and phytoremediation. Acidophiles such as *Euglena mutabilis* and *Chlamydomonas acidophila* are particularly suited for genetic studies since the genomes of their circumneutral model organisms (*E. gracilis* and *C. reinhardtii*) have also been sequenced. To have a close relative for which genetic information is available, including an annotated full-genome sequence and results from transcriptomic studies, greatly facilitates the study of a non-model organism. In addition, both extremophiles are readily cultivated in the laboratory, both in liquid and on solid media, which allows axenic colonies to be isolated, and their genetic responses to different experimental conditions to be examined.

In a recent study regarding arsenic effects on photosynthetic protists, a comparative transcriptomic analysis using the metal hyper-tolerant *E. mutabilis*, and the more sensitive species, *E. gracilis*, was carried out, revealing multiple mechanisms involved in arsenic resistance (Halter *et al.*, 2014). Previous analyses had revealed that the arsenic response of *E. mutabilis* differs markedly from that is observed in *E. gracilis* (Halter *et al.*, 2012). In particular, the accumulation of lower amounts of total arsenic in *E. mutabilis* intracellular compartment may explain its very high tolerance to this toxic metalloid. Such a phenotype results either from a reduced entry of arsenic into the cell, a higher efflux, or both.

Transcripts presenting homologies with Acr3 and an arsenite-activated ATPase related to ArsA, known as transporters and ATPases, respectively, were found in the transcriptome of *E. mutabilis* but not in that of *E. gracilis* (Halter *et al.*, 2014). These transcripts were also observed in the absence of arsenic, suggesting a constitutive expression of these defence mechanisms in *E. mutabilis*. In addition, arsenate has structural analogies with phosphate, and is therefore often incorporated into the cells via low specificity transporters of this latter element (Wysocki and Tamás, 2010). Interestingly, the transcription of genes presenting homologies with specific phosphate transporters preventing this phenomenon was found in *E. mutabilis* but not in *E. gracilis*. This observation suggests that *E. mutabilis* is able to prevent the accumulation of arsenic through low specific phosphate transporters more efficiently than *E. gracilis*. In addition, functions related to phosphonate transport were specifically found in *E. mutabilis* in the presence of arsenic but not in *E. gracilis*. Such organo-phosphoric compounds could correspond to sources other than inorganic phosphate being used in the presence of arsenic, incidentally preventing the accumulation of this toxic metalloid through low-specific phosphate transporters. Additionally, *E. mutabilis* induces the expression of active DNA reparation and protein turnover mechanisms, which allows it to maintain its functional integrity under challenging conditions, suggesting that this protist has acquired specific functions regarding arsenic and has developed an original metabolism to cope with acid mine drainage-related stresses (Halter *et al.*, 2014).

In addition, it has been suggested that acidophile metal resistance strategies do not fully explain why these microorganisms are able to grow in solutions containing highly elevated concentrations of metals. Dopson *et al.* (2014) proposed metal tolerance via complexation of free metals by sulfate ions and passive tolerance to metal influx via an internal positive cytoplasmic transmembrane potential. Acidophilic microorganisms often grow in environments containing high concentrations of sulfate ions that can complex metal cations at acidic pH values. Therefore, the concentrations of free ions that can enter the cytoplasm and consequently challenge acidophiles are significantly lower than the total concentration of the metal. As a consequence, it is possible that extreme metal tolerance in acidophiles is partially a function of free metal ion complexation by sulfate that precludes the metal ion entry into the cell (Dopson *et al.*, 2014). Additionally, acidophiles have an inside positive cytoplasmic transmembrane potential that is reinforced by the dipole potential. This helps to maintain pH homeostasis as protons must travel against a chemiosmotic gradient to enter the cytoplasm (Slonczewski *et al.*, 2009; Baker-Austin and Dopson, 2007). The transmembrane and dipole potentials also provide a hindrance to metal cation influx. If this barrier is sufficiently strong then only a portion of the available free metal ion will penetrate the cell and cause a toxicity response (Dopson *et al.*, 2014).

Photosynthesis in acidic environments

The establishment and maintenance of phototrophic protist biofilms in acidic waters relies on the optimal use of light resources for photosynthesis during the daylight hours, a process particularly sensitive to stressful environmental conditions, such as salinity, pH or the presence of toxic chemicals. There are relatively few reports on photosynthesis in acidic environments in the literature, and most have focused on primary productivity measurements in acidic lakes. It has been reported that low primary productivity is mainly due to metal stress or the availability of phosphate (Spijkerman *et al.*, 2007), when low pH itself does not reduce photosynthetic activity (Gyure *et al.*, 1987). Light is another limiting factor for primary productivity in acidic environments, due to the fact that ferric iron has a strong orange-red pigmentation and large therefore large concentrations of this form of iron can cause significant quantitative and qualitative changes in the light environment in the water column (Souza-Egipsy *et al.*, 2011). Adaptation to low light intensities has been reported for benthic biofilms of diatoms in acidic lakes (Koschorreck and Tittle, 2002) as well as for *E. mutabilis* biofilms in acidic rivers (Souza-Egipsy *et al.*, 2011).

Photo-adaptive capacities related to environmental microhabitat conditions of phototrophic biofilms in the Río Tinto have been examined by using oximetric approaches (Souza-Egipsy *et al.*,

2011). Oximetry allows the calculation of photosynthesis–irradiance curves (P–I curves), which provides a convenient and objective mean for distinguishing between light-limited and light-saturated photosynthesis. The shape and magnitude of the P-I curves reflects the underlying biophysical, biochemical and metabolic processes that regulate photosynthesis (Falkowski and Raven, 1997), being able to estimate the main photosynthetic parameters such as maximum photosynthesis (P_{max}), compensating irradiance (I_c), maximum irradiance (I_{max}), photosynthetic efficiency (α) or photoinhibition rate (β).

In the Río Tinto, all the biofilms analysed (biofilms formed mainly by *Chlorella* spp., *Euglena* spp., *Pinnularia* spp. and *Zygnemopsis* spp.) showed a strong photoinhibition behaviour in their P–I curves except for the *Zygnemopsis* sp. biofilm, which exhibited a photo-saturated P–I curve behaviour (Souza-Egipsy *et al.*, 2011). Generally, photoinhibition in P-I curves is commonly observed in shade-adapted organisms that cannot adequately dissipate the excessive photon flux provided by a high irradiance (Platt *et al.*, 1980). In the Rio Tinto, the biofilms exhibiting photo-inhibition are usually located at the bottom of the riverbed, covered by several centimetres of highly coloured red water. However, *Zygnemopsis* sp. is a filamentous alga usually found at the water surface during the summer, when solar irradiation is extremely high, and in this way can be considered a high light-adapted species. In addition, the analysed species can be considered as low-light or shade-adapted organisms due to their low Ic and Ik values (ca. 20 and 50 µmol photons/m^2/s respectively), which may be related to the fact that they develop under highly coloured waters that affect quantitatively and qualitatively the light available for phototrophic organisms. Photosynthetic activity has been detected even at irradiances as low as 5 µmol photons/m^2/s in the case of the diatom *Pinnularia* spp. These results are congruent with previous data from sediments of acidic lakes, where photosynthetic ability at low light intensities ($< 1.2\,\mu E\ m^2/s$) were found in a benthic biofilm of diatoms suggesting an efficient absorption of red light, the dominant wavelength available in these iron-rich acidic waters, by these organisms (Koschorreck and Tittle, 2002). Maximum photosynthesis values (P_{max}) were also low in comparison with other environments (ca. 25 mg O_2/mg Chl*a*/h), in which rates higher than 200 mg O_2/mg Chl*a*/h are usually reached (Harting *et al.*, 1998). In acidic mining lakes, planktonic primary productivity is usually low probably due to the low phytoplankton biomass (Nixdorf *et al.*, 2003) and the lack of nutrients such as ammonium- and nitrate-N or phosphate (Falagan *et al.*, 2007).

Additionally, all the photosynthetic parameters analysed have shown statistical significant differences among species and sampling locations (Souza-Egipsy *et al.*, 2011). Thus, the *Euglena* biofilms isolated from different locations of Río Tinto showed different photosynthetic values despite being mainly formed by the same phototrophic species. These results could be explained by photo-adaptation processes instead of photo-acclimation procedures (Falkowski and La Roche, 1991). Photo-adaptation refers to changes in the genotype that arise either from mutations or from changes in the distribution of alleles within a gene pool, while photo-acclimation refers to phenotypic adjustments that arise in response to variations of environmental factors. Besides nutrient limitation, the presence of elevated concentrations of heavy metals could also explain the low values of maximum photosynthetic activity, since photosynthesis is one of the most sensitive cellular responses to metal stress conditions (Spijkerman *et al.*, 2007; Hanikenne, 2003). Thus, a significant decrease in the amount of some photosynthetic proteins (RuBisCo, cytrochrome C peroxidase or Photosystem I (PSI) 11 K protein precursor) has been reported in an isolate of *Chlamydomonas* spp. from the Río Tinto when the cells were grown in natural metal-rich water from the river (Cid *et al.*, 2010).

In summary, although considerable work is yet to be done to understand the photosynthetic metabolism under coloured water in acidic environments, the results found in the Rio Tinto have shown that low light adaptation of the phototrophic species could be an important determinant in the competitive success for colonizing these environments.

Overview and future prospects

Most eukaryotes found in extreme environments are unicellular and the main obstacle to their study is the lack of reliable cultivation methods. Recent advances in DNA sequencing technology and statistical analyses have revolutionized the way inferences about the biodiversity and biology of organisms can be made. Molecular ecology is an exciting and expanding area that poses a wide array of questions on biodiversity, ecology, physiology, evolution, conservation and management that molecular biology approaches can address. Although these techniques have revealed an unexpectedly high eukaryotic phylogenetic diversity in all kinds of environments, their impact is even more evident in the field of extreme environments, which are usually more difficult to reproduce in the laboratory. Researchers are now capable of dealing with important questions about microbial diversity that could not be realistically formulated even a few years ago.

References

Aguilera, A., and Amils, R. (2005). Tolerance to cadmium in *Chlamydomonas* sp. (Chlorophyta) strains isolated from an extreme acidic environment, the Tinto River (SW, Spain). Aquat. Toxicol. 75, 316–329.

Aguilera, A., Manrubia, S.C., Gómez, F., Rodríguez, N., and Amils, R. (2006a). Eukaryotic community distribution and its relationship to water physicochemical parameters in an extreme acidic environment, Rio Tinto (SW, Spain). Appl. Environ. Microbiol. 72, 5325–5330.

Aguilera, A., Souza-Egipsy, V., Gómez, F., and Amils, R. (2006b) Development and structure of eukaryotic biofilms in an extreme acidic environment, Río Tinto (SW, Spain). Microb. Ecol. 53, 294–305.

Aguilera, A., Souza-Egipsy, V., San Martín-Úriz, P., and Amils, R. (2008). Extracellular matrix assembly in extreme acidic eukaryotic biofilms and their possible implications in heavy metal adsorption. Aquat. Toxicol. *88*, 257–266.

Amaral-Zettler-Zettler, L., Zettler, E.R., Theroux, S., Palacios, C., Aguilera, A., and Amils, R. (2011). Microbial community structure across the tree of life in the extreme Río Tinto. ISME J. 5, 42–50.

Amils, R., de la Fuente, V., Rodriguez, N., Zuluaga, J., Menendez, N., and Tornero, J. (2007). Composition, speciation and distribution of iron minerals in *Imperata cylindrica*. Plant Physiol. Biochem. 45, 335–340.

Armstrong, G.M. (1981). Studies in the physiology of the fungi-sulfur nutrition, the use of thiosulfate as influenced by hydrogen ion concentration. Ann. Missouri Bot. Garden. *8*, 237–248.

Baffico, G.D., Díaz, M.M., Wenzel, M.T., Koschorreck, M., Schimmele, M., Neu, T.R., and Pedrozo, F. (2004). Community structure and photosynthetic activity of epilithon from a highly acidic (pH<2) mountain stream in Patagania, Argentina. Extremophiles 8, 465–475.

Baker, B.J., Lutz, M.A., Dawson, S.C., Bond, P.L., and Banfield, J.F. (2004). Metabolically active eukaryotic communities in extremely acidic mine drainage. App. Environ. Microbiol. 70, 6264–6271.

Baker-Austin, C., and Dopson, M. (2007). Life in acid: pH homeostasis in acidophiles. Trends Microbiol. *15*, 165–171.

Bell, E.M., and Weithoff, G. (2003). Benthic recruitment of zooplankton in an acidic lake. J. Exp. Mar. Biol. Ecol. *285–286*, 205–219.

Caron, D.A., Countway, P.D., and Brown, M.V. (2004). The growing contributions of molecular biology and immunology to protistan ecology: molecular signatures as ecological tools. J. Euk. Microbiol. *51*, 38–48.

Cid, C., Garcia-Descalzo, L., Casado-Lafuente, V., Amils, R., and Aguilera, A. (2010). Proteomic analysis of the response of an acidophilic strain of *Chlamydomonas* sp. (Chlorophyta) to natural metal-rich water. Proteomics *10*, 2026–2036.

Costerton, J.W., Cheng, K.J., Geesey, G.G., Ladd, T., Nickel, J.C., Dasgupta, M., and Marrie, T.J. (1987). Bacterial biofilms in nature and disease. Ann. Rev. Microbiol. *45*, 1921–1931.

Deneke, R. (2000). Review of rotifers and crustaceans in highly acidic environments of pH<3. Hydrobiologia *433*, 176–172.

Dopson, M., Ossandon, F.J., Lövgren, L., and Holmes, D.S. (2014). Metal resistance or tolerance? Acidophiles confront high metal loads via both abiotic and biotic mechanisms. Front. Microbiol. *5*, 1–4.

Duran, C., Marin, I., and Amils, R. (1999). Specific metal sequestering acidophilic fungi. In Biohydrometallurgy and the Environment, Amils, R., and Ballester, A., eds (Elsevier, Amsterdam, the Netherlands), pp. 234–251.

Ernst, W.H.O. (1998). Effects of heavy metals in plants at cellular and organismic level. In Ecotoxicology, Schürmann, G., and Marker, B., eds (John Wiley and Sons Inc., New York, NY, USA), pp. 587–620.

Falagan, C., Sanchez-España, J., and Johnson, D.B. (2014). New insights into the biogeochemistry of extremely acidic environments revealed by a combined cultivation-based and culture-independent study of two stratified pit lakes. FEMS Microb. Ecol. *87*, 231–243.

Falkowski, P.G., and La Roche, J. (1991). Acclimation to spectral irradiance in algae. J. Phycol. *27*, 8–14.

Falkowski, P.G., and Raven, J.A. (1997). Aquatic Photosynthesis (Blackwell Science, Malden, MA, USA).

Ferris, F.G., Schultze, S., Witten, T.C., Fyfe, W.S., and Beveridge, T.J. (1989). Metal interactions with microbial biofilms in acidic and neutral pH environments. Appl. Environ. Microbiol. *55*, 1249–1257.

Flemming, H.C., Wingender, J., Moritz, R., Borchard, W., and Mayer, C. (1999). Physicochemical properties of biofilms- a short review. In Biofilms in the Aquatic Environment, Keevil, C.W., Goodfree, A., Holt, D., and Dow, C., eds, Royal Society of Chemistry Special Publication No. 242 (Royal Society of Chemistry, Cambridge, UK), pp. 1–12.

Fuente, V., Rodrıguez, N., and Amils, R. (2012). Immunocytochemical analysis of the subcellular distribution of ferritin in *Imperata cylindrical* (L.) Raeuschel, an iron hyperaccumulator plant. Acta Histochem. *114*, 232– 236.

Gadanho, M., and Sampaio, J.P. (2009). *Cryptococcus ibericus* sp. nov., *Cryptococcus aciditolerans* sp. nov. and *Cryptococcus metallitolerans* sp. nov., a new ecoclade of anamorphic basidiomycetous yeast species from an extreme environment associated with acid rock drainage in Sao Domingos pyrite mine, Portugal. Int. J. Syst. Evol. Microbiol. *59*, 2375–2379.

Gadanho, M., Libkind, D., and Sampaio, J.P. (2006). Yeast diversity in the extreme acidic environments of the Iberian Pyrite Belt. Microb. Ecol. *2*, 552–563.

García-Meza, J.V., Barrangue, C., and Admiraal, W. (2005). Biofilm formation by algae as a mechanism for surviving on mine tailings. Environ. Toxicol. Chem. *24*, 573581.

Gimmler, H., Weis, U., Weiss, C., Kugel, H., and Treffny, B. (1989). *Dunaliella acidophila* (Kalina) Masyuk-an alga with a positive membrane potential. New Phytol. *113*, 175–184.

Gimmler, H., Treffny, B., Kowalski, M., and Zimmermann, U. (1991). The resistance of *Dunaliella acidophila* against heavy metals: the importance of the zeta potential. Plant Physiol. *138*, 708–716.

Gross, S., and Robbins, E.I. (2000). Acidophilic and acid-tolerant fungi and yeasts. Hydrobiologia *433*, 91–109.

Guyre, R.A., Konopka, A., Brooks, A., and Doemel, W. (1987). Algal and bacterial activities in acidic strip mine lakes. Appl. Environ. Microbiol., 53, 2069–2076.

Halter, D., Goulhen-Chollet, F., Gallien, S., Casiot, C., Hamelin, J., Gilard, F., Heintz, D., Schaeffer, C., Carapito, C., Van Dorsselaer, A., *et al.* (2012). *In situ* proteo-metabolomics reveals metabolite secretion by the acid mine drainage bio-indicator, *Euglena mutabilis*. ISME J. *6*, 1391–1402.

Halter, D., Andres, J., Plewniak, F., Poulain, J., Da Silva, C., Arsène-Ploetze, F., and Bertin, P.N. (2014). Arsenic hypertolerance in the protist *Euglena mutabilis* is mediated by specific transporters and functional integrity maintenance mechanisms. Environ. Microbiol. *17*, 1941–1949.

Hanikenne, M. (2003). *C. reinhardtii* as a eukaryotic photosynthetic model for studies of heavy metal homeostasis and tolerance. New Phytol. *159*, 331–340.

Harting, P., Wolfstein, K., Lippemeier, S., and Colijn, F. (1998). Photosynthetic activity of natural microbenthos populations measured by fluorescence PAM and 14C-tracer: a comparison. Mar. Ecol. Prog. Ser. *166*, 53–62.

Hirst, C.N., and Jordan, I.A. (2003). Distribution of exopolymeric substances in the littoral sediments of an oligotrophic lake. Microb. Ecol. *46*, 22–32.

Hujslová, M., Kubátová, A., Kostovčík, M., Blanchette, R.A., de Beer, Z.W., Chudíčková, Z.M., and Kolařík, M. (2014). Three new genera of fungi from extremely acidic soils. Mycol. Progress *13*, 819–831 doi: 10.1007/s11557-014-0965-3.

van Hullebusch, E.D., Zandvoort, M.H., and Lens, P.N.L. (2003). Metal immobilisation by biofilms: mechanisms and analytical tools. Rev. Environ. Sci. Bio.Technol. *2*, 9–33.

Hutchinson, N., Nagarkar, S., Aitchison, J.C., and Williams, G.A. (2006). Microspatial variation in marine biofilm abundance on intertidal rock surfaces. Aquat. Microb. Ecol. *42*, 187–197.

Jefree, C.E., and Read, N.D. (1991). Ambient and low-temperature scanning electron microscopy. In Electron Microscopy of Plant Cells, Hall, J.L., and Hawes, C., eds (Academic Press, London, UK), pp. 313–413.

Johnson, D.B. (1998). Biodiversity and ecology of acidophilic microorganisms. FEMS Microbial. Ecol. *27*, 307–317.

Joseph, J.M. (1953). Microbiological study of acid mine waters; preliminary report. Ohio J. Sci. *53*, 123–127.

Kaplan, D., Christiaen, D., and Arad, S. (1987). Binding of heavy metals by algal polysaccharides. In Algal Biotechnology, Stadler, J., Mollion, M.C., Verdus, Y., Karamanos, H., Moran, D., and Christiaen, D., eds (Elsevier Applied Science, London, UK), pp. 179–187.

Koschorrek, M., and Tittel, J. (2002). Benthic photosynthesis in acidic mining lake (pH. 2.6). Limnol. Oceanog. *47*, 1197–1201.

Krembs, C., Eicken, H., Junge, K., and Deming, J.W. (2002). High concentrations of exopolymeric substances in Arctic winter sea ice: implications for the polar ocean carbon cycle and cryoprotection of diatoms. Deep-Sea Res. *49*, 2163–2181.

Lawrence, J.R., Korber, D.R., Hoyle, B.D., Costerton, J.W., and Caldwell, D. (1991). Optical sectioning of microbial biofilms. J. Bacteriol. *173*, 6558–6567.

Lovley, D.R., Phillips, E.J.P., and Lonergan, D.J. (1991). Enzymatic versus noenzymatic mechanisms for Fe(III) reduction acuatic sediments. Environ. Sci. Technol. *25*, 1062–1067.

Macfie, S., Tarmohamed, Y., and Welbourn, P.M. (1994). Effects of cadmium, cobalt, copper, and nickel on growth of the green alga, *Chamydomonas reinhardtii*: the influences of the cell wall and pH. Arch. Environ. Contam. Toxicol. *27*, 454–458.

Mancuso-Nichols, C.A., Guezennec, J., and Bowman, J.P. (2005). Bacterial exopolysaccharides from extreme marine environments with special consideration of the Southern Ocean, Sea Ice, and Deep-Sea hydrothermal vents: A review. Mar. Biotechnol. *7*, 253–271.

Messerli, M.A., Amaral-Zettler-Zettler, L.A., Zettler, E., Junh, S.K., Smith, P.J., and Sogin, L.M. (2005). Life at acidic pH imposes an increased energetic cost for a eukayotic acidophile, J. Exp. Biol. *208*, 2569–2579.

Moser, M., and Weisse, T. (2011a). The most acidified Austrian lake in comparison to a neutralized mining lake. Limnologica *41*, 303–316.

Moser, M., and Weisse, T. (2011b). Combined stress effect of pH and temperature narrows the niche width of flagellates in acid mining lakes. J. Plank. Res. 33, 1023–1032.

Nishikawa, K., and Tominaga, N. (2001). Isolation, growth, ultrastructure, and metal tolerance of the green alga, *Chlamydomonas acidophila* (Chlorophyta). Biosci. Biotechnol. Biochem. *65*, 2650–2656.

Nixdorf, B., Mischke, U., and Lessmann, D. (1998). Chrysophytes and chlamydomonads: pioneer colonists in extremely acidic mining lakes (pH<3) in Lusatia (Germany). Hydrobiologia *369/370*, 315–327.

Nixdorf, B., Krumbeeck, H., Jander, J., and Beulker, C. (2003). Comparison of bacterial and phytoplankton productivity in extremely acidic mining lakes and eutrophic hard water lakes. Acta Oecologica *24*, 281–288.

Nordstrom, D.K., and Southam, G. (1997). Geomicrobiology of sulfide mineral oxidation, In Geomicrobiology: Interactions Between Microbes and Minerals, Vol. 35, Banfield, J.F., and Nealson, K.H., eds

(Mineralogical Society of America, Washington, DC, USA), pp. 361–390.

Norton, T.A., Thompson, R.C., Pope, J., Veltkamp, C.J., Banks, B., Howar, C.V., and Awkns, H. (1998). Using confocal laser scanning microscopy, scanning electron microscopy and phase contrast light microscopy to examine marine biofilms. Aquatic Microb. Ecol. *16*, 199–204.

Oliva, S.R., Mingorance, M.D., and Leidi, E.O. (2012). Tolerance to high Zn in the metallophyte *Erica andevalensis* Cabezudo & Rivera. Ecotoxicology *21*, 2012–2021.

Packroff, G. (2000). Protozooplankton in acidic mining lakes with special respect to ciliates. Hydrobiologia *433*, 157–166.

Platt, T., Gallegos, C.L., and Harrison, W.G. (1980). Photoinhibition of photosynthesis in natural assemblages of marine phytoplankton. J. Mar. Res. *38*, 687–701.

Reeb, V., and Bhattacharya, D. (2010). The thermo-acidophilic *Cyanidiophyceae* (Cyanidiales). In Red Algae in the Genomic Age, Seckbach, J., and Chapman, D.J., eds (Springer Science and Business Media, Dordrecht, Netherlands), pp. 409–426.

Roberts, D.M.L. (1999). Eukaryotic cells under extreme conditions. In Enigmatic Microorganisms and Life in Extreme Environments, Seckbach, J., ed. (Kluwer Academic Publishers, Dordrecht, Netherlands), pp. 165–173.

Rodriguez, N., Menendez, N., Tornero, J., Amils, R., and de la Fuente, V. (2005). Internal iron biomineralization in *Imperata cylindrica*, a perennial grass: chemical composition, speciation and plant localization. New Phytol. *165*, 781–789.

Russo, G., Libkind, D., Sampaio, J.P.,and VanBrock, M.R. (2008). Yeast diversity in the acidic Río Agrio-Lake Caviahue volcanic environment (Patagonia, Argentina). FEMS Microbiol. Ecol. *65*, 415–424.

Sabater, S., Buchaca, T., Cambra, J., Catalan, J., Guasch, H., Ivorra, N., Muñoz, I., Navarro, E., Real, M., and Romani, A. (2003). Structure and function of benthic algal communities in an extremely acid river. Phycology *39*, 481–489.

Schönknecht, G., Chen, W.H., Ternes, C.M., Barbier, G.G., Shrestha, R.P., Stanke, M., Bräutigam, A., Baker, B.J., Banfield, J.F., Garavito, R.M., *et al.* (2013). Gene transfer from bacteria and archaea facilitated evolution of an extremophilic eukaryote. Science *339*, 1207–1210.

Seckbach, J. (1999). Enigmatic Microorganisms and Life in Extreme Environments (Kluwer Academic Publishers, Dordrecht, Netherlands).

Sekler, I., Glaser, H.U., and Pick, U. (1991). Characterization of a plasma membrane H^+-ATPase from the extremely acidophilic alga *Dunaliella acidophila*. J. Membr. Biol. *121*, 51–57.

Sittenfeld, A., Mora, M., Ortega, J.M., Albertazzi, F., Cordero, A., and Roncel, M. (2002). Characterization of a photosynthetic *Euglena* strain isolated from an acidic hot mud pool of a volcanic area of Costa Rica. FEMS Microbiol. Ecol. *42*, 151–161.

Slonczewski, J.L., Fujisawa, M., Dopson, M., and Krulwich, T.A. (2009). Cytoplasmic pH measurement and homeostasis in bacteria and archaea. Adv. Microb. Physiol. *55*, 1–79.

Souza-Egipsy, V., González-Toril, E., Zettler, E., Amaral-Zettler-Zettler, L., Aguilera, A., and Amils, R. (2008). Prokaryotic community structure in algal photosynthetic biofilms from extreme acidic streams in Río Tinto (southwestern Spain). Int. Microbiol. *11*, 251–260.

Souza-Egipsy, V., Altamirano, M., Amils, R., and Aguilera, A. (2011). Photosynthetic performance of phototrophic biofilms in extreme acidic environments. Environ. Microbiol. *13*, 2351–2358.

Spijkerman, E., Barua, D., Gerloff-Elias, A., Kern, J., Gaedke, U., and Heckathorn, S.A. (2007). Stress responses and metal tolerance of *Chlamydomonas acidophila* in metal-enriched lake water and artificial medium. Extremophiles *11*, 551–562.

Stal, L.J., and Défarge, C. (2005). Structure and dynamics of exopolymers in an intertidal diatom biofilm. Geomicrobiol. J. *22*, 341–352.

Stone, A.T. (1997). Reactions of extracellular organic ligands with dissolved metal ions and mineral surfaces. In Geomicrobiology: Interactions Between Microbes and Minerals, Banfield, J.F., and Nealson, K.H., eds (Mineralogical Society of America, Washington, DC, USA), pp. 309–334.

Sunda, W.G., and Huntsman, S.A. (1998). Processes regulating cellular metal accumulation and physiological effects: phytoplankton as model systems. Sci. Total. Environ. *219*, 165–181.

Sutherland, I.W. (2001). The biofilm matrix- an immobilized but dynamic microbial environment. Trends Microbiol. *9*, 222–227.

Weisse, T., Moser, M., Scheffel, U., Stadler, P., Berendonk, T., Weithoff, G., and Berger, H. (2013). Systematics and species-specific response to pH of *Oxytricha acidotolerans* sp. nov. and *Urosomoida* sp. (Ciliophora, Hypotricha) from acid mining lakes. Eur. J. Protistol. *49*, 255–271.

Wong, J.W.C., and Gu, X.Y. (2004). Enhanced heavy metal bioleaching efficiencies from anaerobically digested sewage sludge with coinoculation of *Acidithiobacillus ferrooxidans* ANYL-1 and *Blastoschizomyces capitatus* Y5. Water Sci. Technol. *50*, 83–89.

Wysocki, R., and Tamás, M.J. (2010). How *Saccharomyces cerevisiae* copes with toxic metals and metalloids. FEMS Microbiol. Rev. *34*, 925–951.

Part IV

Acidophile Community Dynamics

Microbial Communities and Interactions in Low-pH Environments

8

D. Barrie Johnson

Abstract

Most of the types of interactions that occur between microorganisms in environments considered to be non-extreme have also been observed in those which are extremely acidic. These include interactions that are both positive (mutualism, commensalism, synergy and syntrophy) and negative (competition, amensalism, predation and pathogenicity) from the perspective of one or more of the co-habiting microorganisms. Carbon flow between autotrophic (chemolithotrophs and phototrophs) and heterotrophic acidophiles is now known to operate in both directions and can help sustain stable consortia of these prokaryotes. By catalysing redox transformations of iron and sulfur, bacteria and archaea can generate electron donors and acceptors used by other species of acidophiles and thereby induce rapid biogeochemical cycling of these elements, a key characteristic of many low-pH environments. Examples of relatively simple and more complex communities that have been studied in acidic streams, pit lakes and anthropogenic (biomining) environments are used to illustrate the importance of microbial interactions at low pH.

Introduction

The English poet John Donne wrote that 'no man is an island', expressing the reality that it is nearly impossible for a human individual to live without in some way interacting with others. The same is true in the in the microbial world. Bacteria, archaea and eukaryotic microorganisms live, sometimes in very close proximity, with others of their own species and with different species, and within these communities they can exhibit a variety of interactions which may have positive, negative or nil effects on their neighbours. The structures of microbial communities and interactions within them have been major areas of microbiological research for many decades. This chapter outlines what has emerged from those studies that have focused on extremely acidic environments, and includes examples of the different types of interactions that have been identified and how these impact the geochemistry of natural and anthropogenic environments. The topic of how acidophiles communicate with each other is dealt with in Chapter 10.

Types of microbial interactions that occur in low-pH environments

The terminologies used to describe the different forms of microbial interactions that have been observed in low-pH environments are listed and defined below. Some, such as grazing of acidophilic bacteria by phagotrophic protozoa, are readily observed, though more often interactions are more clandestine. The list includes most of the types of microbial interactions that have been observed in other environments with one or two exceptions, such as parasitism (one species deriving substrates or nutrients from a living host) which has yet to be confirmed at extremely low pH, though Baker *et al.* (2010) have suggested that some archaeal associations observed in samples from the Richmond Mine at Iron Mountain, California (Chapter 13), may involve parasitism of one group by another. It should also be noted that microbial associations can also be defined in terms of dependency (facultative of obligate), duration (transient or long-term) and specificity (unique or general). The term symbiosis (from the Greek for 'living together') has been used in some contexts to imply that both (or more) partners involved in the association derive some benefit, though one of the earliest definitions (which will be adhered to here) is that it refers in more general terms to 'the living together of unlike organisms.' Microbial interactions observed in low-pH environments include:

1 *Neutral symbioses,* in which neither partner derives any positive or negative effect from sharing an environmental niche, and which tend to be relatively rare.
2 *Positive symbioses,* which are those where one or more partners derive some benefit from sharing an environmental niche. In the most extreme cases, the relationship may be such that one or both of the partners requires the other to remain active and viable (precluding, for example, their growth as pure cultures, though such associations have yet to be identified in extremely acidic environments). Positive symbioses can be further divided into: (a) mutualistic interactions, where both partners derive benefit from their association; (b) commensalism (or facilitation), where only one of the partners is perceived to derive benefit; (c) synergistic interactions that result in the complimentary activities of both (or all) participants being more efficient in, for example, degrading a substrate, than the individual species working alone; (d) syntrophy, where the degradation of a substrate by one species is only made thermodynamically possible through the removal of an end product by another species.
3 *Negative symbioses* (also referred to antagonism), in which the activities of one or both partners sharing an environmental niche are impaired or suffer more serious negative impact. Again these comprise several types of interaction: (a) competition, where populations of co-habiting species are limited by their communal requirement of an electron donor, electron acceptor or inorganic nutrient (nitrogen, phosphorus, carbon dioxide etc.) which is present in limited quantities. This is the most widespread interaction in extremely acidic as in other environments; (b) amensalism, in which one species can repress another, e.g. via the production of toxin(s); (c) predation, where one organism is ingested by another.

Microbial interactions involving carbon flow in extremely acidic environments

As elsewhere in the biosphere, microbial communities in extremely acidic environments depend on primary producers that fix inorganic carbon. However, while solar energy is an important energy source for primary production in some low-pH sites, the energy derived from the oxidation of inorganic electron donors is the main or sole driver of carbon fixation in many others, such as subterranean water bodies, and within bioheaps and stirred tanks used in biomining operations. Table 8.1 lists some species of autotrophic acidophiles that have been described. Microalgae are the dominant phototrophs in extremely acidic environments (Chapter 7), and cyanobacteria (and anaerobic phototrophic bacteria) either rare or absent. Acidophilic chemo-autotrophs include many different species of bacteria and archaea. These derive the energy required to fix CO_2 from oxidizing ferrous iron, reduced forms of sulfur (including elemental sulfur) or hydrogen. Some species use only one of these electron donors (e.g. *Leptospirillum* spp., which oxidize only ferrous iron, and *Acidithiobacillus thiooxidans,* which oxidizes only reduced sulfur), some use two (e.g. *Acidithiobacillus caldus,* which oxidizes both reduced sulfur and hydrogen, and *Sulfolobus metallicus,* which oxidizes ferrous iron and reduced sulfur) while other use all three (such as *Acidithiobacillus ferrooxidans, Acidithiobacillus ferridurans* and *Sulfobacillus thermosulfidooxidans*). While the oxidation of ferrous iron is, on thermodynamic grounds, necessarily coupled to oxygen reduction in low-pH environments, oxidation of reduced sulfur and hydrogen can also be coupled to ferric iron reduction, and occur in anoxic acidic environments.

Heterotrophic acidophiles include bacteria (e.g. *Acidiphilium* and *Acidocella* spp.) and archaea (e.g. *Ferroplasma* and *Thermoplasma* spp.) that use mostly soluble low-molecular-weight organic compounds, and phagotrophic acidophilic protozoa (e.g. *Eutreptia* sp.) and rotifera (e.g. *Elosa woralii*) that graze smaller acidophiles, such as bacteria. While many acidophilic prokaryotes are obligately heterotrophic, others use organic carbon preferentially but will switch to an autotrophic mode when suitable compounds are unavailable. Examples of the latter are *Acidiphilium acidophilum, Sulfobacillus*

Table 8.1 Examples of phototrophic and chemolithotrophic autotrophs* that have been reported in extremely acidic environments

Phototrophs	Chemolithotrophs
Chlorophytes	Iron-oxidizers
Chlamydomonas acidophila	*Leptospirillum* spp.
Chlorella spp.	'*Fv. myxofaciens*'
Dunaliella acidophila	Sulfur-oxidizers
Rhodophytes	*At. thiooxidans*
Cyanidioschyzon merolae	*At. albertensis***
Diatoms	Sulfur/hydrogen-oxidizers
Eunotia spp.	*At. caldus*
Pinnularia spp.	*Hydrogenobaculum acidophilum*
Euglenoids	Iron/sulfur-oxidizers
Euglena mutabilis	*At. ferrivorans****
Euglena gracilis	*Acidiferrobacter thiooxydans*
	Sulfolobus metallicus
	Acidihalobacter prosperus
	Iron/sulfur/hydrogen-oxidizers
	At. ferrooxidans
	At. ferridurans
	At. ferriphilus

*Obligate autotrophs; **hydrogen oxidation status unclear; ***most species do not oxidize hydrogen.

spp. and *Acidianus brierleyi*. The term 'mixotrophy' has sometimes been used to describe this, and also for prokaryotes that use inorganic electron donors and organic carbon (as C-source). To avoid ambiguity, the term 'facultative autotroph' is better used for species, like those just mentioned, that can use both organic and inorganic carbon.

The first evidence for mutualistic interactions involving carbon flow between different species of acidophilic bacteria came from the isolation of *Thiobacillus acidophilus* (later renamed *Acidiphilium acidophilum*) from a supposedly pure culture of (*Acidi*)*thiobacillus ferrooxidans* grown on ferrous iron liquid medium (Guay and Silver, 1974). *A. acidophilum* is a facultative autotroph, but since it does not use ferrous iron as an electron donor the inference was that it was using some of the small molecular weight substances that accumulate as exudates from living cells and lysates of moribund and dead cells of chemolithotrophic acidophiles. Schnaitman and Lundgren (1965) had previously reported that '*Ferrobacillus ferrooxidans*' (now the type strain of *At. ferrooxidans*) released a large proportion of the CO_2 that it fixed into its growth medium, and identified pyruvic acid as one of the small molecular weight organic compounds. The obligately heterotrophic acidophile, *Acidiphilium cryptum*, which was isolated from various mine sites was so named by Harrison (1981) because of its tendency to grow 'hidden' in cultures of acidophilic chemolithotrophs. Johnson and Kelso (1983) reported that two of the four supposedly axenic cultures of *At. ferrooxidans* maintained by the National Collections of Industrial and Marine Bacteria (UK) were contaminated with *Acidiphilium*-like heterotrophic bacteria. Ňancucheo and Johnson (2010) found that glycolic acid was produced as a metabolic waste product by three species of biomining bacteria (*At. ferrooxidans*, *Acidithiobacillus caldus* and *Leptospirillum ferriphilum*), and accounted for up to 24% of the total dissolved organic carbon (DOC) found in cell-free media of pure cultures. Only some strains of *Firmicutes* (chiefly *Sulfobacillus* spp.) were found to be able to metabolize glycolic acid, and it was highly toxic to many other species, such as *L. ferriphilum* and *Acidithiobacillus ferrivorans*. Ňancucheo and Johnson (2010) suggested that exudation of glycolic acid, which was thought to arise from the hydrolysis of phosphoglycolate formed, at least in the case of the acidithiobacilli, from RuBisCO acting as an oxygenase, helped explain the presence of *Sulfobacillus* spp. in continuous flow stirred tanks that are used to bio-process mineral concentrates (described later in this chapter).

While one-way flow of (organic) carbon was readily implicated as the reason why heterotrophic acidophiles could grow in supposedly organic-carbon free liquid media, the proof of this as a truly mutualistic relationship was more difficult to establish. Concentrations of DOC are invariably greater in axenic cultures of chemolithotrophic acidophiles than in mixed cultures containing heterotrophic species (such as *Acidiphilium*) and catabolism of organic compounds by the latter can lower the mortality rates of chemolithotrophic bacteria (like *At. ferrooxidans*) in stored cultures (Table 8.2), suggesting a positive feedback mechanism whereby the heterotrophic species 'detoxifies' the liquid medium for the chemolithotroph. The same hypothesis formed the basis of the design and formulation of 'overlay' solid media, in which heterotrophic acidophiles are deliberately incorporated into the lower layer of a two-layer gel in order to remove both potential toxins produced as exudates and as acid-hydrolysis products of the gelling agent (agar or agarose), thereby promoting the growth of *Leptospirillum* spp. and other 'organic sensitive' species of acidophiles (Johnson and Hallberg, 2007). Proof that the positive feedback also involved assimilation of CO_2 generated by the heterotrophic partner came from an elegant study carried out by Kermer *et al.* (2012). They used protein-based stable isotope probing (SIP) to investigate two-way carbon transfer in a mixed culture of *At. ferrooxidans* and *A. cryptum*, grown in a liquid medium containing tetrathionate and ^{13}C-labelled galactose. The heterotroph was shown to metabolize the galactose, releasing $^{13}CO_2$ which was taken up and assimilated by *At. ferrooxidans*. The autotroph in turn secreted ^{13}C-labelled organic substrates (mostly peptides) which were assimilated by *A. cryptum*.

As indicated in Table 8.1, acidophilic microalgae are also primary producers in many extremely acidic environments and are probably the major carbon-fixing biota in many surface-located niches, such as acidic pit lakes and mine drainage streams. There have been relatively few studies describing carbon flow between phototrophic and heterotrophic acidophiles, but in one of these Ňancucheo and Johnson (2012) examined DOC in axenic cultures of the micro-alga *Chlorella protothecoides* var. *acidicola* and the phototrophic protozoan *Euglena mutabilis*, isolated from abandoned copper mines in Spain and Wales, respectively. Growth of the phototrophs in liquid media was paralleled with increasing concentrations of DOC. Glycolic acid was again identified as a significant (12–14%) component of the total DOC but, in contrast to the chemolithotrophic acidophiles monosaccharides (glucose, fructose and mannitol) were also identified in cell-free liquors. The sugars were rapidly metabolized by heterotrophic acidophiles (*Acidiphilium*, *Acidocella* and *Acidobacterium*) and were also identified as substrates that were used by species of acidophilic sulfate-reducing bacteria that were found in the same mine sites (e.g. Rowe *et al.*, 2007).

Table 8.2 Dissolved organic carbon in pure cultures of *At. ferrooxidans* and in mixed cultures containing *Acidiphilium* sp. and its effect on long-term survival of the autotroph in substrate-depleted cultures

	DOC (mg/l)		*At. ferrooxidans*/ml		*Acidiphilium*/ml	
	Day 30	Day 160	Day 30	Day 160	Day 30	Day 160
*At. ferrooxidans*ᵀ (axenic culture)	2.4	7.1	2.4×10^7	$<10^2$	–	–
*At. ferrooxidans*ᵀ/*Acidiphilium* sp. (mixed culture)	0.4	1.7	2.4×10^7	3.7×10^4	1.1×10^5	4.0×10^5
At. ferrooxidans (ATCC 13728; axenic culture)	2.5	5.6	3.8×10^6	$<10^2$	–	–
At. ferrooxidans (ATCC 13728/*Acidiphilium* sp. (mixed culture)	0.5	1.1	3.8×10^6	8.2×10^4	3.0×10^5	1.0×10^7

Microbial interactions involving redox transformations of iron and sulfur in extremely acidic environments

Iron and sulfur are frequently two of the most abundant elements found in extremely acidic environments, for reasons described in Chapter 1, though in solfatara and some low-pH subterranean environments (e.g. within the Frasassi cave complex in Italy) concentrations of iron may be relatively small. Both elements have multiple oxidation states – three in the case of iron (0, +2 and +3, though zero-valent iron is only found in some anthropogenic environments) and nine in the case of sulfur (ranging from −2 to +6). Redox reactions of iron in natural environments involve transformations between iron(II) (ferrous iron) and iron(III) (ferric iron). While these can be observed in most terrestrial and aquatic environments, they play a particularly important role in low-pH sites, due to ferrous iron being far more stable in acidic liquors (and therefore spontaneous chemical oxidation being far less pronounced) and ferric iron being far more soluble, particularly at pH <2.5. This facilitates both ferrous iron acting as an electron donor and ferric iron acting as an electron acceptor for acidophilic prokaryotes, and is probably the reason why these physiological traits are far more widespread among species of acidophiles than in other microbial groups. Another contrast between low-pH and circumneutral-pH environments is that the redox potential (E_H) of the ferrous/ferric couple is several hundred millivolts more positive in the former. The value often quoted for low-pH waters is +770 mV, though this value assumes that neither ferrous nor ferric iron is complexed, which is not the case in most natural or anthropogenic acidic environments, where ferric iron occurs in a variety of soluble sulfate [e.g. $Fe(SO_4)_2^-$] and hydroxyl [e.g. $Fe(OH)^{2+}$] complexes, and, consequently, the E_H value is less positive at (depending on pH and sulfate concentration) ~ +690 to +720 mV (Welham *et al.*, 2000; Johnson *et al.*, 2012). Even so, this is still a relatively highly positive value, and has the implications that (i) only molecular oxygen can act as an electron acceptor coupled to iron oxidation at low pH (the E_H of O_2/H_2O couple is +1.12 V at pH 2) and (ii) using ferric iron as an alternative electron acceptor to oxygen in anoxic or micro-aerobic sites does not carry a great thermodynamic penalty.

Some species of acidophilic prokaryotes can only oxidize ferrous iron (e.g. *Leptospirillum* spp., and *S. metallicus*), others only reduce ferric iron (e.g. *Acidiphilium* and *Acidocella* spp.) while many species can do both (e.g. *At. ferrooxidans*, *At. ferridurans*, *At. ferrivorans*, *Sulfobacillus* spp., *Ferrimicrobium acidiphilum*, *Acidimicrobium ferrooxidans* and *Ferroplasma* spp.; Johnson *et al.*, 2012). Ferrous iron is an important energy source in many extremely acidic environments, and is used as such by obligate and facultative autotrophs, and also by some species of obligately heterotrophic acidophiles. The ferric iron generated may, depending on pH and other factors, remain in solution or hydrolyse, forming minerals such as jarosites [e.g. $KFe_3(SO_4)_2(OH)_6$] and schwertmannite [$Fe_8O_8(SO_4)(OH)_6$]. Soluble ferric iron can diffuse into the oxygen-depleted zones of stream sediments where it can be reduced back to ferrous and thereafter diffuse into the overlying aerated waters. Ferric iron minerals can form ochre-like accumulations on stream beds, the lower layers of which adjacent to the anoxic sediments can be subjected to reductive dissolution, again releasing soluble ferrous iron. Both scenarios explain the dynamic cycling of iron which is seen in many extremely acidic environments. It should be noted though that microbially mediated ferric iron reduction requires an electron donor, which can be organic carbon (derived from phototrophic and chemolithotrophic primary producers, as described above), reduced sulfur or elemental sulfur (from partial oxidation of sulfide minerals), or hydrogen (formed, for example, via the acid dissolution of metals, such as relics of mining). Cycling of iron may involve more than one species (e.g. the iron-oxidizer *At. ferrooxidans* and the iron-reducer *Acidiphilium*) or a single species (e.g. *Sulfobacillus*) that can catalyse both reactions. Ferric iron reduction in acidic waters can also be mediated by hydrogen sulfide (produced by sulfidogenic bacteria and archaea), by reacting with sulfide minerals (e.g. Equation 8.1) and by solar energy (Johnson *et al.*, 2012). Diez-Ercilla *et al.* (2009) showed that the kinetics of photo-reduction of ferric iron were zero order and were a function of temperature and the intensity of solar radiation. Photo-chemical regeneration of ferrous iron explains how significant populations of iron-oxidizing chemolithotrophic acidophiles can be maintained in some situations, even when their primary energy source, such as pyrite, is depleted. A

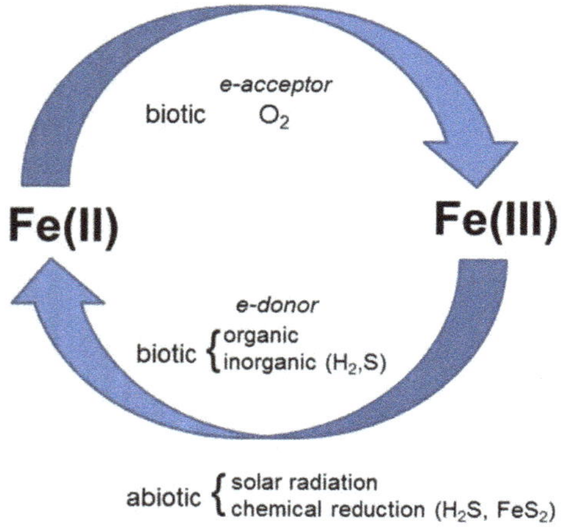

Figure 8.1 Oxido-reduction of iron in extremely acidic environments. Ferrous iron oxidation is almost exclusively mediated by iron-oxidizing prokaryotes, and is coupled to the reduction of molecular oxygen whereas ferric iron reduction can be mediated by a various biotic and abiotic mechanisms.

generalized picture of iron cycling in an extremely acidic water body is shown in Fig. 8.1.

Microbial interactions involving oxido-reduction of sulfur are also thought to be of great significance in extremely acidic environments and, although there is a great deal of literature describing the dissimilatory oxidation of sulfur by a large number of species of acidophilic archaea and bacteria (*Sulfolobus* spp., *Acidithiobacillus* spp. etc.), relatively little is known about microorganisms that catalyse the reduction of sulfur, and especially sulfate, at low pH. Oxyanions in which sulfur occurs in varying oxidation states are produced directly or indirectly by biological activity in acidic environments that contain sulfur and/or sulfide minerals. These are ultimately oxidized to sulfate (S oxidation state of +6), a soluble anion that, as described for ferric iron at low pH, can freely diffuse into anoxic sediments where it can act as the terminal electron acceptor for acidophilic and acid-tolerant sulfate reducing bacteria (aSRB). Relatively few strains of aSRB have been reported (reviewed in Dopson and Johnson, 2012) and all isolates so far are members of the phylum *Firmicutes*. Sulfate-reducing prokaryotes generate sulfide which, at low pH, is present almost exclusively as gaseous H_2S, rather than soluble anionic HS^- or S^{2-}. If metals that form very insoluble sulfides, such as copper, are present in the sediment, these react with the H_2S, producing metal sulfides (e.g. CuS) in reactions that generate protons and help maintain the acidity of the anoxic zones. Such a scenario was reported in an acidic channel draining an abandoned copper mine in south-west Spain (Rowe *et al.*, 2007). However, in the absence of significant amounts of chalcophilic metals that form sulfides at low pH, dissimilatory sulfate reduction, which is a strongly proton-consumptive reaction at low pH, will cause the local pH to increase to ~5, at which point ferrous sulfide (FeS) will form and buffer the anoxic zone at a much higher pH than the overlying water body. Any hydrogen sulfide gas which diffuses into the aerobic water layer can act as the electron donor for the many extremely (e.g. *Acidithiobacillus*) and moderately (such as *Thiomonas*) acidophilic species that populate low-pH streams, perpetuating the sulfur cycle in these environments. Examples of situations where interconnected redox transformations of sulfur occur in sites that are spatially removed from each other and of widely different pH values can be found in concrete sewers and some sulfur-rich subterranean environments (Dopson and Johnson, 2012). In both cases the sites of dissimilatory sulfide/sulfur oxidation and far more acidic than those where dissimilatory sulfate reduction occurs.

Other examples of positive interactions between acidophiles (synergy and syntrophy)

Microbial interactions can result in the complimentary activities of both (or all) participants being more efficient in, for example, degrading a substrate, than the individual species working alone. These are referred to as synergistic, and may be either facultative or obligate associations. An example of the former is oxidative dissolution of pyrite by mixed cultures of *L. ferrooxidans* and *At. thiooxidans* at mesophilic temperatures (~20–35°C), and by *L. ferriphilum* and *At. caldus* at more elevated temperatures (up to ~45°C). *Leptospirillum* spp. are iron-oxidizing chemolithotrophs that cannot oxidize sulfur, while *At. thiooxidans* and *At. caldus* oxidize elemental sulfur and reduced sulfur anions, but not ferrous iron. Pyrite (FeS_2) is an acid-insoluble sulfide mineral that can be oxidized by ferric iron produced by ferrous iron-oxidizing leptospirilli in a reaction (via the 'thiosulfate pathway'; Schippers and Sand, 1999) that generates protons (Equation 8.1):

$$FeS_2 + 6Fe^{3+} + 3H_2O \rightarrow 7Fe^{2+} + S_2O_3^{2-} + 6H^+ \quad (8.1)$$

The thiosulfate produced in this reaction is oxidized by ferric iron (Equation 8.2):

$$S_2O_3^{2-} + 2\,Fe^{3+} + H_2O \rightarrow S^0 + SO_4^{2-} + 2\,Fe^{2+} + 2H^+ \quad (8.2)$$

These combined reactions generate eight protons/hydronium ions for each pyrite oxidized. However, the subsequent re-oxidation of the nine ferrous irons produced uses nine protons (Equation 8.3), making the overall process net consumptive of protons.

$$9Fe^{2+} + 9H^+ + 2.25O_2 \rightarrow 9Fe^{3+} + 4SH_2O \quad (8.3)$$

Hydrolysis of the ferric iron generated buffers the solution pH at ~ 2.4, but this is above that of the growth optima of *Leptospirillum* spp. (generally ~1.5–2.0). However, in mixed cultures with sulfur-oxidizing prokaryotes, the elemental sulfur produced in Equation 8.2 is oxidized to sulfuric acid (Equation 8.4):

$$S^0 + 1.5\,O_2 + H_2O \rightarrow 2H^+ + SO_4^{2-} \quad (8.4)$$

This maintains the more extreme pH favoured by the leptospirilli allowing them to generate sulfur intermediates more rapidly than in axenic cultures.

Since pure cultures of *Leptospirillum* spp. are also able to accelerate the oxidative dissolution of pyrite but *At. caldus* and *At. thiooxidans* cannot, this is an example of facultative synergy, at least from the perspective of the leptospirilli. In contrast, the interaction of the heterotrophic iron-oxidizer *Ferrimicrobium acidiphilum* and *At. thiooxidans* is an obligate synergistic relationship when metabolizing pyrite (Bacelar-Nicolau and Johnson, 1999). *Fm. acidiphilum* is a heterotrophic iron-oxidizer, requiring organic carbon such as yeast extract, to grow. *At. thiooxidans* can supply organic carbon (as exudates and lysates) that also promotes the growth of *Fm. acidiphilum. At. thiooxidans* does not oxidize pyrite, though it can grow on the sulfur intermediates produced by ferric iron attack on the mineral (Equation 8.1). The synergistic interaction between these two acidophiles therefore involves carbon flow as well as redox transformations of iron and sulfur, as illustrated in Fig. 8.2.

Syntrophic relationships involving interactions between sulfate-reducing bacteria and a partner microorganism have mostly been described in circumneutral pH, anaerobic environments, and frequently involve inter-species transfer of hydrogen. Hydrogen is a common end product of fermentative metabolism, and the oxidation of this gas, coupled to sulfate reduction, can result in a change in the overall free energy (ΔG) of a catabolic reaction and allow an otherwise thermodynamically unfeasible intermediary reaction to proceed. A

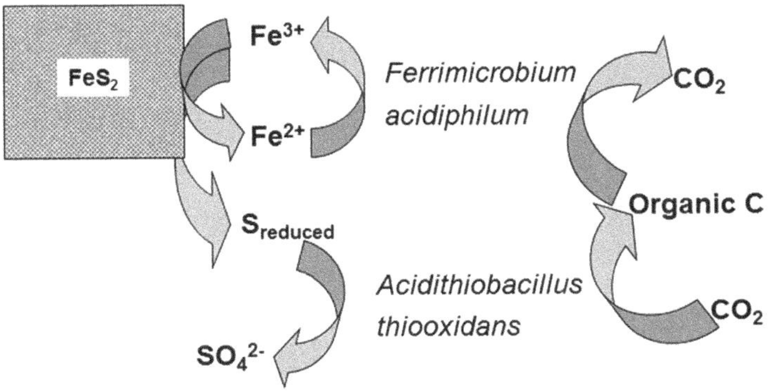

Figure 8.2 Synergistic interactions between *Ferrimicrobium acidiphilum* and *Acidithiobacillus thiooxidans* which allow the mixed culture to grow on pyrite. Neither bacterium can grow on pyrite in pure culture (Bacelar-Nicolau and Johnson, 1999).

syntrophic association involving *Desulfosporosinus acididurans* (an aSRB) and the type strain *Acidocella aromatica* was described by Kimura *et al.* (2006). Acetic acid accumulates in pure cultures of *D. acididurans* in equimolar proportion to the amount of glycerol oxidized. In contrast, in mixed cultures containing *Ac. aromatica* the appearance of acetic acid is transient, and more sulfide is produced per mole of glycerol metabolized. *Ac aromatica* does not grow on glycerol, but can catabolize acetic acid under anaerobic conditions. The syntrophic association proposed involved the initial partial oxidation (coupled to sulfate reduction) of glycerol to acetic acid by of *D. acididurans*, followed by the degradation of acetic acid to hydrogen and carbon dioxide by *Ac. aromatica* and, finally, oxidation of this hydrogen (also coupled to sulfate reduction) by *D. acididurans* (Fig. 8.3). Two of these reactions have negative Δ*G* values, though the acetoclastic reaction has a positive Δ*G* value and does not occur spontaneously. However, as is the case with other syntrophic associations, tight coupling of the reactions allow the complete oxidation of glycerol to CO_2 (which is an exergonic reaction) to proceed. Supporting evidence for this syntrophic interaction came from the observations that *D. acididurans* could grow on acetic acid in co-culture with *Ac. aromatica* though not in pure culture, and that this aSRB could use hydrogen, as well as glycerol, as an electron donor.

Negative interactions between acidophilic microorganisms

Negative interactions between different species of acidophiles can take many forms. For example, generation of sulfuric acid by *Acidithiobacillus* spp. can lower the pH of an environmental niche (or a laboratory culture) to the level at which growth of more moderately acidophilic sulfur-oxidizers (e.g. *Thiomonas* spp.) or heterotrophs (e.g. *Acidobacterium* spp.) is completely inhibited. Competition is, however, probably the most widespread negative microbial interaction in extremely acidic, as in many other environments. Acidophiles compete for electron donors, electron acceptors and nutrients. Phosphorus is often a limiting macro-nutrient in low-pH environments due to its low bio-availability resulting from adsorption of $H_2PO_4^-$ onto positively charge ferric iron colloids and minerals, and strategies developed by acidophiles to growing under phosphate-limited conditions have been

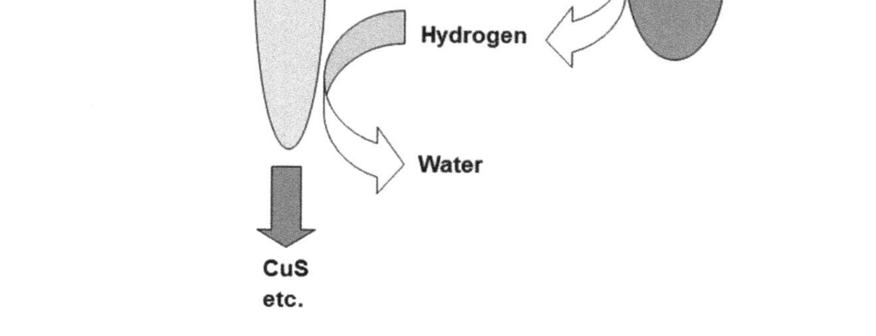

Figure 8.3 Syntrophic interaction between the acid-tolerant sulfate-reducing bacterium *Desulfosporosinus acididurans* and the heterotrophic acidophile *Acidocella aromatica*. The key reaction in this association is the fermentation of acetic acid to hydrogen, catalysed by *Ac. aromatica*. Although this is an energetically unfavourable reaction (as indication by the positive Δ*G* value shown) the rapid consumption of hydrogen allows the net reaction to proceed (Δ*G*= −199 kJ/mol).

described (e.g. Vera *et al.*, 2008). Competition between *At. ferrooxidans* and *L. ferrooxidans* for their common substrate (ferrous iron) has been studied in detail, due to the perceived importance of these two acidophiles in commercial mineral processing and in the genesis of acid mine drainage. Although it was once thought that *At. ferrooxidans* is always the dominant iron-oxidizing bacterium in acidic ferruginous environments, this was mostly based on the observation that this iron-oxidizer is far more easily isolated, and invariably dominates ferrous iron cultures used to enrich iron-oxidizers from mesophilic acidic environments. Other approaches, e.g. direct plating on overlay media and, in particular, cultivation-independent methods, have frequently highlighted the fact that *L. ferrooxidans* is more numerous (e.g. Schrenk *et al.*, 1998; Walton and Johnson, 1992). *At. ferrooxidans*, and the closely related species *At. ferridurans*, can grow more rapidly than *Leptospirillum* spp., and therefore tend to dominate environments (and enrichment culture liquors) that contain elevated concentrations of ferrous iron. On the other hand, *L. ferrooxidans* has greater affinity for ferrous iron and greater tolerance of ferric iron (Norris *et al.*, 1988) and these factors give this acidophile a competitive advantage in other situations, for example in waters containing small concentrations of ferrous iron and/or elevated concentrations of ferric iron. In terms often used to differentiate heterotrophic bacteria, *At. ferrooxidans* may be considered to be a copiotroph while *L. ferrooxidans* is an oligotroph. Another consequence of having greater affinities for ferrous iron is that redox potentials in iron-rich liquors that contain leptospirilli can greatly exceed those that contain only iron-oxidizing acidithiobacilli, which has important consequences for mineral leaching (Rawlings *et al.*, 1999). Other factors, such as temperature, pH and concentrations of dissolved metals, are also important in influencing competitive interactions between *At. ferrooxidans* and *L. ferrooxidans*. Strains of *At. ferrooxidans* are, for example, far more copper tolerant than those of *L. ferrooxidans*, and therefore dominant in copper-rich solutions, such as some mineral leachates. However, strains of the thermo-tolerant species *L. ferriphilum* are at least as copper tolerant as *At. ferrooxidans*, and sometimes more so, and therefore are often found in large numbers of copper leachates, e.g. in some biomining operations. At the other end of the temperature spectrum, it is now known that cold, acidic iron-rich environments are more likely to be dominated by the psychrotolerant species *At. ferrivorans* and '*Ferrovum myxofaciens*' than by *Leptospirillum* spp. or other *Acidithiobacillus* spp. (e.g. Dold *et al.*, 2013; Kay *et al.*, 2013).

In one of the few other reports on competition between two iron-oxidizing acidophiles, Okibe and Johnson (2004) described the microbial kinetics and dissolution of pyrite at 45°C by a mixed culture of thermo-tolerant iron-oxidizing bacteria, the facultative autotroph *Am. ferrooxidans* and the obligate autotroph *L. ferriphilum*. Neither of these bacteria are able to oxidize sulfur, and a parallel controlled pH/temperature bioreactor was set up which also included the thermo-tolerant sulfur-oxidizer *At. caldus*. At pH 1.5 in the *At. caldus*-free bioreactor, numbers of the two iron-oxidizers (estimated using fluorescent *in situ* hybridization) remained similar during mineral bioleaching, but when the pH of the bioreactor was lowered to 1.2, *L. ferriphilum* emerged as the dominant bacterium. In contrast, in the presence of *At. caldus* numbers of *Am. ferrooxidans* were greater than those of *L. ferriphilum* at both pH 1.5 and 1.2. The reason for this was thought to be the additional organic carbon available for the heterotrophically inclined *Am. ferrooxidans* from exudates and lysates of autotrophic *At. caldus* that used the reduced sulfur produced by ferric iron attack on the pyrite (Equation 8.1). Because of much higher free energy available from oxidizing reduced sulfur (ΔG^0 −507 kJ/mol in the case of elemental sulfur) compared with oxidizing ferrous iron (ΔG^0 −30 kJ/mol at pH 2), *At. caldus* was present in larger numbers than the combined total of the iron-oxidizers in the three-membered bacterial consortium.

The fact that some bacterial predators (unicellular flagellates, ciliates and amoeboid protozoa, and multicellular rotifers) can also live in extremely acidic environments and laboratory cultures has been known for many years. Rotifers have, for example, been observed feeding on acid streamer communities using their wheel-like ciliated corona to draw vortices of bacteria into their mouth structures (Johnson, 2009). Ehrlich (1963) noted that a *Eutreptia*-like flagellate could grow in cultures of acidophilic bacteria enriched from metal mine waters. Another *Eutreptia*-like flagellate studied by McGinness and Johnson (1992) was shown to be obligately acidophilic, with a pH range of 1.8 to 4.5. This flagellate was found to graze a variety of acidophilic bacteria, including *At. ferrooxidans*, *L. ferrooxidans* and *A. cryptum*, but some degree of selectivity was reported, with the highly motile iron-oxidizer *L. ferrooxidans* being less effectively grazed than the non-motile acidophile *At. ferrooxidans*, and consequently mixed cultures of the two bacteria were dominated by *L. ferrooxidans* when the protozoan was present. Filamentous and aggregate growth by some acidophiles also appeared to give them some protection from predation by the flagellate. The *Eutreptia* was highly tolerant of soluble iron (both ferrous and ferric) but sensitive to other metals such as copper, silver and molybdenum. Johnson and Rang (1993) later described *in vitro* grazing of acidophilic bacteria by two other flagellates, a *Cinetochilum*-like ciliate and a *Vahlkampfia*-like amoeba. In each case, population dynamics followed classic predator-prey population dynamics. Attempts at cultivating the protozoa-bacteria cultures in pyrite, as opposed to ferrous iron media, all resulted in the elimination of the protozoa. This was shown to be due to the fine-grain size (< 61 mm) of the pyrite used in initial experiments, and that when coarser-grain (61–200 mm) particles were used, all four protozoa (and the earlier *Eutreptia* flagellate) were able to grow effectively. It was hypothesized that the protozoa were unable to discriminate between bacteria and bacterial-size pyrite particles, and inadvertent ingestion of the latter resulted in the death of the eukaryotes. Johnson and Rang (1993) also noted that dramatic reductions of numbers of planktonic iron-oxidizing bacteria due to protozoan grazing did not necessarily result in decreased rates of pyrite dissolution, possibly because of the overriding influence of mineral-oxidizing bacteria attached to the pyrite, which were not grazed.

There have been reports of more clandestine interactions occurring between species of acidophiles, reasons for which are currently not known. Okibe and Johnson (2004) found that the presence of a strain of the heterotrophic iron-oxidizing archaeon *Ferroplasma acidiphilum* suppressed oxidation of pyrite by *L. ferriphilum* in both shake flask and bioreactor cultures, though this was not the case when *At. caldus* was added to the consortium mix. Bellenberg *et al.* (2014) showed that, while both *L. ferrooxidans* and a strain of *Acidiferrobacter* grew successfully in pure culture, these bacteria did not grow in non-amended mixed cultures. Addition of synthetic N-acyl-homoserine lactones (see Chapter 10) to the mixed cultures relieved the inhibition, and these were noted to be dominated by *Acidiferrobacter*. The same authors also reported that *At. thiooxidans* cells attached more efficiently to pyrite when the mineral was pre-colonized by iron-oxidizers. These positive and negative interactions merit further investigation, particularly as they are likely to impact mineral processing biotechnologies.

Microbial community dynamics: examples from extremely acidic environmental and industrial habitats

The diversity of acidophiles that can be found in any low-pH environment can vary enormously, from stratified pit lakes where juxtaposed zones varying greatly in some important physico-chemical parameters (e.g. concentrations of dissolved oxygen and carbon, redox potential, access to solar energy) and therefore provide discrete niches for colonization by many different species of acidophiles, to continuous flow stirred tank reactors used for biomining, where controlled temperature and pH and fast throughout of mineral slurries result in long-term colonization by relatively few species.

The following section describes the diversities and interactions of microbial populations that occur in three contrasting extremely acidic environments, and how these impact the geochemistry of these environments. The omission of descriptions of interactions between viruses and prokaryotic and eukaryotic microorganisms is due to the fact that, in extremely low-pH environments, this has been little researched and is poorly understood, in contrast to most other most other 'normal' and even extreme environments.

Acid streamers and other macroscopic growths

Acid streamers are filamentous macroscopic growths found in flowing waters such as streams draining metal and coal mines. These can be very extensive (> 100 m^3 streamer biomass has been reported in the abandoned Cae Coch pyrite mine in Wales, Kimura *et al.*, 2011) and were among the first biological materials to be described in extremely acidic environments (e.g. Lackey, 1938). Related to acid streamer growths are (i) acid mats, which are more dense in texture and often found below growths of acid streamers; (ii) acid slimes: thick, macroscopic biofilms that grow on moist surfaces of exposed rock faces and in acidic pools; (iii) suspended macroscopic growths attached to mine and cave roofs and pit props (small-scale growths sometimes described as snottites and larger structures as microbial stalactites).

Snottite-like growths have been reported in the Grotta Grande del Vento–Grotta del Flume (Frasassi) cave complex, Italy. This is an example of a sulfidic cave system formed when sulfide-rich groundwaters contact oxygen within subterranean carbonate rock strata and the sulfuric acid formed by (microbiological) oxidation of the sulfide causes the dissolution of the carbonate minerals. Areas of extremely low pH (0–1) in Frasassi have been found to be colonized by viscous microbial biofilms. Molecular analysis of these has shown that they have limited biodiversity, and are dominated by bacteria related to the mesophilic sulfur-oxidizing autotroph *At. thiooxidans*. Other bacteria identified include bacteria related to Gram-positive *Sulfobacillus* and *Acidimicrobium* spp., and archaea related to *Ferroplasma* (Macalady *et al.*, 2007).

Most other streamer-like growths that have been reported are associated directly or indirectly with sulfidic ore bodies, and grow in acidic waters that are enriched with soluble iron as well as sulfur. The most extensive macroscopic growths (streamers, slimes and stalactites) of acidophiles reported are those within a pyrite mine in north Wales that has been abandoned for ~ 100. Early research on these acidophilic communities (e.g. Johnson *et al.*, 1979) was limited by the lack of effective cultivation techniques and also predated the advent of nucleic acid-based molecular tools that bypass the need for cultivation. More recently the site was revisited and a comprehensive account of the biodiversity of the macroscopic growths and water bodies (ponds and the main drainage stream in the mine, which have pH values of 1.8–2.3) described by Kimura *et al.* (2011). Among the findings were that the dominant bacteria in the macroscopic growths were *Betaproteobacteria* (*'Fm. myxofaciens'*, and occasionally a (presumed) acidophilic *Gallionella* sp.) and that smaller numbers of *Alphaproteobacteria* (*Acidiphilium* spp. and a novel *Sphingomonas* sp.), *Gammaproteobacteria* (*At. ferrivorans* and *Metallibacterium scheffleri*), *Nitrospira* (*L. ferrooxidans*), *Firmicutes* (distantly related to *Sulfobacillus*) and *Actinobacteria* (*Ferrimicrobium*-like) were detected and, in most cases, isolated. Archaea were found to be less abundant than bacteria, possibly due to the constant low temperature (8–9°C) within the mine. Smaller-scale microbial stalactites were found in an abandoned pyrite mine (pH 2.6, 15°C) in the Harz mountains (Germany) and shown to be dominated by iron-oxidizing *Leptospirillum* and *'Ferrovum'*-like bacteria, and smaller numbers of heterotrophic *Acidiphilium* were also found to be present (Ziegler *et al.*, 2009). In contrast, the very warm (35–45°C) and extremely acidic (pH 0–2) waters within the abandoned Richmond Mine (Chapter 14) contain acid streamers and slimes composed primarily of thermo-tolerant *Leptospirillum* spp. (the major primary producers within the mine) and the iron-oxidizing heterotrophic archaeon *'Ferroplasma acidarmanus'*, with mesophilic acidophiles (such as *At. ferrooxidans*) confined to the cooler areas of the mine (Bond *et al.*, 2000).

In most cases, microbial and molecular analysis of acid streamers and related structures are single 'snapshots' of their composition at their time of sampling and give no indication how these may have changed over time. In 2003, a new channel was constructed to drain ground-water (pH 2.5, 11 ± 1°C) that accumulated in the deep mine workings of an abandoned copper mine (Mynydd Parys) in Wales. Acid streamers were found to appear within months following commissioning of the new drain channel, and the evolution of these growths was followed over the next nine years using a combined cultivation-based and molecular approach (Kay *et al.*, 2013). Although there were some fluctuations in the composition of the streamer growths, especially for species present in relatively low abundance, the dominant prokaryotes were all iron-oxidizing bacteria, two psychrotolerant autotrophic species (*'Fv. myxofaciens'* and *At. ferrivorans*) which together accounted for ~80% of the streamer biomass, and a novel genus of filamentous, iron-oxidizing heterotrophic *Actinobacteria* (*'Acidithrix ferrooxidans'*). Growth of the latter was considered to be sustained by organic carbon derived from the chemolithotrophs. In the summer months, the surfaces of some acid streamers in the channel were green in colour, due to colonization by *E. mutabilis*, which would constitute a second, more season-dependent, source of organic carbon into the oligotrophic stream.

A contrasting acid streamer/mat community was observed in another channel draining an abandoned copper mine (Cantareras), located in the Iberian Pyrite Belt (IPB) in south-west Spain (Rowe *et al.*, 2007). Here the acidic (pH 2.6) stream displayed seasonal temperature changes, reaching > 30°C in the summer months. This, and the greater light intensity, promoted extensive growths of micro-algae (*Chlamydomonas* and a filamentous alga identified as a *Zygnema* sp.) and *Euglena*, which dominated the biomass of green/turquoise-coloured streamer filaments which overlaid microbial mats (about 12 cm thick). In contrast to the acid streamers described above, the dominant prokaryotes in the mat layer were iron-reducing heterotrophic bacteria, though some autotrophic species (e.g. '*Fv. myxofaciens*') were also identified. The lowest layers of the mats were dark grey in colour and contained non-classified species of aSRB. The hypothesis suggested for the contrasting biomass composition of Cantareras and other acid streamer/mat communities was that there was far more input of organic carbon into the Spanish stream, as a consequence of greater (micro-algal) primary production. This organic carbon was used by both iron- and sulfate-reducing bacteria, which generated ferrous iron and hydrogen sulfide (the latter promoting the biomineralization of copper sulfide, which caused the coloration of the deepest mat zone) in the oxygen-depleted acid mats, whereas oxidative processes (generating ferric iron and sulfate) were prevalent in the aerated surface waters (Fig. 8.4).

Acidic pit lakes

Acidic pit lakes form in all but the most arid global zones, as a consequence of opencast mining of coals and metals. Both opencast and deep mines are necessarily dewatered during their active lives, but once mining ceases the water pumps are shut down and the opencast voids become in-filled with precipitation and ground-waters. The time taken for the water rebound to reach its zenith varies greatly from location to location, from months to many years, depending on climatic, hydrologic and topographic factors. Extremely acidic lakes may

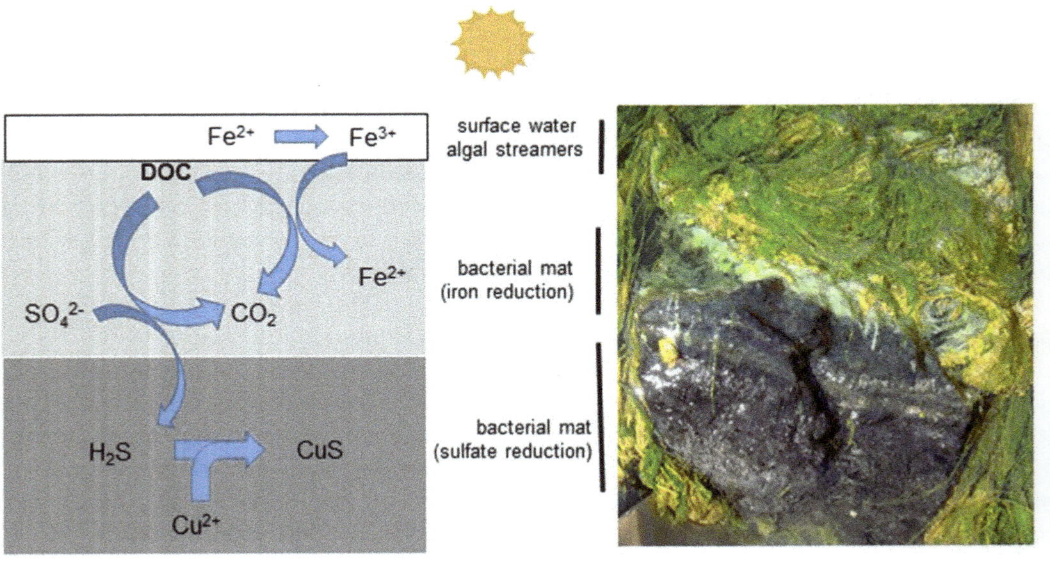

Figure 8.4 Biogeochemistry of an acid streamer/mat community in a drainage channel at an abandoned copper mine (Cantareras, Spain). Dissolved organic carbon (DOC) originating from the main primary producers in the stream (acidophilic micro-algae) fuels the reduction of ferric iron and sulfate in the sub-surface bacterial mats, while iron oxidation occurs in the oxygenated surface stream water. Soluble copper in the mine water combines with biogenic hydrogen sulfide to form black CuS, which accumulates in the anaerobic bacterial mats. The picture on the right shows the filamentous surface algal streamers and the dense subsurface bacterial mats.

also form in craters in volcanic areas. Natural and anthropogenic acidic lakes are also discussed in Chapter 10.

Extensive opencast mining of lignite coal in Germany and some other central European countries has left a legacy of a large number of pit lakes, many of which are acidic, and some extremely so (Geller *et al.*, 1998). The main sources of acidity are residual sulfide minerals such as pyrite, and sulfur present in the waste material produced which, from their exposure to both water and air, are susceptible to biologically catalysed oxidative reactions that generate protons (Chapter 1). Some (acid-table) silicate minerals dissolve in the low-pH solutions generated. Pit lakes from coal mining operations are enriched with sulfate, but the only metals that tend to occur in elevated concentrations are iron (from pyrite and other iron sulfides) and aluminium (from silicate mineral dissolution).

In contrast, pit lakes that form in opencast voids from metal mining, where the ore minerals are sulfides, can contain, in addition to iron and aluminium, elevated concentrations of many other transition metals (e.g. copper and zinc) and metalloids (e.g. arsenic), and therefore are more hostile to most life forms, but also have potential for greater biogeochemical complexity than coal mine pit lakes. Over 20 pit lakes resulting from metal mining in the IPB have been reported. These vary in size, depth, age and water composition, though they share a common geological framework. Most of these are meromictic and display a permanent stratification pattern, with the water column composed of an upper oxic layer (the mixolimnion), a bottom anaerobic layer (the monimolimnon) and a transitional chemocline separating the two zones. The microbial ecology of some of these pit lakes has been recently investigated. Santofimia *et al.* (2013) described the geochemistry and microbiology in two acidic pit lakes (Nuestra Señora del Carmen and Concepción), and identified bacteria involved in the oxido-reduction of iron and others not previously reported in acidic mine waters. These researchers presented a model accounting for the inter-connected carbon and iron cycles in the Nuestra Señora del Carmen pit lake. Falagan *et al.* (2014) investigated two other IPB pit lakes, Cueva de la Mora and Guadiana, that have water depths of ~40 m and 60 m, respectively, and which had previously been reported to have the most dramatic vertical gradients of pH and water chemistry of those found in pit lakes in the IPB. The main redox couple in both lakes was the ferrous/ferric couple, though evidence was also presented of sulfur cycling in the lakes which had major impacts on the bioavailability of some chalcophilic metals such as copper. A major apparent anomaly in the pit lakes examined by Falagan *et al.* (2014) was that the greatest accumulation of algal biomass (and therefore of primary production) was found at depths where photosynthetically active radiation (PAR) was <2% of that close to the water surface, a phenomenon which was thought due to the highly variable bioavailability of phosphorus in these water bodies. In contrast to inorganic nitrogen, where concentrations were fairly uniform throughout the water columns, phosphate was present in far greater concentrations in the anaerobic monimolimnion than in the aerobic mixolimnion, as a consequence of ferrous iron oxidation in the upper aerated water, producing ferric iron which induced the formation of highly insoluble ferric phosphate. Both the size and the diversity of the microbial communities in the chemocline were far greater than those found in waters located below and above this zone, and consequently the geochemical cycling of carbon, iron and sulfur were all much more dynamic within this relatively narrow stratified zone (~2 m thick in Cueva de la Mora) than in the rest of the water column. A diagram depicting the geochemical transformations of iron, sulfur and carbon in Cueva de la Mora, and some of the main microbial players involved in these reactions is shown in Fig. 8.5.

Engineered biomining environments

Some acidophilic microorganisms are used to facilitate the extraction and recovery from precious and base metals from mineral ores and concentrates, a process usually referred to as 'biomining' and described in detail in Chapter 17. Different engineered environments are created in biomining operations, of which irrigated systems (bioheaps and low-grade ore dumps) and aerated stirred tank reactors are currently the most important. More details of the microbiology of biomining environments are given in Chapter 11, while the following section considers what is currently known of the nature and interactions of microorganisms in bioheaps and stirred tanks used for mineral processing (Rawlings and Johnson, 2007).

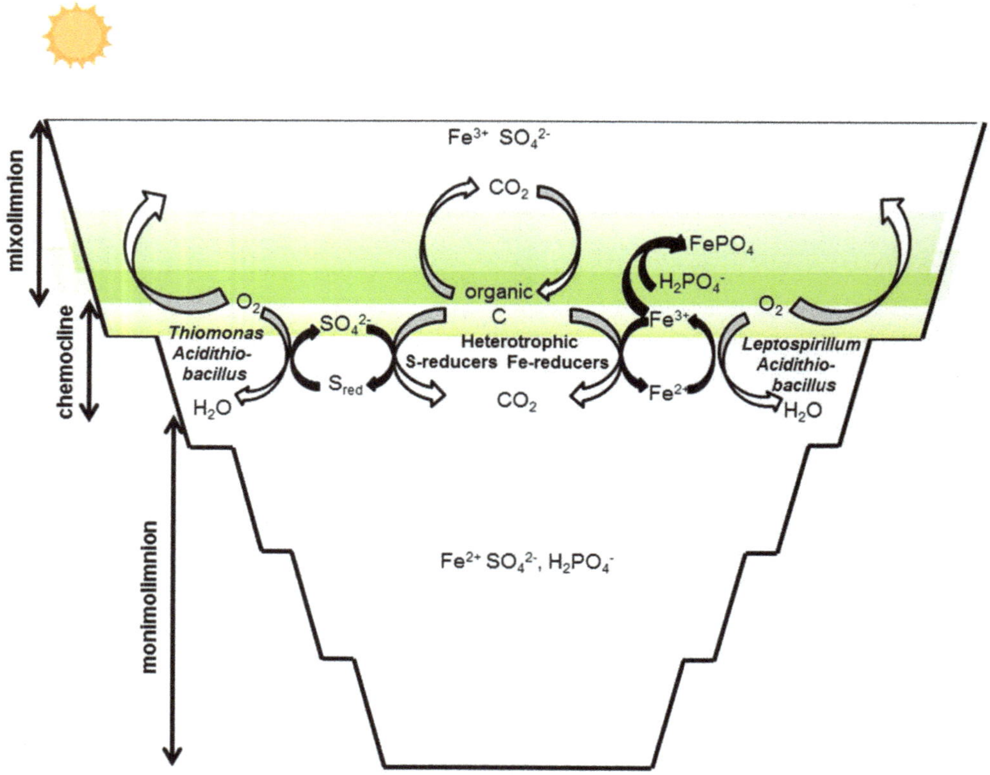

Figure 8.5 Carbon, iron and sulfur cycling in the meromictic Cueva de la Mora pit lake in the Iberian Pyrite Belt. The bioavailability of phosphate dictates the position and depth of the chemocline, where biogeochemical transformations are most intense (modified from Falagan *et al.*, 2013)

Biomining operations are, necessarily, non-sterile systems and, although inoculation with defined microbial consortia may be carried out, these may be accompanied or even usurped by other acidophilic microorganisms introduced, for example, in the mineral ore or concentrate being processed. Bioheaps are constructed to pre-determined dimensions, mostly using graded ores, and are usually irrigated from above with acidic liquors and aerated from below to provide both the oxygen and carbon dioxide required by autotrophic mineral-oxidizing microorganisms, which are the key players in the process. Inevitably even the most carefully engineered bioheaps are highly heterogeneous, and contain micro-environments that vary greatly in terms of such fundamental physico-chemical determinants as pH, temperature and concentrations of dissolved oxygen and solutes. These allow a bioheap to be colonized in the short- and long-terms by very diverse microorganisms, not all (or even most, in some cases) of which have active roles in mineral dissolution. In addition to spatial heterogeneity, bioheaps also display temporal changes in their physico-chemical conditions, which may involve increasing mean temperatures as a result of increasing rates of sulfide mineral oxidation and thermal insulation as heaps mature or are stacked upon each other, and also changes in solution chemistries (redox potentials, dissolved metals, etc.). The net result is that the microbial ecology of bioheaps and other irrigated biomining operations is highly complex. Much of what has been implied about this is based on the microbiology of leach liquors draining bioheaps and dumps (e.g. Demergasso *et al.*, 2005), though how truly reflective this is of the microbial populations attached to solid phases where mineral dissolution is actually occurring has been questioned. Some principles have been established, such as the importance of cell attachment to prevent washout in percolating liquors, the fact that rapid cell growth cycles is of minor importance compared to stirred tanks (described below), and that microbial biodiversity (possibly increased by using carefully designed microbial inocula) is an important factor in promoting the colonization of different micro-environments in a bioheap (Rawlings and Johnson, 2007).

In contrast to bioheaps and mineral dumps, continuous-flow stirred tank reactors are operated at set pH and temperatures and controlled aeration, and physico-chemical conditions within a single tank are highly homogenous. Consequently, the microbial communities in a stirred tank tend to be dominated by relatively few species (typically two to four) of bacteria and archaea, although smaller numbers of other microorganisms may be present. However, conditions change from tank to tank in a continuous flow series as mineral oxidation becomes increasingly extensive, and this can have a significant impact on the relative numbers of indigenous prokaryotes (Okibe *et al.*, 2003). One of the main challenges faced by microorganisms in any continuous flow set up is the possibility of being washed out in the through-flow of process liquor/slurry. In some other types of reactors, carrier materials that are retained in the reactors are provided for microbial attachment and biofilm formation. However, in stirred tanks used to bioprocess mineral concentrates, solid particles (the degrading minerals) are continuously washed out in flowing slurries. This process therefore selects for those microorganisms that are able to grow most rapidly and efficiently in the tanks under the conditions imposed, using whatever substrates are available. In the case of sulfide mineral oxidation, the latter are primarily ferrous iron and reduced sulfur present in the mineral, though organic carbon originating from the primary producers is an important secondary source, particularly in downstream tanks set up in series.

Microbiological analysis that has been carried out on continuous flow stirred tanks operated under acidic conditions but at different temperatures and using different mineral concentrates has confirmed that these all harbour limited diversity of prokaryotic acidophiles (reviewed in Rawlings and Johnson, 2007). However, in all cases reported three distinct groups of microorganisms have been identified: (i) iron-oxidizing acidophiles, classed as the primary microorganisms as these catalyse the regeneration of the main oxidant of sulfide minerals (ferric iron); (ii) reduced sulfur-oxidizing acidophiles (secondary microorganisms) that generate sulfuric acid from the sulfur intermediates generated by ferric iron attack on sulfide minerals; (iii) carbon-degrading acidophiles (obligate or facultative heterotrophic prokaryotes, classed

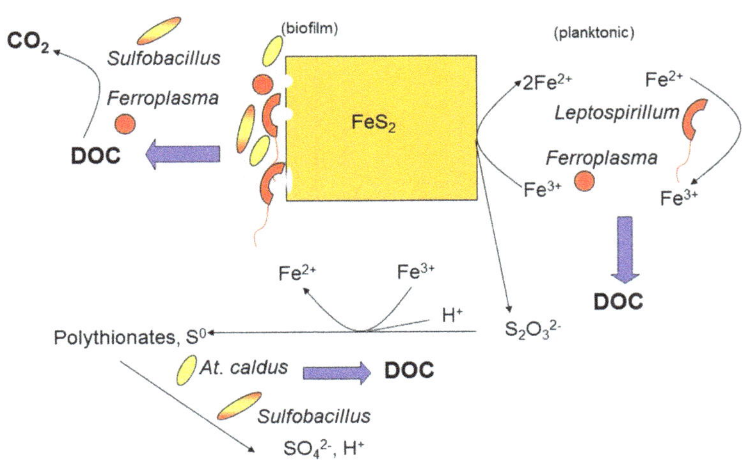

Figure 8.6 Schematic representation of a microbial consortium bioleaching a pyritic mineral concentrate in a continuous flow stirred tank, operated at 40–45°C and at pH ~ 1.5. The bacteria and archaea included at those identified by Okibe *et al.* (2003) and their roles in redox transformations of iron and sulfur, and in carbon flow, are indicated (DOC, dissolved organic carbon).

as tertiary microorganisms) that metabolize the organic carbon from lysates and exudates of the autotrophic species of iron- and sulfur-oxidizers present. The actual identities of acidophiles within each of these three categories depends mostly on the temperature at which the stirred tank is operated. For example, *L. ferriphilum* is frequently found as the most abundant autotrophic iron-oxidizer (and *At. caldus* as the dominant sulfur-oxidizer) in stirred tanks operated at 35–45°C, while at ~80°C thermoacidophilic archaea, such as *Metallosphaera* are the iron-oxidizers present. Often the carbon-metabolizing acidophiles that have been identified probably contribute to some extent to primary mineral dissolution, such as *Ferroplasma* and *Acidiplasma* which also oxidize iron (but not sulfur) and *Sulfobacillus* spp. that can oxidize ferrous iron and reduced sulfur. Fig. 8.6 summarizes the interactions that occur between acidophilic microorganisms in a continuous feed stirred tank oxidizing sulfide minerals and operated at ca. 40 to 45°C.

Studies on microbial communities that occur in man-made and natural extremely acidic environments have revealed a complex network of interactions involving many different species. The application of 'omics' to acidophilic biofilms, acid streamers and the like as greatly enhanced our understanding of these from a genetic perspective. The potential of harnessing positive, and in some cases negative, microbial interactions offers intriguing opportunities for future biotechnological applications of acidophile microbiology.

References

Bacelar-Nicolau, P., and Johnson, D.B. (1999). Leaching of pyrite by acidophilic heterotrophic iron-oxidizing bacteria in pure and mixed cultures. Appl. Environ. Microbiol. *65*, 585–590.

Baker, B.J., Comolli, L.R., Dick, G.J., Hauser, L.J., Hyatt, D., Dill, B.D., Land, M.L., VerBerkmoes, N.C., Hettich, R.L., and Banfield, J.F. (2010). Enigmatic, ultrasmall, uncultivated Archaea. Proc. Natl. Acad. Sci. U.S.A. *107*, 8806–8811.

Bellenberg, S., Díaz, M., Noël, N., Sand, W., Poetsch, A., Guiliani, N., and Vera, M. (2014), Biofilm formation, communication and interactions of mesophilic leaching bacteria during colonization of pyrite and sulfur surfaces. Res. Microbiol. *165*, 773–781.

Bond, P.L., Smriga, S.P., and Banfield, J.F. (2000). Phylogeny of microorganisms populating a thick, subaerial, predominantly lithotrophic biofilm at an extreme acid mine drainage site. App. Environ. Microbiol. *66*, 3842–3849.

Demergasso, C.S., Galeguillos, P.A., Escudero, L.V., Zepeda, V.J., Castillo, D., and Casamayor, E.O. (2005), Molecular characterization of microbial populations in a low-grade copper ore bioleaching test heap. Hydrometallurgy *80*, 241–253.

Diez Ercilla, M., López Pamo, E., and Sánchez España, J. (2009). Photoreduction of Fe(III) in the acidic mine pit lake of San Telmo (Iberian Pyrite Belt): field and experimental work. Aquat. Geochem. *15*, 391–419.

Dold, B., Gonzalez-Toril, E., Aguilera, A., Lopez-Pamo, E., Cisternas, M.E., Bucchi, F., and Amils, R. (2013). Acid rock drainage and rock weathering in Antarctica: important sources for iron cycling in the Southern Ocean. Environ. Sci. Technol. *47*, 6129–6136.

Dopson, M., and Johnson, D.B. (2012). Biodiversity, metabolism and applications of acidophilic sulfur-metabolizing micro-organisms. Environ. Microbiol. *14*, 2620–2631.

Falagan, C., Sanchez-Espana, J., and Johnson, D.B. (2014). New insights into the biogeochemistry of extremely acidic environments revealed by a combined cultivation-based and culture-independent study of two stratified pit lakes. FEMS Microbiol. Ecol. *87*, 231–243.

Geller, W., Klapper, H., and Salomons, W., eds. (1998). Acidic Mining Lakes (Springer-Verlag, Berlin, Germany).

Guay, R., and Silver, M. (1974). *Thiobacillus acidophilus* sp. nov.: isolation and some physiological characteristics. Can. J. Microbiol. *21*, 281–288.

Harrison, A.P. Jr. (1981). *Acidiphilium cryptum* gen. nov., sp. nov., heterotrophic bacterium from acidic mineral environments. Int. J. Syst. Bacteriol. *31*, 327–332.

Johnson, D.B. (2009). Extremophiles: acid environments. In Encyclopaedia of Microbiology, Schaechter, M., ed. (Elsevier, Oxford, UK), pp. 107–126.

Johnson, D.B., and Kelso, W.I. (1983). Detection of heterotrophic contaminants in cultures of *Thiobacillus ferrooxidans* and their elimination by subculturing in media containing copper sulphate. J. Gen. Microbiol. *123*, 2969–2972.

Johnson, D.B., and Hallberg, K.B. (2007). Techniques for detecting and identifying acidophilic mineral-oxidising microorganisms. In Biomining, Rawlings, D.E., and Johnson, D.B., eds (Springer-Verlag, Heidelberg, Germany), pp. 237–262.

Johnson, D.B., and Rang, L. (1993). Effects of acidophilic protozoa on populations of metal-mobilising bacteria during the leaching of pyritic coal. J. Gen. Microbiol. *139*, 1417–1423.

Johnson, D.B., Kelso, W.I., and Jenkins, D.A. (1979). Bacterial streamer growth in a disused pyrite mine. Environ. Pollut. *18*, 107–118.

Johnson, D.B., Kanao, T., and Hedrich, S. (2012). Redox transformations of iron at extremely low pH: fundamental and applied aspects. Front. Microbiol. 3, 96.

Kay, C.M., Rowe, O.F., Rocchetti, L., Coupland, K., Hallberg, K.B., and Johnson, D.B. (2013). Evolution of microbial "streamer" growths in an acidic, metal-contaminated stream draining an abandoned underground copper mine. Life 3, 189–210.

Kermer, R., Hedrich, S., Taubert, M., Baumann, S., Schlömann, M., Johnson, D.B., von Bergen, M., and

Seifert, J. (2012). Elucidation of carbon transfer in a mixed culture of *Acidiphilium cryptum* and *Acidithiobacillus ferrooxidans* using protein-based stable isotope probing. J. Integr. OMICS *2*, 37–45.

Kimura, S., Hallberg, K.B., and Johnson, D.B. (2006). Sulfidogenesis in low pH (3.8 – 4.2) media by a mixed population of acidophilic bacteria. Biodegradation *17*, 57–65.

Kimura, S., Bryan, C.G., Hallberg, K.B., and Johnson, D.B. (2011). Biodiversity and geochemistry of an extremely acidic, low temperature subterranean environment sustained by chemolithotrophy. Environ. Microbiol. *13*, 2092–2104.

Lackey, J.B. (1938). Aquatic life in waters polluted by acid mine waste. Public Health Rep. *54*, 740–746.

Löhr, A.J., Laverman, A.M., Braster, M., van Straalen, N.M., and Röling, W.F.M. (2006). Microbial communities in the world's largest acidic volcanic lake, Kawah Ijen in Indonesia, and in the Banyupahit river originating from it. Microbial Ecol. *52*, 609–618.

Macalady, J.L., Jones, D.S., and Lyon, E.H. (2007). Extremely acidic, pendulous cave wall biofilms from the Frasassi cave system, Italy. Environ. Microbiol. *9*, 1402–1414.

McGinness, S., and Johnson, D.B. (1992). Grazing of acidophilic bacteria by a flagellate protozoan. Microbial Ecol. *23*, 75–86.

Ňancucheo, I., and Johnson, D.B. (2010). Production of glycolic acid by chemolithotrophic iron- and sulfur-oxidizing bacteria and its role in delineating and sustaining acidophilic sulfide mineral-oxidizing consortia. Appl. Environ. Microbiol. *76*, 461–467.

Ňancucheo, I., and Johnson, D.B. (2012). Acidophilic algae isolated from mine-impacted environments and their roles in sustaining heterotrophic acidophiles. Front. Microbiol. *3*, 235.

Norris, P.R., Barr, D.W., and Hinson, D. (1988). Iron and mineral oxidation by acidophilic bacteria: affinities for iron and attachment to pyrite.. In Biohydrometallurgy: Proceedings of the International Symposium, Warwick 1987, Norris, P.R., and Kelly, D.P., eds (Science and Technology Letters, Kew, UK), pp. 43–59.

Okibe, N., and Johnson, D.B. (2004), Biooxidation of pyrite by defined mixed cultures of moderately thermophilic acidophiles in pH-controlled bioreactors: the significance of microbial interactions. Biotechnol. Bioeng. *87*, 574–583.

Okibe, N., Gericke, M., Hallberg, K.B., and Johnson, D.B. (2003). Enumeration and characterization of acidophilic microorganisms isolated from a pilot plant stirred tank bioleaching operation. Appl. Environ. Microbiol. *69*, 1936–1943.

Rawlings, D.E., and Johnson, D.B. (2007). The microbiology of biomining: development and optimization of mineral-oxidizing microbial consortia. Microbiology *153*, 315–324.

Rawlings, D.E., Tributsch, H., and Hansford, G.S. (1999). Reasons why *Leptospirillum*-like species rather than *Thiobacillus ferrooxidans* are the dominant iron-oxidizing bacteria in many commercial processes for the biooxidation of pyrite and related ores. Microbiology *145*, 5–13.

Rowe, O.F., Sánchez-España, J., Hallberg, K.B., and Johnson, D.B. (2007). Microbial communities and geochemical dynamics in an extremely acidic, metal-rich stream at an abandoned sulfide mine (Huelva, Spain) underpinned by two functional primary production systems. Environ. Microbiol. *9*, 1761–1771.

Santofimia, E., González-Toril, E., López-Pamo, E., Gomariz, M., Amils, R., and Aguilera, A. (2013). Microbial diversity and its relationship to physicochemical characteristics of the water in two extreme acidic pit lakes from the Iberian Pyrite Belt (SW Spain). PLoS One *8*, e66746.

Schippers, A., and Sand, W. (1999). Bacterial leaching of metal sulfides proceeds by two indirect mechanisms via thiosulfate or via polysulfides and sulfur. Appl. Environ. Microbiol. *65*, 319–321.

Schnaitman, C., and Lundgren, D.G. (1965). Organic compounds in the spent medium of *Ferrobacillus ferrooxidans*. Can. J. Microbiol. *11*, 23–27.

Schrenk, M.O., Edwards, K.J., Goodman, R.M., Hamers, R.J., and Banfield, J.F. (1998). Distribution of *Thiobacillus ferrooxidans* and *Leptospirillum ferrooxidans*: implications for generation of acid mine drainage. Science *279*,1519–1522.

Vera, M., Pagliai, F., Guiliani, N., and Jerez, C.A. (2008). The chemolithoautotroph *Acidithiobacillus ferrooxidans* can survive under phosphate-limiting conditions by expressing a C-P lyase operon that allows it to grow on phosphonates. Appl. Environ. Microbiol. *74*, 1829–1835.

Walton, K.C., and Johnson, D.B. (1992). Microbiological and chemical characteristics of an acidic stream draining a disused copper mine. Environ. Pollut. *76*, 169–175.

Welham, N.J., Malatt, K.A., and Vukcevic, S. (2000). The effect of solution speciation on iron-sulphur-arsenic-chloride systems at 298 K. Hydrometallurgy *57*, 209–223.

Zeigler, S., Ackermann, S., Majzian, J., and Gescher, J. (2009). Matrix composition and community structure analysis of a novel bacterial leaching community. Environ. Microbiol. *11*, 2329–2338.

Biofilm Formation by Acidophile Bacteria and Archaea

9

Alvaro Orell and Nicolas Guiliani

Abstract

The ability of microbes to grow as surface-associated communities, or biofilms, is seen as a lifestyle widely spread in all biomes on planet earth, including extremely acidic ecosystems. Within these harsh and constantly changing habitats, the biofilm mode of growth is thought to offer the microbial community a regular condition so as to cope with the adverse scenery. Though acidophilic biofilm communities are mainly dominated by bacterial species, archaeal and eukaryotic populations appear to play crucial roles, broadening metabolic diversity and maintaining the community structure.

Environmental signals and molecular mechanisms that underlay the biofilm lifestyle have been largely characterized for bacteria that colonize human cavities. In contrast, the genetic basis that enable environmental acidophiles to form and develop biofilms are far from comprehended. Nonetheless, the molecular mechanisms that govern the biofilm developmental processes in a few model acidophiles have begun to be elucidated. This chapter describes what is currently known on the cellular and molecular mechanisms that promote biofilm formation of bacteria and archaea inhabiting extremely acidic ecosystems.

Introduction

Microbial communities attached to surfaces have been observed using different techniques since the beginning of the study of microbiology. Nonetheless, for almost three centuries of reductionist and operationally oriented thinking, single cells were considered as the dominant form of prokaryotic life. Accordingly, this was the driving force to generate theories, practices and finally knowledge about the lifestyle of microbial life forms. Indeed, it was not until the last quarter of the 20th century that the first article introducing the term 'biofilm', and describing the microbial sessile mode of growth, was published (Mack *et al.*, 1975). In recent years, Parsek and Greenberg (2005) introduced the term 'sociomicrobiology' and initiated a new and exciting field of study focused at deciphering the social activities developed by bacteria, thus linking biofilm formation as a community lifestyle with the so called cell–cell communication system, quorum sensing.

Biofilm formation as a social phenomenon is a complex biological process that has been shown to be regulated in several ways through distinctive molecular mechanisms. Most efforts focused at understanding biofilm formation and its regulation have emerged from studies using model bacterial pathogens such as *Pseudomonas aeruginosa*, *Vibrio cholerae* and *Escherichia coli*. Conversely, and despite extensive evidence reporting the occurrence of acidophile microorganisms in biofilms, slimes and other macroscopic growth formations on mineral surfaces, streams and water bodies (see Chapters 1 and 11), the comprehension of the molecular mechanisms that allow acidophiles to develop into sessile communities is yet in its infancy.

The development of microbial biofilms on minerals plays a relevant role in the generation of Acid Rock/Mine Drainage (AR/MD) in natural and man-made environments. AR/MD is a worldwide problem of water contamination due to the production of sulfuric acid and the release of ferric iron and other toxic heavy metals ions through weathering of metal ores, e.g. pyrite (FeS_2). On the other hand, the leaching activity has been advantageously employed in biomining industrial processes for the recovery of valuable metals from sulfidic ores, such as copper or gold. Hence, it is evident that understanding the regulatory mechanisms that lead acidophilic bacteria and archaea to develop into biofilms communities are relevant research avenues for both controlling the AM/RD generation process and improving bioleaching rates for the biomining industry.

Analyses of biofilm microbial communities from subterranean extremely acidic mine waters, acidic streamers and other macrofilamentous structures growing in low-pH environments from extant geographical regions have begun to provide first insights into the biology and ecology of biofilms formed by acidophilic prokaryotes (reviewed in Johnson *et al.*, 2012). Most of these studies have focused on the identification of the microorganisms present and the characterization of their phylogenetic diversity and metabolic capacities. A general observation emerging from these studies is that the phylogenetic diversity of bacteria, archaea and eukaryotes that inhabit these extremely acidic ($pH \leq 2$) and metal-rich environments is relatively low (e.g. Garcia-Moyano *et al.*, 2007). In addition to the challenging conditions within acidic environments that directly affect the species richness, the abundances of the different members is also thought to be affected by the presence of several classes of protists that act as predators.

Acidophilic biofilms communities are mainly dominated by bacterial species. For instance, *Leptospirillum* spp. dominate the Richmond Mine (Iron Mountain, CA, USA; see Chapter 13) AMD biofilms, where *Leptospirillum rubarum* is reported as the primary colonizer (e.g. Wilmes *et al.*, 2009). On the other hand, several studies of biofilms from the Rio Tinto ecosystem (Spain) revealed that the dominant species within the macro-filamentous structures are *Acidithiobacillus ferrooxidans, L. ferrooxidans* and *Acidiphilium* spp. (e.g. González-Toril *et al.*, 2010). In addition, recent metagenomic and metatranscriptomic analysis from other AMD sites (Liljeqvist *et al.*, 2015; Chen *et al.*, 2014) and terrestrial hot springs (e.g. Menzel *et al.*, 2015) identified *Acidithiobacillus* spp. as the predominant player in acidophile communities structures that develop in temperature ranges of 6–10°C and 76–90°C. Nonetheless, archaea-dominated acidophilic biofilm communities (e.g. Justice *et al.*, 2012) as well as algal photosynthetic biofilms (Chapter 7) have been also found to occur in both the Iron Mountain and Rio Tinto ecosystems.

Due to the difficulties associated with genetic manipulation of acidophilic bacteria and archaea, understanding the mechanisms operating at the cellular and molecular level during the biofilm mode of growth displayed by microorganisms inhabiting extreme acidic environments has been largely delayed. Nonetheless, recent data regarding the structural and genetic components assisting biofilm formation of model acidophiles have begun to emerge (Castro *et al.*, 2015; Orell *et al.*, 2013a). This chapter focuses on current knowledge on the molecular mechanisms underlying biofilm formation by acidophilic bacteria and archaea such as *Acidithiobacillus* spp. (Fig. 9.1A) and *Sulfolobus* spp. (Fig. 9.1B).

Molecular understanding of prokaryotic and acidophilic biofilm

Bacteria

The most and best-studied molecular mechanisms controlling biofilm development in bacteria are quorum sensing and the secondary messenger bis-(3′–5′)-cyclic dimeric guanosine monophosphate signalling pathway (c-di-GMP). Several reports have demonstrated that both pathways are linked to regulate exopolysaccharide (EPS) production and biofilm formation in several bacterial species (e.g. Kozlova *et al.*, 2011; Ueda and Wood, 2009). Bacteria are capable to sense cell density through quorum sensing systems and environmental signals through c-di-GMP pathway, and to choose accordingly between two different possible lifestyles: planktonic state, as individual and motile cells versus adhered state, as social and sessile cells conforming a biofilm structure.

Quorum sensing

Bacteria are able to communicate with each other by producing and secreting small molecules called autoinducers. This process is known as quorum sensing (QS). It is an important mechanism for the timing of collective behaviours through the regulation of population density-dependent cellular processes, such as the production of virulence factors, motility, EPS production and biofilm

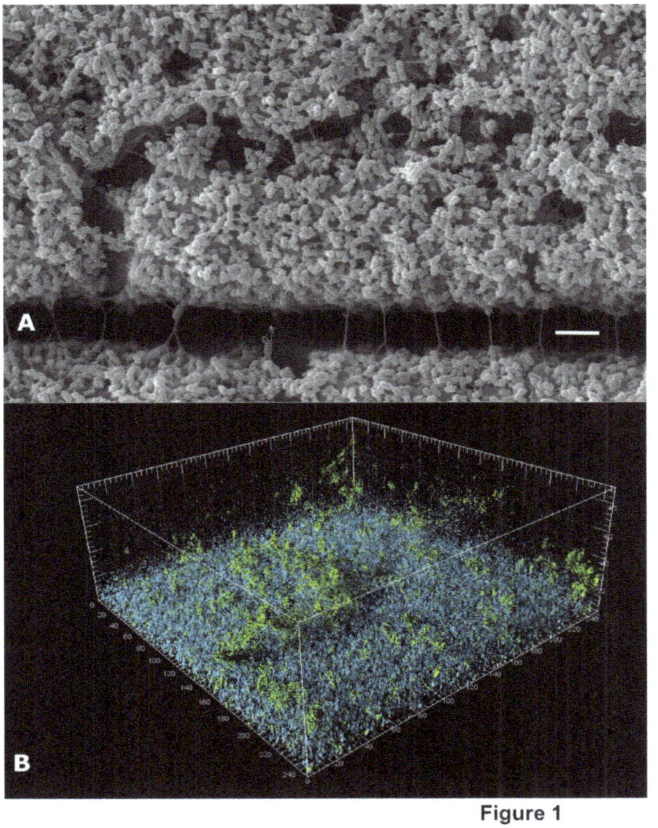

Figure 1

Figure 9.1 Microscopy visualization of biofilm developed by acidophile bacteria and archaea. (A) Scanning electronic microscopy of biofilm formed by *Acidithiobacillus thiooxidans* cells adhered on sulfur coupons (unpublished); Size bar represents 3 mm. (B) Confocal laser scanning microscopy analysis of 3-day-old biofilms formed by the Crenarchaea *Sulfolobus acidocaldarius* MW001 on polystyrene surfaces. Cells were stained with DAPI (blue channel) and fluorescently labelled lectins ConA (green channel) and IB4 (yellow channel) to visualize glucose/mannose residues and a-galactosyl-residues, respectively. An overlay image of all three channels is shown. Numbers indicate length in μm (unpublished).

formation (Parsek and Greenberg, 2005). Different QS systems have been described in Gram-positive and Gram-negative bacteria (Stevens *et al.*, 2011; Bassler and Losick, 2006).

In Gram-negative bacteria, the major QS system involves at least three different molecular elements (Stevens *et al.*, 2011): (i) N-acyl homoserine lactones (AHLs), also called autoinducer type 1 (AI-1); (ii) the *luxI*-like gene coding for the AHL synthase; (iii) the *luxR*-gene coding for a transcriptional regulator which binds AHL molecules and modulates the expression of different target genes accordingly. AHLs are synthesized by LuxI-like proteins that catalyse the amide bond formation between S-adenosyl methionine and the acyl chain of an specific acyl-Acyl carrier protein (acyl-ACP). Specificity of such AHL synthases is directly related with the structural conformation of the enzyme and the available cellular pool of acyl-ACP. LuxR-like proteins are transcription factors composed by two structural and functional subdomains. Indeed, the amino-terminal part binds the cognate AHL, whereas the carboxy-terminal region is involved in DNA binding through an HTH motif. The homodimeric LuxR–AHL complexes bind their specific promoter sequences (*lux*-boxes) generally inducing the expression of QS-regulated genes, though some repressors have also been identified.

QS modulates the expression of many genes through global transcriptional regulators belonging to the R-protein family (Antunes *et al.*, 2007). R-proteins can coordinate directly the expression of several genes; however, many of their effects on global gene expression are likely mediated indirectly by other regulatory genes (Gilbert *et al.*, 2009). In *P. aeruginosa*, the QS regulon encompasses a complex regulatory response, affecting the expression (either positively or negatively) of approximately 300 genes. However, the size of the QS regulon may cover between 3% and 8% of the genome, depending on the experimental strategy and bacterial model used for analysis (Cantero *et al.*, 2006; Vasil, 2003).

As phenotypes such as pathogenicity and biofilm formation are known to be QS controlled in an expanding number of Gram-negative bacteria, several laboratories have focused on the characterization of AHL analogues capable of interfering with the bacterial cell–cell communication (e.g. Stevens *et al.*, 2011). Synthetic AHL analogues have been tested allowing the identification of different chemical families of QS-quenching molecules, as well as super-activators of QS (Geske *et al.*, 2008). These molecules have opened a broad new spectrum of biotechnological applications in which improvement and/or inhibition of QS-controlled functions may be required. These QS modulators can be exploited for proteomic and transcriptomic studies to identify proteins or genes that compose a specific QS regulon (Eberl and Riedel, 2011; Wenbin *et al.*, 2011).

QS mediated by AHL/R-protein binary complex is mainly considered to be involved in intra-species communication. However, there have been several reports showing that in mixed biofilms formed by phylogenetically close Gram-negative species, some interspecies communication may occur (e.g. Riedel *et al.*, 2001). In addition, AHL-signalling also mediates inter-kingdom communication between bacterial and eukaryotic cells (e.g. Gonzalez and Venturi, 2013).

C-di-GMP pathway

Bis-(3′-5′)-cyclic dimeric guanosine monophosphate (c-di-GMP) is a universal bacterial second messenger (reviewed in Römling *et al.*, 2013). Studies on the innate immune response and more recently on social life of dictyostelid amoebas have revealed that c-di-GMP signalling also operates in eukaryotes (Schaap, 2013; Burdette *et al.*, 2011). To date, no evidence for c-di-GMP mediated signalling has been yet presented in archaea. In bacteria, this second messenger controls physiological processes allowing the switch between a free-living planktonic lifestyle and a sedentary biofilm-related lifestyle (reviewed in Römling *et al.*, 2013; Boyd and O'Toole, 2012; Hengge, 2009). Indeed, c-di-GMP regulates surface attachment in many bacteria through control of the biosynthesis of pili and flagellum, EPS production and flagella motor speed. It is now widely accepted that low levels of c-di-GMP promote motility, while high levels promote sessile cellular states. Sessility is generally accomplished by binding c-di-GMP to specific effector proteins that induce inhibition of flagella motility and/or activation of specific polysaccharide biosynthetic machineries such as pelicle polysaccharides (PEL) and alginate in *P. aeruginosa* and cellulose in several Gram-negative bacteria. Indeed, recently the study of *E. coli* macro-colony biofilms and its wrinkled structures revealed how the c-di-GMP signalling is a key player for the spatial arrangements of several polymeric molecules such as flagella, amyloid curli fibres and cellulose (Serra *et al.*, 2013).

C-di-GMP is synthesized from two GTP molecules by the diguanylate cyclase enzyme (DGC) and it can be degraded by two families of phosphodiesterases (PDEs). DGC proteins are characterized by a GGDEF domain whereas PDE proteins can present EAL or HD-GYP domains. Most of the GGDEF domains (>60%) possess an RxxD sequence motif involved in allosteric feedback inhibition by a product (dimer of c-di-GMP) named I-site. Key residues for DGC and PDE activities of GGDEF and EAL domains respectively have been identified (Tchigvintsev *et al.*, 2010; Schirmer and Jenal, 2009).

The complexity of c-di-GMP metabolism depends on the bacterial species. For instance, *Mycobacterium smegmatis* has a single bifunctional protein with both GGDEF and EAL domains, whereas *Vibrio cholerae* possesses 61 domain proteins related to c-di-GMP metabolism, Therefore, understanding the relationship between the number and type of genes involved in c-di-GMP metabolism, the number and type of c-di-GMP responsive genes, and the environment of the bacteria has become of relevance.

The identification of new c-di-GMP effector proteins is growing fast since the characterization of PilZ (Amikam and Galperin, 2006), but it is still an open challenge in several bacterial species to fully decipher the complex network of metabolic enzymes and effectors. Depending of its molecular acting and its specific c-di-GMP binding site, different classes of c-di-GMP effector-proteins have been identified so far (reviewed in Römling *et al.*, 2013; Boyd and O'Toole, 2012). These effector proteins can affect protein complexes involved in flagella motility and biosynthesis of EPS such as alginate, PEL and cellulose, as well the transcriptional regulation of these genes. In addition to c-di-GMP effector proteins, two classes of riboswitch are also involved in c-di-GMP regulation, and different kinds of evidence point to the existence of as yet unidentified specific c-di-GMP-dependent effectors (reviewed in Boyd and O'Toole, 2012; Schirmer and Jenal, 2009).

The structure and c-di-GMP binding site of PilZ and PelD effector proteins has been well characterized in a number bacterial species. Two highly conserved sequence motifs (RxxxR and D/NxSxxG) located in the N-terminal part of PilZ domain and involved in c-di-GMP binding have been identified. However, to date, the diversity of cellular functions regulated by PilZ domain proteins have complicated further functional studies. Regarding PelD, the c-di-GMP binding site is identical to the I-site (RxxD) from known GGDEF domains.

Quorum sensing and c-di-GMP pathway in acidophilic bacteria

Molecular elements involved in QS systems have been identified in prokaryotes living in different extreme environments (Fig. 9.2A). In acidic niches, molecular players of QS systems have been characterized in biomining bacterial models such as *Acidithiobacillus* (Valenzuela *et al.*, 2007; Farah *et al.*, 2005; Rivas *et al.*, 2005) and *Leptospirillum* spp. (Moreno-Paz *et al.*, 2010). Early thin-layer chromatography experiments have indicated that

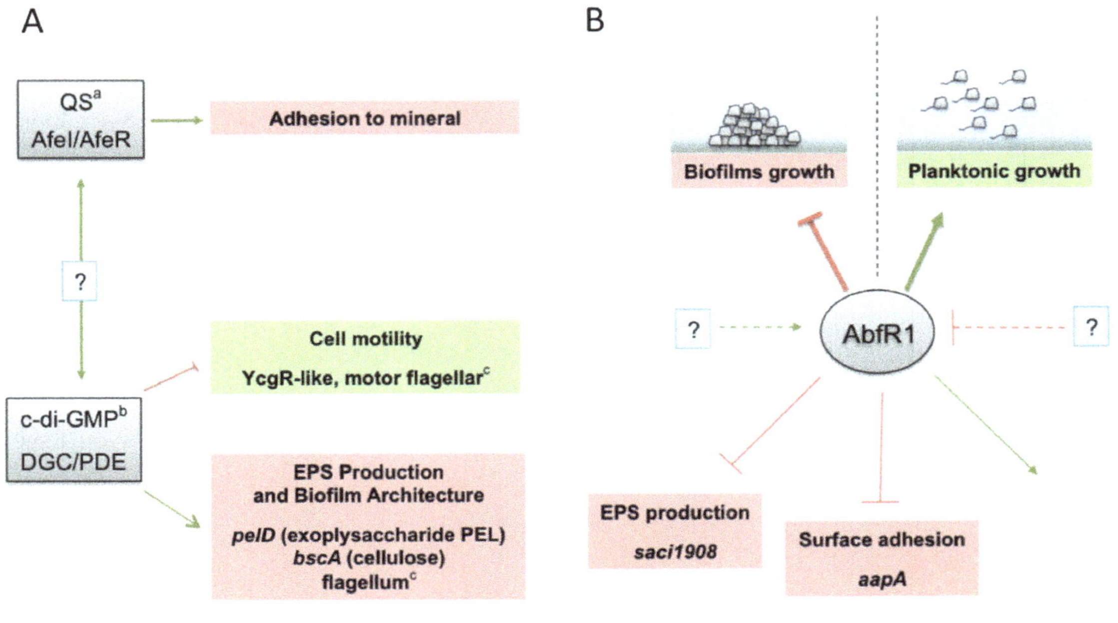

Figure 9.2 Molecular regulatory mechanisms of biofilm formation by acidophilic archaea and bacteria. Green and red arrows represent activation and repression, respectively; question marks indicate still unknown regulation mechanisms. Identified genes that are involved in distinctive cellular processes are shown in italics. (A) A quorum sensing system (QS) has so far been identified in *Acidithiobacillus* iron/sulfur-oxidizing species and *Leptospirrillum* sp. (a); A c-di-GMP pathway has been identified in all characterized members of *Acidithiobacillus* genus (b) and several key players for biofilm architecture are currently under study (personal communication). DGC, diguanylate cyclase; PDE, phosphodiesterase; flagellum participates in either cellular motility or promoting different biofilm architectures (c), except in *At. ferrooxidans* as it does not possess genes for flagellum synthesis. (B) Schematic representation of the archaeal biofilm regulator 1 (AbfR1) transcriptional regulatory network. AbfR1 is proposed to act as a biofilm transcriptional repressor, controlling the expression of genes: *flaB*, *flaX*, *appA* and *saci1908*.

at least two strains of *Leptospirillum* sp. type I do not produce AHL-signalling molecules (Ruiz *et al.*, 2008). However, analysis of available metagenomic data from the Iron Mountain AMD biofilm allowed the identification of a locus for QS type AI-1 in *Leptospirillum* sp. type III (Ruiz *et al.*, 2008). On the other hand, transcriptomic assays performed with a genomic microarray of *L. ferrooxidans* RT32a (former L3.2) and total RNA obtained from environmental samples of filamentous biofilms in the Rio Tinto also suggested that QS type AI-2 is involved in biofilm formation regulation in *Leptospirillum* spp. (Moreno-Paz *et al.*, 2010). AHL signals have been identified in the supernatant of different strains of *Acidithiobacillus thiooxidans* (Ruiz *et al.*, 2008), though no genes related to QS were identified in the draft genomes of either *At. thiooxidans* or *Acidithiobacillus caldus* (Valdés *et al.*, 2008). To date the type strain of *At. ferrooxidans* is the only acidophile species known so far, in which a functional QS system has been studied and characterized (Farah *et al.*, 2005; Rivas *et al.*, 2005). It involves a canonical QS locus made up of the genes *afeI* and *afeR* encoding, respectively, the AHL synthase AfeI and the transcriptional regulator AfeR. AfeI is capable of synthesizing at least nine different AHLs that differ for large of acyl-chains and substitutions at carbon 3 (Valenzuela *et al.*, 2007). Therefore it was proposed that *At. ferrooxidans* should be capable of interspecies communication with other acidophiles, and González *et al.* (2013) demonstrated that AHLs with large acyl chain promote cell attachment to minerals of both *At. ferrooxidans* and *Leptospirillum ferriphilum* in pure culture. Moreover, several AHLs with medium size acyl chain (C8 and C10-AHL) increase the capacity of *At. thiooxidans* to colonize solid substrates such as sulfur coupons, and the addition of different AHL-mixtures affects bioleaching efficiency of *At. ferrivorans* SS3, *Acidiferrobacter* sp. SPIII/3, and *L. ferrooxidans* DSM 2391 (Bellenberg *et al.*, 2014).

On the other hand, due to the lack of available genetic tools to build QS-null mutant strains, early studies on the role of QS system in *At. ferrooxidans* made use of synthetic analogous of AHLs (Valenzuela *et al.*, 2007). Using the same line of thought, Wenbin *et al.* (2011) used the synthetic QS blocker (5Z)-4-bromo-5-(bromomethylene)-2(5H)-furanone and found that copper tolerance in *At. ferrooxidans* involved QS regulation. Other kinds of functional studies, including transcriptional experiments, cell adhesion assays and electron microscopy analyses, have revealed that a specific tetrazolic compound increases the transcriptional levels of *afeI* and promotes colonization of sulfur coupons by *At. ferrooxidans* (Mamani *et al.*, 2013).

Together, these data indicate that QS plays, directly or indirectly, a role in biofilm formation in the acidithiobacilli. However, further studies are necessary to decipher which genes and pathways are regulated by QS in *Acidithiobacillus* spp. It has been reported that capsular EPS and several proteins involved in its biosynthesis and export are overexpressed in *At. ferrooxidans* cells adhered on pyrite (Vera *et al.*, 2013; Bellenberg *et al.*, 2012). In addition, using fluorescence microscopy analysis, González *et al.* (2013) demonstrated that EPS biosynthesis is induced by QS signalling molecules in *At. ferrooxidans* cells attached to solid strata.

Nicolas Guiliani's laboratory first suggested the existence of functional c-di-GMP pathways in acidophilic species belonging to the *Acidithiobacillus* genus (Castro *et al.*, 2009; Ruiz *et al.*, 2007). In 2010, Moreno-Paz and colleagues provided supporting evidence for the occurrence and increased expression of several genes encoding proteins with domains involved in c-di-GMP metabolism in *Leptospirillum* spp. from Rio Tinto. However, further functional characterization of c-di-GMP pathways in acidophiles has been strongly restricted by the capacity to develop null mutant strains for genes involved in c-di-GMP metabolism and signal transduction. Despite this, insights into the c-di-GMP metabolism in *Acidithiobacillus* spp. have been obtained over the last decade. With only five genes encoding proteins with EAL (1) and GGDEF/EAL (4) domains and two genes encoding PilZ effector proteins, *At. ferrooxidans* possesses the c-di-GMP pathway with the lowest complexity currently known among species of this genus. Higher c-di-GMP levels observed in attached cells of *At. ferrooxidans* (Ruiz *et al.*, 2012), *At. caldus* (Castro *et al.*, 2015) and *At. thiooxidans* (Diaz *et al.*, 2013) compared to planktonic cells, suggest that the c-di-GMP pathway may also dictate biofilm formation in these acidophiles. On the other hand, the c-di-GMP effector protein PelD and the *pel* operon involved in the biosynthesis of PEL exopolysaccharide are present only in sulfur-oxidizing *Acidithiobacillus* compared to iron/sulfur-oxidizing suggesting a pivotal role in biofilm formation in both species (Castro *et al.*, 2015; Diaz *et al.*, 2013). Recently, the development of a DGC-defective mutant strain of *At. caldus*, whose intracellular levels of c-di-GMP were strongly decreased, reveals that c-di-GMP pathway is involved in the regulation of swarming motility and adherence to elemental sulfur surfaces (Castro *et al.*, 2015). In addition, preliminary results obtained by comparing the wild type ATCC 19377 strain of *At. thiooxidans* with a c-di-GMP effector null mutant, strongly indicate that the c-di-GMP pathway may be involved in the regulation of biofilm architecture in acidophile species (unpublished) (Fig. 9.2A).

Archaea

Over the past 30 years, efforts directed to understand the microbial biofilm lifestyle have mainly been restricted to bacteria of medical importance and its relationship with persistent and chronic infections. Conversely, studies on sessile communities formed by microorganisms belonging to the *Archaea* domain have been largely ignored. Archaeal species are well known for their ability to inhabit extreme environments such as hot springs, deep-sea hydrothermal vents, salt lakes, sulfuric pools, etc., and it is widely recognized that metabolic and physiological features of archaea imply a major contribution on the biogeochemical cycling of essential elements on earth (Justice *et al.*, 2012). Likewise, evidence is increasing to suggest that archaea (mostly methanogens), via their syntrophic interactions with bacteria, build up biofilm communities that form part of the human microbioma (Bang *et al.*, 2014). Though archaea are found to ubiquitously colonize planet earth ecosystems, it is startling how limited molecular details are known on their sessile mode of growth.

Archaeal biofilms in acidic environments

Acid mine drainage sites (AMD) have probably been the most extensively studied acidic ecosystems to date (Chapters 1, 9, 10 and 13). Despite the often extreme acidity and high concentrations of toxic metals, a diverse range of autotrophic and heterotrophic archaea and bacteria can thrive within these environments, often forming mixed-community biofilms. Occasionally, these biofilms communities exhibit relatively low microbial diversity as frequently no more than five genera can be distinguished. Nonetheless, within some sites, where extreme physic-chemical conditions turn to be milder, biodiversity of biofilms/streamers in most AMD can be much greater. For instance, within Iron Mountain (CA), *Leptospirillum* group II corresponds to the predominant bacterial taxon, encompassing around 75% of the microbial fraction within some of the biofilms, and archaea accounts for ~10%, mostly represented by *Ferroplasma* (Comolli *et al*, 2009; Baker *et al.*, 2006). Their abundance increases with biofilm age and thickness, eventually dominating the top layer in stratified, mature biofilms (Wilmes *et al.*, 2009). On the other hand, archaea dominate the submerged suboxic biofilms in this system, acting as relevant players in carbon and nutrient cycling within these sessile communities (Justice *et al*, 2012). Besides *Ferroplasma*, a deeply branching clade of uncultivated euryarchaea referred to as ARMAN (archaeal Richmond Mine acidophilic nanoorganisms) have also been detected in the Iron Mountain biofilms (Comolli *et al.* 2009; Baker *et al.*, 2006). Other uncultivated archaea, including methanogens, have also been detected in other subterranean AMD-impacted environments (e.g. Johnson, 2012).

Some geothermal springs are moderately or extremely acidic, and host thermoacidophilic microorganisms, some of which may be the principal primary producers in these ecosystems. The Yellowstone National Park (USA) geothermal ecosystem has received particular attention. Unlike the biofilm communities of AMD environments, acidic ferric iron mat-biofilms in Yellowstone acidic hot springs were found to be principally composed of archaea, although bacterial diversity increases below 60°C (Kozubal *et al.*, 2012). Biofilms formed on the surface of these iron oxide mats exhibited high diversity of crenarchaeal representatives from the *Sulfolobales*, *Thermoproteales*, and *Desulfurococcales*, with *Metallosphaera yellowstonensis* being the dominant species.

Studies on biofilm formation by acidophilic archaea

The first acidophilic archaeal biofilm was described for the strain '*Ferroplasma acidarmanus*' Fer1

(Baker-Austin *et al.*, 2010). This mesophile organism formed flat multi-layered biofilms on glass and pyrite surfaces as well as long filamentous biofilms in AMD waters. Proteomic profiling of these biofilms revealed the up-regulation of proteins involved in adaptation to anoxia indicating the existence of anaerobic zones within the '*F. acidarmanus*' biofilm structure (Baker-Austin *et al.*, 2010). Several detailed studies have also been carried out in thermoacidophilic *Sulfolobus* spp. Biofilm formation was first described to occur in *Sulfolobus acidocaldarius*, *Sulfolobus solfataricus* and *Ssulfolobus tokodaii*. These strains exhibited very diverse biofilm morphologies: simple carpet-like structures in *S. solfataricus* and highly dense tower-like arrangements in *S. acidocaldarius* (Koerdt *et al.*, 2010) (Fig. 9.1B). Among *Sulfolobus* spp., *S. acidocaldarius* was shown to be the one that engages most readily in biofilm growth (Koerdt *et al.*, 2010). Additionally, pH and temperature changes promote the formation of biofilms in these organisms, suggesting that the sessile lifestyle is protective against environmental insults and changes. During biofilm development *S. acidocaldarius*, *S. solfataricus and S. tokodaii* produced an extracellular matrix containing glucose, galactose, mannose and N-acetylglucosamine residues.

Seeking to uncover a common biofilm genetic program, a comparative genome-wide proteomic/transcriptomic study was carried out on biofilms formed by *S. acidocaldarius*, *S. solfataricus* or *S. tokodaii* (Koerdt *et al.*, 2011). Intriguingly, expression profiling of biofilms versus planktonic cells for these three closely related species revealed evidence of species-specific responses, where ca. 15% of the *S. acidocaldarius* genes were differentially expressed, whereas *S. solfataricus* and *S. tokodaii* displayed only 3.4% and 1% of changes, respectively. The regulated genes encompassed a variety of basic cellular functions, comprising energy production/conversion, amino acid metabolism, lipid and carbohydrate metabolism, transport related functions and cell surface modifications. Despite the notably different transcriptomic and proteomic responses displayed by each species, one relevant common response gene was uncovered – the Lrs14 candidate transcriptional regulator, which is hypothesized to play a key role in the regulation of biofilm development in *Sulfolobus* (see below).

Structural components of acidophilic archaeal biofilms: cellular surface appendages and EPS production

Studies using thermoacidophilic crenarchaeal model *Sulfolobus* spp. have provided significant information on the structural components that contribute to the biofilm mode of growth in archaea, although the picture is still far from clear. *Sulfolobus* cell surfaces house a variety of type IV pili-like structures. The archaellum is the functional analogue appendage of the bacterial flagellum. Archaella were described as mediating surface adherence for *S. solfataricus* and other neutrophilic archaea (Zolghadr *et al.*, 2010). In *S. solfataricus* archaella deletion mutants were unable to adhere to surfaces.

The *Sulfolobales* possess other type-IV pili, including the UV inducible ups pilus which has proved to be essential in surface adhesion (Zolghadr *et al.*, 2010). The ups pili are highly induced after UV damage and lead to strain specific cell aggregation and subsequent DNA exchange (Ajon *et al.*, 2011). In addition to the archaellum and the ups pili *S. acidocaldarius* exhibits a third type IV pilus, the aap pili or archaeal adhesive pili (Henche *et al.*, 2012a). The influence on biofilms maturation of these diverse type IV pili structures has been investigated in *S. acidocaldarius*. For instance, the archaella does not affect biofilm formation, but plays a relevant role in cell dispersal from the biofilm after eight days of maturation. Conversely, the deletion of both the aap pili and the ups pili led to major structural biofilm impairments (Henche *et al.*, 2012b). *S. acidocaldarius* mutant cells lacking the aap pili developed very compact biofilms, whereas the deletion of the ups pili resulted in a fragile biofilm structure, unevenly distributed over the surface (Henche *et al.*, 2012b). Accordingly, the fact that both structures impacted in a rather opposing ways may reflect their complementary roles in optimal cell density and distance control during *S. acidocaldarius* biofilms development.

The production of EPS within acidophilic archaeal biofilms has been successfully studied using fluorescence-labelled lectins that bind selectively to terminal sugars residues. First studies showed that glass- and mica-attached *S. solfataricus* cells produced large amounts of EPS in between cells, which were composed of mannose, galactose and N-acetylglucosamine (Zolghadr *et al.*, 2010). Further studies revealed changes in the

EPS composition of archaeal biofilms during their development and maturation. In three-days old biofilms formed by *S. solfataricus*, *S. acidocaldarius* and *S. tokodaii*, the predominant sugar residues corresponded to glucosyl and mannosyl, whereas in six- days old mature biofilms signals for galactosyl and N-acetyl-glucosamine residues increased (Koerdt *et al.*, 2010). Interestingly the signals for galactosyl and N-acetyl-glucosamine residues were mainly observed on the surface of the biofilm structures formed by *S. acidocaldarius* and *S. tokodaii* (Koerdt *et al.*, 2010). It has also been reported that three other acidophilic archaea, *Ferroplasma acidiphilum*, *Sulfolobus metallicus* and a novel isolate of *Acidianus*, produced various EPS glycoconjugates when grown either on pyrite or elemental sulfur surfaces. A change of substratum induced different EPS glycoconjugates and biofilm structures of *Acidianus* sp. cells, and major differences in the EPS binding patterns (either tightly or loosely bound) were detected according to the substrate (Zhang *et al.*, 2015).

To date, the EPS biosynthesis pathways of archaea remain undescribed. However, enzymes with potential roles in archaeal EPS formation have been identified. One of these is the a-mannosidase form *S. solfataricus* as its homologous overexpression efficiently reduced the amount of carbohydrate and the protein concentration of the isolated EPS (Koerdt *et al.*, 2012). On the other hand, *S. acidocaldarius* genome contains a gene locus encoding 11 glycosyltransferases (GTs) and other genes coding for bacterial-like enzymes involved in the modification and polymerization of sugars. Via genetic epistasis analysis it was shown that the deletion of one of the genes in this cluster abolished EPS production, thus suggesting that its gene product plays a relevant role in the EPS biosynthetic pathways of *S. acidocaldarius* (Orell *et al.*, 2013a). The gene in question encodes a putative membrane protein, exhibiting 15 transmembrane regions, and no function assigned so far.

Besides exopolysaccharides, the archaeal EPS also consists of extracellular DNA (eDNA), as demonstrated for biofilms formed by different *Sulfolobus* species. These biofilms were stained with 7-hydroxy-9H-(1,3-dichloro-9,9-dimethylacridin-2-one) (DDOA) which is a membrane-impermeable DNA-intercalating dye. eDNA was mainly located in regions of dense cell aggregates (Koerdt *et al.*, 2010). However, the degradation of eDNA by Dnase I-treated 3-day-old biofilms of *Sulfolobus* did not show any effects on the biofilm structure (Koerdt *et al.*, 2010).

Regulation of archaeal biofilm formation: preliminary evidence

The regulatory network that governs the transition from planktonic (free-living) lifestyle to a biofilm mode of growth is just beginning to be elucidated for biofilm-forming archaea. Fundamental differences found with the bacterial biofilm-forming signalling pathways have made its understanding particularly intriguing and challenging. As described above, much of the community-wide behaviour in bacteria (e.g. biofilms formation), is regulated through a tightly coordinated interplay between the extracellular cell–cell communication system QS and the c-di-GMP mediated signalling.

Very few reports on archaeal cell–cell communication are available. On one hand, it has been shown that neutrophilic methanogenic and halophilic archaea can activate AHL bioreporters (Tommonaro *et al.*, 2012; Zhang *et al.*, 2012), suggesting that they are able to produce AHL-like QS inducers. Like Gram-negative bacteria, *Methanosaeta harundinacea* 6Ac strain, *Methanosarcina mazei* and *Methanothermobacter thermautotrophicus* produce a type of carboxylated AHLs via the synthetase encoding gene *filI*, a bacterial-like *luxI* orthologue (Zhang et. al, 2012). However a QS-based biofilm developmental process for these methano-archaeal strains has yet to be evaluated. On the other hand, the possibility of AHL-based QS in hydrothermal systems appears rather improbable. This is mainly due to (i) the intrinsic thermolability of AHL molecules and (ii) the absence of orthologues to mesophilic AHL synthetases in the genomic sequences of thermo-acidophiles (Nichols *et al.*, 2009). Despite it, the thermoacidophilic archaea *S. solfataricus* and *Sulfolobus islandicus* were shown to produce lactonase enzymes, which are involved in the hydrolysis of AHLs, most likely preventing cell–cell QS communication. This suggests that these archaeal lactonases might play a role in niche competition against bacterial species. In contrast, a boronated furanone-based QS autoinducer (AI-2) was found to be produced by the neutrophilic thermophile *P. furiosus*. As proposed, the AI-2 molecule is synthesized through a combination of biotic and abiotic reaction steps (Nichols *et al.*, 2009). Regulatory implications of AI-2 mediated signalling have not yet been investigated in any biofilm-forming thermoacidophilic archaea.

As described above, c-di-GMP is a well studied bacterial second messenger that promotes biofilm formation via mechanisms that can differ even among related microorganisms. Thus far, the occurrence of c-di-GMP has not been described for any archaeon (Baker-Austin *et al.*, 2010), thereby indicating different intracellular signal transduction pathways for biofilm development in the *Archaea* domain. On the other hand, recently discovered c-di-AMP, is involved in processes that allow *Bacillus subtilis* cells to manage extreme membrane and cell wall stress (Witte *et al.*, 2008). Emerging data also indicate that increased cellular levels of c-di-AMP stimulate biofilm development in certain bacteria (Corrigan *et al.*, 2011). Bioinformatic analyses have revealed the occurrence of DAC (diadenylyl cyclase) domain-containing genes in a subset of archaeal genomes, suggesting that c-di-AMP synthesis and signalling occurs in both the bacterial and archaeal domains (Römling, 2008). Whether c-di-AMP might also play a role in the regulation of archaeal biofilm formation remains to be assessed.

More recently, two novel biofilm-related transcriptional regulators have been identified in the thermoacidophilic crenarchaea *S. acidocaldarius*, providing the first insights into the transcriptional regulation of archaeal biofilm development (Orell *et al.*, 2013b; Koerdt *et al.*, 2011). These archaea-specific regulators markedly influenced biofilm formation. Functional studies have suggested that one of these Lrs14-like regulator (for leucine-responsive regulator of *Sulfolobus*) acts as a biofilm global activator in *S. acidocaldarius*, since its absence led to notable impairment of biofilm development. In contrast, the AbfR1 regulator (for archaeal biofilm regulator 1) was proposed to function as a biofilm repressor, since its deletion favoured EPS production and lead to the formation of thick biofilms (Fig. 9.2B). Remarkably, in *S. acidocaldarius abfR1* deletion mutant strain, the expression of the adhesive pili (*aap*) encoding genes was found to be up regulated, whereas transcript levels of the motility structure genes, the archaellum (*fla*), were down-regulated rendering the mutant strain non-motile. The AbfR1 protein was demonstrated to bind to the promoter regions of *fla* and *aap* genes, presumably controlling the expression of both cell surface structures (Orell *et al*, 2013b). The dual role of AbfR1 in controlling cell motility and formation of biofilms is reminiscent of *E. coli* CsgD biofilm master regulator. CsgD was shown to directly repress genes coding for flagella components and activate the synthesis of EPS, thereby regulating the switch from planktonic growth to the biofilm lifestyle (Pesavento *et al.*, 2008; Zogaj *et al.*, 2001). Unlike the *E. coli* CsgD regulator, AbfR1 from *S. acidocaldarius* appears to work as an activator of the archaellum synthesis and a repressor of the genes required for EPS biosynthesis and biofilm formation (Orell *et al*, 2013b) (Fig. 9.2B).

Concluding remarks

From studies using neutrophilic bacteria models, it is now well recognized that microbial biofilm development involves coordinated events leading to well-defined and distinct phenotypes that follow a tightly regulated genetic program. As described in this chapter, further studies are clearly needed in order to better understand what are the molecular mechanisms governing the transitional switch from planktonic to biofilm lifestyles in acidophilic bacteria and archaea.

So far, most of the available molecular information on biofilm formation and its regulation in acidic habitats have been obtained assessing pure cultures of either the mesophile-acidophile bacterium *At. ferrooxidans* or the thermoacidophilic crenarchaea *S. acidocaldarius*. Although these model organisms have shed substantial light into the biofilm physiology of acidophilic bacteria and archaea, the use of mixed cultures experimental approaches so as to mimic acidophilic-biofilms communities occurring in nature, seems imperative. Besides binary systems, such as bacteria/bacteria and bacteria/archaea, the inclusion in artificial biofilm experimental systems of acidophilic eukaryotes as predators (e.g. protozoa) might provide a more comprehensive understanding of the complexity and dynamics of natural biofilm communities in extreme acidic niches. Additionally, the study of moderately thermophiles and thermo-tolerant acidophilic microorganisms might contribute to broaden our understanding of the molecular pathways that promote both development and community dynamics within acidophile biofilms. Due to its intriguing physiological versatility, and because it has been found in several natural and industrial leaching communities, moderately thermophilic and thermo-tolerant species of the ferrous-iron- and mineral-sulfide-oxidizing Gram-positive acidophilic bacterial genus *Sulfobacillus* might represent suitable candidates.

Though several structural components that build up acidophilic archaeal biofilms have been successfully identified, using the thermoacidophile crenarchaeaote *S. acidocaldarius* as model, fundamental questions related to the archaeal biofilm physiology remain to be elucidated, for instance (i) do archaea inhabiting extremely acidic environments communicate with each other by means of either intra- or inter-species cell to cell communication system? and (ii) do they use c-di-nucleotides-based signal transduction pathways such as c-di-AMP to regulate biofilm formation? So far, the first insights into transcriptional regulation of archaeal biofilm development point in a different direction. As mentioned, two novel biofilm-related transcriptional regulators have been identified: Lrs14-like and AbfR1. These archaea-specific regulators markedly influenced biofilm formation; the former one has been suggested to act like a biofilm master activator in *S. acidocaldarius*, whereas AbfR1 regulator has been proposed to function as a biofilm repressor in the same archaeal model. Accordingly, it seems that the evidently opposing modes of action of these two archaeal regulators, by either enhancing or impairing biofilm formation, might be providing a suitable system to further unravel and reconstruct the biofilm transcriptional regulatory network of archaea.

Acknowledgement

FONDECYT Grants 11130229 (AO) and 1120295 (NG) for financial support.

References

Ajon, M., Frols, S., van Wolferen, M., Stoecker, K., Teichmann, D., Driessen, A.J., Grogan, D.W., Albers, S.V., and Schleper. C. (2011). UV-inducible DNA exchange in hyperthermophilic archaea mediated by type IV pili. Mol. Microbiol. *82*, 807–817.

Amikam, D., and Galperin, M.Y. (2006). PilZ domain is part of the bacterial c-di-GMP binding protein. Bioinformatics 22, 3–6.

Antunes, L.C., Schaefer, A.L., Ferreira, R.B., Qin, N., Stevens, A.M., Ruby, E.G., and Greenberg, E.P. (2007). Transcriptome analysis of the *Vibrio fischeri* LuxR-LuxI regulon. J. Bacteriol. *189*, 8387–8391.

Baker, B.J., Tyson, G.W., Webb, R.I., Flanagan, J., Hugenholtz, P., Allen, P., Allen, E.E., and Banfield, J.F., (2006). Lineages of acidophilic archaea revealed by community genomic analysis. Science *314*, 1933–1935.

Baker-Austin, C., Potrykus, J., Wexler, M., Bond, P.L., and Dopson, M. (2010). Biofilm development in the extremely acidophilic archaeon *'Ferroplasma acidarmanus'* Fer1. Extremophiles *14*, 485–491.

Bang, C., Ehlers, C., Orell, A., Prasse, D., Spinner, M., Gorb, S.N., Albers, S.V., and Schmitz, R.A. (2014). Biofilm formation of mucosa-associated methanoarchaeal strains. Front. Microbiol. *5*, 353–362.

Bassler, B.L., and Losick, R. (2006). Bacterially speaking. Cell. *25*, 237–246.

Bellenberg, S., Leon-Morales, C.F., Sand, W., and Vera, M., (2012). Visualization of capsular polysaccharide induction in *Acidithiobacillus ferrooxidans*. Hydrometallurgy *129–130*, 82–89.

Bellenberg, S., Díaz, M., Noël, N., Sand, W., Poetsch, A., Guiliani, N., and Vera, M. (2014). Biofilm formation, communication and interactions of leaching bacteria during colonization of pyrite and sulfur surfaces. Res. Microbiol. *165*, 773–781.

Boyd, C.D., and O'Toole, G.A. (2012). Second messenger regulation of biofilm formation: breakthroughs in understanding c-di-GMP effector systems. Annu. Rev. Cell Dev. Biol. *28*, 439–462.

Burdette, D.L., Monroe, K.M., Sotelo-Troha, K., Iwig, J.S., Eckert, B., Hyodo, M., Hayakawa, Y., and Vance, R.E. (2011). STING is a direct innate immune sensor of cyclic di-GMP. Nature *478*, 515–518.

Cantero, L., Palacios, J.M., Ruiz-Argueso T., and Imperial, J. (2006). Proteomic analysis of quorum sensing in *Rhizobium leguminosarum* biovar viciae UPM791. Proteomics *6*, S97–S106.

Castro, M., Ruiz, L.M., Barriga, A., Jerez, C.A., Holmes, D., and Guiliani, N. (2009). C-di-GMP pathway in biomining bacteria. Adv. Mater. Res. *71–73*, 223–226.

Castro, M., Deane, S.M., Ruiz, L., Rawlings, D.E., and Guiliani, N. (2015). Diguanylate cyclase null mutant reveals that C-Di-GMP pathway regulates the motility and adherence of the extremophile bacterium *Acidithiobacillus caldus*. PLoS One *10*, e0116399.

Chen, L.X., Hu, M., Huang, L.N., Hua, Z.S., Kuang, J.L., Li, S.J., and Shu, W.S. (2014). Comparative metagenomic and metatranscriptomic analyses of microbial communities in acid mine drainage. ISME J. *9*, 1579–1592.

Comolli, L.R., Baker, B.J., Downing, K.H., Siegerist, C.E., and Banfield, J.F. (2009). Three-dimensional analysis of the structure and ecology of a novel, ultra-small archaeon. ISME J. *3*, 159–167.

Corrigan, R.M., Abbott, J.C., Burhenne, H., Kaever, V., and Grundling, A. (2011). c-di-AMP is a new second messenger in *Staphylococcus aureus* with a role in controlling cell size and envelope stress. PLoS Pathog. *7*, e1002217.

Diaz, M., Copaja, S., and Guiliani, N. (2013). Functional analysis of c-di-GMP pathway in biomining bacteria *Acidithiobacillus thiooxidans*. Adv. Mater. Res. *825*, 133–136.

Dopson, M., and Johnson, D.B. (2012). Biodiversity, metabolism and applications of acidophilic sulfur-metabolizing microorganisms. Environ. Microbiol. *14*, 2620–2631.

Eberl, L., and Riedel, K. (2011). Mining quorum sensing regulated proteins – Role of bacterial cell-to-cell communication in global gene regulation as assessed by proteomics. Proteomics *11*, 3070–3085.

Farah, C., Vera, M., Morin, D., Haras, D., Jerez, C.A., and Guiliani, N. (2005). Evidence for a functional quorum-sensing type AI-1 system in the extremophilic bacterium *Acidithiobacillus ferrooxidans*. Appl. Environ. Microbiol. *71*, 7033–7040.

Garcia-Moyano, A., González-Toril, E., Aguilera, A., and Amils, R. (2007). Prokaryotic community composition and ecology of floating macroscopic filaments from an extreme acidic environment, Río Tinto (SW, Spain). Syst. Appl. Microbiol. *30*, 601–614.

Geske, G.D., O'Neill, J.C., and Blackwell, H.E. (2008). Expanding dialogues: from natural autoinducers to non-natural analogues that modulate quorum sensing in Gram-negative bacteria. Chem. Soc. Rev. *37*, 1432–1447.

Gilbert, K.B., Kim, T.H., Gupta, R., Greenberg, E.P., and Schuster, M. (2009). Global position analysis of the *Pseudomonas aeruginosa* quorum-sensing transcription factor LasR. Mol. Microbiol. 73, 1072–1085.

González, A., Bellenberg, S., Mamani, S., Ruiz, L., Echeverría, A., Soulère, L., Doutheau, A., Demergasso, C., Sand, W., Queneau, Y., *et al.* (2013). AHL signaling molecules with a large acyl chain enhance biofilm formation on sulfur and metal sulfides by the bioleaching bacterium *Acidithiobacillus ferrooxidans*. Appl. Microbiol. Biotechnol. *97*, 3729–3737.

Gonzalez, J.F., and Venturi, V. (2013). A novel widespread interkingdom signaling circuit. Trends Plant Sci. *18*,167–174.

González-Toril, E., Aguilera, A., Rodriguez, N., Fernández-Remolar, D., Gómez, F., Diaz, E., García-Moyano, A., Sanz, J.L., and Amils, R. (2010). Microbial ecology of Río Tinto, a natural extreme acidic environment of biohydrometallurgical interest. Hydrometallurgy *104*, 329–333.

Henche, A.L., Ghosh, A., Yu, X., Jeske, T., Egelman, E., and Albers, S.V. (2012a). Structure and function of the adhesive type IV pilus of *Sulfolobus acidocaldarius*. Environ. Microbiol. *14*, 3188–3202.

Henche, A.L., Koerdt, A., Ghosh, A., and Albers, S.V. (2012b). Influence of cell surface structures on crenarchaeal biofilm formation using a thermostable green fluorescent protein. Environ. Microbiol. *14*, 779–793.

Hengge, R. (2009). Principles of c-di-GMP signalling in bacteria. Nat. Rev. Microbiol. 7, 263–273.

Johnson, D.B. (2012). Geomicrobiology of extremely acidic subsurface environments. FEMS Microbiol. Ecol. *81*, 2–12.

Justice, N.B., Pan, C., Mueller, R., Spaulding, S.E., Shah, V., Sun, C.L., Yelton, A.P., Miller, C.S., Thomas, B.C., Shah, M., *et al.* (2012). Heterotrophic archaea contribute to carbon cycling in low-pH, suboxic biofilm communities. Appl. Environ. Microbiol. *78*, 8321–8330.

Koerdt, A., Godeke, J., Berger, J., Thormann, K.M., and Albers, S.V. (2010). Crenarchaeal biofilm formation under extreme conditions. PLoS One *5*, e14104.

Koerdt, A., Orell, A., Pham, T.K., Mukherjee, J., Wlodkowski, A., Karunakaran, E., Biggs, C.A., Wright, P.C., and Albers, S.V. (2011). Macromolecular Fingerprinting of *Sulfolobus Species* in Biofilm: A Transcriptomic and Proteomic Approach Combined with Spectroscopic Analysis. J. Proteome Res. *10*, 4105–4119.

Koerdt, A., Jachlewski, S., Ghosh, A., Wingender, J., Siebers, B., and Albers, S.V. (2012). Complementation of *Sulfolobus solfataricus* PBL2025 with an α-mannosidase: effects on surface attachment and biofilm formation. Extremophiles *16*,115–1125.

Kozlova, E.V., Khajanchi, B.K., Sha, J., and Chopra, A.K. (2011). Quorum sensing and c-di-GMP-dependent alterations in gene transcripts and virulence-associated phenotypes in a clinical isolate of *Aeromonas hydrophila*. Microb. Pathog. *50*, 213–223.

Kozubal, M., Macur, R.E., Jay, Z.J., Beam, J.P., Malfatti, S.A., Tringe, S.G., Kocar, B.D., Borch, T., and Inskeep, W.P. (2012). Microbial iron cycling in acidic geothermal springs of yellowstone national park: integrating molecular surveys, geochemical processes, and isolation of novel fe-active microorganisms. Front. Microbiol. *3*, e109.

Liljeqvist, M., Ossandon, F.J., González, C., Rajan, S., Stell, A., Valdes, J., Holmes, D.S., and Dopson, M. (2015). Metagenomic analysis reveals adaptations to a cold-adapted lifestyle in a low-temperature acid mine drainage stream. FEMS Microbiol. Ecol. *91*, pii: fiv011.

Mack, W.N., Mack, J.P., and Ackerson, A.O. (1975). Microbial film development in a trickling filter. Microb. Ecol. *2*, 215–226.

Mamani, S., Denis, Y., Moinier, D., Sabbah, M., Soulère, L., Queneau, Y., Bonnefoy, V., and Guiliani, N. (2013). Characterization of the quorum sensing regulon in *Acidithiobacillus ferrooxidans*. Adv. Mater. Res. *825*, 129–132.

Menzel. P., Gudbergsdóttir., S.R., Rike, A.G., Lin, L., Zhang, Q., Contursi, P., Moracci, M., Kristjansson, J.K., Bolduc, B., Gavrilov, S., *et al.* (2015). Comparative metagenomics of eight geographically remote terrestrial hot springs. Microb. Ecol. *70*, 411–424.

Moreno-Paz, M., Gómez, M.J., Arcas, A., and Parro, V. (2010). Environmental transcriptome analysis reveals physiological differences between biofilm and planktonic modes of life of the iron oxidizing bacteria *Leptospirillum* spp. in their natural microbial community. BMC Genomics *11*, 404.

Nichols, J.D., Johnson, M.R., Chou, C.J., and Kelly, R.M. (2009). Temperature, not LuxS, mediates AI-2 formation in hydrothermal habitats. FEMS Microbiol. Ecol. *68*, 173–181.

Orell, A., Fröls, S., and Albers, S.V. (2013a). Archaeal biofilms: the great unexplored. Annu. Rev. Microbiol. *67*, 337–354.

Orell, A., Peeters, E., Vassen, V., Jachlewski, S., Schalles, S., Siebers, B., and Albers, S.V. (2013b). Lrs14 transcriptional regulators influence biofilm formation and cell motility of Crenarchaea. ISME J. *7*, 1886–1898.

Parsek, M.R., and Greenberg, E.P. (2005). Sociomicrobiology: the connections between quorum sensing and biofilms. Trends Microbiol. *13*, 27–33.

Pesavento, C., Becker, G., Sommerfeldt, N., Possling, A., Tschowri, N., Mehlis, A., and Hengge, R. (2008). Inverse regulatory coordination of motility and curli-mediated adhesion in *Escherichia coli*. Genes Dev. *22*, 2434–2446.

Riedel, K., Hentzer, M., Geisenberger, O., Huber, B., Steidle, A., Wu, H., Høiby, N., Givskov, M., Molin, S., and Eberl, L. (2001). N-acylhomoserine-lactone-mediated communication between *Pseudomonas aeruginosa* and *Burkholderia cepacia* in mixed biofilms. Microbiology *147*, 3249–3262.

Rivas, M., Seeger, M., Holmes, D.S., and Jedlicki, E. (2005). A Lux-like quorum sensing system in the extreme acidophile *Acidithiobacillus ferrooxidans*. Biol. Res. *38*, 283–288.

Römling, U. (2008). Great times for small molecules: c-di-AMP, a second messenger candidate in Bacteria and Archaea. Sci. Signal. *1*, 39.

Römling, U., Galperin, M.Y., and Gomelskyc, M. (2013). Cyclic di-GMP: the first 25 years of a universal bacterial second messenger Microbiol. Mol. Biol. Rev. *77*, 1–52.

Ruiz, L.M., Sand, W., Jerez, C.A., and Guiliani, N. (2007). C-di-GMP pathway in *Acidithiobacillus ferrooxidans*: analysis of putative diguanylate cyclases (DGCs) and phosphodiesterases (PDEs) bifunctional proteins. Adv. Mater. Res. *20–21*, 551–556.

Ruiz, L.M., Valenzuela, S., Castro, M., Gonzalez, A., Frezza, M., Soulère, L., Rohwerder, T., Queneau, Y., Doutheau, A., Sand, W., *et al.* (2008). AHL communication is a widespread phenomenon in biomining bacteria and seems to be involved in mineral-adhesion efficiency. Hydrometallurgy *94*, 133–137.

Ruiz, L.M., Castro, M., Barriga, A., Jerez, C.A., and Guiliani, N. (2012). The extremophile *Acidithiobacillus ferrooxidans* possesses a c-di-GMP signalling pathway that could play a significant role during bioleaching of minerals. Lett. Appl. Microbiol. *54*, 133–139.

Schirmer, T., and Jenal, U. (2009). Structural and mechanistic determinants of c-di-GMP signaling. Nat. Rev. Microbiol. *7*, 724–735.

Serra, D.O., Richter, A.M., and Hengge, R. (2013). Cellulose as an architectural element in spatially structured *Escherichia coli* biofilms. J. Bacteriol. *195*, 5540–5554.

Simm, R., Morr, M., Kader, A., Nimtz, M., and Römling, U. (2004). GGDEF and EAL domains inversely regulate cyclic di-GMP levels and transition from sessility to motility. Mol. Microbiol. 53, 1123–1134.

Stevens, A.M., Queneau, Y., Soulère, L., von Bodman, S., and Doutheau, A. (2011). Mechanisms and synthetic modulators of AHL-dependent gene regulation. Chem. Rev. *111*, 4–27.

Tchigvintsev, A., Xu, X., Singer, A., Chang., C., Brown, G., Proudfoot, M., Cui, H., Flick, R., Anderson, W.F., Joachimiak, A., *et al.* (2010). Structural insight into the mechanism of c-di-GMP hydrolysis by EAL domain phosphodiesterases. J. Mol. Biol. *402*, 524–538.

Tommonaro, G., Abbamondi, G.R., Iodice, C., Tait, K., and De Rosa, S. (2012). Diketopiperazines produced by the halophilic archaeon, *Haloterrigena hispanica*, activate AHL bioreporters. Microb Ecol. *63*, 490–495.

Ueda, A., and Wood, T.K. (2009). Connecting quorum sensing, c-di-GMP, pel polysaccharide, and biofilm formation in *Pseudomonas aeruginosa* through tyrosine phosphatase TpbA (PA3885). PLoS Pathog. *5*, e1000483.

Valdés, J., Pedroso, I., Quatrini, R., and Holmes, D.S. (2008). Comparative genome analysis of *Acidithiobacillus ferrooxidans, A. thiooxidans* and *A. caldus*: insights into their metabolism and ecophysiology. Hydrometallurgy *94*, 180–184.

Valenzuela, S., Banderas, A., Jerez, C.A., and Guiliani, N. (2007). Cell–cell communication in bacteria: a promising new approach to improve bioleaching efficiency? In Microbial Processing of Metal Sulfides, Donati, E., and Sand, W., eds (Springer, Dordrecht, Netherlands), pp. 253–264.

Vasil, M.L. (2003). DNA microarrays in analysis of quorum sensing: strengths and limitations. J. Bacteriol. *185*, 2061–2065.

Vera, M., Krok, B., Bellenberg, S., and Sand, W., and Poetsch, A. (2013). Shotgun proteomics study of early biofilm formation process of *Acidithiobacillus ferrooxidans* ATCC 23270 on pyrite. Proteomics *13*, 1133–1144.

Wenbin, N., Dejuan, Z., Feifan, L., Lei, Y., Peng, C., Xiaoxuan, Y., and Hongyu, L. (2011). Quorum-sensing system in *Acidithiobacillus ferrooxidans* involved in its resistance to Cu2+. Lett. Appl. Microbiol. *53*, 84–91.

Wilmes, P., Remis, J.P., Hwang, M., Auer, M., Thelen, M.P., and Banfield, J.F. (2009). Natural acidophilic biofilm communities reflect distinct organismal and functional organization. ISME J. *3*, 266–270.

Witte, G., Hartung, S., Buttner, K., and Hopfner, K.P. (2008). Structural biochemistry of a bacterial checkpoint protein reveals diadenylate cyclase activity regulated by DNA recombination intermediates. Mol. Cell. *30*, 167–178.

Zhang, G., Zhang, F., Ding, G., Li, J., Guo, X., Zhu, J., Zhou, L., Cai, S., Liu, X., Luo, Y., *et al.* (2012). Acyl homoserine lactone-based quorum sensing in a methanogenic archaeon. ISME J. *6*, 1336–1344.

Zhang, R.Y., Neu, T.R., Bellenberg, S., Kuhlicke, U., Sand, W., and Vera, M. (2015). Use of lectins to *in situ* visualize glycoconjugates of extracellular polymeric substances in acidophilic archaeal biofilms. Microb. Biotechnol. *8*, 448–461.

Zogaj, X., Nimtz, M., Rohde, M., Bokranz, W., and Römling, U. (2001). The multicellular morphotypes of *Salmonella typhimurium* and *Escherichia coli* produce cellulose as the second component of the extracellular matrix. Mol. Microbiol. *39*, 1452–1463.

Zolghadr, B., Klingl, A., Koerdt, A., Driessen, A.J., Rachel, R., and Albers, S.V. (2010). Appendage-mediated surface adherence of *Sulfolobus solfataricus*. J. Bacteriol. *192*, 104–110.

Distribution of Acidophilic Microorganisms in Natural and Man-made Acidic Environments

10

Sabrina Hedrich and Axel Schippers

Abstract

Acidophilic microorganisms can thrive in both natural and man-made environments. Natural acidic environments comprise hydrothermal sites on land or in the deep sea, cave systems, acid sulfate soils and acidic fens, as well as naturally exposed ore deposits (gossans). Man-made acidic environments are mostly mine sites including mine waste dumps and tailings, acid mine drainage and biomining operations. The biogeochemical cycles of sulfur and iron, rather than those of carbon and nitrogen, assume centre stage in these environments. Ferrous iron and reduced sulfur compounds originating from geothermal activity or mineral weathering provide energy sources for acidophilic, chemolithotrophic iron- and sulfur-oxidizing bacteria and archaea (including species that are autotrophic, heterotrophic or mixotrophic) and, in contrast to most other types of environments, these are often numerically dominant in acidic sites. Anaerobic growth of acidophiles can occur via the reduction of ferric iron, elemental sulfur or sulfate. While the activities of acidophiles can be harmful to the environment, as in the case of acid mine drainage, they can also be used for the extraction and recovery of metals, as in the case of biomining. Considering the important roles of acidophiles in biogeochemical cycles, pollution and biotechnology, there is a strong need to understanding of their physiology, biochemistry and ecology.

Introduction

Environments inhabited by acidophiles can be of natural origin where acidic conditions have existed for many years, such as volcanic or geothermal areas, or environments where acidity has arisen due to human activities, such as mining of metals and coal. In such environments elemental sulfur and other reduced inorganic sulfur compounds (RISCs) are formed from geothermal activities or the dissolution of minerals. Weathering of metal sulfides due to their exposure to air and water leads to their degradation to protons (acid), RISCs and metal ions such as ferrous and ferric iron, copper, zinc etc. (see Chapter 1).

RISCs and metal ions are abundant at acidic sites where most of the acidophilic microorganisms thrive using iron and/or sulfur redox reactions. In contrast to biomining operations such as heaps and bioleaching tanks, which are often aerated to enhance the activities of mineral-oxidizing prokaryotes, geothermal and other natural environments, can harbour a more diverse range of acidophiles including obligate anaerobes.

Populations in natural acidic environments

Environments where acid is formed naturally without the influence of mining include geothermal terrestrial sites (solfatara) that occur at active volcanoes, deep-sea hydrothermal systems, and naturally exposed sulfide ore deposits. These sites are of great interest when searching for organisms to be used, for example in biomining operations, as it is often the case that these environments have existed for many years and therefore indigenous microorganisms are potentially adapted to high metal and salt concentrations, extreme temperature and low pH (Table 10.1).

Table 10.1 Physico-chemical parameters and prokaryotic diversity of selected natural environments where acidophiles have been detected

Site	pH	Temperature (°C)	*Bacteria*	*Archaea*	Reference
Geothermal sites					
Yellowstone National Park, Wyoming, USA	2.0–4.4	30–90	*Fx. thermotolerans*, *Sulfobacillus*-like, *Am. ferrooxidans*-like, *Methylobacterium*-like, *Acidisphaera* sp., *Alicyclobacillus*-like, *H. acidophilus*, *Thiomonas* sp.	*Ad. brierleyi*, *Sulfolobales*, *Thermoproteales*, *Desulforococales*, '*Geoarchaeota*', *Thaumarchaeota*, Metallosphaera-like, *Marinithermus* sp.	Johnson *et al.* (2003), Inskeep *et al.* (2004, 2010), Kozubal *et al.* (2013), Beam *et al.* (2014)
Montserrat, Lesser Antilles	1.6–7.4	24–99	*At. caldus*, *At. ferrooxidans*. *L. ferrooxidans*, *Am. ferrooxidans*, *Sulfobacillus* spp., *Acidiphilium*-like, *Alicyclobacillus* spp., *Desulfosporosinus* sp.	*Sulfolobus* spp., *Acidianus* spp., *Ferroplasma*-like	Atkinson *et al.* (2000), Burton and Norris (2000)
St. Lucia, Lesser Antilles	2.0	42.6	*At. caldus*, *Sulfobacillus* spp.	*Acidianus* spp., unknown *Eukaryota*	Stout *et al.* (2009)
Palaeochori Bay, Milos	2.0	50–60	*At. caldus*, '*Acidithiomicrobium*', *Acidithiobacillus* spp., *Sulfobacillus* spp.	*Acidianus* spp., *Ferroplasma*-like	Norris (2011)
Vulcano, Italy	2.0–3.5	35–75	*Acidithiobacillus* spp., *T. prosperus*	*Tp. volcanium*, *Ad. brierleyi*, *Ad. inferno*	Simmons and Norris (2002)
Mutnovsky volcano, Kamchatka	3.5–4.0	70	*At. caldus*, *Thermotogae*	*Thaumarchaeota*, *Acidilobus* sp., *Vulcanisaeta* sp., *Thermoplasma* sp.	Wemheuer *et al.* (2013)
Hveragerji, Iceland Salar de Gorbea	2.5–3.5 1.1–4.8	81–90 8–20	NR *Sulfobacillus* spp., *Alicyclobacillus* spp., *Acidisoma* sp., *Alkalibacter* sp., *Proteo-* and *Actinobacteria*, *Cyanobacteria*	*Sulfolobales*, *Thermoproteales*, crenarchaeotal group I.1b *Thermoplastales*, *Thermococcales*, *Halobacteriales*, *Methanosarcinales and Methanobacteriales*, *Thermoproteales*	Kvist *et al.* (2007) Davis-Belmar *et al.* (2013), Escudero *et al.* (2013)
Copahue volcanic area including Rio Agrio, Argentina	<1.0–4.0	8–85	*At. ferrooxidans*, *At. thiooxidans*, *Leptospirillum* spp., *At. caldus*-like, *Sulfobacillus* spp.	*Ferroplasma* spp., *Ad. copahuensis*, *Sulfolobus* spp.	Urbieta *et al.* (2012, 2014a)
Deep sea hydrothermal vents					
deep sea vent, Eastern Lau Spreading Centre	3–6	50–80	NR	'*Aciduliprofundum boonei*'	Reysenbach *et al.* (2006)
Caves					
Frasassi cave, Italy	0–1.5	13	*At. thiooxidans*, *Sulfobacillus* spp., *Am. ferrooxidans*	*Ferroplasma* spp., '*G-plasma*'	Macalady (2007), Jones (2012)
Cueva de Villa Luz, Mexico	1.4	28	*Acidithiobacillus* spp., *Am. ferrooxidans*	NR	Hose *et al.* (2000)

Table 10.1 Continued

Site	pH	Temperature (°C)	*Bacteria*	*Archaea*	Reference
Gossan					
Citronen Fjord, Greenland	1.9–2.5	2–5	*Acidithiobacillus*-like, heterotrophs	NR	Langdahl and Ingvorsen (1997)
Antarctica	3.2–4.5	0–4	*T. plumbophilus*, *At. ferrivorans*	NR	Dold *et al.* (2013)

Abbreviations: *Acidianus* (*Ad.*), *Acidilobus* (*Al.*), *Acidimicrobium* (*Am.*), *Acidiphilium* (*A.*), *Acidisphaera* (*As.*), *Acidithiobacillus* (*At.*), '*Acidithiomicrobium*' (*Atm.*), *Alicyclobacillus* (*Alb.*), *Ferrithrix* (*Fx.*), *Ferroplasma* (*Fp.*), *Hydrogenobaculum* (*H.*), *Leptospirillum* (*L.*), *Metallosphaera* (*M.*), *Picrophilus* (*P.*), *Sulfisphaera* (*Ss.*), *Sulfobacillus* (*Sb.*), *Sulfolobus* (*S.*), *Thermoplasma* (*Tp.*), *Thiobacillus* (*T.*).
NR, not reported.

Hydrothermal environments

Solfatara are characterized by high temperatures (up to 100°C) and elevated concentrations of sulfur (and RISCs), hydrogen and often also soluble metals and arsenic. When elemental sulfur, formed by the condensation of volcanic gases such as H_2S and SO_2, is oxidized by acidophilic prokaryotes, sulfuric acid is formed which results in low-pH environments. Owing to the low pH and high temperatures in volcanic areas, acid-labile minerals are dissolved and release elevated concentration of transition metals and metalloids. The nature of hydrothermal sites is dictated by the subterranean geology and water flow. In particular, the degree of mixing of deep hydrothermally heated water (buffered in the neutral–alkaline region by CO_2/HCO_3^-) with cold shallow groundwater (strongly acidified by microbial and chemical oxidation of sulfur compounds) determines pH and temperature of the environment (Atkinson, 2000). The microbiology of acidic geothermal areas has been studied for more than four decades (e.g. Johnson *et al.*, 2003; Brock, 2001; 1978) and microbial communities in these environments host a variety of deeply rooted known and unknown *Archaea*, *Bacteria* and *Eukarya*.

One of the most intensively studied natural acidic hydrothermal environment is Yellowstone National Park (YNP; Wyoming, USA) which houses the most numerous and diverse geothermal terrestrial systems on Earth with widely varying geochemical properties, such as temperature, pH, and concentrations of dissolved ions and oxygen.

The first sulfur-metabolizing archaeon, subsequently named as *Acidianus brierleyi*, was isolated by James Brierley in 1965 from geothermal sites in Yellowstone (Brierley, 1973). Further cultivation-based studies on acidophiles from YNP resulted in the enrichment of novel thermophilic iron- and/or sulfur-metabolizing and heterotrophic bacteria (Johnson *et al.*, 2003), including the isolation of the type strain of the acidophilic, moderately thermophilic species *Ferrithrix thermotolerans* (Johnson *et al.*, 2009) and *Acidicaldus organivorans* (Johnson *et al.*, 2006).

While a number of described acidophilic archaeal species have been detected in samples of YNP, several studies have shown that the hot springs of YNP are a source of novel and deeply branching archaea (e.g. Beam *et al.*, 2014; Inskeep *et al.*, 2010, 2004; Table 10.1). Archaea in an acidic iron oxide geothermal spring were further analysed by Kozubal *et al.* (2013) and proposed as a new candidate phylum within the domain *Archaea* referred to as '*Geoarchaeota*' or 'novel archaeal group 1 (NAG1)'. These organisms were found to contain pathways necessary for the catabolism of peptides and complex carbohydrates and genes involved in the metabolism of oxygen. Furthermore Beam *et al.* (2014) applied metagenome sequence analysis to study thermophilic populations (65–72°C) at acidic iron oxide- and sulfur-rich sediment environments of YNP. These deeply rooted *Thaumarchaeota* were proposed to be chemo-organotrophic and couple growth to the reduction of oxygen or nitrate in iron oxide habitats, or sulfur in hypoxic (low-oxygen) sulfur sediments. Possible carbon sources for these archaea include aromatic compounds, complex carbohydrates, oligopeptides and amino acids or carbon dioxide. No evidence for the oxidation of ammonia was obtained from the *de novo* sequence assemblies.

Some of the geothermal springs within YNP often contain elevated concentrations of arsenic (1–150 mg/l) which occurs predominantly as arsenious acid [As(III)] when the water is discharged, but is rapidly oxidized downstream from the spring source. The oxidation of As(III) to arsenate (As(V)) can occur through abiotic and biotic processes and, although biological oxidation is typically much faster (Cullen and Reimer, 1989), only a few studies have examined the organisms that carry out this process. Studies on the microbial communities of acid-sulfate-chloride springs within YNP (pH ~3.1, T = 53–74°C) indicated the dominance of hydrogen-oxidizing *Hydrogenobaculum acidophilum*, *Desulphurella* and smaller numbers of *Acidimicrobium* and *Thiomonas* (Inskeep *et al.*, 2004). Archaeal sequences of characterized members in these springs were related to thermophilic, metal-mobilizing *Metallosphaera* and *Marinithermus*, but the majority were unknown crenarchaeotes and euryarchaeotes, most closely related to sequences from marine hydrothermal vents and which are proposed to be responsible for the oxidation of As(III) in the spring (Jackson *et al.*, 2001). D'Imperio *et al.* (2007) described the isolation of a novel arsenite-oxidizing *Acidicaldus* species and documented arsenite oxidation inhibition by hydrogen sulfide and its influence on the distribution of arsenite-oxidizing chemolithotrophs.

Another interesting and intensively studied hydrothermal site is located on the Caribbean island of Montserrat (Atkinson *et al.*, 2000; Burton and Norris, 2000; Table 10.1), the topography of which is dominated by the active Chances Peak volcano (914 m) and the eroded peaks of three extinct volcanoes, Centre Hills (676 m), Silver Hill (403 m) and South Soufrière Hills (756 m).

Microbial studies on Montserrat samples identified the thermophilic sulfur-oxidizer *Acidithiobacillus caldus*, which is known to grow over a wide temperature range, as the dominant organism at moderate to higher temperature (Burton and Norris, 2000). Cultivation studies furthermore indicated that iron-/sulfur-oxidizing *Acidithiobacillus ferrooxidans* dominated over iron-oxidizing *Leptospirillum* spp. at moderate temperatures, while the latter was dominant in samples of 50–58°C. These low to moderate temperature site populations were also accompanied by heterotrophic *Acidiphilium*-like spp., *Sulfobacillus* spp., *Acidimicrobium ferrooxidans*, *Alicyclobacillus* species, and an unknown actinobacterium (RIV 14). 16S rRNA gene sequences of two potentially moderate sulfur-reducing genera (one of the *Desulfurella* group) were detected in the same pools, while sulfate-reducing bacteria could be isolated from moderate pools (30°C, pH 3.2) one of which (strain M1) was later assigned as the type strain of the novel species *Desulfosporosinus acididurans* (Sánchez-Andrea *et al.*, 2015). Higher temperature pools of Montserrat contained clone sequences related to *Ferroplasma*-like organisms, while the dominant group clustered in the *Crenarchaeota*, distantly related to classified *Sulfolobus* spp. and *Acidianus* spp. Members of the thermoacidophilic crenarchaeotal genus *Sulfolobus* were isolated from sites of 65–98°C including one isolate representing a new species which has also been isolated from the Azores (P.R. Norris, unpublished data).

Two more sites with similar microbiological composition such as Montserrat have been studied. One is located on the island St. Lucia, where volcanic, acidic sulfur-springs occur (pH of 2.0 and 42.6°C) and the other one is a geothermal site at Palaeochori Bay of the island of Milos in the Aegean Sea (T = 55–60°C). Both microbial communities were dominated by *At. caldus* and *Acidianus* spp. and samples from St. Lucia also contained low numbers of *Sulfobacillus* spp. and unknown *Euryarchaeota* (Stout *et al.*, 2009; Table 10.1). The Milos samples additionally harboured unknown *Ferroplasma*/*Acidiplasma*-like clone sequences as well as a novel sulfur-oxidizing actinobacterium '*Acidithiomicrobium*' (also isolated from Milos) and a novel *Acidithiobacillus* sp. V1 (Norris *et al.*, 2011).

Copahue-Caviahue, a geothermal area in Argentina, is dominated by the Copahue Volcano which is responsible for the thermal activity in this area. Several acid ponds, pools and hot springs occur in this region, some of which are the source of the acidic river Rio Agrio that discharges in the acidic Caviahue lake. The various habitats harbour a broad range of acidophilic, moderate to thermophilic microorganisms as they vary in their physico-chemical characteristics (mainly pH and temperature).

Molecular analysis of the microbial river community (Urbieta *et al.*, 2012) revealed the presence of moderate, sulfur-oxidizing bacteria (*Acidithiobacillus thiooxidans* and *Acidithiobacillus albertensis*) and moderately thermophilic iron- and sulfur-oxidizing bacteria (*Alicyclobacillus* spp. and *Sulfobacillus* spp.) ubiquitous along the river. Iron-oxidizing bacteria (*Leptospirillum* spp. and *Ferrimicrobium* spp.) and archaea (*Ferroplasma* spp.) were present at the source of the river where iron concentrations were much lower than further downstream. The biodiversity of various ponds studied in the Copahue area (Urbieta *et al.*, 2014a; Table 10.1) is determined by the temperature, resulting in archaea (novel *Sulfolobales* spp. and *Thermoplasmatales* spp.) colonizing higher-temperature ponds, whereas ponds of moderate temperature are colonized by sulfur-oxidizing bacteria, e.g. *Thiomonas* spp., *Acidithiobacillus* spp., *Hydrogenobaculum* spp. and *Acidiphilium* spp. Also a novel *Acidianus* species, *Acidianus copahuensis*, whose genome has recently been sequenced, has been isolated from hot springs of Copahue (Urbieta *et al.*, 2014b). The presence of sulfate-reducing bacteria in anaerobic sediments of an acidic hot spring of Copahue was confirmed by the isolation of several strains related to the sulfate-reducer '*Desulfobacillus acidavidus*' strain CL4 from these sediments (Willis *et al.*, 2013). Several eukaryotes, such as yeasts and filamentous fungi, have also been detected in the geothermal area (Chiacchiarini *et al.*, 2009). In microbial terms, Copahue is a very diverse environment and analyses have shown that the distribution of microorganisms in this system is determined by pH, temperature and conductivity, rather than mineral chemistry as it is the case in Yellowstone National Park (Urbieta *et al.*, 2014a).

Numerous hydrothermal vents are also located on the Kamchatka Peninsula in the northeast of Russia. These vents constantly release geothermal gases and fluids dominated by N_2, CO_2 and H_2; CH_4 and H_2S also frequently occur. The hot springs found in this area have multiple origins, including meteoric and magmatic water. Active hydrothermal springs with a large variety in chemical composition, temperature, and pH are located in the area of the Uzon Caldera, Geyser Valley near the volcanoes Karymskii and Mutnovsky. Similar sites can also be found in Iceland, which is located on the Mid-Atlantic Ridge and harbours a zone of active rifting and is characterized by active volcanoes, fissure swarms, numerous normal faults, and high-temperature geothermal fields. In the famous Hveragerði high-temperature geothermal field, located about 50 km southwest of Reykjavik, geothermal manifestations consist of fumaroles and hot springs, which have different values of temperatures and pH.

Studies on the prokaryotic diversity of the sediment samples near the Mutnovsky volcano (pH 3.5–4.0, $T = 70°C$) revealed the dominance of sulfur-oxidizing *At. caldus* and *Thermotogae* (accompanied by small numbers of *Desulfurella*, *Kosmotoga* and *Hydrogenobaculum* spp.) and sequences clustering with the thaumarchaeal 'terrestrial hot spring group' as well as minor numbers of *Acidilobus* spp., *Vulcanisaeta* spp. and uncultivated *Thermoplasma* spp. (Wemheuer *et al.*, 2013; Table 10.1). The type strain of the hyperthermophile *Acidilobus aceticus* has previously been isolated from the same area and described as a strictly anaerobic acidophile with a heterotrophic lifestyle (Prokofeva *et al.*, 2000).

The archaeal composition of samples from Hveragerði (Iceland) hot springs (pH 2.5, $T = 81–90°C$) was more diverse, comprising members of the acidophilic, sulfur-oxidizing *Sulfolobales*, *Thermoproteales* and uncultured members of the crenarchaeal group I.1b (Kvist *et al.*, 2007). Further studies on archaeal communities of water and mud samples from hot springs in Kamchatka and Iceland detected archaeal ammonia monooxygenase (*amoA*) genes in agreement with considerable *in situ* nitrification rates (Reigstadt, 2010). The wide temperature and pH range ($T = 38–97°C$ and pH 2.5–7) at which ammonium-oxidizing archaea (AOA) have been detected, together with the geographically disparate sampling locations, provide evidence for an active role of hyperthermophilic AOA in nitrogen cycling in hot spring microbial communities throughout the world.

Natural acidic, saline lakes are relatively rare compared to alkaline salt lakes, but can be found in Australia and the Atacama region of northern Chile. The Salar de Gorbea and Salar Ignorado are basins in Chile where the oxidation of native volcanic sulfur occurrences leads to the release of sulfuric acid and strong hydrothermal alteration of the country rocks lowers the buffering capacity and thereby cause these habitats to be acidic. The water bodies of Salar de Gorbea are characterized by high salinity (100–300% total dissolved solids) and acidic pH (1.1–4.8) with temperatures between 8 and 19.5°C. The dominant bacteria in these samples were members of the *Proteo*-and *Actinobacteria, Cyanobacteria* as well as *Firmicutes* (e.g. *Sulfobacillus, Alicyclobacillus*) (Davis-Belmar *et al.*, 2013; Escudero *et al.*, 2013). Cultivation based studies revealed the presence of novel strains of the genus *Acidisoma* (Davis-Belmar *et al.*, 2013). The archaeal community of Salar de Gorbea was dominated by various members of the *Euryarchaeota* (*Thermoplastales, Thermococcales, Halobacteriales, Methanosarcinales and Methanobacteriales*) and minor numbers of *Crenarchaeota* (*Thermoproteales*) (Escudero *et al.*, 2013).

Salt-tolerant acidophiles appear to be relatively uncommon. One such is the acidophilic iron-oxidizer '*Thiobacillus prosperus*' (Huber and Stetter, 1989), which was isolated from the Mediterranean island of Volcano (located to the north of Sicily; Table 10.1) and which can grow in media containing up to 6% NaCl. 16S rRNA gene clone libraries of samples from Volcano sites were dominated by a novel *Acidithiobacillus* sp. at 35–45°C, while novel iron-oxidizing bacteria related to '*T. prosperus*' were isolated but not detected in gene banks (Simmons and Norris, 2002). Acidophilic archaea at Volcano included *Acidianus* spp. and *Thermoplasma* spp., all of which displayed a broad tolerance (up to 4%) to sodium chloride (Simmons and Norris, 2002).

While acidophilic prokaryotic communities in hydrothermal areas have been studied quite intensively, eukaryotes in geothermal sites have received far less attention. In one study, eukaryotes observed in Nymph Creek, YNP (pH 2.5, $T=40$°C), comprised *Cyanidium*-like algae, the acidophilic diatom *Pinnularia*, the amoeba *Tetramitus thermoacidophilus and Naegleria* spp., flagellates and novel members of the amoeba *Vampyrellidae* (Amaral-Zettler *et al.*, 2013).

Deep-sea hydrothermal vents

The walls of deep-sea hydrothermal vent deposits (chimneys) with their steep chemical and thermal gradients provide a wide range of microhabitats for microorganisms. *In situ* measurement of pH at the exterior of such a deposit wall on the East Pacific Rise reported pH values of 4.1 at 120°C (von Damm *et al.*, 1995). Calculated pH values based on various flow rates and exchange of fluids within the pores can range from 3 to 6 at temperatures of 50–80°C (Le Bris *et al.*, 2005). Acidity is also produced locally in the chimneys by the precipitation of minerals such as pyrite and chalcopyrite. The majority of thermophiles isolated from deep-sea vents are, however, neutrophiles or acid-tolerant organisms. A notable exception is the thermoacidophilic sulfur- and ferric iron-reducing heterotroph '*Aciduliprofundum boonei*', a member of the 'deep-sea hydrothermal vent *Euryarchaeota* 2' (DHVE2) group, which grows between pH 3.3 and 5.8 (Reysenbach *et al.*, 2006; Table 10.1). Further thermoacidophilic strains related to '*Acp. boonei*' could be isolated from vents along the East Pacific Rise and Mid-Atlantic Ridge, showing that the DHVE 2 are widespread in various hydrothermal vents and constitute a major part of the archaeal population (Flores *et al.*, 2012).

Cave systems

Another example of natural-acidic environments are cave systems, such as the Lechugilla Cave (New Mexico, USA; Hill, 1995), Frasassi cave (Italy; Macalady *et al.*, 2007) and Cueva de la Villa Luz (Mexico; Hose *et al.*, 2000) (Table 10.1). These develop in subterranean environments containing sulfide-rich ground-waters where, under aerobic conditions, microbial oxidation of sulfide to sulfuric acid leads to extensive dissolution of carbonate rock strata. One of the most comprehensively studied sites of this type, in terms of microbiology and chemical parameters, is the sulfidic Frasassi cave in Italy where biofilms ('snottites') of pH 0–1 have been described (Macalady *et al.*, 2007). Community analysis of theses snottites revealed a limited biodiversity with sulfur-oxidizing *At. thiooxidans* as the most abundant bacterium, and smaller numbers of *Sulfobacillus* spp., *Acidimicrobium ferrooxidans* and archaea related to the uncultivated 'G-plasma' clade of the *Thermoplasmatales* (Jones *et al.*, 2012). Also some protists and fungal filaments were observed via microscopy in the Frasassi cave samples (Macalady *et al.*, 2007).

The hypogenic cave Cueva de Villa Luz in southern Mexico also comprises extremely acidic microenvironments (pH 0.1–3.0) and sulfur-rich springs and shows elevated hydrogen sulfide emission. The microbial composition of water drips from snottites is similar to the Frasassi community with sulfur-oxidizing *Acidithiobacillus* spp. as key players and smaller numbers of moderately thermophilic

Am. ferrooxidans mediating redox reactions within the cave (Hose *et al.*, 2000).

Acid sulfate soils and acidic fens

Acid sulfate soils (ASS) form predominantly when anoxic marine sediments or salt-marshes rich in pyrite and other sulfide minerals are aerated, e.g. when drained for agricultural or industrial purposes, and oxidation reactions generate proton acidity and mobilize metals. ASS are widespread in coastal regions and cause a serious environmental risk due to severe soil acidity and acid metal-rich run-off effecting adjacent flora and fauna. As described earlier, microorganisms catalyse the oxidation of sulfidic minerals and acid production and are therefore the key drivers in the formation of acid sulfate soils. Initial studies on the formation of ASS by Arkesteyn (1980) resulted in the isolation of iron-/sulfur-oxidizing *At. ferrooxidans* and sulfur-oxidizing *At. thiooxidans* from the acidifying soil material but were not attributed to contribute to the initial pH decrease to pH 4. Sulfur-oxidizing acidophiles were also detected in a buried potential acid sulfate soil layer of a Japanese paddy field which were proposed to contribute to ASS formation by oxidizing sulfur compounds to sulfate and thereby producing acidity (Ohba and Owa, 2005). Wu *et al.* (2013) studied the zones of the Risöfladan experimental field, Finland, which had been drained for more than 40 years. Bacteria present in the plough and oxidized layers were related to known acidophilic heterotrophs and iron- and sulfur-oxidizers previously found in acidic, metal- and sulfur-containing environments. Enrichment cultures of partial oxidized soil at pH 3.0 included iron-/sulfur-oxidizing *At. ferrivorans, At. ferrooxidans, Sulfobacillus* spp. and *Thiomonas* spp..

Naturally exposed ore deposits

Acidophilic microorganisms also occur naturally in the uppermost weathered zones of sulfide ore deposits (gossans; Table 10.1). Such deposits have been mainly exploited by mining and remain only in remote areas e.g. in high altitudes. Langdahl and Ingvorsen (1997) studied the microbiology of gossan material at the Citronen Fjord in North Greenland, which was characterized by subzero temperatures and therefore a potential habitat for psychrophilic microorganisms. They successfully enriched *Acidithiobacillus*-like organisms and heterotrophic acidophiles from this environment. The bacteria assimilated ^{14}C-labelled bicarbonate and glucose below 0°C, but were found to be only cold-tolerant rather than psychrophilic.

Acid rock drainage (ARD), resulting from the microbially catalysed oxidation of sulfide minerals, becomes increasingly significant in polar regions as global warming causes increased glacier melting and thereby the exposing of more sulfide-rich rock strata to air. Schwertmannite present in glacier ice and icebergs of the Antarctica and Artic was the first indication of pyrite oxidation in this area. Dold *et al.* (2013) investigated the microbial community causing ARD on the Antarctic landmass, which was exclusively dominated by bacteria especially related to *Thiobacillus plumbophilus* and *At. ferrivorans*, both of which are psychrotolerant. Other bacteria in these Antarctic samples were species typically detected in acidic environments including *Frateuria*-like species, *Acidisphaera, Actinobacteria* and *Acidobacteria*.

Populations in mine-impacted environments

The best-studied low-pH environments are man-made and mostly associated with the mining of metals and coals. Pyrite becomes exposed to air (oxygen) and water during mining and oxidizes to sulfuric acid and ferric iron. However, the prime oxidant of pyrite is ferric iron, and sulfate is only the final product with the highest oxidation state, meaning that sulfur compound intermediates such as thiosulfate occur in the oxidation pathway (Vera *et al.*, 2013; see also Chapters 1 and 8).

Organisms in such habitats mainly catalyse carbon, iron and sulfur transformations and are widespread among the domains *Bacteria* and *Archaea*, and some *Eukarya*. In general, members of the *Proteobacteria* phylum are by far the most common representatives of *Bacteria* in mine-impacted environments followed by *Nitrospirae, Actinobacteria, Firmicutes* and *Acidobacteria*, while members of the *Bacteriodetes*, and Candidate division TM7 occur only in low numbers (Chapter 1).

In contrast to geothermal sites, which mainly comprise sulfur-oxidizing microorganisms that do not oxidize ferrous iron, few of these organisms are detected in AMD and one explanation might be that reduced sulfur compounds resulting from pyrite oxidation can also be oxidized to sulfate by ferric iron (Druschel *et al.*, 2003) and are thus not available as substrate for acidophiles. The numerically dominant microorganisms in AMD sites are often iron-oxidizing microorganisms (many of which also oxidize sulfur), as iron is often the dominant metal in these environments. A well-known iron-oxidizer frequently found in acidic mine waters is *At. ferrooxidans* (Colmer *et al.*, 1950), which is also capable of oxidizing reduced sulfur compounds and hydrogen, and of ferric iron respiration under anaerobic conditions. *At. ferrooxidans* (and related iron-oxidizing acidithiobacilli) is mostly found in AMD-environments of low pH and moderate temperature. Another common iron-oxidizing bacterium in acidic environments is *L. ferrooxidans*, which has a higher affinity for ferrous iron and a greater tolerance of ferric iron than *At. ferrooxidans*, and therefore correlates with changing ferrous iron concentrations in the mine waters. Heterotrophic acidophilic bacteria, mostly belonging to the *Alphaproteobacteria* class (e.g. *Acidiphilium*, *Acidocella*, *Acidisphaera*), also play an important role in the carbon cycle of mine waters. Archaea have also been reported to be less abundant than bacteria in mine-impacted environments as most of them are (moderate) thermophiles and are more frequently found in geothermal areas. One archaeon, however, which is often found in association with AMD is the mesophilic, iron-oxidizer *Ferroplasma acidiphilum* (Golyshina *et al.*, 2000). Eukaryotes often only account for a minor portion of the microbial community in some AMD environments, but have been studied to a great extent e.g. in the Rio Tinto, Spain (e.g. González-Toril *et al.*, 2003) and to some extent at the Richmond Mine, California, USA (e.g. Edwards *et al.*, 1999).

Microorganisms in these environments often occur as biofilms on mineral surfaces, microbial stalactites, snottites, macroscopic growths in streams as well as in the planktonic phase of acidic water bodies, streams and pit lakes. Examples of the most intensively studied AMD environments are listed in Table 10.2 and some relevant and microbial diverse sites are further described below. Additional examples from extremely acidic environmental and industrial habitats are given in Chapters 9 and 17.

Microbial populations at various mine sites

Richmond Mine, Iron Mountain

One of the most intensively studied and most extreme acidic site is the Richmond Mine at Iron Mountain in California, comprising natural underground water pools of negative pH which are the lowest pH environments discovered yet (Nordstrom *et al.*, 2000). The Richmond Mine harbours a broad range of biodiversity and has been the source of several new species of iron-oxidizing prokaryotes (e.g. Tyson *et al.*, 2004; Edwards *et al.*, 1999; Table 10.2). Oxidative dissolution of pyrite and other sulfide minerals occurs at greatly accelerated rates within Iron Mountain causing increased temperatures within the mine (Nordstrom *et al.*, 2000).

The higher temperature sites of the mine have been found to be dominated by sulfur-oxidizing *Sulfobacillus* spp., while low-pH sites of the same temperature range were colonized by the iron-oxidizing archaeon '*Ferroplasma acidarmanus*' (Bond *et al.*, 2000). Higher pH and cooler temperature areas are mainly dominated by *At. ferrooxidans*, while *Leptospirillum* spp. are widely distributed throughout the mine. All three groups of leptospirilli, *L. ferrooxidans* ('Group I'), *L. ferriphilum* and '*L. rubarum*' ('Group II'), '*Leptospirillum ferrodiazotrophum*' ('Group III'), are present in the Richmond Mine and recently a fourth group '*Leptospirillum* group IV UBA BS' has been detected (Aliaga Goltsmann *et al.*, 2013). Underground slime biofilms and acid streamer growths within the Richmond Mine were dominated by *Leptospirillum* spp. and '*Ferroplasma acidarmanus*', accompanied by the facultative heterotroph *Ferrimicrobium acidiphilum* and *Deltaproteobacteria* (Edwards et al., 1999).

Genomic (Tyson *et al.*, 2004) and proteomic (Ram *et al.*, 2005) studies of biofilm communities within Iron Mountain carried out by the Banfield group have provided further insights into the nature of the microbial communities and motivated revision of the *Leptospirillum* and *Ferroplasma* taxonomy (e.g. *Leptospirillum* group IV, *Ferroplasma* group II). One particularly finding revealed by community genomic analysis was the presence of novel archaeal lineages designated as ARMAN (Archaeal Richmond Mine Acidophilic Nanoorganisms; Baker *et al.*, 2006), which form a deep branch within the *Euryarchaeota*. The cells of

Table 10.2 Characteristics of selected mine-impacted sites, and dominant microorganisms detected

	Richmond Mine, Iron Mountain	Parys Mountain (Mynydd Parys)	Cae Coch	Drei Kronen und Ehrt	Dyffryn Adda (draining Mynydd Parys)	Cantereras	Rio Tinto
Location	California (USA)	North Wales	North Wales	Germany	North Wales	Spain	Spain
Type of mine	Copper	Copper	Pyrite	Pyrite	Copper	Copper	Pyrite
Microbial growth	Mine water Biofilms Streamer growth	Underground lake Drapes	Water pools Water droplets Streamer growth Stalactites	Stalactites	Mine water Streamer growth	Mine water Streamer growth	Mine water Biofilms Streamer growth Sediments
pH	<0–2.0	2.3	1.8–2.3	2.6	2.5	2.5–2.8	2.2–4.7
Temperature (°C)	30–50	8–9	8–9	15	11	–	17.9
Dominant bacteria	*Leptospirillum* spp. *Sulfobacillus* spp. *At. ferrooxidans*	Iron-oxidizing acidithiobacilli *Acidobacteriaceae*	*At. ferrivorans* '*Fv. myxofaciens*' *Gallionella* sp.	*Leptospirillum* sp. '*Ferrovum*' sp.	*At. ferrivorans* '*Fv. myxofaciens*' '*Atx. ferrooxidans*'	*At. ferrivorans* *Acidobacteriaceae* *Acidiphilium* spp.	*L. ferrooxidans* *At. ferrooxidans* *Acidiphilium* spp.
Dominant archaea	'*Fp. acidarmanus*' ARMAN group	Euryarchaeota	Euryarchaeota	ARMAN-group *Thermoplasmatales*	Euryarchaeota	Euryarchaeota	*Ferroplasma* spp. *Thermoplasma* spp.
Eukarya	Algae Fungi	–	–	–	*Euglena mutabilis*	*Chlamydomonas acidophila*	Algae Protists Yeasts Fungi

ARMAN-members are of extremely small size and sequences have been detected in various locations of the Richmond Mine and also other habitats (e.g. the Drei Kronen und Ehrt mine; Ziegler *et al.*, 2013). The Richmond Mine, similar to the Rio Tinto in Spain, is also one of the best studied habitats of acidophilic eukaryotes, and harbours species of, for example, red algae, fungi and protists (*Vahlkampfiidae, Dothideomycetes* and *Eurotiomycetes;* Baker *et al.*, 2009, 2004). A more in-depth account of the microbiology of the metagenomics of the Richmond Mine site can be found in Chapter 14.

Cae Coch pyrite mine

In contrast to the relatively high temperature Richmond Mine site, two low-temperature and extreme acidic sites in north Wales, UK, Cae Coch mine and Mynydd Parys mine, have been extensively studied by Johnson and colleagues. The abandoned Cae Coch pyrite mine contains several water accumulations in form of several small pools and one single drainage stream. Microbial growth occurs as biofilms on the walls, stalactites, streamer growths within the drainage stream and in water droplets on the ceiling and stalactites. The microbial community of the water droplets is relatively simple and dominated by the iron-oxidizing *At. ferrivorans* and smaller numbers of iron-oxidizing *L. ferrooxidans* (Table 10.2). The streamer and slime communities, in contrast, are more diverse and vary between the different sampling locations, but are mostly dominated by the streamer-forming iron-oxidizer '*Ferrovum myxofaciens*', which has been detected as a dominant bacterium in a number of acidic environments (Johnson *et al.*, 2014). '*Fv. myxofaciens*' often occurs as macroscopic streamer growths due to its ability to produce copious amounts of extracellular polymeric substances (EPS). This bacterium, like *L. ferrooxidans,* appears to be only capable of growth by ferrous iron oxidation and fixes carbon dioxide and nitrogen (Johnson *et al.*, 2014). The dominance of '*Fv. myxofaciens*' in the stream correlates well with the occurrence of *At. ferrivorans* in the mine, as both species are psychrotolerant and capable of growth at low temperatures as is characteristic of the Cae Coch mine (9 +/− 1°C). One of the stalactites analysed was predominantly colonized by bacteria closely related to the neutrophilic iron-oxidizer *Gallionella ferruginea*. There have been a number of other reports on the occurrence of these proposed acidophilic/acid-tolerant iron-oxidizing *Gallionella* spp. in several mine-impacted environments (e.g. Fabisch *et al.*, 2013; Heinzel *et al.*, 2009). Minor numbers of *Alphaproteobacteria* (*Acidiphilium* spp. and a novel *Sphingomonas*-isolate), *Gammaproteobacteria* (*At. ferrivorans* and bacteria related to the unclassified species 'WJ2'), *Actinobacteria* (*Ferrimicrobium*-like bacteria), *Nitrospirae* (*L. ferrooxidans*) and *Firmicutes* were also detected in the streamer and slime growths. Archaeal 16S rRNA genes related to a single, novel euryarchaeotal species were detected in one pool characterized by lower pH and higher dissolved solutes than in the other sampled water bodies. With *At. ferrivorans, 'Fv. myxofaciens', L. ferrooxidans* and the proposed iron-oxidizing *Gallionella* spp. identified as the dominant organisms, iron oxidation seems to be the main metabolism in this pyrite mine (Johnson, 2012).

Mynydd Parys copper mine

The flooded underground copper mine, Mynydd Parys (Parys Mountain), north Wales, represents a similar extreme acidic subsurface environment to the Cae Coch mine, though the underground lake is essentially anoxic. When the mine was partially dewatered, due to concerns that a concrete dam could fail and result in the flooding and severe pollution of a nearby coastal town, an opportunity to study the microbial community within the mine arose. The subterranean waters were dominated by iron-oxidizing *Acidithiobacillus* spp., and even so the exact species was not determined at this time, the dominance of *At. ferrivorans* is most likely due to the low temperatures and subsequent studies (Coupland and Johnson, 2004). As in the Cae Coch mine, *L. ferrooxidans, Gallionella*-related species, heterotrophic acidophiles (*Acidiphilium, Acidobacterium, Acidisphaera*) and *Fm. acidophilum* were present in the samples. Archaeal 16S rRNA gene sequences detected in the Parys Mountain underground lake were, however, very different to the Cae Coch archaea with clones related to potential methanogens and only minor numbers of *Thermoplasma*/*Ferroplasma*-like species (Coupland and Johnson, 2004; Table 10.2). Macroscopic growths hanging from pit props within the mine had a different microbial composition and were dominated by the heterotrophic iron-bacteria *Ferrimicrobium, Acidimicrobium* and the iron-reducing genus

Actinobacterium as well as a variety of archaea similar to those detected in the streamers (Coupland and Johnson, 2008). Owing to the limited oxygen ingress reported for Mynydd Parys, activities of iron-oxidizing microorganisms seemed to be low, but ferric iron reduction catalysed by autotrophic and heterotrophic bacteria, as well as due to reaction with pyrite and other residual sulfide minerals, led to ferrous iron being the dominant form in the subterranean lake (Johnson, 2012).

Other mine sites

Similar structures and microbial communities as described for the two former mine sites were also found in the Drei Kronen und Ehrt pyrite mine in the Harz mountains in Germany. Ziegler *et al.* (2009) studied small microbial stalactites and identified the iron-oxidizers *L. ferrooxidans* and '*Fv. myxofaciens*' as dominant bacteria and smaller numbers of *Acidiphilium* spp.. They also detected archaea related to the ARMAN-group, previously detected in the Richmond Mine, and uncultivated members of the *Thermoplasmatales* (Ziegler *et al.*, 2013; Table 10.2).

Mine sites with a more moderately acidic to circumneutral pH, as a habitat for moderate acidophilic and acid-tolerant microorganisms, have also been studied. One of these is the former Wheal Jane tin mine in Cornwall (England) which harbours similar microorganism as the Parys mine but is dominated by moderate acidophilic iron-oxidizing species closely related to *Halothiobacillus neapolitanus* (Hallberg and Johnson, 2003).

Various other mine sites across the world, which will not be discussed in detail, have been analysed for their microbial composition: e.g. metal mines in China (e.g. Hao *et al.*, 2010; He *et al.*, 2007), the Königsstein uranium mine in Germany (Seifert *et al.*, 2008) and the Los Rueldos mercury mine in Spain (Mendez-Garcia *et al.*, 2014).

Acidic streams draining abandoned mines

Dyffryn Adda

Closely connected to the Mynydd Parys copper mine microbiology described above is the microbial colonization of a stream bed, the Dyffryn Adda, into which AMD draining the abandoned mine was diverted in 2003. Kay *et al.* (2013) described a study of this stream carried out for 9 years following its inception (Table 10.2). A few months after it carried AMD, the stream became heavily colonized with macroscopic streamer growths, which continued to occupy a large part of the stream over the entire study period. The prokaryotic community of the streamer growths was dominated by the autotrophic iron-oxidizers *At. ferrivorans* and '*Fv. myxofaciens*' and a novel heterotrophic iron-oxidizer ('*Acidithrix ferrooxidans*'). As time progressed acidophiles, such as iron-reducing heterotrophs of the genus *Acidiphilium*, *Acidobacterium*, *Acidocella* and *Metallibacterium*, as well as the autotrophic iron-oxidizers *At. ferrooxidans* and *L. ferrooxidans*, were detected in the stream. Archaeal species found in the Dyffryn Adda were all distantly related to known *Euryarchaeota*. The surface of the stream becomes occasionally colonized by the algae *Euglena mutabilis*, which might be one cause of the varying abundance of heterotrophic bacteria within the stream as the algae provide organic carbon that can sustain their growth (Ňancucheo and Johnson, 2012). The relatively constant physico-chemical parameters supported a stable microbial community in the stream over the years; with most organisms originating from the underground mine water of Parys Mountain. The only notable difference between the Parys mine underground water and the Dyffryn Adda stream is the dominance of '*Fv. myxofaciens*' in the stream, which was not detected within the mine itself, possibly because it is an obligate aerobe.

Cantareras

Similar physico-chemical parameters to the waters draining Mynydd Parys was reported in a stream draining the small abandoned Cantareras copper mine in south-west Spain (Table 10.2). The microbial community inhabiting stream water draining the mine and the stream sediment was found to be dominated by the psychrophilic iron-oxidizer *At. ferrivorans* and heterotrophic bacteria (Rowe *et al.*, 2007). The microbial mat-like structures were populated by acidophilic algae (*Chlamydomonas acidophila*). Even so the mat was dominated by iron-oxidizing bacteria, high concentrations of ferrous iron could be detected, which is explained by the presence of heterotrophic iron-reducing microorganisms (*Acidiphilium* spp., *Acidobacteriaceae*) whose growth was postulated to be supported by organic carbon released by the algae. Another acidophilic alga, *Chlorella protothecoides* var. *acidicola*, was isolated from the mat at Cantareras (Nancucheo and Johnson, 2012) and had an optimum growth pH of 2.5 tolerating elevated concentrations of various transition metals. Novel acidophilic sulfate-reducing bacteria (designated as '*Desulfobacillus acidavidus*') were detected and isolated from the anaerobic, sulfate-rich bottom layer of the Cantareras stream (Rowe *et al.*, 2007). Archaeal sequences related to the *Euryarchaeota* could be detected in the mine water and the streamer growths of Cantareras.

Rio Tinto

The Rio Tinto is a long (ca. 100 km) and extremely acidic (pH 2.0–2.5), river, which originates from the Peña de Hierro in south-west Spain and enters the Atlantic Ocean at Huelva (Table 10.2). The microbial communities of the Tinto river have intensively been studied by Amils and colleagues, and are described in more detail in Chapter 17. The biogeochemical transformations of iron dominates the microbial ecology of the Tinto river, acting as both electron donor and electron acceptor for chemoautotrophic (*L. ferrooxidans, At. ferrooxidans*) and heterotrophic (*Acidiphilium* spp., *Fm. acidiphilum* and a moderately acidophilic iron-oxidizer related to strain WJ2) acidophiles (Gonzalez-Toril *et al.*, 2003). Bacteria involved in the sulfur redox cycle in Rio Tinto include *At. ferrooxidans, At. thiooxidans* and the sulfate-reducing *Desulfosporosinus* spp.. Macroscopic growth occurring in some parts of the river are devoid of iron-metabolizing acidophiles, predominantly colonized by *Alphaproteobacteria* (including *Sphingomonas*-like bacteria) with smaller numbers of *Betaproteobacteria, Actinobacteria* and *Firmicutes* (Lopez-Archilla *et al.*, 2001). Archaea related to the genera *Ferroplasma* and *Thermoplasma* have also been detected in Rio Tinto, although they account for only a small proportion of the total prokaryotic population. The most remarkable trait of the Tinto river is, however, the unexpectedly high diversity of eukaryotes detected (Aguilera, 2013). The algae are the main primary producers in the river and are mainly represented by diatoms and *Euglena mutabilis* (Lopez-Archilla *et al.*, 2001). Further eukaryotes detected in the Rio Tinto are other photosynthetic algae, protists, yeasts and filamentous fungi (Aguilera, 2013), which are described in detail in Chapter 7.

Acidic pit lakes

Pit lakes form as a result of opencast mining of coal or metals when the abandoned voids become filled with groundwater. Oxidation of sulfide minerals in adjacent rocks or dumps leads to acidification of these lakes and release of sulfate, iron and other metals (Geller *et al.*, 1998). Due to the high iron content of pit lakes, ferric minerals are precipitated and form sediments which often leads to the adsorption of phosphorus and lead to oligotrophic conditions developing in most acid pit lakes (Kleeberg and Grüneberg, 2005). Acidic pit lakes can be found in mining areas of e.g. China, Australia, Poland, Spain and Germany, whereby the two later areas have been studied most intensively.

Many pit lakes in Germany resulting from lignite coal mining have an acidic pH (2.0–4.0) and are rich in sulfate and iron with minor amounts of toxic metals (Geller *et al.*, 1998), while concentrations of nutrients such as nitrate, phosphate, and ammonium, are very low. The maximum depth of the lakes is 11 m and they receive most of their inflow by a ditch connecting the lakes and only secondarily from the aquifer. Most of the pit lakes have an anoxic hypolimnion during summer stratification and a complete mixture of the water column occurs during winter (Blodau *et al.*, 1998). The phytoplankton community commonly observed in these lakes is dominated by the phytoflagellates *Ochromonas*, *Chlamydomonas* and *Gymnodinium* (Beulker *et al.*, 2003) which is influenced by the availability of inorganic carbon. Microbial oxidation of iron occurring in the oxic zones of these lakes results in the sedimentation of ferric iron minerals (e.g. schwertmannite, goethite) serving as electron acceptor for acidophilic iron reducing bacteria (*Acidiphilium* spp., *Acidobacterium* spp.) under oxygen-limited conditions. The pH of the lakes increases with depth causing the microbial communities (including Fe(III)-reducing microorganisms) to be more heterogeneous than those in the surface waters. The acidic sediments of 'lake 77' in the Lower Lusatia area (Germany) were found to be dominated by moderately acidophilic *Acidobacterium* spp., which were able to catalyse the reductive dissolution of ferric iron minerals (Blöthe *et al.*, 2008). *Acidithiobacillus* spp. and *Acidocella* spp. were more dominant in the acidic sediments, while *Bacillus* spp. and *Alicyclobacillus* spp. were isolated from moderate acidic sediments of the lake and *Acidiphilium* spp. could be found in almost all sediment depths (Lu *et al.*, 2010; Blöthe *et al.*, 2008). Acidic conditions in the upper sediments of the lakes are stabilized by the cycling of iron which restricts fermentative and sulfate-reducing activities, but as reactive iron decreases and pH increases with sediment depth, sulfate-reducers (e.g. *Desulfosporosinus* spp.) have been detected (Koschorrek, 2008). Enhanced sulfate-reduction occurred in the sediments of the acidic mining lake '*Gruenewalder Lauch*' where the sediment was covered by a thick layer of periphytic algae separating the sediment (pH 6) from the acidic water (pH 3.1) and provided large amounts of organic carbon as electron donors (Koschorreck *et al.*, 2007). Sulfate-reduction in lake sediments below pH 4.5 has also been reported for other environments (e.g. Koschorreck *et al.*, 2007; Gyure *et al.*, 1990; Satake, 1977) and is usually related to increased influx of organic matter and the inhibition of competing iron-reducing bacteria and methanogens (Koschorreck, 2008).

Pit lakes located in Spain are the consequence of metal mining rather than coal mining, and are biochemically more complex as they can contain various transition metals (e.g. Zn) and metalloids in addition to iron and sulfate. There are more than 25 pit lakes in the Iberian Pyrite Belt (IPB), which hosts one of the world's largest accumulations of mine wastes and AMD, including the Rio Tinto. The lakes in the IPB are characterized by low pH and high concentration of heavy metals, but each lake presents unique characteristics. The microbiology of two acidic, sulfate- and metal-rich pit lakes in the IPB, Nuestra Señora del Carmen and Concepción, was described by Santofimia *et al.* (2013). In Nuestra Señora del Carmen, a chemically and thermally stratified pit lake consisting of two different layers, they detected typical AMD microorganisms involved in iron redox-cycling. The upper layer of the lake was dominated by heterotrophic *Acidiphilium* spp. and accompanied by iron-oxidizing *Leptospirillum* spp. and *Planctomycetes*. Eukaryotic sequences detected in the upper layer belonged to the genus *Chlamydomonas* and filamentous species and diatoms were observed by microscopy. The lower layer (15 m) of the lake, which exhibited a slightly reduced environment and equal concentrations of ferrous and ferric iron, was more diverse and mainly represented by *Acidithiobacillus* spp., *Acidimicrobiaceae*, both facultative iron-reducers, leptospirilli, *Actinobacteria*, and archaeal members of the *Thermoplasmata*. The microbial diversity in the Concepción pit lake was more diverse, but dissolved iron concentrations were low and therefore less typical AMD microorganisms could be detected. In the lower layers, however, where dissolved iron concentrations increased, iron-oxidizing '*Fv. myxofaciens*' and potential iron-reducing *Acidiphilium* spp., *Acidimicrobaceae* and *Acidobacteria* were present. Possibly because of the low concentrations of sulfate and transition metals in this lake, members of AMD atypical genera (*Erwinia*, *Legionella* and *Halomonas*) were also detected.

Falagan *et al.* (2014) studied two other meromictic stratified pit lakes in the IPB, Cueva de la Mora and Guadiana, which were claimed to have the most dramatic vertical pH gradients and water chemistry of pit lakes in the region. Cueva de la Mora has three layers, the oxygenated mixolimnion, the chemocline and the more anoxic monimolimnon. Most interesting was the observed enhanced accumulation of algal biomass well below the lake surface, which was thought likely due to the highly variable bioavailability of phosphorus in these water bodies. Acidophilic micro-algae were reported to be the primary producers in the lake, providing oxygen and organic carbon and thereby supporting growth of heterotrophic acidophiles (e.g. *Acidobacteriaceae*), which were dominant in the mixolimnion and the chemocline. The chemocline was quite diverse and populated by iron-oxidizing *L. ferrooxidans*, iron-/sulfur-oxidizing and facultative iron-reducing *At. ferrooxidans* and sulfate-reducing *Desulfomonile* sp. Based on chemical parameters and the various microbial metabolisms present, the chemocline seems to be the zone where dynamics of carbon, iron and sulfur redox transformations were most dynamic. Archaeal sequences detected in the water column and sediments of this lake belonged to the *Eukaryota* and few sequences to the recently proposed *Thaumarchaeota*, possibly mediating the transformation of nitrogen.

Mine water treatment systems

Mine waters pose serious threats to the environmental, owing to their high metal and sulfate loads and (often) their low pH, and therefore require treatment prior to release in receiving water courses. The ability of bacteria found in acidic environments to mediate iron and sulfur redox reactions can be advantageous in mine water treatment to remove toxic metals and sulfate and raise pH.

A pilot passive treatment plant was constructed at the Wheal Jane tin mine in Cornwall (England) to remediate acidic (pH 3.4), metal-rich water draining the mine utilizing naturally occurring microorganisms (Hallberg and Johnson, 2005). The treatment plant consisted of three separate composite systems, comprising a series of aerobic wetlands for iron oxidation and precipitation, a compost bioreactor for removing chalcophilic metals and to generate alkalinity, and rock filter ponds for removing soluble manganese and organic carbon. Moderately acidophilic iron-oxidizing bacteria (related to *Halothiobacillus neapolitanus*) and later heterotrophic acidophiles were the dominant cultivatable bacteria in the Wheal Jane AMD. Heterotrophic acidophiles (*Acidiphilium* spp., *Acidobacterium*-like) and smaller numbers of moderately and extremely acidophilic iron-oxidizing bacteria (*Acidithiobacillus* spp., *Leptospirillum* spp.) were isolated from the surface waters and sediments of the constructed aerobic wetlands. The dominant microbial isolate in waters draining the anaerobic compost bioreactors was an iron- and sulfur-oxidizing moderate acidophile closely related to *Thiomonas intermedia*. Acidophiles enumerated at the Wheal Jane plant only accounted for up to 25% of the total microbial population.

The Carbondale constructed wetland in Waterloo Township (Ohio, USA) was constructed to treat coal mine drainage of more or less constant pH (2.0–3.9), metal and sulfate concentrations. Each wetland cell was layered with mushroom or manure compost followed by a limestone layer. The microbial community of the oxidized surface of the wetland was dominated by iron–sulfur oxidizing *Acidithiobacillus* spp. and accompanied by iron-oxidizing *Actinobacteria* and '*Fv. myxofaciens*' (Nicomrat *et al.*, 2008).

Another example is an acid mine water treatment plant in the Lusatia area (Germany) operated at pH 3.0 and continuously fed with mine water pH 4.5 containing an elevated concentration of iron (and sulfate) which is microbially oxidized to form schwertmannite. The iron-oxidizing community in the plant, which proofed to be very stable (Heinzel *et al.*, 2009), was dominated by the iron-oxidizing betaproteobacterium '*Ferrovum*' and often accompanied by an acidophilic organism related to the neutrophilic iron-oxidizer *Gallionella ferruginea* (Heinzel *et al.*, 2009). Despite being inoculated with more familiar iron-oxidizing acidophiles (*At. ferrooxidans* and *L. ferrooxidans*) at the beginning of the plant operation, '*Ferrovum*' and the *Gallionella*-related bacterium prevailed in the plant.

Mine heaps and waste rock dumps

From an industrial viewpoint, sulfidic mine deposits arrangements can be divided into commercial heap and dump bioleaching operations, and waste rock dumps and tailings which may generate acid mine drainage (AMD). Heap and dump bioleaching is used extensively around the world for copper extraction with an estimated 15% of the world's copper produced (Brierley and Brierley 2013; Schippers *et al.* 2014; see also Chapter 17).

Waste rock originates from mining and comprises of rocks of different size in which the metal content is too low for an economically feasible ore processing. Tailings are the fine grained remains of ore processing which are dumped as a slurry (Fig. 10.1). Both kinds of mine waste often contain high amounts of iron sulfide minerals (pyrite or pyrrhotite) and often produce AMD due to chemical and microbial metal sulfide oxidation. Over a period of several years or decades, an oxidized zone with depleted sulfide content, low pH, and enrichment of secondary minerals develops above an unoxidized zone with unaltered material in tailings dumps (Fig. 10.2; Korehi *et al.*, 2013; Schippers *et al.*, 2010).

The biogeochemical processes in sulfidic waste rock dumps and tailings are similar to those in commercial heap bioleaching operations. At low pH <4, the biological metal sulfide oxidation by acidophilic, chemolithoautotrophic iron- and sulfur-oxidizing bacteria and archaea, dominates over chemical oxidation. These prokaryotes can dissolve pyrite very efficiently, and the biological pyrite oxidation rate has shown to be two orders of magnitude or greater than the chemical rate. At pH below 2.5, ferric iron is soluble and serves as a more efficient oxidant for metal sulfides than molecular oxygen. By reacting with a metal sulfide, ferric iron is being reduced to ferrous, the substrate for Fe(II)-oxidizing acidophiles. In addition, ferric iron can act as electron acceptor for acidophiles that oxidize RISCs (or organic carbon) in the absence of oxygen. Elemental sulfur and RISCs are formed

Figure 10.1 Depositing sulfidic mine tailings as slurry on top of the tailings dump (ore processing plant in the background).

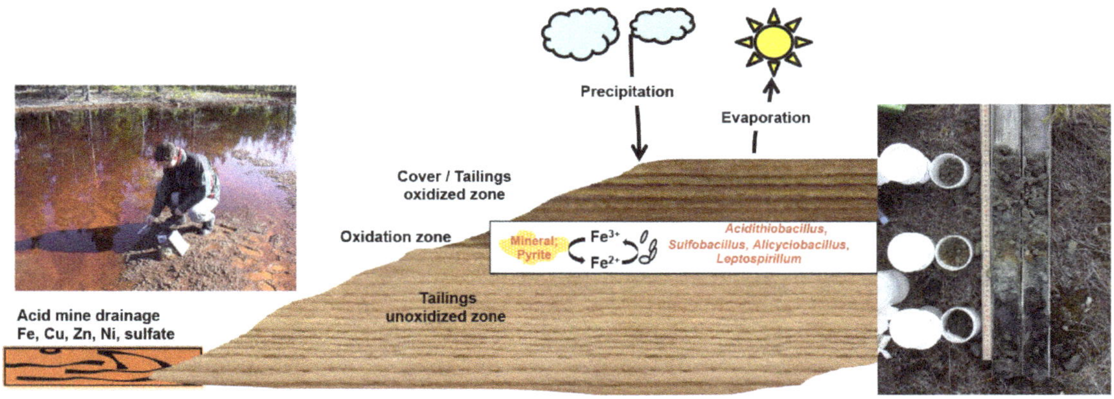

Figure 10.2 Acid mine drainage releasing sulfidic mine tailings with developed different zones due to microbially driven pyrite oxidation (modified from Korehi *et al*., 2013).

as intermediates in the metal sulfide oxidation process with sulfate as the most oxidized sulfur form (Schippers *et al.*, 2010, 2014; Vera *et al.*, 2013).

Since concentrations of dissolved organic carbon are usually very low in mine heaps and dumps, microbial metal sulfide oxidation is the dominant biogeochemical process at low pH and chemolithotrophic, acidophilic iron- and sulfur-oxidizing bacteria are often the dominant members of the microbial community. In addition, ferric oxyhydroxides can be dissolved by ferric iron-reducing microorganisms, with the consequent release of adsorbed metals and metalloids (such as As). In contrast, if sulfate reduction is occurring, metals may precipitate as metal sulfides within the dump (Schippers *et al.*, 2010).

As the main geochemical processes in sulfidic mine dumps and heaps are catalysed by microorganisms, analysis of microbial communities is vital to understand the scale and nature of these transformations. Microbial analysis helps understand which organisms are responsible for these processes, where in a dump or heap a particular process takes place, where zones of high or low microbial activity occur, and how the kinetics of particular processes, e.g. pyrite oxidation or sulfate reduction, can be influenced by stimulating or inhibiting microbial activity. Monitoring microbiology is also useful for evaluating the success of AMD countermeasures (Sand *et al.*, 2007) or heap leaching operations (Remonsellez *et al.*, 2009).

In the Escondida mine in northern Chile, the extensive industrial copper heap leaching operation has been monitored by analysing 16S rRNA gene copy numbers of acidophiles in the pregnant leach solution over time. *At. ferrooxidans* was the most abundant organism during the first part of the leaching cycle, while the abundance of *L. ferriphilum* and *Fp. acidiphilum* increased with age of the heap. *At. thiooxidans* remained at similar levels throughout the leaching cycle, and *Firmicutes* group showed a low and a patchy distribution in the heap. *Acidiphilium*-like bacteria reached their highest abundance corresponding to the amount of autotrophs (Remonsellez *et al.*, 2009).

The microbial diversity in sulfidic mine dumps and heaps comprises aerobic and anaerobic species which are autotrophic (CO_2-fixation) or heterotrophic (C_{org} as carbon source) as well as lithotrophic (inorganic compounds as energy source) or organotrophic (C_{org} as energy source). A comprehensive list of microorganisms on the genus level from the three domains of life, *Bacteria*, *Archaea* and *Eukarya*, which have been detected in mine dumps or heaps by cultivation and molecular biological techniques, is shown in Tables 10.3 and 10.4. These lists are based on the results of published papers. Important physiological properties relevant to biogeochemical oxidation and reduction processes are given for each genus.

Most microbial genera detected in dumps and heaps belong to the domain *Bacteria* and comprise acidophilic species. The *Archaea* are mainly thermophiles and all of them are acidophiles. *Eukarya* detected in these environments include algae, fungi, yeasts and protozoa. The iron- and sulfur-oxidizing bacteria or archaea are responsible for metal sulfide oxidation. These microorganisms have been widely detected in mine waste tailings located in different climate zones. Ferric iron and oxidized sulfur produced by these *Bacteria* or *Archaea* may be used as terminal electron acceptors by obligately and facultatively anaerobic acidophiles under anoxic conditions (Schippers *et al.*, 2010).

In addition to exploring the microbial diversity of sulfidic mine dumps and heaps, the quantification of particular organisms, e.g. acidophilic iron-oxidizers, in different areas and depth layers of a dump is of interest especially for monitoring purposes. Primarily cultivation techniques, i.e. solid media and most-probable-number (MPN), have been used to enumerate prokaryotes involved in oxidation and reduction processes in sulfidic mine dumps (Schippers *et al.*, 2010).

Far more studies have been carried out for mine tailings than for waste rock dumps. A study by Schippers *et al.* (2010) compared the numbers of relevant groups of microorganisms in selected waste rock and tailings dumps with different metal contents, pH values and located in eight countries in different climate zones. The maximum number of total cells ranged between 10^6 and 10^8 cells per gram waste rock or tailings material, and acidophilic iron- and sulfur-oxidizing bacteria represented a significant proportion of these. Large numbers of these organisms were detected in waste rock dumps and tailings all over the world. Neutrophilic sulfur-oxidizing bacteria oxidize the sulfur compounds formed by chemical sulfide oxidation at circumneutral pH. These bacteria were detected in most of the cases in similar numbers to acidophilic representatives. Acidophilic heterotrophic bacteria can facilitate the continued growth of autotrophic iron- and sulfur-oxidizing bacteria, and have also been detected in some cases. In waste rock dumps, maximum numbers of heterotrophic acidophiles were in the range of 10^3–10^6 cells/g dry sample, and tailings contained lesser numbers.

A quantification of the microbial communities in four different mine tailings was carried out by Kock and Schippers (2008) and Korehi *et al.* (2013) using both cultivation and molecular techniques. Depth profiles of cell numbers showed that the compositions of the microbial communities are significantly different at the four sites, and varied greatly between zones of oxidized and non-oxidized tailings. qPCR data indicated that *Bacteria* dominated over *Archaea* and *Eukarya* and that the genus *Acidithiobacillus* represented the most abundant iron- and sulfur-oxidizers detected. *Leptospirillum* and *Sulfobacillus* were detected in smaller numbers and not in all tailings dumps. Anaerobes such

Table 10.3 Bacteria detected in mine dumps or heaps and their physiological properties. Genus names are listed in alphabetical order. Genus names in bold means that the genus comprises acidophilic species

	Oxidation of					Reduction of	
Genus	Pyrite	Other MS	Fe(II)	Sulfur	C_{org}	Fe(III)	Sulfate
Bacteria							
Acidimicrobium	+		+		+		
Acidiphilium				+	+	+	
Acidisphaera					+		
Acidithiobacillus	+	+	+	+		+	
Acidobacterium					+		
Acidovorax					+		
Acinetobacter					+		
Alicyclobacillus	+	+	+	+	+	+	
Amycolatopsis					+		
Arthrobacter					+		
Bacillus					+	+	
Bradyrhizobium					+		
Brevibacterium					+		
Caulobacter					+		
Cellulomonas					+		
Chryseobacterium					+		
Comamonas					+		
Corynebacterium					+		
Desulfobacter					+	+	+
Desulfobacterium					+	+	+
Desulfobotulus					+		+
Desulfobulbus					+	+	+
Desulfococcus					+		+
Desulfonema					+		+
Desulfosarcina					+		+
Desulfosporosinus					+		+
Desulfotomaculum					+	+	+
Desulfovibrio					+	+	+
Enhydrobacter					+		
Ferribacterium					+	+	
Ferrimicrobium	+	+	+		+	+	
Ferritrophicum			+				
Gallionella			+				
Gemmatimonas							
Geobacter					+	+	
Hydrogenophaga							
Hyphomicrobium					+		
Legionella					+		
Leptospirillum	+	+	+				
Leptothrix					+		
Meiothermus					+		
Methylobacterium					+		

Table 10.3 Continued

Genus	Oxidation of					Reduction of	
	Pyrite	Other MS	Fe(II)	Sulfur	C_{org}	Fe(III)	Sulfate
Methylococcus					+		
Microbacterium					+		
Microvirgula					+		
Nocardiopsis					+		
Ochrobactrum					+		
Paenibacillus					+		
Peredibacter					+		
Pseudomonas				+	+	+	
Ralstonia					+		
Rubrobacter					+		
Sedimentibacter					+		
Sphingomonas					+		
Subtercola					+		
Sulfobacillus	+	+	+	+	+	+	
Thiobacillus		+		+	+		
Thiomonas		+		+	+		
Thiovirga		+		+			

MS, metal sulfide other than pyrite; sulfur, inorganic sulfur compounds; +, physiological property shown (modified from Schippers *et al.*, 2010).

Table 10.4 Archaea and Eukarya detected in mine dumps or heaps and their physiological properties. Genus names are listed in alphabetical order. Genus names in bold means that the genus comprises acidophilic species

Genus	Oxidation of					Reduction of	
	Pyrite	Other MS	Fe(II)	Sulfur	C_{org}	Fe(III)	Sulfate
Archaea							
Acidianus	+	+	+	+	+		
Ferroplasma	+	+	+		+		
Metallosphaera	+	+	+	+	+		
Sulfolobus	+	+	+	+	+		
Sulfurisphaera				+	+		
Thermogymnomonas					+		
Eukarya							
Amoeba					+		
Cladosporium					+		
Euglena					+		
Eutrepia					+		
Hormidium							
Penicillium					+		
Rhodotorula					+		
Trichosporon					+		
Ulothrix							

MS, metal sulfide other than pyrite; sulfur, inorganic sulfur compounds; +, physiological property shown (modified from Schippers *et al.*, 2010).

as sulfate-reducers or the neutrophilic iron- and manganese-reducing *Geobacteraceae* were detected in lesser abundance.

Several tailings studies have reported the predominance of iron- and sulfur-oxidizing acidophiles at low pH, but microbial communities at the moderately acidic oxidation stage (between the initial circumneutral to alkaline pH and the strong acidic final stage) have only been studied for few mine tailings sites (Liu *et al.*, 2014; Chen *et al.*, 2013). Microbial communities detected were found to be very different to those in tailings of < pH 3. For example microbial communities in lead/zinc mine tailings samples in China with a pH range of 1.8–7.5 revealed a predominance of *Proteobacteria,* including the hydrogen- and sulfur-oxidizing genera *Hydrogenophaga, Thiovirga* and *Thiobacillus,* respectively, in the circumneutral samples at the initial weathering stage, while gene sequences related to the acidophilic, iron-oxidizing genera *Ferroplasma* (*Euryarchaeota*), *Acidithiobacillus* (*Proteobacteria*), *Leptospirillum* (*Nitrospira*) and *Sulfobacillus* (*Firmicutes*) were mainly detected in samples of lower pH and more intense weathering and iron precipitations (Chen *et al.*, 2013). The finding that the composition of microbial communities in sulfidic mine tailings is mainly controlled by pH was also shown for copper mine tailings (Liu *et al.*, 2014) and supported by experimental evidence (Chen *et al.*, 2014).

Conclusions

Acidophiles can thrive in both natural and man-made environments. The most ancient sites where acidophilic life has developed over many decades represent geothermal areas which display high temperatures and are rich in sulfur compounds. Due to these physico-chemical parameters, such sites are mostly colonized by organisms that metabolize reduced sulfur compounds, and archaea dominate over bacteria at high temperatures. Mine-impacted environments, in contrast, often contain more soluble iron than reduced sulfur compounds and are therefore predominantly inhabited by microorganisms that can catalyse redox transformations of iron. Microbial growths, occurring as filamentous 'acid streamers' – growths in flowing waters, thick and often dense 'acid mats' – and 'pipes'/'snotites' – pendulous growths attached to mine roofs, are characteristic of acidic environments. The microbial composition of these growths is determined by the physico-chemical parameters of the site, resulting in warm and extremely acidic streams being dominated by iron-oxidizing bacteria (*Leptospirillum* spp.) and archaea (*'Fp. acidarmanus'*), while in higher pH and more moderate mine waters other iron-oxidizing bacteria (*Acidithiobacillus* spp. and *'Fv. myxofaciens'*) are more abundant. Numbers of heterotrophic species are often relatively low in acidic environments, as dissolved organic carbon levels are usually extremely low. However, extraneous inputs of dissolved organic carbon and that originating from indigenous acidophilic algae in acidic streams and pit lakes promotes the growth of heterotrophic bacteria, which may then the outnumber chemolithotrophic bacteria. The microbial ecology of extremely acidic environments can therefore be both complex and dynamic, and provides an exciting area for microbiological and molecular-based research.

Acknowledgement

S. Hedrich was supported by the German Federal Ministry of Education and Research (BMBF, Grant ID 033RF001).

References

Aguilera, A. (2013). Eukaryotic organisms in extreme acidic environments, the Río Tinto case. Life 3, 363–374.

Aliaga Goltsman, D.S., Dasari, M., Thomas, B.C., Shah, M.B., VerBerkmoes, N.C., Hettich, R.L., and Banfield, J.F. (2013). New group in the leptospirillum clade: Cultivation-independent community genomics, proteomics, and transcriptomics of the new species "*Leptospirillum* group IV UBA BS". Appl. Environ. Microbiol. 79, 5384–5393.

Amaral-Zettler, L.A. (2013). Eukaryotic diversity at pH extremes. Front. Microbiol. 3, 441.

Arkesteyn, G.J.M.W. (1980). Pyrite oxidation in acid sulphate soils: The role of microorganisms. Plant Soil 54, 119–134.

Atkinson, T., Cairns, S., Cowan, D.A., Danson, M.J., Hough, D.W., Johnson, D.B., Norris, P.R., Raven, N., Robinson, C., Robson, R., *et al.* (2000). A microbiological survey of Montserrat Island hydrothermal biotopes. Extremophiles 4, 305–313.

Baker, B.J., Tyson, G.W., Webb, R.I., Flanagan, J., Hugenholtz, P., Allen, E.E., and Banfield, J.F. (2006). Lineages of acidophilic archaea revealed by community genomic analysis. Science *314*, 1933–1935.

Baker, B.J., Tyson, G.W., Goosherst, L., and Banfield, J.F. (2009). Insights into the diversity of eukaryotes in acid mine drainage biofilm communities. Appl. Environ. Microbiol. 75, 2192–2199.

Beam, J.P., Jay, Z.J., Kozubal, M.A., and Inskeep, W.P. (2014). Niche specialization of novel *thaumarchaeota* to oxic and hypoxic acidic geothermal springs of Yellowstone national park. ISME J. *8*, 938–951.

Beulker, C., Lessmann, D., and Nixdorf, B. (2003). Aspects of phytoplankton succession and spatial distribution in an acidic mining lake (Plessa 117 Germany). Acta Oecol. *24*, 25–31.

Blodau, C., Hoffmann, S., Peine, A., and Peiffer, S. (1998). Iron and sulfate reduction in the sediments of acidic mine lake 116 (Brandenburg, Germany): rates and geochemical evaluation. Water Air Soil Pollut. *108*, 249–270.

Blöthe, M., Akob, D.M., Kostka, J.E., Göschel, K., Drake, H.L., and Küsel, K.(2008). pH gradient-induced heterogeneity of Fe(III)-reducing microorganisms in coal mining-associated lake sediments. Appl. Environ. Microbiol. *74*, 1019–1029.

Bond, P.L., Druschel, G.K., and Banfield, J.F. (2000). Comparison of acid mine drainage microbial communities in physically and geochemically distinct ecosystems. Appl. Environ. Microbiol. *66*, 4962–4971.

Brierley, C.L., and Brierley, J.A. (1973). A chemoautotrophic and thermophilic microorganism isolated from an acid hot spring. Can. J. Microbiol. *19*, 183–188.

Brierley, C.L., and Brierley, J.A. (2013). Progress in bioleaching: part B: applications of microbial processes by the minerals industries. Appl. Microbiol. Biotechnol. *97*, 7543–7552.

Brock, T.D. (1978). Thermophilic Microorganisms and Life at High Temperatures (Springer-Verlag, New York, NY, USA).

Brock, T.D. (2001). The origins of research on thermophiles. In Thermophiles: Biodiversity, Ecology and Evolution, Reysenbach, A.L., and Voytek, A. eds (Kluwer Academic/Plenum, New York, NY, USA), pp. 1–9.

Burton, N.P., and Norris, P.R. (2000). Microbiology of acidic, geothermal springs of Montserrat: Environmental rDNA analysis. Extremophiles *4*, 315–320.

Chen, L.X., Li, J.T., Chen, Y.T., Huang, L.N., Hua, Z.S., Hu, M., and Shu, W.S. (2013). Shifts in microbial community composition and function in the acidification of a lead/ zinc mine tailings. Environ. Microbiol. *15*, 2431–2444.

Chen, Y., Li, J., Chen, L., Hua, J., Huang, L., Liu, J., Xu, B., Liao, B., and Shu, W. (2014). Biogeochemical processes governing natural pyrite oxidation and release of acid metalliferous drainage. Environ. Sci. Technol. *48*, 5537–5545.

Chiacchiarini, P., Lavalle, L., Giaveno, A., and Donati, E. (2009). Acidophilic microorganisms from geothermal Copahue Volcano system. Assessment of biotechnological applications. Adv. Mat. Res. *71–73*, 87–91.

Colmer, A.R., Temple, K.L., and Hinkle, M.E. (1950). An iron-oxidizing bacterium from the acid drainage of some bituminous coal mines. J. Bacteriol. *59*, 317–328.

Coupland, K., and Johnson, D.B. (2004). Geochemistry and microbiology of an impounded subterranean acidic water body at Mynydd Parys, Anglesey, Wales. Geobiology 2, 77–86.

Coupland, K., and Johnson, D.B. (2008). Evidence that the potential for dissimilatory ferric iron reduction is widespread among acidophilic heterotrophic bacteria. FEMS Microbiol. Lett. *279*, 30–35.

Cullen, W.R., and Reimer, K.J. (1989). Arsenic speciation in the environment. Chem. Rev. *89*, 713–764.

Davis-Belmar, C.S., Pinto, E., Demergasso, C., and Rautenbach, G. (2013). Proteo and Actinobacteria diversity at a sulfide, salt and acid -rich lake in the north of Chile. Adv. Mat. Res. *825*, 37–41.

D'Imperio, S., Lehr, C.R., Breary, M., and McDermott, T.R. (2007). Autecology of an arsenite chemolithotroph: Sulfide constraints on function and distribution in a geothermal spring. Appl. Environ. Microbiol. *73*, 7067–7074.

Dold, B., Gonzalez-Toril, E., Aguilera, A., Lopez-Pamo, E., Cisternas, M.E., Bucchi, F., and Amils, R. (2013). Acid rock drainage and rock weathering in antarctica: Important sources for iron cycling in the southern ocean. Environ. Sci. Technol. *47*, 6129–6136.

Druschel, G.K., Hamers, R.J., and Banfield, J.F. (2003). Kinetics and mechanism of homogeneous polythionate oxidation at low pH with oxygen and ferric iron. Geochim. Cosmochim. Ac. *67*, 4457–4469.

Edwards, K.J., Gihring, T.M., and Banfield, J.F.. (1999). Seasonal variations in microbial populations and environmental conditions in an extreme acid mine drainage environment. Appl. Environ. Microbiol. *65*, 3627–3632.

Escudero, L.V., Bijman, J., Chong, G., Pueyo, J.J., and Demergasso, C. (2013). Geochemistry and microbiology in an acidic, high altitude (4,000 m) salt flat. high andes, Northern Chile. Adv. Mat. Res. *825*, 28–32.

Fabisch, M., Beulig, F., Akob, D.M., and Küsel, K. (2013). Surprising abundance of *Gallionella*-related iron oxidizers in creek sediments at pH 4.4 or at high heavy metal concentrations. Front. Microbiol. *4*, 390.

Falagan, C., Sanchez-Espana, J., and Johnson, D.B. (2014). New insights into the biogeochemistry of extremely acidic environments revealed by a combined cultivation-based and culture-independent study of two stratified pit lakes. FEMS Microbiol. Ecol. *87*, 231–243.

Flores, G.E., Wagner, I.D., Liu, Y., and Reysenbach, A.L. (2012). Distribution, abundance, and diversity patterns of the thermoacidophilic "deep-sea hydrothermal vent euryarchaeota 2". Front. Microbiol. 3, 47.

Geller, W., Klapper, H., and Salomons, W. (Eds.), 1998. Acidic Mining Lakes: Acid Mine Drainage, Limnology, and Reclamation (Springer Verlag, Berlin, Heidelberg, Germany).

Golyshina, O.V., Pivovarova, T.A., Karavaiko, G.I., Kondrat'eva, T.F., Moore, E.R.B., Abraham, W.R., Lunsdorf, H., Timmis, K.N., Yakimov, M.M., and Golyshin, P.N. (2000). *Ferroplasma acidiphilum* gen. nov., sp. nov., an acidophilic, autotrophic, ferrous-iron-oxidizing, cell-wall-lacking, mesophilic member of the *Ferroplasmaceae fam. nov.*, comprising a distinct lineage of the *Archaea*. Int. J. Syst. Evol. Microbiol. *50*, 997–1006.

González-Toril, E., Llobet-Brossa, E., Casamayor, E.O., Amann, R., and Amils, R. (2003). Microbial ecology of an extreme acidic environment, the Tinto River. Appl. Environ. Microbiol. *69*, 4853–4865.

Gyure, R.A., Konopka, A., Brooks, A., and Doemel, W. (1990). Microbial sulfate reduction in acidic (pH 3) strip-mine lakes. FEMS Microb. Ecol. 73, 193–202.

Hallberg, K.B., and Johnson, D.B. (2003). Novel acidophiles isolated from moderately acidic mine drainage waters. Hydrometallurgy *71*, 139–148.

Hallberg, K.B., and Johnson, D.B. (2005.) Microbiology of a wetland ecosystem constructed to remediate mine drainage from a heavy metal mine. Sci. Total Environ. *338*, 53–66.

Hao, C., Wang, L., Gao, Y., Zhang, L., and Dong, H. (2010). Microbial diversity in acid mine drainage of Xiang Mountain sulfide mine, Anhui Province, China. Extremophiles *14*, 465–474.

He, Z., Xiao, S., Xie, X., Zhong, H., Hu, Y., Li, Q., and Qiu, G. (2007). Molecular diversity of microbial community in acid mine drainages of Yunfu sulfide mine. Extremophiles *11*, 305–314.

Heinzel, E., Hedrich, S., Janneck, E., Glombitza, F., Seifert, J., and Schlömann, M. (2009). Bacterial diversity in a mine water treatment plant. Appl. Environ. Microbiol. *75*, 858–861.

Hill, C.A. (1995). Sulfur redox reactions: hydrocarbons, native sulfur, Mississippi Valley-type deposits, and sulfuric acid karst in the Delaware Basin, New Mexico and Texas. Environ. Geol. *25*, 16–23.

Hose, L.D., Palmer, A.N., Palmer, M.V., Northup, D.E., Boston, P.J., and DuChene, H.R. (2000). Microbiology and geochemistry in a hydrogen-sulphide-rich karst environment. Chem. Geol. *169*, 399–423.

Huber, H., and Stetter, K.O. (1989). *Thiobacillus prosperus* sp. nov., represents a new group of halotolerant metal-mobilizing bacteria isolated from a marine geothermal field. Arch. Microbiol. *151*, 479–485.

Inskeep, W.P., Macur, R.E., Harrison, G., Bostick, B.C., and Fendorf, S. (2004). Biomineralization of As(V)-hydrous ferric oxyhydroxide in microbial mats of an acid-sulfate-chloride geothermal spring, Yellowstone National Park. Geochim. Cosmochim. Acta *68*, 3141–3155.

Inskeep, W.P., Rusch, D.B., Jay, Z.J., Herrgard, M.J., Kozubal, M.A., Richardson, T.H., Macur, R.E., Hamamura, N., Jennings, R., Fouke, B.W., *et al.* (2010). Metagenomes from high-temperature chemotrophic systems reveal geochemical controls on microbial community structure and function. PLoS One *5*, e9773.

Jackson, C.R., Langner, H.W., Donahoe-Christiansen, J., Inskeep, W.P., and McDermott, T.R. (2001). Molecular analysis of microbial community structure in an arsenite-oxidizing acidic thermal spring. Environ. Microbiol. *3*, 532–542.

Johnson, D.B. (2012). Geomicrobiology of extremely acidic subsurface environments. FEMS Microbiol. Ecol. *81*, 2–12.

Johnson, D.B., Okibe, N., and Roberto, F.F. (2003). Novel thermo-acidophilic bacteria isolated from geothermal sites in Yellowstone National Park: physiological and phylogenetic characteristics. Archiv. Microbiol. *180*, 60–68.

Johnson, D.B., Stallwood, B., Kimura, S., and Hallberg, K.B. (2006). Characteristics of *Acidicaldus organivorus, gen. nov.*, sp. *nov.*; a novel thermo-acidophilic heterotrophic *Proteobacterium*. Arch. Microbiol. *185*, 212–221.

Johnson, D.B., Bacelar-Nicolau, P., Okibe, N., Thomas, A., and Hallberg, K.B. (2009). *Ferrimicrobium acidiphilum* gen. nov., sp. nov. and *Ferrithrix thermotolerans* gen. nov., sp. nov.: heterotrophic, iron-oxidizing, extremely acidophilic actinobacteria. Int. J. Syst. Evol. Microbiol. *59*, 1082–1089.

Johnson, D.B., Hallberg, K.B., and Hedrich, S. (2014). Uncovering a microbial enigma: isolation and characterization of the streamer-generating, iron-oxidizing, acidophilic bacterium "*Ferrovum myxofaciens*". Appl. Environ. Microbiol. *80*, 672–680.

Jones, D.S., Albrecht, H.L., Dawson, K.S., Schaperdoth, I., Freeman, K.H., Pi, Y., Pearson, A., and Macalady, J.L. (2012). Community genomic analysis of an extremely acidophilic sulfur-oxidizing biofilm. ISME J. *6*, 158–170.

Kay, C.M., Rowe, O.F., Rocchetti, L., Coupland, K., Hallberg, K.B., and Johnson, D.B. (2013). Evolution of microbial "streamer" growths in an acidic metal-contaminated stream draining an abandoned underground copper mine. Life 3, 189–211.

Kleeberg, A., and Grüneberg, B. (2005). Phosphorus mobility in sediments of acid mining lakes, Lusatia, Germany. Ecol. Eng. *24*, 89–100.

Kock, D., and Schippers, A. (2008). Quantitative microbial community analysis of three different sulfidic mine tailings dumps generating acid mine drainage. Appl. Environ. Microbiol. *74*, 5211–5219.

Korehi, H., Blöthe, M., Sitnikova, M.A., Dold, B., and Schippers, A. (2013). Metal mobilization by iron- and sulfur-oxidizing bacteria in a multiple extreme mine tailings in the Atacama Desert, Chile, Environ. Sci. Technol. *47*, 2189–2196.

Koschorreck, M. (2008). Microbial sulphate reduction at a low pH. FEMS Microbiol. Ecol. *64*, 329–342.

Koschorreck, M., Bozau, E., Frömmichen, R., Geller, W., Herzsprung, P., and Wendt-Potthoff, K. (2007). Processes at the sediment water interface after addition of organic matter and lime to an acid mine pit lake mesocosm. Environ. Sci. Technol. *41*, 1608–1614.

Kozubal, M.A., Romine, M., Jennings, R.D., Jay, Z.J., Tringe, S.G., Rusch, D.B., Beam, J.P., McCue, L.A., and Inskeep, W.P. (2013). *Geoarchaeota*: A new candidate phylum in the *Archaea* from high-temperature acidic iron mats in Yellowstone National Park. ISME J. 7, 622–634.

Kvist, T., Ahring, B.K., and Westermann, P. (2007). Archaeal diversity in Icelandic hot springs. FEMS Microbiol. Ecol. 59, 71–80.

Langdahl, B.R., and Ingvorsen, K. (1997). Temperature characteristics of bacterial iron solubilisation and 14C assimilation in naturally exposed sulfide ore material at Citronen fjord, north Greenland (831N). FEMS Microbiol. Ecol. *23*, 275–283.

Le Bris, N., Zbinden, M., and Gaill, F. (2005). Processes controlling the physicochemical micro-environments associated with Pompeii worms. Deep Sea Res. *52*, 1071–1083.

Liu, J., Hua,Z.-S., Chen,Z.-N., Kuang, J.-L., Li, S.J., Shu, W.-S., and Huang, L.-N. (2014). Correlating microbial diversity patterns with geochemistry in an extreme and heterogeneous mine tailings environment. Appl. Environ. Microbiol. *80*, 3677–3686.

López-Archilla, A.I., Marín, I., and Amils, R. (2001). Microbial community composition and ecology of an acidic aquatic environment: the Tinto River, Spain. Microb. Ecol. *41*, 20–35.

Lu, S., Gischkat, S., Reiche, M., Akob, D.M., Hallberg, K.B., and Küsel, K. (2010). Ecophysiology of Fe-cycling bacteria in acidic sediments. Appl. Environ. Microbiol. *76*, 8174–8818.

Macalady, J.L., Jones, D.S., and Lyon, E.H. (2007). Extremely acidic, pendulous cave wall biofilms from the Frasassi cave system, Italy. Environ. Microbiol. *9*, 1402–1414.

Méndez-García, C., Mesa, V., Sprenger, R.R., Richter, M., Diez, M.S., Solano, J., Bargiela, R., Golyshina, O.V., Manteca, A., Ramos, J.L., *et al.* (2014). Microbial stratification in low pH oxic and suboxic macroscopic growths along an acid mine drainage. ISME J. *8*, 1259–1274.

Ñancucheo, I., and Johnson, D.B. (2012). Acidophilic algae isolated from mine-impacted environments and their roles in sustaining heterotrophic acidophiles. Front. Microbiol. *3*, 1–8.

Nicomrat, D., Dick, W.A., Dopson, M., and Tuovinen, O.H. (2008). Bacterial phylogenetic diversity in a constructed wetland system treating acid coal mine drainage. Soil Biol. Biochem. *40*, 312–321.

Nordstrom, D.K. (2000). Advances in the hydrogeochemistry and microbiology of acid mine waters. Int. Geol. Rev. *42*, 499–515.

Norris, P.R., Davis-Belmar, C.S., Brown, C.F., and Calvo-Bado, L.A. (2011). Autotrophic, sulfur-oxidizing actinobacteria in acidic environments. Extremophiles *15*, 155–163.

Ohba, H., and Owa, N. (2005). Vertical distribution of physicochemical properties and number of sulfur-oxidizing bacteria in the buried layer of soil profiles with marine-reduced sulfur compounds. Soil Sci. Plant Nut. *51*, 379–388.

Prokofeva, M.I., Miroshnichenko, M.L., Kostrikina, N.A., Chernyh, N.A., Kuznetsov, B.B., Tourova, T.P., and Bonch-Osmolovskaya, E.A. (2000). *Acidilobus aceticus gen. nov.*, sp. *nov.*, a novel anaerobic thermoacidophilic archaeon from continental hot vents in Kamchatka. Int. J. Syst. Evol. Microbiol. *50*, 2001–2008.

Ram, R.J., Verberkmoes, N.C., Thelen, M.P., Tyson, G.W., Baker, B.J., Blake, R.C. II, Shah, M., Hettich, R.L., and Banfield, J.F. (2005). Community proteomics of a natural microbial biofilm. Science *308*, 1915–1920.

Reigstad, L.J., Jorgensen, S.L., and Schleper, C. (2010). Diversity and abundance of Korarchaeota in terrestrial hot springs of Iceland and Kamchatka. ISME J. *4*, 346–356.

Remonsellez, F., Galleguillos, F., Moreno-Paz, M., Parro, V., Acosta, M., and Demergasso, C. (2009). Dynamic of active microorganisms inhabiting a bioleaching industrial heap of low-grade copper sulfide ore monitored by real-time PCR and oligonucleotide prokaryotic acidophile microarray. Microbial. Biotechnol. *2*, 613–624.

Reysenbach, A.-L., Liu, Y., Banta, A.B., Beveridge, T.J., Kirshtein, J.D., Schouten, S., Tivey, M.K., Von Damm, K.L., and Voytek, M.A. (2006). A ubiquitous thermoacidophilic archaeon from deep-sea hydrothermal vents. Nature *442*, 444–447.

Rowe, O.F., Sánchez-España, J., Hallberg, K.B., and Johnson, D.B. (2007). Microbial communities and geochemical dynamics in an extremely acidic, metal-rich stream at an abandoned massive sulfide mine (Huelva, Spain) underpinned by two primary production systems. Environ. Microbiol. *9*, 1761–1771.

Sánchez-Andrea, I., Stams, A.J.M., Hedrich, S., Ñancucheo, I., and Johnson, D.B. (2015). *Desulfosporosinus acididurans* sp. nov.: an acidophilic sulfate-reducing bacterium isolated from acidic sediments. Extremophiles *19*, 39–47.

Sand, W., Jozsa, P.-G., Kovacs, Zs.-M., Săsăran, N., and Schippers, A. (2007). Long-term evaluation of acid rock drainage mitigation measures in large lysimeters. J. Geochem. Explor. *92*, 205–211.

Santofimia, E., González-Toril, E., López-Pamo, E., Gomariz, M., Amils, R., and Aguilera, A. (2013). Microbial diversity and its relationship to physicochemical characteristics of the water in two extreme acidic pit lakes from the Iberian Pyrite Belt (SW Spain). PLoS One *8*, e66746.

Satake, K. (1977). Microbial sulphate reduction in a volcanic acid lake having pH 1.8 to 2. Jap. J. Limnol. *38*, 33–35.

Schippers, A., Breuker, A., Blazejak, A., Bosecker, K., Kock, D., and Wright, T.L. (2010). The biogeochemistry and microbiology of sulfidic mine waste and bioleaching dumps and heaps, and novel Fe(II)-oxidizing bacteria. Hydrometallurgy *104*, 342–350.

Schippers, A., Hedrich, S., Vasters, J., Drobe, M., Sand, W., and Willscher, S. (2014). Biomining: metal recovery from ores with microrganisms. Adv. Biochem. Eng. Biotechnol. *141*, 1–47.

Seifert, J., Erler, B., Seibt, K., Rohrbach, N., Arnold, J., Schlömann, M., Kassahun, A., and Jenk, U. (2008). Characterization of the microbial diversity in the abandoned uranium mine Königstein. In Uranium, Mining and Hydrogeology, Merkel, B., and Hasche-Berger, A., eds (Springer, Berlin, Germany), pp. 733–742.

Sen, A.M., and Johnson, D.B. (1999). Acidophilic sulphate-reducing bacteria: candidates for bioremediation of acid mine drainage. In Biohydrometallurgy and the Environment Toward the Mining of the 21st Century, Amils, R., and Ballester, A., eds (Process Metallurgy 9A. Elsevier, Amsterdam, the Netherlands), pp. 709–718.

Simmons, S., and Norris, P.R. (2002). Acidophiles of saline water at thermal vents of Vulcano, Italy. Extremophiles *6*, 201–207.

Stout, L.M., Blake, R.E., Greenwood, J.P., Martini, A.M., and Rose, E.C. (2009). Microbial diversity of boron-rich volcanic hot springs of St. Lucia, Lesser Antilles. FEMS Microbiol. Ecol. *70*, 402–412.

Tyson, G.W., Chapman, J., Hugenholtz, P., Allen, E.E., Ram, R.J., Richardson, P.M., Solovyev, V.V., Rubin, E.M., Rokhsar, D.S., and Banfield, J.F. (2004). Community structure and metabolism through reconstruction of microbial genomes from the environment. Nature *428*, 37–43.

Urbieta, M.S., González Toril, E., Aguilera, A., Giaveno, M.A., and Donati, E. (2012). First prokaryotic biodiversity assessment using molecular techniques of an acidic river in Neuquén, Argentina. Microbial Ecol. *64*, 91–104.

Urbieta, M.S., González-Toril, E., Giaveno, M.A., Aguilera-Bazán, A., and Donati, E.R. (2014a). Relevance of sulphur oxidising species and new archaea in Copahue geothermal ponds in Argentina. Syst. Appl. Microbiol. *37*, 429–441.

Urbieta, M.S., Rascovan, N., Castro, C., Revale, S., Giaveno, M.A., Vazquez, M., and Donati, E.R. (2014b). Draft genome sequence of the novel thermoacidophilic archaeon *Acidianus copahuensis* strain ALE1, isolated from the Copahue volcanic area in Neuquén, Argentina. Genome Announc. *2*, e00259–14.

Vera, M., Schippers, A., and Sand, W. (2013). Progress in bioleaching: fundamentals and mechanisms of bacterial metal sulfide oxidation—part A. Appl. Microbiol. Biotechnol. *97*, 7529–7541.

Von Damm, K.L. (1995). Controls on the chemistry and temporal variability of seafloor hydrothermal fluids. In Physical, Chemical, Biological, and Geological Interactions within Seafloor Hydrothermal Systems, Humphris, S., Lupton, J., Mullineaux, L., and Zierenberg, R., eds (American Geophysical Union, Washington, DC, USA), pp. 222–247.

Wemheuer, B., Taube, R., Akyol, P., Wemheuer, F., and Daniel, R. (2013). Microbial diversity and biochemical potential encoded by thermal spring metagenomes derived from the Kamchatka peninsula. Archaea *2013*, 136714.

Willis, G., Hedrich, S., Ñancucheo, I., Johnson, D.B., and Donati, E. (2013). Microbial diversity in acidic anaerobic sediments at the geothermal Caviahue-Copahue system, Argentina. Adv. Mat. Res. *825*, 7–10.

Wu, X., Wong, Z.L., Sten, P., Engblom, S., Österholm, P., and Dopson, M. (2013). Microbial community potentially responsible for acid and metal release from an Ostrobothnian acid sulfate soil. FEMS Microbiol. Ecol. *84*, 555–563.

Ziegler, S., Ackermann, S., Majzlan, J., and Gescher, J. (2009). Matrix composition and community structure analysis of a novel bacterial pyrite leaching community. Environ. Microbiol. *11*, 2329–2338.

Ziegler, S., Dolch, K., Geiger, K., Krause, S., Asskamp, M., Eusterhues, K., Kriews, M., Wilhelms-Dick, D., Goettlicher, J., Majzlan, J. *et al.* (2013). Oxygen-dependent niche formation of a pyrite-dependent acidophilic consortium built by archaea and bacteria. ISME J. *7*, 1725–1737.

Part V

Molecular Biology of Acidophiles

Progress in Acidophile Genomics

11

Juan Pablo Cárdenas, Raquel Quatrini and David S. Holmes

Abstract

By March 2015, 161 genomes of acidophilic microorganisms had been published, including 126 permanent draft and closed genomes and 35 partial genomes reconstructed from metagenomic data, distributed between *Archaea* (71), *Bacteria* (86) and *Eukarya* (4). This provides a rich source of latent data that can be exploited for understanding the biology of this group of organisms and for advancing biotechnological applications. Genomic data are already yielding valuable insights into the genome architecture and cellular metabolism of acidophiles, allowing the construction of useful models to probe their evolution and ecophysiology both in naturally occurring acidic environments and in industrial operations including biomining.

Introduction

The advent of genomics and the possibility of genome-wide comparative and evolutionary analyses have revolutionized the field of microbiology. This is particularly true for studies on acidophiles where genomics have proven to be a rich source of information that, in many cases, has proven difficult to access via biochemical and genetic studies given difficulties in their culturing and manipulation.

An early review of acidophile genomes focused on comparative genomics and provided a very useful overview of metabolic functions and evolution (Angelov and Liebl, 2007). Several other mini-reviews discussed the genomics of acidophiles but these have focused on microorganisms involved in the biological recovery of copper and other metals under acidic conditions (bioleaching and biooxidation; Chapter 16) (Bonnefoy and Holmes, 2012; Cardenas *et al.*, 2010; Jerez, 2008; Holmes and Bonnefoy, 2007; Quatrini *et al.*, 2007b; Valenzuela *et al.*, 2006). More recent commentaries touched upon acidophile genomics in the context of iron oxidation (Altermann, 2014; Hedrich *et al.*, 2011).

The present chapter describes and updates the genomic information discussed in the above reviews and extends it to describe more recently published genomes from diverse acidic environments. The chapter does not cover metagenomics (Chapter 13), except in the instances where metagenomic data have been used to partially reconstruct acidophile genomes; nor are genomes of viruses or plasmids included (Chapter 12); also, it does not discuss functional genomic applications such as proteomics (Chapter 14) and transcriptomics.

One of the ultimate goals of genome science is to predict the overall cellular organization from molecular to higher levels, eventually leading to useful models of interactions of microorganisms with their environment including other microbial co-inhabitants. This goal can be approached by systematic comparison of genomes of related organisms adapted to both similar and different environments, thereby identifying corresponding trends in genomic content.

Timeline of genome sequencing projects

According to the IMG database (https://img.jgi.doe.gov/cgi-bin/w/main.cgi), as of March 2015, there were 25,253 bacterial, 527 archaeal and 190 eukaryotic genomes completely sequenced or with permanent draft sequences. Of these, 126 genomes were derived from acidophiles, representing 0.48%

of the total. This low percentage is partly due a bias towards sequencing microorganisms of medical relevance which represent more than 50% of the projects. If instead, one takes into account only genomes from environmental samples, the percentage of sequenced acidophile genomes is 10-fold higher (4.8%). However, since the biodiversity of acidic niches is known to be relatively low in comparison with some other environmental ecosystems such as soil and marine environments, the variety of sequenced acidophiles may provide a better representation of the members present in acidic environments than might be expected based on these raw numbers, at least in the prokaryotic group. The most genomes per genus are from *Sulfolobus* (total 27) and the acidithiobacilli (total 10), representatives of acidophilic hot-springs and AMD/bioleaching operations, respectively.

The first complete genome of an acidophilic prokaryote, the archaeon *Thermoplasma acidophilum* DSM1728 (Ruepp *et al.*, 2000) and the first draft genome of a bacterial acidophile *Acidithiobacillus ferrooxidans* (ATCC 23270^T; Selkov *et al.*, 2000) were published in 2000. Between 2001 and the end of 2006 only four more acidophile genome sequences were published (Fig. 11.1). From 2007 until 2012 there was an approximately linear increase in the number of genomes sequenced per year.

Diversity of acidophiles with sequenced genomes

Of the 126 permanent draft and complete acidophile genomes, 55 (44%) are Archaea, 67 (53) are Bacteria and only 4 (3%) are Eukaryotes of which 3 are red algae (Schönknecht *et al.*, 2013; Qiu *et al.*, 2013; Matsuzaki *et al.*, 2004) and one is a yeast (project GOLD Gp0110612) (see Appendix). The paucity of acidophile genomes from Eukarya reflects the relatively recent recognition of this group as important members of acidophilic communities and also, to some extent, the greater effort and expense required to sequence more complex genomes.

Phylogenetic distribution and associated metadata of sequenced genomes

Sequenced acidophiles include representatives widely distributed on the phylogenetic tree of life

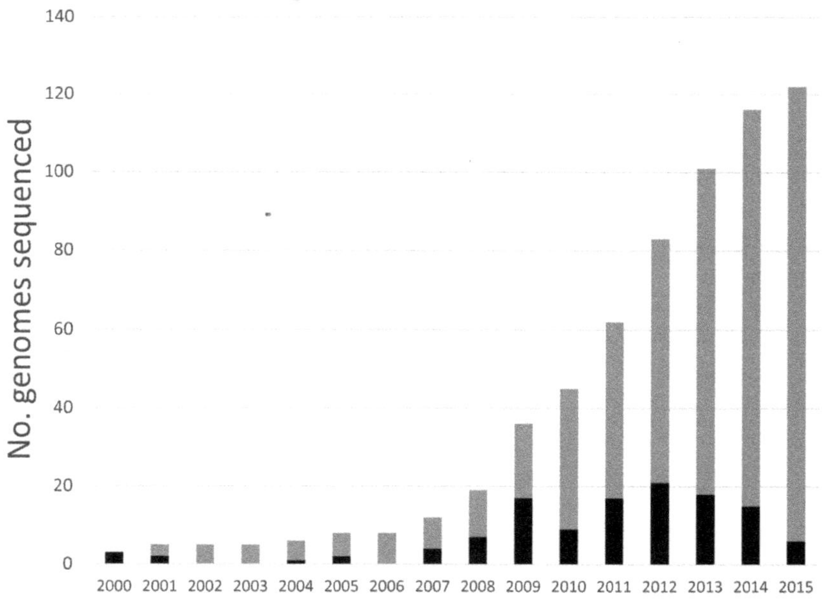

Figure 11.1 The number of acidophile genomes sequenced by year up to March, 2015. The dark bars represent the number of genomes sequenced in any particular year; grey bars represent the accumulative number sequenced genomes.

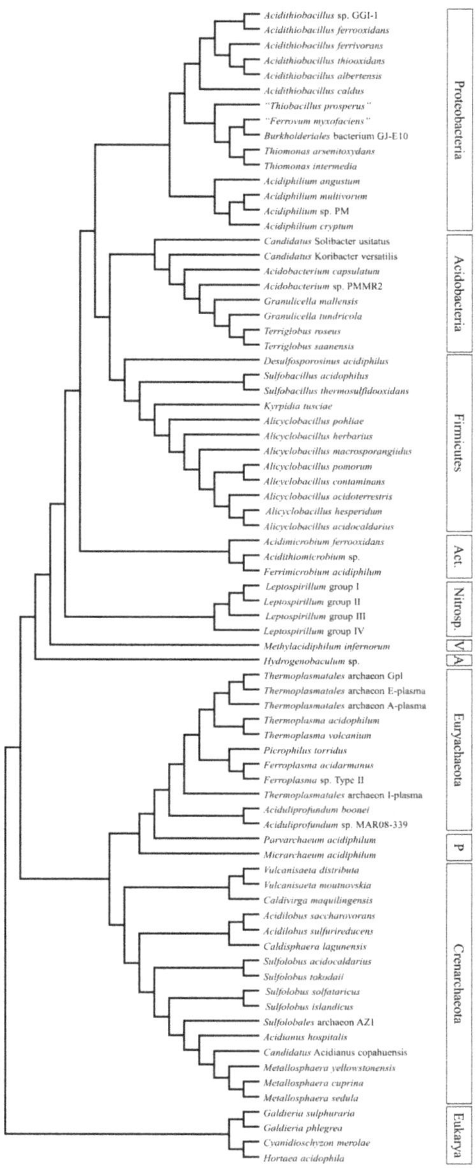

Organism	(1)	(2)	(3)	(4)	(5)	(6)	(7)	(8)	(9)	(10)
Acidithiobacillus sp. GGI-1	1	E	M	OA	FAe	+?	+?	+	CBB	-
Acidithiobacillus ferrooxidans	3	E	M	OA	FAe	+	+	+	CBB	-
Acidithiobacillus ferrivorans	1	E	M#	OA	FAe	+	+	+	CBB	-
Acidithiobacillus thiooxidans	2	E	M	OA	FAe	-	+	-	CBB	B
Acidithiobacillus albertensis	1	E	M	OA	FAe	-	+	-	CBB	B
Acidithiobacillus caldus	2	E	MT	OA	FAe	-	+	-	CBB	B
"Thiobacillus prosperus"	1	E	M	OA	FAe	+	+	+	CBB	B
"Ferrovum myxofaciens"	1	E	M#	OA	FAe	+	-	+	CBB	B
Burkholderiales bacterium GJ-E10	1	E	M	OA?	FAe?	+	-	+	CBB	B
Thiomonas arsenitoxydans	1	M	M	FA	FAe	-	+	-	CBB	B
Thiomonas intermedia	1	M	M	FA	FAe	-	+	-	CBB	B
Acidiphilium angustum	1	M	M	H	FAe	-	-	-	CBB	B
Acidiphilium multivorum	1	M	M	H	FAe	-	-	-	CBB	B
Acidiphilium sp. PM	1	M	M	FA	FAe	-	-	-	CBB	B
Acidiphilium cryptum	1	M	M	H	FAe	-	-	-	CBB	B
Candidatus Solibacter usitatus	1	M	M	H	A	-	-	-	n.a	B
Candidatus Koribacter versatilis	1	M	M	H	A	-	-	-	n.a	B
Acidobacterium capsulatum	1	M	M	H	A	-	-	-	n.a	B
Acidobacterium sp. PMMR2	1	M	M	H	A	-	-	-	n.a	B
Granulicella mallensis	1	M	M	H	A	-	-	-	n.a	-
Granulicella tundricola	1	M	M	H	A	-	-	-	n.a	B
Terriglobus roseus	1	M	M	H	A	-	-	-	n.a	B
Terriglobus saanensis	1	M	M	H	A	-	-	-	n.a	B
Desulfosporosinus acidiphilus	1	M	M	H	An	-	-	+	n.a	B
Sulfobacillus acidophilus	2	E	T	M	FAe	+	+	-	CBB	B
Sulfobacillus thermosulfidooxidans	3	E	T	M	FAe	+	+	-	CBB	B
Kyrpidia tusciae	1	M	T	FA	A	-	-	-	CBB	B
Alicyclobacillus pohliae	1	M	T	H	A	-	-	-	n.a	B
Alicyclobacillus herbarius	1	M	T	H	A	-	-	-	n.a	B
Alicyclobacillus macrosporangiidus	1	M	T	H	A	-	-	-	n.a	B
Alicyclobacillus pomorum	1	M	MT	H	A	-	-	-	n.a	B
Alicyclobacillus contaminans	1	M	T	H	A	-	-	-	n.a	B
Alicyclobacillus acidoterrestris	1	M	T	H	A	-	-	-	n.a	B
Alicyclobacillus hesperidum	1	M	T	H	A	-	-	-	n.a	B
Alicyclobacillus acidocaldarius	3	M	T	H	A	-	-	-	n.a	B
Acidimicrobium ferrooxidans	1	E	MT	FA	A	+	-	-	CBB	B
Acidithiomicrobium sp.	1	E	MT	FA	A	+	+	-	CBB	B
Ferrimicrobium acidiphilum	1	E	MT	H	A	+	-	-	CBB	B
Leptospirillum group I	1	E	MT	OA	A	+	-	+	rTCA	B
Leptospirillum group II	5	E	MT	OA	A	+	-	±	rTCA	B
Leptospirillum group III	1	E	MT	OA	A	+	-	+	rTCA	B
Leptospirillum group IV	1	E	MT	OA	A	+	-	+	rTCA	B
Methylacidiphilum infernorum	1	E	T	OA	A	-	-	+	CBB	-
Hydrogenobaculum sp.	5	E	T	OA	A	-	+	-	rTCA	B
Thermoplasmatales archaeon Gpl	1	E	MT	?	?	?	?	-	?	-
Thermoplasmatales archaeon E-plasma	1	E	MT	?	?	?	?	-	?	A
Thermoplasmatales archaeon A-plasma	1	E	MT	?	?	?	?	-	?	A
Thermoplasma acidophilum	1	E	T	H	FAe	-	-	-	n.a	A
Thermoplasma volcanium	1	E	T	H	FAe	-	-	-	n.a	A
Picrophilus torridus	1	H	T	H	FAe	-	-	-	n.a	-
Ferroplasma acidarmanus	2	E	MT	M	FAe	+	-	-	n.a	-
Ferroplasma sp. Type II	1	E	MT	M?	FAe?	+?	-	-	n.a	-
Thermoplasmatales archaeon I-plasma	1	E	MT	?	?	?	?	-	?	-
Aciduliprofundum boonei	1	M	T	H	An	-	-	-	n.a	A
Aciduliprofundum sp. MAR08-339	1	M	T	H	An	-	-	-	n.a	A
Parvarchaeum acidiphilum	2	E	MT?	?	?	?	?	-	?	-
Micrarchaeum acidiphilum	1	E	MT?	?	?	?	?	-	?	-
Vulcanisaeta distributa	1	M	HT	H	An	-	-	-	n.a	-
Vulcanisaeta moutnovskia	1	M	HT	H	An	-	-	-	n.a	-
Caldivirga maquilingensis	1	M	HT	H	An	-	-	-	n.a	-
Acidilobus saccharovorans	1	M	HT	H	An	-	-	-	n.a	A
Acidilobus sulfurireducens	1	M	HT	H	An	-	-	-	n.a	A
Caldisphaera lagunensis	1	M	T	H	An	-	-	-	n.a	-
Sulfolobus acidocaldarius	4	E	T	FA	A	-	+	-	m3HP	A
Sulfolobus tokodaii	1	E	T	FA	A	+	+	-	m3HP	A
Sulfolobus solfataricus	2	E	T	FA	A	-	+	-	m3HP	A
Sulfolobus islandicus	20	E	T	FA	A	-	+	-	m3HP	A
Sulfolobales archaeon AZ1	1	E	T?	FA?	A?	?	?	-	m3HP	-
Acidianus hospitalis	1	E	T	FA	A	+	+	-	m3HP	-
Candidatus Acidianus copahuensis	1	E	T	FA	FAe	+	+	-	m3HP	-
Metallosphaera yellowstonensis	1	E	T	FA	A	+	+	-	m3HP	A
Metallosphaera cuprina	1	E	T	FA	A	+	+	-	m3HP	-
Metallosphaera sedula	1	E	T	FA	A	+	+	-	m3HP	A
Galdieria sulphuraria	1	E	MT	FA	A	-	-	-	CBB	-
Galdieria phlegrea	1	E	MT	FA	A	-	-	-	CBB	-
Cyanidioschyzon merolae	1	E	MT	FA	A	-	-	-	CBB	-
Hortaea acidophila	1	E	M	H	A	-	-	-	n.a	-

Figure 11.2 Phylogenetic distribution of sequenced permanent draft and closed acidophile genomes based on small subunit ribosomal RNA gene sequences for (A) Bacteria and (B) Archaea and Eukarya. Habitat metadata for each organism and their predicted genomic features, derived from the literature are also provided. Numbers at top of columns: (1) Number of genomes sequenced per species; (2) Optimum acidity by range: H: hyper-acidophile (pH < 1.0), E: extreme acidophile (pH 1.0–3.0), M: moderate acidophile (pH > 3.0); (3) Optimum temperature by range: M: mesophile (< 40°C), MT: meso-thermophile (40–55°C), T: thermophile (55–80°C), H: hyperthermophile (> 80°C); (4) trophy: H: heterotroph, OA: obligate autotroph, F-A: facultative autotroph, M: mixotroph; (5) oxygen requirements: A: aerobe, An: anaerobe, F-A: facultative aerobe; (6) Fe(II)-oxidizing capabilities; (7) sulfur-oxidizing capabilities; (8) nitrogen-fixing capabilities where + is positive and – is negative; (9) carbon fixation pathway where CBB = Calvin cycle, rTCA = reverse TCA cycle and m3HP = modified 3-hydroxypropionate cycle; (10) flagella genes detected where B = bacterial type and A = archaeal type. Abbreviations used: *Detected in genome, but without experimental validation; # = psychrotolerant; ? = unknown. Sequences were extracted from genome projects and aligned by MAFFT (Katoh *et al.*, 2002). The tree was made in MEGA 6, using Maximum Likelihood method (Kimura-2-parameter, G+) (Tamura *et al.*, 2013). The tree also includes the phylum affiliation for each clade. Abbreviations: Act, *Actinobacteria*; A, *Aquificae*; V, *Verrucomicrobia*; Nitro, *Nitrospirae*; P, *Parvarchaeota*.

and include *Archaea*, *Bacteria* and *Eukarya*, providing a first glimpse of the genomics of acidophilic life over a range of environmental conditions (Fig. 11.2). They span a wide diversity of lifestyles and include organisms that are psychrotolerant, mesophilic, moderately and extremely thermophilic; there are also some moderate acidophiles (optimum growth above pH 3.5) that are hyperthermophilic. In addition, the wide diversity involves different trophies including heterotrophs, autotrophs and mixotrophs, and have different oxygen requirements (obligate aerobes, facultative and obligate anaerobes) and some are halotolerant. This genomic data can now be mined to reveal evolutionary tendencies of acidophiles including horizontal gene transfer (Chapter 12) and associations of genes and pathways with the acidophilic environment (Chapters 2–4). For example, several studies have already detected a tendency for extracellular proteins from acidophiles to have high (basic) isoelectric points (Duarte *et al.*, 2011; 2009; Hou *et al.*, 2008) and to have a number of mechanisms for survival at very low pH (Chapter 3 and references therein). Most of these studies were carried out using individual genomes or on a relatively small sample of genomes. With the new, more substantial genomic data, these studies can be revisited, consolidating the evidence and perhaps allowing the detection of additional trends associated with extremely low pH (< pH 1) versus pH around 3 or with specific phylogenetic branches of the acidophiles. This would help address questions related to the evolution of life in acidophilic conditions.

Genomes partially reconstructed from metagenomic data

Deep sequencing of metagenomes from relatively low-diversity acid mine drainages has permitted the partial reconstruction of 16 archaeal and 19 bacterial genomes of the dominant organisms in these environments (Table 11.1). This approach is especially useful when the metagenome sequences can be compared to genomes from cultured strains revealing new insights into the genetic diversity of environmental strains including genomic rearrangements, deletions and insertions of genes and genomic islands and the abundance and location of single nucleotide polymorphisms (see references in Table 11.1).

Comparative genomics

With the increasing number of sequenced acidophile genomes, significant progress is being made in the understanding of the evolution of acidophiles using comparative genomics. In perhaps what is the most extensive comparative genomic study of acidophiles, the roles that geography and genome evolution play in speciation of *Sulfolobus islandicus* have been explored (Cadillo-Quiroz *et al.*, 2012; Reno *et al.*, 2009). Patterns of homologous gene flow in sympatric speciation indicated that certain regions of the genome were more likely than others to be involved in strain differentiation. Although these studies focused on one archaeal genus, the findings may well be applicable to other organisms and other environments, helping us address the challenging problem of understanding microbial speciation and the events that lead to it.

Comparative genomics has also been useful for identifying differences between genomes from cultured isolates and environmental sequences. Genome information of environmental '*Ferroplasma acidarmanus*'-like populations showed important differences compared with the genome of strain fer1 (Allen *et al.*, 2007). These include differences in mobilizable genome regions and heterogeneity of single nucleotide polymorphisms, reflecting population diversity and adaptation to environmental fluctuations. In a study comparing the *At. ferrivorans* type strain genome with matching metagenomic sequences from a low-temperature, acidic copper mine drainage, several differences were observed in gene content and arrangement, including gene clusters present only in the isolate genome potentially encoding a copper efflux P-type ATPase system and a polyphosphate kinase involved in polyphosphate mediated copper resistance (Gonzalez *et al.*, 2014).

A comparative genomics study of 17 acidophile genomes focused on mechanisms of CO_2 fixation and other metabolic and physiological features that were used to generate an initial model for time–course interactions of acidophilic organisms in bioleaching operations (Valdés *et al.*, 2010). A comparison of two strains of *At. ferrooxidans* (ATCC 23270^T and ATCC 53993) revealed the presence of unique genome islands in both strains even though they were 100% identical at the 16S rRNA gene level (Holmes *et al.*, 2009; Levicán *et al.*, 2009). One of these islands was subsequently shown to

Table 11.1 List of acidophiles with partial genome sequences reconstructed from metagenomic data

NCBI accession no.	Organism	Reference
AOMP00000000-AOMQ00000000	*Acidithiobacillus ferrivorans*-like	Gonzalez *et al.* (2014)
4565622.3 – 4565632.3*	*Acidithiobacillus ferrooxidans*-like (FKB1)	Hua *et al.* (2014)
	Acidithiobacillus thiooxidans-like (FKB2)	
	'*Leptospirillum rubarum*'-like (FKB3)	
	Leptospirillum ferrodiazotrophum-like (FKB4)	
	Acidiphilium cryptum-like (FKB5)	
	Alicyclobacillus acidocaldarius-like (FKB6)	
	'*Ferrovum*' sp. (FKB7)	
	Candidatus Micrarchaeum acidiphilum ARMAN-2 (FKA8)	
	Candidatus Parvarchaeum acidiphilum ARMAN-4 (FKA9)	
	Candidatus Parvarchaeum acidophilus ARMAN-5 (FKA10)	
	Picrophilus torridus-like (FKA11)	
ACVJ00000000	*Candidatus* Micrarchaeum acidiphilum ARMAN-2	Baker *et al.* (2010)
ADCE00000000	*Candidatus* Parvarchaeum acidiphilum ARMAN-4	
ADHF00000000	*Candidatus* Parvarchaeum acidophilus ARMAN-5	
ATMF00000000	*Thermoplasmatales* archaeon A-plasma	Yelton *et al.* (2013)
ACXL00000000	*Thermoplasmatales* archaeon E-plasma	
ATDV00000000	*Thermoplasmatales* archaeon Gpl	
ATMD00000000	*Thermoplasmatales* archaeon I-plasma	
ATDU00000000	*Ferroplasma* sp. Type II	
AAWO00000000	'*Leptospirillum rubarum*'	Goltsman *et al.* (2009)
ACNP00000000	*Leptospirillum ferrodiazotrophum*	
ASRH00000000	*Sulfolobales* Archaeon AZ1	Servin-Garciduenas and Martinez-Romero (2014)
AIJM00000000	*Leptospirillum* sp. Group II 'C75'	Denef and Banfield (2012)
AURA00000000	*Leptospirillum* sp. Group IV 'UBA BS'	Goltsman *et al.* (2013)
AADL00000000	*Leptospirillum* sp. Group II '5-way CG'	Tyson *et al.* (2004)
	Leptospirillum sp. Group III	
	'*Ferroplasma acidarmanus*' Type I	
	Ferroplasma sp. Type II	
	Thermoplasmatales archaeon Gpl	
SRR191843 #	*Sulfobacillus*-like AMDSBA1	Justice *et al.* (2014)
	Sulfobacillus-like AMDSBA2	
	Sulfobacillus-like AMDSBA3	
	Sulfobacillus-like AMDSBA4	
	Sulfobacillus-like AMDSBA5	

*G-RAST accession number; #NCBI Sequence Read Archive accession, genomic information available at: http://ggkbase.berkeley.edu

encode copper resistance traits (Orellana and Jerez, 2011) and the other biofilm formation and maturation traits (Bustamante *et al.*, 2012). Similarly, a prophage (Tapia *et al.*, 2012) and other potentially mobile elements were detected in the type strain of *At. caldus* (ATCC 51756) and not in the closely related strain SM-1 (Acuña *et al.*, 2013). Comparative genomics has also been useful in revealing differences in the way acidophiles carry out iron transport and metabolism (Osorio *et al.*, 2008), biofilm formation (Núñez *et al.*, 2011), energy metabolism (Levican *et al.*, 2012), oxidative stress (Cárdenas *et al.*, 2012) and several characteristics of native strains of *At. thiooxidans* and *Sulfobacillus thermosulfidooxidans* of importance to bioleaching (Guo *et al.*, 2014; Travisany *et al.*, 2014)

Metabolic reconstructions and genome-scale models

An increasing number of metabolic models have been constructed based on bioinformatic predictions of genome data. This is a particularly useful approach because, in many cases, these organisms are difficult to grow and handle, making bioinformatic predictions of gene repertoires and metabolic potential an important way to peer inside the 'black box' of acidophile ecophysiology.

An early study of the *At. ferrooxidans* type strain genome predicted a pathway for sulfur assimilation and suggested how it could be connected with hydrogen metabolism and nitrogen fixation (Valdes *et al.*, 2003). Additional models for *At. ferrooxidans*^{T} metabolism were subsequently generated (Barreto *et al.*, 2003) including carbon metabolism (Appia-Ayme *et al.*, 2006) and iron uptake/homeostasis (Osorio *et al.*, 2008; Quatrini *et al.*, 2005, 2007a). These models were subsequently incorporated and extended to produce a more complete picture of *At. ferrooxidans* metabolism using genomic analysis (Valdés *et al.*, 2008) and functional genomic approaches (Quatrini *et al.*, 2009).

Models for carbon and energy metabolism have been made for other strains of *At. ferrooxidans*, and also for *At. thiooxidans* and *Leptospirillum* group II (Levicán *et al.*, 2008) and *At. caldus* (You *et al.*, 2011a). Models for energy metabolism include the reconstruction of sulfur oxidation pathways for *At. thiooxidans* (Yin *et al.*, 2014) and *At. caldus* (Mangold *et al.*, 2011), and the electron transfer chains involved in iron oxidation pathways in *Leptospirillum* (Levicán *et al.*, 2012) and the *Firmicute*, *Sulfobacillus thermosulfidooxidans* (Guo et al., 2014). The anaerobic metabolism of *At. ferrooxidans* has been studied using basic metabolic reconstruction (Osorio *et al.*, 2009), in combination with proteomics and transcriptomics (Osorio *et al.*, 2013). Those investigations found a potential pathway of Fe(III) reduction and differential patterns in central carbon metabolism, including carbon fixation (up-regulated in anaerobiosis) and glycolysis (up-regulated in aerobiosis).

The metabolic reconstruction from the genome also involved in some cases the analysis of regulons. In *At. ferrooxidans*, for example, the bioinformatic prediction and experimental validation of Fur targets allowed a preliminary model of the Fur regulatory network to be deduced (Quatrini *et al.*, 2007a). Other studies using genome-wide evidence of *At. ferrooxidans*, predicted the AfeR regulon regulated by quorum sensing (Banderas and Guiliani, 2013) and the reconstruction of the ArsR and MerR regulons, involved in responses to arsenic- and heavy metal-induced stresses (Hodar *et al.*, 2012).

The above examples illustrate the power of genome biology to predict gene repertoires, metabolic potential and first glimpses of potential regulatory connections. They represent an important first step in the construction of genome-scale models that can be used to mathematically model metabolic networks. However, to date the only example of a full genome-scale model for an acidophile is the reconstruction of a metabolic network in the crenarchaeote *Sulfolobus solfataricus* (Ulas *et al.*, 2012). This model includes 718 metabolic reactions, 58 transport/exchange processes and 705 unique metabolites, constructed by using genomic and experimental data, in addition to literature-based curation. The model permitted mathematical simulations to be carried out for network behaviour using 35 different carbon sources and under both heterotrophic and autotrophic conditions. The model defined a set of essential metabolic functions by gene deletion simulation using the minimization of metabolic adjustment (MOMA) method.

Since the elaboration of genome-scale models requires deep knowledge of the organism (Thiele and Palsson, 2010), the use of this technique remains difficult for many acidophiles at this time. However, stoichiometric models for *At. ferrooxidans* metabolic flux analysis (Hold *et al.*, 2009) and reduced inorganic sulfur compound oxidation in *At. thiooxidans* (Bobadilla Fazzini *et al.*, 2013) have been proposed. In addition, there is a metabolomic study (Chapter 15) of strains 'Wenelen' and 'Licanantay' of *At. ferrooxidans* and *At. thiooxidans* respectively, revealing variations of substrate preference and/or growth phase differences and supporting the conjecture that glutathione plays a role in extracellular sulfur oxidation (Martínez *et al.*, 2013).

Perspectives

More genomes are coming

There is likely to be a surge in the number of genome projects of acidophiles, not only because lower sequencing costs have put genome sequencing within reach of most investigators, but also due to the increasing appreciation of the importance of acidophiles for understanding environmental metal cycling, the production of acid mine drainage and natural acidic drainage; also for their role in industrial metal recovery and corrosion processes and as potential sources of new enzymes. There is also a growing recognition of their fundamental relevance for building models of early evolutionary metabolisms and for their burgeoning importance in extremophile research and astrobiology.

But is more genome information useful?

Additional genome information can be plundered for the discovery and harnessing of untapped genetic information, enhancing biotechnological applications and facilitating our understanding of the metabolism of acidophiles. More genomes will enable a deeper understanding of the partitioning of gene functions between the core and dispensable genomes in selected acidophile groups, using the powerful tools of comparative genomics. This, in turn, will facilitate the identification of genes and genome islands that have been acquired by horizontal gene transfer, enabling models of evolution of niche specificity to be built. More genomes will also allow the search for signatures of acidophilia to be undertaken and the evolution of acidophilia with more robustness due to greater sample depth.

New acidophile genomes will probably be added to poorly sampled branches of the phylogenetic tree fleshing out its foliage. New genomes are also expected to come and from ultra-deep sampling of populations from closely related species, and from new low abundance representatives uncovered by metagenomics and single cell sequencing projects.

Single cell genomics

Partial reconstruction of several acidophile genomes has been carried out using metagenomic sequencing data and this method is likely to yield many more partial genomes in the future. This approach works best using model genomes from a cultured specimen as a templates and it is difficult to reconstruct genomes of unknown organisms even in environments with relatively low organismal diversity such as acid mine drainages. However, single cell genomics can recover bacterial and archaeal genomes directly from uncultured environmental samples and is a powerful technique for identifying new phylogenetic relationships, interdomain gene transfers and genome adaptation to environmental conditions, all of which would be challenging to identify in standard metagenomic studies (Lasken, 2012).

Genomes of eukaryotic acidophiles

Only four genomes of acidophilic Eukaryotes have been published but it is expected that more such genomes will be forthcoming. Not only are Eukaryotes major components of many acidophilic niches but one group has recently been shown to have received many bacterial and archaeal genes by horizontal gene transfer (Schönknecht *et al.*, 2013), adding a weighty contribution to the accumulating body of evidence of inter-kingdom gene transfer. Information from additional genomes of acidophilic Eukaryotes is likely to illuminate further the wide range of evolutionary forces that can act on genomes and will help expand the known horizons of gene transfer beyond the bounds of prevailing orthodoxy.

Acknowledgements

The authors thank Ana Moya-Beltrán, Francisco Issotta, Francisco Ossandon and María Yanquepe for help in data compilation and Fondecyt 1130683, 1140048 and a Basal Grant PFB-16 for financial support.

References

Acuña, L.G., Cardenas, J.P., Covarrubias, P.C., Haristoy, J.J., Flores, R., Nunez, H., Riadi, G., Shmaryahu, A., Valdes, J., Dopson, M., *et al.* (2013). Architecture and gene repertoire of the flexible genome of the extreme acidophile *Acidithiobacillus caldus*. PLoS One *8*, e78237.

Allen, E.E., Tyson, G.W., Whitaker, R.J., Detter, J.C., Richardson, P.M., and Banfield, J.F. (2007). Genome dynamics in a natural archaeal population. Proc. Natl. Acad. Sci. U.S.A. *104*, 1883–1888.

Altermann, E. (2014). Invited Commentary: Lubricating the rusty wheel, new insights into iron oxidising bacteria through comparative genomics. Front. Microbiol. *5*, 386.

Angelov, A., and Liebl, W. (2007). Genomics of acidophiles. In Physiology and Biochemistry of Extremophiles, Gerday, C., and Glansdorff, N., eds (ASM Press, Washington, DC, USA), pp. 279–292.

Appia-Ayme, C., Quatrini, R., Denis, Y., Denizot, F., Silver, S., Roberto, F., Veloso, F., Valdés, J., Pablo Cárdenas, J., Esparza, M., *et al.* (2006). Microarray and bioinformatic analyses suggest models for carbon metabolism in the autotroph *Acidithiobacillus ferrooxidans*. Hydrometallurgy *83*, 273–280.

Baker, B.J., Comolli, L.R., Dick, G.J., Hauser, L.J., Hyatt, D., Dill, B.D., Land, M.L., Verberkmoes, N.C., Hettich, R.L., and Banfield, J.F. (2010). Enigmatic, ultrasmall, uncultivated Archaea. Proc. Natl. Acad. Sci. U.S.A. *107*, 8806–8811.

Banderas, A., and Guiliani, N. (2013). Bioinformatic prediction of gene functions regulated by quorum sensing in the bioleaching bacterium *Acidithiobacillus ferrooxidans*. Int. J. Mol. Sci. *14*, 16901–16916.

Barreto, M., Quatrini, R., Bueno, S., Arriagada, C., Valdes, J., Silver, S., Jedlicki, E., and Holmes, D.S. (2003). Aspects of the predicted physiology of *Acidithiobacillus ferrooxidans* deduced from an analysis of its partial genome sequence. Hydrometallurgy *71*, 97–105.

Bobadilla Fazzini, R.A., Cortes, M.P., Padilla, L., Maturana, D., Budinich, M., Maass, A., and Parada, P. (2013). Stoichiometric modeling of oxidation of reduced inorganic sulfur compounds (Riscs) in *Acidithiobacillus thiooxidans*. Biotechnol. Bioeng. *110*, 2242–2251.

Bonnefoy, V., and Holmes, D.S. (2012). Genomic insights into microbial iron oxidation and iron uptake strategies in extremely acidic environments. Environ. Microbiol. *14*, 1597–1611.

Cadillo-Quiroz, H., Didelot, X., Held, N.L., Herrera, A., Darling, A., Reno, M.L., Krause, D.J., and Whitaker, R.J. (2012). Patterns of gene flow define species of thermophilic Archaea. PLoS Biol. *10*, e1001265.

Cárdenas, J.P., Valdés, J., Quatrini, R., Duarte, F., and Holmes, D.S. (2010). Lessons from the genomes of extremely acidophilic bacteria and archaea with special emphasis on bioleaching microorganisms. Appl. Microbiol. Biot. *88*, 605–620.

Cárdenas, J.P., Moya, F., Covarrubias, P., Shmaryahu, A., Levicán, G., Holmes, D.S., and Quatrini, R. (2012). Comparative genomics of the oxidative stress response in bioleaching microorganisms. Hydrometallurgy *127–128*, 162–167.

Denef, V.J., and Banfield, J.F. (2012). In situ evolutionary rate measurements show ecological success of recently emerged bacterial hybrids. Science *336*, 462–466.

Duarte, F., Araya-Secchi, R., González, W., Perez-Acle, T., González-Nilo, D., and Holmes, D.S. (2009). Protein function in extremely acidic conditions: molecular simulation studies of a predicted aquaporin and a voltage gated potassium channel in *Acidithiobacillus ferrooxidans*. Adv. Mat. Res. *71–73*, 211–214.

Duarte, F., Sepulveda, R., Araya, R., Flores, S., Perez-Acle, T., Gonzales, W., Gonzales, D., Neshich, G., Neshich, I., Mazoni, I., *et al.* (2011). Mechanisms of protein stabilization at very low pH. Paper presented at: 19th International Biohydrometallurgy Symposium (Changsha, China).

Goltsman, D.S., Denef, V.J., Singer, S.W., VerBerkmoes, N.C., Lefsrud, M., Mueller, R.S., Dick, G.J., Sun, C.L., Wheeler, K.E., Zemla, A., *et al.* (2009). Community genomic and proteomic analyses of chemoautotrophic iron-oxidizing "*Leptospirillum rubarum*" (Group II) and "*Leptospirillum ferrodiazotrophum*" (Group III) bacteria in acid mine drainage biofilms. Appl. Environ. Microb. *75*, 4599–4615.

Goltsman, D.S., Dasari, M., Thomas, B.C., Shah, M.B., VerBerkmoes, N.C., Hettich, R.L., and Banfield, J.F. (2013). New group in the *Leptospirillum* clade: cultivation-independent community genomics, proteomics, and transcriptomics of the new species "*Leptospirillum* group IV UBA BS". Appl. Environ. Microb. *79*, 5384–5393.

Gonzalez, C., Yanquepe, M., Cardenas, J.P., Valdes, J., Quatrini, R., Holmes, D.S., and Dopson, M. (2014). Genetic variability of psychrotolerant *Acidithiobacillus ferrivorans* revealed by (meta)genomic analysis. Res. Microbiol. *165*, 726–734.

Guo, X., Yin, H., Liang, Y., Hu, Q., Zhou, X., Xiao, Y., Ma, L., Zhang, X., Qiu, G., and Liu, X. (2014). Comparative genome analysis reveals metabolic versatility and environmental adaptations of *Sulfobacillus thermosulfidooxidans* strain ST. PLoS One *9*, e99417.

Hedrich, S., Schlömann, M., and Johnson, D.B. (2011). The iron-oxidizing proteobacteria. Microbiology *157*, 1551–1564.

Hodar, C., Moreno, P., di Genova, A., Latorre, M., Reyes-Jara, A., Maass, A., Gonzalez, M., and Cambiazo, V. (2012). Genome wide identification of *Acidithiobacillus ferrooxidans* (ATCC 23270) transcription factors and comparative analysis of ArsR and MerR metal regulators. Biometals *25*, 75–93.

Hold, C., Andrews, B.A., and Asenjo, J.A. (2009). A stoichiometric model of *Acidithiobacillus ferrooxidans* ATCC 23270 for metabolic flux analysis. Biotechnol. Bioeng. *102*, 1448–1459.

Holmes, D.S., and Bonnefoy, V. (2007). Genetic and bioinformatic insights into iron and sulfur oxidation mechanisms of bioleaching organisms. In Biomining, Rawlings, D., and Johnson, D.B., eds (Springer, Berlin, Heidelberg, Germany), pp. 281–307.

Holmes, D.S., Cárdenas, J.P., Valdés, J., Quatrini, R., Esparza, M., Osorio, H., Duarte, F., Lefimil, C., and Jedlicki, E. (2009). Comparative genomics begins to unravel the ecophysiology of bioleaching. Adv. Mat. Res. *71–73*, 144–150.

Hou, S., Makarova, K.S., Saw, J.H., Senin, P., Ly, B.V., Zhou, Z., Ren, Y., Wang, J., Galperin, M.Y., Omelchenko, M.V., *et al.* (2008). Complete genome sequence of the extremely acidophilic methanotroph isolate V4, *Methylacidiphilum infernorum*, a representative of the bacterial phylum *Verrucomicrobia*. Biol. Direct 3, 26.

Hua, Z.-S., Han, Y.-J., Chen, L.-X., Liu, J., Hu, M., Li, S.-J., Kuang, J.-L., Chain, P.S.G., Huang, L.-N., and Shu, W.-S. (2014). Ecological roles of dominant and rare prokaryotes in acid mine drainage revealed by metagenomics and metatranscriptomics. ISME J. *9*, 1280–1294.

Jerez, C.A. (2008). The use of genomics, proteomics and other OMICS technologies for the global understanding of biomining microorganisms. Hydrometallurgy *94*, 162–169.

Justice, N., Norman, A., Brown, C., Singh, A., Thomas, B., and Banfield, J. (2014). Comparison of environmental and isolate *Sulfobacillus* genomes reveals diverse carbon, sulfur, nitrogen, and hydrogen metabolisms. BMC Genomics *15*, 1107.

Katoh, K., Misawa, K., Kuma, K., and Miyata, T. (2002). MAFFT: a novel method for rapid multiple sequence alignment based on fast Fourier transform. Nucleic Acids Res. *30*, 3059–3066.

Lasken, R.S. (2012). Genomic sequencing of uncultured microorganisms from single cells. Nat. Rev. Micro. *10*, 631–640.

Levicán, G., Ugalde, J., Ehrenfeld, N., Maass, A., and Parada, P. (2008). Comparative genomic analysis of carbon and nitrogen assimilation mechanisms in three indigenous bioleaching bacteria: predictions and validations. BMC Genomics *9*, 581.

Levicán, G., Katz, A., Valdés, J., Quatrini, R., Holmes, D.S., and Orellana, O. (2009). A 300 kb genome segment, including a complete set of tRNA genes, is dispensable for *Acidithiobacillus ferrooxidans*. Adv. Mat. Res. *71–73*, 187–190.

Levicán, G., Gómez, M.J., Chávez, R., Orellana, O., Moreno-Paz, M., and Parro, V. (2012). Comparative genomic analysis reveals novel facts about *Leptospirillum* spp. cytochromes. J. Mol. Microb. Biotech. *22*, 94–104.

Mangold, S., Valdes, J., Holmes, D.S., and Dopson, M. (2011). Sulfur metabolism in the extreme acidophile *Acidithiobacillus caldus*. Front. Microbiol. *2*, 17.

Martínez, P., Galvez, S., Ohtsuka, N., Budinich, M., Cortes, M.P., Serpell, C., Nakahigashi, K., Hirayama, A., Tomita, M., Soga, T., *et al.* (2013). Metabolomic study of Chilean biomining bacteria *Acidithiobacillus ferrooxidans* strain Wenelen and *Acidithiobacillus thiooxidans* strain Licanantay. Metabolomics *9*, 247–257.

Matsuzaki, M., Misumi, O., Shin, I.T., Maruyama, S., Takahara, M., Miyagishima, S.Y., Mori, T., Nishida, K., Yagisawa, F., Nishida, K., *et al.* (2004). Genome sequence of the ultrasmall unicellular red alga *Cyanidioschyzon merolae* 10D. Nature *428*, 653–657.

Núñez, H., Garcés, K., Cárdenas, J.P., Covarrubias, P., Shmaryahu, A., Demergasso, C., Holmes, D.S., and Quatrini, R. (2011). Biofilm formation capabilities in the bioleaching microbial consortium: a comparative genomics approach. Paper presented at: 19th International Biohydrometallurgy Symposium (Changsha, China).

Orellana, L.H., and Jerez, C.A. (2011). A genomic island provides *Acidithiobacillus ferrooxidans* ATCC 53993 additional copper resistance: a possible competitive advantage. Appl. Microbiol. Biot. *92*, 761–767.

Osorio, H., Martinez, V., Nieto, P.A., Holmes, D.S., and Quatrini, R. (2008). Microbial iron management mechanisms in extremely acidic environments: comparative genomics evidence for diversity and versatility. BMC Microbiol. *8*, 203.

Osorio, H., Cárdenas, J.P., Valdés, J.H., and Holmes, D.S. (2009). Prediction of FNR regulated genes and metabolic pathways potentially involved in anaerobic growth of *Acidithiobacillus ferrooxidans*. Adv. Mat. Res. *71–73*, 195–198.

Osorio, H., Mangold, S., Denis, Y., Ñancucheo, I., Esparza, M., Johnson, D.B., Bonnefoy, V., Dopson, M., and Holmes, D.S. (2013). Anaerobic sulfur metabolism coupled to dissimilatory iron reduction in the extremophile *Acidithiobacillus ferrooxidans*. Appl. Environ. Microb. 79, 2172–2181.

Qiu, H., Price, D.C., Weber, A.P.M., Reeb, V., Chan Yang, E., Lee, J.M., Kim, S.Y., Yoon, H.S., and Bhattacharya, D. (2013). Adaptation through horizontal gene transfer in the cryptoendolithic red alga *Galdieria phlegrea*. Curr. Biol. 23, R865–R866.

Quatrini, R., Jedlicki, E., and Holmes, D. (2005). Genomic insights into the iron uptake mechanisms of the biomining microorganism *Acidithiobacillus ferrooxidans*. J. Ind. Microbiol. Biot. *32*, 606–614.

Quatrini, R., Lefimil, C., Veloso, F.A., Pedroso, I., Holmes, D.S., and Jedlicki, E. (2007a). Bioinformatic prediction and experimental verification of Fur-regulated genes in the extreme acidophile *Acidithiobacillus ferrooxidans*. Nucleic Acids Res. *35*, 2153–2166.

Quatrini, R., Valdés, J., Jedlicki, E., and Holmes, D. (2007b). The use of bioinformatics and genome biology to advance our understanding of bioleaching microorganisms. In Microbial Processing of Metal Sulfides, Donati, E., and Sand, W., eds. (Springer, Houten, Netherlands), pp. 221–239.

Quatrini, R., Appia-Ayme, C., Denis, Y., Jedlicki, E., Holmes, D.S., and Bonnefoy, V. (2009). Extending the models for iron and sulfur oxidation in the extreme acidophile *Acidithiobacillus ferrooxidans*. BMC Genomics *10*, 394.

Reno, M.L., Held, N.L., Fields, C.J., Burke, P.V., and Whitaker, R.J. (2009). Biogeography of the *Sulfolobus islandicus* pan-genome. Proc. Natl. Acad. Sci. U.S.A. *106*, 8605–8610.

Ruepp, A., Graml, W., Santos-Martinez, M.L., Koretke, K.K., Volker, C., Mewes, H.W., Frishman, D., Stocker, S., Lupas, A.N., and Baumeister, W. (2000). The genome sequence of the thermoacidophilic scavenger *Thermoplasma acidophilum*. Nature *407*, 508–513.

Schönknecht, G., Chen, W.-H., Ternes, C.M., Barbier, G.G., Shrestha, R.P., Stanke, M., Bräutigam, A., Baker, B.J., Banfield, J.F., Garavito, R.M., *et al.* (2013). Gene transfer from bacteria and archaea facilitated evolution of an extremophilic eukaryote. Science *339*, 1207–1210.

Selkov, E., Overbeek, R., Kogan, Y., Chu, L., Vonstein, V., Holmes, D., Silver, S., Haselkorn, R., and Fonstein, M. (2000). Functional analysis of gapped microbial genomes: amino acid metabolism of *Thiobacillus ferrooxidans*. Proc. Natl. Acad. Sci. U.S.A. *97*, 3509–3514.

Servin-Garciduenas, L.E., and Martinez-Romero, E. (2014). Draft genome sequence of the *Sulfolobales* Archaeon AZ1, obtained through metagenomic analysis of a mexican hot spring. Genome Announc. *2*, e00164-14.

Tamura, K., Stecher, G., Peterson, D., Filipski, A., and Kumar, S. (2013). MEGA6: Molecular Evolutionary Genetics Analysis version 6.0. Mol. Biol Evol. *30*, 2725–2729.

Thiele, I., and Palsson, B.O. (2010). A protocol for generating a high-quality genome-scale metabolic reconstruction. Nat. Protoc. *5*, 93–121.

Travisany, D., Cortes, M.P., Latorre, M., Di Genova, A., Budinich, M., Bobadilla-Fazzini, R.A., Parada, P., Gonzalez, M., and Maass, A. (2014). A new genome of *Acidithiobacillus thiooxidans* provides insights into adaptation to a bioleaching environment. Res. Microbiol. *165*, 743–752.

Tyson, G.W., Chapman, J., Hugenholtz, P., Allen, E.E., Ram, R.J., Richardson, P.M., Solovyev, V.V., Rubin, E.M., Rokhsar, D.S., and Banfield, J.F. (2004). Community structure and metabolism through reconstruction of microbial genomes from the environment. Nature *428*, 37–43.

Ulas, T., Riemer, S.A., Zaparty, M., Siebers, B., and Schomburg, D. (2012). Genome-scale reconstruction and analysis of the metabolic network in the hyperthermophilic archaeon *Sulfolobus solfataricus*. PLoS One *7*, e43401.

Valdés, J., Veloso, F., Jedlicki, E., and Holmes, D. (2003). Metabolic reconstruction of sulfur assimilation in the extremophile *Acidithiobacillus ferrooxidans* based on genome analysis. BMC Genomics *4*, 51.

Valdés, J., Pedroso, I., Quatrini, R., Dodson, R.J., Tettelin, H., Blake, R. 2nd, Eisen, J.A., and Holmes, D.S. (2008). *Acidithiobacillus ferrooxidans* metabolism: from genome sequence to industrial applications. BMC Genomics *9*, 597.

Valdés, J., Cárdenas, J.P., Quatrini, R., Esparza, M., Osorio, H., Duarte, F., Lefimil, C., Sepulveda, R., Jedlicki, E., and Holmes, D.S. (2010). Comparative genomics begins to unravel the ecophysiology of bioleaching. Hydrometallurgy *104*, 471–476.

Valenzuela, L., Chi, A., Beard, S., Orell, A., Guiliani, N., Shabanowitz, J., Hunt, D.F., and Jerez, C.A. (2006). Genomics, metagenomics and proteomics in biomining microorganisms. Biotechnol. Adv. *24*, 197–211.

Yelton, A.P., Comolli, L.R., Justice, N.B., Castelle, C., Denef, V.J., Thomas, B.C., and Banfield, J.F. (2013). Comparative genomics in acid mine drainage biofilm communities reveals metabolic and structural differentiation of co-occurring archaea. BMC Genomics *14*, 485.

Yin, H., Zhang, X., Li, X., He, Z., Liang, Y., Guo, X., Hu, Q., Xiao, Y., Cong, J., Ma, L., *et al.* (2014). Whole-genome sequencing reveals novel insights into sulfur oxidation in the extremophile *Acidithiobacillus thiooxidans*. BMC Microbiol. *14*, 179.

Appendix

Organism	Status	Size (kbp)	# CDS	%GC	# rRNA	# tRNA	NCBI accession	Reference
Archaea								
Acidianus hospitalis W1	C	2138	2329	34.1	1	46	NC_015518	You *et al.*, 2011b
Acidilobus saccharovorans 345-15 [T‡]	C	1496	1499	57.2	1	45	NC_014374	Mardanov *et al.*, 2010.
Acidilobus sulfurireducens [‡]	D	1649	1864	57.5	1	46	2504136013*	§
Aciduliprofundum boonei T469 [T‡]	C	1487	1550	39.2	1	46	NC_013926	Reysenbach and Flores, 2008
Aciduliprofundum sp. MAR08-339 [‡]	C	1437	1531	44.1	1	46	NC_019942	§
Caldisphaera lagunensis DSM 15908 [T‡]	C	1547	1528	30.0	1	46	NC_019791	§
Caldivirga maquilingensis IC-167 [T‡]	C	2077	1963	43.1	1	45	NC_009954	§
Candidatus Acidianus copahuensis ALE1	D	2278	2332	35.6	1	52	JFZT00000000	Urbieta *et al.*, 2014
Candidatus Micrarchaeum acidiphilum ARMAN-2	D	999	1033	47.2	nd	35	ACVJ00000000	Baker *et al.*, 2010
Candidatus Parvarchaeum acidiphilum ARMAN-4	D	800	916	34.8	nd	39	ADCE00000000	Baker *et al.*, 2010
Candidatus Parvarchaeum acidophilus ARMAN-5	D	921	1046	34.1	nd	42	ADHF00000000	Baker *et al.*, 2010
Ferroplasma acidarmanus fer1 [T]	C	1935	1695	36.5	1	45	NC_021592	Allen *et al.*, 2007
Ferroplasma acidarmanus Type I	D	1476	1663	37.0	nd	39	AADL00000000	Yelton *et al.*, 2013
Ferroplasma sp. Type II	D	1823	2543	38.0	nd	45	ATDU00000000	Yelton *et al.*, 2013
Metallosphaera cuprina Ar-4 [T]	C	1840	2029	42.0	1	45	NC_015435	Liu *et al.*, 2011
Metallosphaera sedula DSM 5348 [T]	C	2192	2293	46.2	1	46	NC_009440	Auernik *et al.*, 2008
Metallosphaera yellowstonensis MK1	D	2871	3356	47.6	1	46	AHKJ00000000	§
Picrophilus torridus DSM 9790 [T]	C	1546	1551	36.0	1	47	NC_005877	Futterer *et al.*, 2004
Sulfolobales archaeon AZ1	D	1798	1975	46.9	1	25	ASRH00000000	Servin-Garciduenas and Martinez-Romero, 2014.
Sulfolobus acidocaldarius DSM 639 [T]	C	2226	2224	36.7	1	49	NC_007181	Chen *et al.*, 2005.
Sulfolobus acidocaldarius N8	C	2176	2275	36.7	1	30	NC_020246	Mao and Grogan, 2012
Sulfolobus acidocaldarius Ron12/I	C	2224	2317	36.7	1	30	NC_020247	Mao and Grogan, 2012

Organism	Status	Size (kbp)	# CDS	%GC	# rRNA	# tRNA	NCBI accession	Reference
Sulfolobus acidocaldarius SUSAZ	C	2062	2146	36.3	1	49	NC_023069	§
Sulfolobus islandicus HVE10/4	C	2655	2720	35.1	1	45	NC_017275	Guo *et al.*, 2011
Sulfolobus islandicus L.D.8.5	C	2722	2916	35.3	1	46	NC_013769	Reno *et al.*, 2009
Sulfolobus islandicus L.S.2.15	C	2736	2737	35.1	1	48	NC_012589	Reno *et al.*, 2009
Sulfolobus islandicus LAL14/1	C	2465	2601	35.1	1	45	NC_021058	Jaubert *et al.*, 2013
Sulfolobus islandicus M.14.25	C	2609	2608	35.1	1	46	NC_012588	Reno *et al.*, 2009
Sulfolobus islandicus M.16.27	C	2692	2615	35.0	1	46	NC_012632	Cadillo-Quiroz *et al.*, 2012
Sulfolobus islandicus M.16.2	D	2645	2629	35.0	1	46	AHJK00000000	Cadillo-Quiroz *et al.*, 2012
Sulfolobus islandicus M.16.22	D	2681	2607	35.2	1	45	AHJN00000000	Cadillo-Quiroz *et al.*, 2012
Sulfolobus islandicus M.16.23	D	2601	2555	35.1	1	45	AHJO00000000	Cadillo-Quiroz *et al.*, 2012
Sulfolobus islandicus M.16.30	D	2608	2595	35.0	1	45	AHJP00000000	Cadillo-Quiroz *et al.*, 2012
Sulfolobus islandicus M.16.4	C	2587	2657	35.0	1	46	NC_012726	Reno *et al.*, 2009
Sulfolobus islandicus M.16.40	D	2665	2591	35.0	1	44	AHJQ00000000	Cadillo-Quiroz *et al.*, 2012
Sulfolobus islandicus M.16.43	D	2590	2735	35.0	1	47	AHJR00000000	Reno *et al.*, 2009
Sulfolobus islandicus M.16.46	D	2690	2600	35.0	1	46	AHJS00000000	Cadillo-Quiroz *et al.*, 2012
Sulfolobus islandicus M.16.47	D	2637	2595	35.1	1	46	AHJT00000000	Cadillo-Quiroz *et al.*, 2012
Sulfolobus islandicus M.16.12	D	2680	2613	35.2	1	46	AHJL00000000	Cadillo-Quiroz *et al.*, 2012
Sulfolobus islandicus M.16.13	D	2640	2552	35.0	1	45	AHJM00000000	Cadillo-Quiroz *et al.*, 2012
Sulfolobus islandicus REY15A	C	2523	2644	35.3	1	45	NC_017276	Guo *et al.*, 2011
Sulfolobus islandicus Y.G.57.14	C	2702	2902	35.4	1	50	NC_012622	Reno *et al.*, 2009
Sulfolobus islandicus Y.N.15.51	C	2812	2850	35.3	1	48	NC_012623	Reno *et al.*, 2009
Sulfolobus solfataricus 98/2 ^{T}	C	2669	2679	35.8	1	45	NC_017274	§
Sulfolobus solfataricus 98/2 ^{T}	D	2680	2679	35.2	1	45	ACUK00000000	§
Sulfolobus solfataricus P2 ^{T}	C	2992	2978	35.8	1	46	NC_002754	She *et al.*, 2001
Sulfolobus tokodaii str. 7 ^{T}	C	2695	2826	32.8	1	46	NC_003106	Kawarabayasi *et al.*, 2001

Organism	Status	Size (kbp)	# CDS	%GC	# rRNA	# tRNA	NCBI accession	Reference
Thermoplasma acidophilum DSM 1728 [T]	C	1565	1527	45.9	1	45	NC_002578	Ruepp *et al.*, 2000
Thermoplasma volcanium GSS1 [T]	C	1585	1561	39.9	1	45	NC_002689	Kawashima *et al.*, 2000.
Thermoplasmatales archaeon A-plasma	D	1990	2280	46.2	nd	nd	ATMF00000000	Yelton *et al.*, 2013
Thermoplasmatales archaeon E-plasma	D	1657	1669	38.4	nd	nd	ACXL00000000	Yelton *et al.*, 2013
Thermoplasmatales archaeon Gpl	D	1827	1923	37.9	nd	nd	ATDV00000000	Yelton *et al.*, 2013
Thermoplasmatales archaeon I-plasma	D	1688	1696	44.3	nd	nd	ATMD00000000	Yelton *et al.*, 2013
Vulcanisaeta distributa DSM 14429 [T‡]	C	2374	2543	45.4	1	46	NC_014537	Mavromatis *et al.*, 2010b
Vulcanisaeta moutnovskia 768-28 [T‡]	C	2299	2320	42.4	1	42	NC_015151	Gumerov *et al.*, 2011
Bacteria								
Acidimicrobium ferrooxidans DSM 10331 [T]	C	2158	1964	68.3	2	46	NC_013124	Clum *et al.*, 2009
Acidiphilium angustum ATCC 35903 [T‡]	D	4175	4175	63.3	3	48	JNJH00000000	§
Acidiphilium cryptum JF-5 [‡]	C	3389	3063	68.0	2	47	NC_009484	§
Acidiphilium multivorum AIU301 [T‡]	C	3749	3448	67.6	2	48	NC_015186	§
Acidiphilium sp. PM [‡]	D	3929	3929	66.4	1	45	AFPR00000000	San Martin-Uriz *et al.*, 2011
Acidithiobacillus albertensis [T]	D	3107	3178	52.4	1	41	nd	§
Acidithiobacillus caldus ATCC 51756 T	C	2778	2944	61.4	1	47	CP005986	Valdes *et al.*, 2009.
Acidithiobacillus caldus SM-1	C	2932	2876	61.3	2	47	NC_015850	You *et al.*, 2011a
Acidithiobacillus ferrivorans CF27	D	3430	3430	56.4	1	71	CCCS000000000	Talla *et al.*, 2014
Acidithiobacillus ferrivorans SS3	C	3208	3093	56.6	2	47	NC_015942	Liljeqvist *et al.*, 2011
Acidithiobacillus ferrooxidans ATCC 23270 [T]	C	2982	2982	58.8	1	81	NC_011761	Valdes *et al.*, 2008
Acidithiobacillus ferrooxidans ATCC 23270 [T]	D	nd	nd	nd	nd	nd	Gp0000798#	Selkov *et al.*, 2000
Acidithiobacillus ferrooxidans ATCC 53993	C	2885	2885	58.9	2	46	NC_011206	§
Acidithiobacillus sp. GGI-221	D	3177	3177	58.6	1	75	AEFB00000000	§
Acidithiobacillus thiooxidans A01	D	3820	3820	53.1	1	117	AZMO00000000	Yin *et al.*, 2014

Organism	Status	Size (kbp)	# CDS	%GC	# rRNA	# tRNA	NCBI accession	Reference
Acidithiobacillus thiooxidans ATCC 19377 [T]	D	3020	3020	53.2	1	45	AFOH00000000	Valdes *et al.*, 2011
Acidithiobacillus thiooxidans CLST	D	3975	3727	48.8	1	70	nd	§
Acidithiobacillus thiooxidans Licanantay	D	3938	3938	52.8	nd	85	JMEB00000000	Travisany *et al.*, 2014
Acidithiomicrobium sp. JTC04	D	2189	2250	52.7	1	45	nd	§
Acidithiomicrobium sp. P2	D	2297	2409	52.5	4	90	nd	§
Acidobacterium capsulatum ATCC 51196 [T‡]	C	4127	4127	60.5	1	45	NC_012483	Ward *et al.*, 2009
Acidobacterium sp. PMMR2 [‡]	D	2520	3687	56.5	1	46	JIAL00000000	§
Alicyclobacillus acidocaldarius DSM 446 [T‡]	C	3019	3019	62.3	6	63	NC_013205	Mavromatis *et al.*, 2010a
Alicyclobacillus acidocaldarius LAA1 [‡]	D	2943	2943	61.7	1	63	ACCS00000000	§
Alicyclobacillus acidocaldarius Tc-4-1 [‡]	C	3124	3124	61.5	6	63	NC_017167	Chen *et al.*, 2011
Alicyclobacillus acidoterrestris ATCC 49025 [T‡]	D	4064	4064	52.2	1	131	AURB00000000	Shemesh *et al.*, 2013
Alicyclobacillus contaminans DSM 17975 [T‡]	D	3271	3271	58.2	7	74	AUCA00000000	§
Alicyclobacillus herbarius DSM 13609 [T‡]	D	3308	3308	57.2	3	68	AUMH00000000	§
Alicyclobacillus hesperidum URH17-3-68 [‡]	D	2967	2967	53.8	2	71	AKZN00000000	§
Alicyclobacillus macrosporangiidus CPP55 [‡]	D	4092	4092	62.3	6	54	JNIL00000000	§
Alicyclobacillus pomorum DSM 14955 [T‡]	D	3398	3398	53.7	3	96	AXAR00000000	§
Burkholderiales bacterium GJ-E10	C	3276	3087	66.3	1	43	AP014683	Fukushima *et al.*, 2015
Candidatus Koribacter versatilis Ellin345 [‡]	C	5650	5650	58.4	1	47	NC_008009	Ward *et al.*, 2009
Candidatus Solibacter usitatus Ellin6076 [‡]	C	9966	9966	61.9	2	52	NC_008536	Ward *et al.*, 2009
Desulfosporosinus acidiphilus SJ4 [T‡]	C	4927	4927	42.1	9	66	NC_018068	Pester *et al.*, 2012
Effusibacillus pohliae DSM 22757 [T‡]	D	3364	3364	54.8	5	71	AQXL00000000	§
Ferrimicrobium acidiphilum DSM 19497 [T]	D	2929	2832	55.3	2	46	JQKF00000000	§

Organism	Status	Size (kbp)	# CDS	%GC	# rRNA	# tRNA	NCBI accession	Reference
Ferrovum myxofaciens P3G [T]	D	2702	2702	54.3	1	43	JPOQ00000000	Moya-Beltran *et al.*, 2014
Granulicella mallensis MP5ACTX8 [T‡]	C	6238	6238	57.9	1	47	NC_016631	§
Granulicella tundricola MP5ACTX9 [T‡]	C	4309	4309	60.5	1	46	NC_015064	Rawat *et al.*, 2014
Hydrogenobaculum sp. 3684	C	1553	1579	34.8	3	45	CP002726	Romano *et al.*, 2013
Hydrogenobaculum sp. HO	C	1553	1553	34.8	2	45	NC_020411	Romano *et al.*, 2013
Hydrogenobaculum sp. SHO	C	1553	1574	34.8	2	45	CP002809	Romano *et al.*, 2013
Hydrogenobaculum sp. SN	C	1553	1553	34.8	2	45	NC_020814	Romano *et al.*, 2013
Hydrogenobaculum sp. Y04AAS1	C	1560	1560	34.8	2	45	NC_011126	Reysenbach *et al.*, 2009
Kyrpidia tusciae DSM 2912 [‡]	C	3385	3385	59.1	3	60	NC_014098	Klenk *et al.*, 2011
Leptospirillum ferriphilum DSM 14647 [T]	D	2406	2406	54.1	nd	44	JPGK00000000	Cardenas *et al.*, 2014
Leptospirillum ferriphilum ML-04	C	2406	2406	54.6	2	48	NC_018649	Mi *et al.*, 2011
Leptospirillum ferriphilum YSK	C	2332	2332	54.4	2	47	CP007243	§
Leptospirillum ferrodiazotrophum	D	2796	2658	57.7	1	*45*	ACNP01000000	Goltsman *et al.*, 2009
Leptospirillum ferrooxidans C2-3	C	2560	2560	50.1	3	51	NC_017094	Fujimura *et al.*, 2012
Leptospirillum rubarum	D	2637	2601	53.3	1	48	AAWO00000000	Goltsman *et al.*, 2009
Leptospirillum sp. Group II 'C75'	D	2605	2605	54.4	1	nd	AIJM00000000	Denef and Banfield, 2012
Leptospirillum sp. Group IV 'UBA BS'	D	1522	1522	59.0	1	18	AURA00000000	Goltsman *et al.*, 2013
Leptospirillum sp. SPCL	D	2554	2858	54.2	1	45	nd	§
Methylacidiphilum infernorum V4 [T]	C	2287	2287	45.5	1	46	NC_010794	Hou *et al.*, 2008
Sulfobacillus acidophilus DSM 10332 [T]	C	3473	3473	56.8	3	53	NC_016884	Anderson *et al.*, 2012
Sulfobacillus acidophilus TPY	C	3551	3551	56.8	5	52	NC_015757	Li *et al.*, 2011
Sulfobacillus thermosulfidooxidans CBAR13	D	3828	3994	48.9	4	52	nd	§
Sulfobacillus thermosulfidooxidans Cutipay	D	3859	3859	49.3	6	50	ALWJ00000000	Travisany *et al.*, 2012
Sulfobacillus thermosulfidooxidans DSM 9293 [T]	D	3861	3861	49.6	6	50	2506210005*	Not referenced

Organism	Status	Size (kbp)	# CDS	%GC	# rRNA	# tRNA	NCBI accession	Reference
Sulfobacillus thermosulfidooxidans ST	D	3325	3325	48.3	1	47	AWEW00000000	Guo *et al.*, 2014
Terriglobus roseus DSM 18391 T‡	C	5228	5228	60.3	2	53	NC_018014	§
Terriglobus saanensis SP1PR4 T‡	C	5095	5095	57.3	1	48	NC_014963	Rawat *et al.*, 2012
Thiobacillus prosperus DSM 5130 T	D	3323	3323	64.5	1	48	JQSG00000000	Ossandon *et al.*, 2014
Thiomonas arsenitoxydans strain 3As ‡	C	3739	3739	63.8	1	43	NC_014145	Arsene-Ploetze *et al.*, 2010
Thiomonas intermedia K12 T‡	C	3396	3396	63.9	1	45	NC_014153	§
Eukarya								
Cyanidioschyzon merolae	C	16356	6170	~55	2	37	NC_010127 - NC_010146	Matsuzaki *et al.*, 2004
Galdieria phlegrea	D	nd	2178	nd	nd	nd	SRR953992 &	Qiu *et al.*, 2013
Galdieria sulphuraria	D	13710	7174	37.7	nd	81	ADNM00000000	Schönknecht *et al.*, 2013
Hortaea acidophila	D	nd	nd	nd	nd	nd	Gp0110612 #	§

List of permanent draft and finished, acidophile genomes with associated metadata. Abbreviations used: T = type strain; ‡ = moderate acidophile; nd = no data; * = sequence only available in IMG-JGI where the IMG Taxon ID value is provided; # = sequence not available and the GOLD project ID is provided; & = NCBI Sequence Read Archive (SRA) accession; § = no primary genome paper available; # rRNA = number of genes encoding the 16SrRNA (18SrRNA in the case of Eukarya); # tRNA = number of genes encoding tRNAs.

References

Allen, E.E., Tyson, G.W., Whitaker, R.J., Detter, J.C., Richardson, P.M., and Banfield, J.F. (2007). Genome dynamics in a natural archaeal population. Proc. Natl. Acad. Sci. U.S.A. *104*, 1883–1888.

Anderson, I., Chertkov, O., Chen, A., Saunders, E., Lapidus, A., Nolan, M., Lucas, S., Hammon, N., Deshpande, S., Cheng, J.F., *et al.* (2012). Complete genome sequence of the moderately thermophilic mineral-sulfide-oxidizing firmicute *Sulfobacillus acidophilus* type strain (NAL(T)). Stand. Genomic Sci. *6*, 1–13.

Arsene-Ploetze, F., Koechler, S., Marchal, M., Coppee, J.Y., Chandler, M., Bonnefoy, V., Brochier-Armanet, C., Barakat, M., Barbe, V., Battaglia-Brunet, F., *et al.* (2010). Structure, function, and evolution of the *Thiomonas* spp. genome. PLoS Genet. *6*, e1000859.

Auernik, K.S., Maezato, Y., Blum, P.H., and Kelly, R.M. (2008). The genome sequence of the metal-mobilizing, extremely thermoacidophilic archaeon *Metallosphaera sedula* provides insights into bioleaching-associated metabolism. Appl. Environ. Microb. *74*, 682–692.

Baker, B.J., Comolli, L.R., Dick, G.J., Hauser, L.J., Hyatt, D., Dill, B.D., Land, M.L., Verberkmoes, N.C., Hettich, R.L., and Banfield, J.F. (2010). Enigmatic, ultrasmall, uncultivated Archaea. Proc. Natl. Acad. Sci. U.S.A. *107*, 8806–8811.

Cadillo-Quiroz, H., Didelot, X., Held, N.L., Herrera, A., Darling, A., Reno, M.L., Krause, D.J., and Whitaker, R.J. (2012). Patterns of gene flow define species of thermophilic Archaea. PLoS Biol. *10*, e1001265.

Cardenas, J.P., Lazcano, M., Ossandon, F.J., Corbett, M., Holmes, D.S., and Watkin, E. (2014). Draft Genome Sequence of the Iron-Oxidizing Acidophile *Leptospirillum ferriphilum* Type Strain DSM 14647. Genome Announc. *2*, e01153–14.

Clum, A., Nolan, M., Lang, E., Glavina Del Rio, T., Tice, H., Copeland, A., Cheng, J.F., Lucas, S., Chen, F., Bruce, D., *et al.* (2009). Complete genome sequence of *Acidimicrobium ferrooxidans* type strain (ICP). Stand. Genomic Sci. *1*, 38–45.

Chen, L., Brugger, K., Skovgaard, M., Redder, P., She, Q., Torarinsson, E., Greve, B., Awayez, M., Zibat, A., Klenk, H.P., *et al.* (2005). The genome of *Sulfolobus acidocaldarius*, a model organism of the Crenarchaeota. J Bacteriol. *187*, 4992–4999.

Chen, Y., He, Y., Zhang, B., Yang, J., Li, W., Dong, Z., and Hu, S. (2011). Complete genome sequence of *Alicyclobacillus acidocaldarius* strain Tc-4–1. J Bacteriol. *193*, 5602–5603.

Denef, V.J., and Banfield, J.F. (2012). In situ evolutionary rate measurements show ecological success of recently emerged bacterial hybrids. Science *336*, 462–466.

Fujimura, R., Sato, Y., Nishizawa, T., Oshima, K., Kim, S.W., Hattori, M., Kamijo, T., and Ohta, H. (2012). Complete genome sequence of *Leptospirillum ferrooxidans* strain C2-3, isolated from a fresh volcanic ash deposit on the island of Miyake, Japan. J Bacteriol. *194*, 4122–4123.

Fukushima, J., Tojo, F., Asano, R., Kobayashi, Y., Shimura, Y., Okano, K., and Miyata, N. (2015). Complete Genome Sequence of the Unclassified Iron-Oxidizing, Chemolithoautotrophic *Burkholderiales* Bacterium GJ-E10, Isolated from an Acidic River. Genome Announc. *3*, e01455–14.

Futterer, O., Angelov, A., Liesegang, H., Gottschalk, G., Schleper, C., Schepers, B., Dock, C., Antranikian, G., and Liebl, W. (2004). Genome sequence of *Picrophilus torridus* and its implications for life around pH 0. Proc. Natl. Acad. Sci. U.S.A. *101*, 9091–9096.

Goltsman, D.S., Denef, V.J., Singer, S.W., VerBerkmoes, N.C., Lefsrud, M., Mueller, R.S., Dick, G.J., Sun, C.L., Wheeler, K.E., Zemla, A., *et al.* (2009). Community genomic and proteomic analyses of chemoautotrophic iron-oxidizing "*Leptospirillum rubarum*" (Group II) and "*Leptospirillum ferrodiazotrophum*" (Group III) bacteria in acid mine drainage biofilms. Appl. Environ. Microb. *75*, 4599–4615.

Goltsman, D.S., Dasari, M., Thomas, B.C., Shah, M.B., VerBerkmoes, N.C., Hettich, R.L., and Banfield, J.F. (2013). New group in the *Leptospirillum* clade: cultivation-independent community genomics, proteomics, and transcriptomics of the new species "*Leptospirillum* group IV UBA BS". Appl. Environ. Microb. *79*, 5384–5393.

Gumerov, V.M., Mardanov, A.V., Beletsky, A.V., Prokofeva, M.I., Bonch-Osmolovskaya, E.A., Ravin, N.V., and Skryabin, K.G. (2011). Complete genome sequence of "*Vulcanisaeta moutnovskia*" strain 768-28, a novel member of the hyperthermophilic crenarchaeal genus Vulcanisaeta. J Bacteriol. *193*, 2355–2356.

Guo, L., Brugger, K., Liu, C., Shah, S.A., Zheng, H., Zhu, Y., Wang, S., Lillestol, R.K., Chen, L., Frank, J., *et al.* (2011). Genome analyses of Icelandic strains of *Sulfolobus islandicus*, model organisms for genetic and virus-host interaction studies. J Bacteriol. *193*, 1672–1680.

Guo, X., Yin, H., Liang, Y., Hu, Q., Zhou, X., Xiao, Y., Ma, L., Zhang, X., Qiu, G., and Liu, X. (2014). Comparative genome analysis reveals metabolic versatility and environmental adaptations of *Sulfobacillus thermosulfidooxidans* strain ST. PloS One *9*, e99417.

Hou, S., Makarova, K.S., Saw, J.H., Senin, P., Ly, B.V., Zhou, Z., Ren, Y., Wang, J., Galperin, M.Y., Omelchenko, M.V., *et al.* (2008). Complete genome sequence of the extremely acidophilic methanotroph isolate V4, *Methylacidiphilum infernorum*, a representative of the bacterial phylum *Verrucomicrobia*. Biol. Direct *3*, 26.

Jaubert, C., Danioux, C., Oberto, J., Cortez, D., Bize, A., Krupovic, M., She, Q., Forterre, P., Prangishvili, D., and Sezonov, G. (2013). Genomics and genetics of *Sulfolobus islandicus* LAL14/1, a model hyperthermophilic archaeon. Open Biol. *3*, 130010.

Kawarabayasi, Y., Hino, Y., Horikawa, H., Jin-no, K., Takahashi, M., Sekine, M., Baba, S., Ankai, A., Kosugi, H., Hosoyama, A., *et al.* (2001). Complete genome sequence of an aerobic thermoacidophilic crenarchaeon, *Sulfolobus tokodaii* strain7. DNA Res. *8*, 123–140.

Kawashima, T., Amano, N., Koike, H., Makino, S., Higuchi, S., Kawashima-Ohya, Y., Watanabe, K., Yamazaki, M., Kanehori, K., Kawamoto, T., *et al.* (2000). Archaeal adaptation to higher temperatures revealed by genomic sequence of *Thermoplasma volcanium*. Proc. Natl. Acad. Sci. U.S.A. *97*, 14257–14262.

Klenk, H.P., Lapidus, A., Chertkov, O., Copeland, A., Del Rio, T.G., Nolan, M., Lucas, S., Chen, F., Tice, H., Cheng, J.F., *et al.* (2011). Complete genome sequence of the thermophilic, hydrogen-oxidizing *Bacillus tusciae* type strain (T2) and reclassification in the new genus, *Kyrpidia* gen. nov. as *Kyrpidia tusciae* comb. nov. and emendation of the family *Alicyclobacillaceae* da Costa and Rainey, 2010. Stand. Genomic Sci. *5*, 121–134.

Li, B., Chen, Y., Liu, Q., Hu, S., and Chen, X. (2011). Complete genome analysis of *Sulfobacillus acidophilus* strain TPY, isolated from a hydrothermal vent in the Pacific Ocean. J Bacteriol. *193*, 5555–5556.

Liljeqvist, M., Valdes, J., Holmes, D.S., and Dopson, M. (2011). Draft genome of the psychrotolerant acidophile *Acidithiobacillus ferrivorans* SS3. J Bacteriol. *193*, 4304–4305.

Liu, L.J., You, X.Y., Zheng, H., Wang, S., Jiang, C.Y., and Liu, S.J. (2011). Complete genome sequence of *Metallosphaera cuprina*, a metal sulfide-oxidizing archaeon from a hot spring. J Bacteriol. *193*, 3387–3388.

Mao, D., and Grogan, D. (2012). Genomic evidence of rapid, global-scale gene flow in a *Sulfolobus* species. ISME J. *6*, 1613–1616.

Mardanov, A.V., Svetlitchnyi, V.A., Beletsky, A.V., Prokofeva, M.I., Bonch-Osmolovskaya, E.A., Ravin, N.V., and Skryabin, K.G. (2010). The genome sequence of the crenarchaeon *Acidilobus saccharovorans* supports a new order, Acidilobales, and suggests an important ecological role in terrestrial acidic hot springs. Appl. Environ. Microb. *76*, 5652–5657.

Matsuzaki, M., Misumi, O., Shin, I.T., Maruyama, S., Takahara, M., Miyagishima, S.Y., Mori, T., Nishida, K., Yagisawa, F., Nishida, K., *et al.* (2004). Genome sequence of the ultrasmall unicellular red alga *Cyanidioschyzon merolae* 10D. Nature *428*, 653–657.

Mavromatis, K., Sikorski, J., Lapidus, A., Glavina Del Rio, T., Copeland, A., Tice, H., Cheng, J.F., Lucas, S., Chen, F., Nolan, M., *et al.* (2010a). Complete genome sequence of *Alicyclobacillus acidocaldarius* type strain (104-IA). Stand. Genomic Sci. *2*, 9–18.

Mavromatis, K., Sikorski, J., Pabst, E., Teshima, H., Lapidus, A., Lucas, S., Nolan, M., Glavina Del Rio, T., Cheng, J.F., Bruce, D., *et al.* (2010b). Complete genome sequence of *Vulcanisaeta distributa* type strain (IC-017). Stand. Genomic Sci. *3*, 117–125.

Mi, S., Song, J., Lin, J., Che, Y., Zheng, H., and Lin, J. (2011). Complete genome of *Leptospirillum ferriphilum* ML-04 provides insight into its physiology and environmental adaptation. J Bacteriol. *49*, 890–901.

Moya-Beltran, A., Cardenas, J.P., Covarrubias, P.C., Issotta, F., Ossandon, F.J., Grail, B.M., Holmes, D.S., Quatrini, R., and Johnson, D.B. (2014). Draft Genome Sequence of the Nominated Type Strain of "*Ferrovum myxofaciens*," an Acidophilic, Iron-Oxidizing Betaproteobacterium. Genome Announc. *2*, e00834–14.

Ossandon, F.J., Cardenas, J.P., Corbett, M., Quatrini, R., Holmes, D.S., and Watkin, E. (2014). Draft Genome

Sequence of the Iron-Oxidizing, Acidophilic, and Halotolerant "*Thiobacillus prosperus*" Type Strain DSM 5130. Genome Announc. *2*, e01042–14.

Pester, M., Brambilla, E., Alazard, D., Rattei, T., Weinmaier, T., Han, J., Lucas, S., Lapidus, A., Cheng, J.F., Goodwin, L., *et al.* (2012). Complete genome sequences of *Desulfosporosinus orientis* DSM765^T, *Desulfosporosinus youngiae* DSM17734^T, *Desulfosporosinus meridiei* DSM13257^T, and *Desulfosporosinus acidiphilus* DSM22704^T. J Bacteriol. *194*, 6300–6301.

Qiu, H., Price, D.C., Weber, A.P.M., Reeb, V., Chan Yang, E., Lee, J.M., Kim, S.Y., Yoon, H.S., and Bhattacharya, D. (2013). Adaptation through horizontal gene transfer in the cryptoendolithic red alga *Galdieria phlegrea*. Curr. Biol. *23*, R865-R866.

Rawat, S.R., Mannisto, M.K., Starovoytov, V., Goodwin, L., Nolan, M., Hauser, L., Land, M., Davenport, K.W., Woyke, T., and Haggblom, M.M. (2012). Complete genome sequence of *Terriglobus saanensis* type strain SP1PR4(T), an Acidobacteria from tundra soil. Stand. Genomic Sci. *7*, 59–69.

Rawat, S.R., Mannisto, M.K., Starovoytov, V., Goodwin, L., Nolan, M., Hauser, L., Land, M., Davenport, K.W., Woyke, T., and Haggblom, M.M. (2014). Complete genome sequence of *Granulicella tundricola* type strain MP5ACTX9(T), an Acidobacteria from tundra soil. Stand. Genomic Sci. *9*, 449–461.

Reno, M.L., Held, N.L., Fields, C.J., Burke, P.V., and Whitaker, R.J. (2009). Biogeography of the *Sulfolobus islandicus* pan-genome. Proc. Natl. Acad. Sci. U.S.A. *106*, 8605–8610.

Reysenbach, A.L., and Flores, G.E. (2008). Electron microscopy encounters with unusual thermophiles helps direct genomic analysis of *Aciduliprofundum boonei*. Geobiology *6*, 331–336.

Reysenbach, A.L., Hamamura, N., Podar, M., Griffiths, E., Ferreira, S., Hochstein, R., Heidelberg, J., Johnson, J., Mead, D., Pohorille, A., *et al.* (2009). Complete and draft genome sequences of six members of the Aquificales. J Bacteriol. *191*, 1992–1993.

Romano, C., D'Imperio, S., Woyke, T., Mavromatis, K., Lasken, R., Shock, E.L., and McDermott, T.R. (2013). Comparative genomic analysis of phylogenetically closely related *Hydrogenobaculum* sp. isolates from Yellowstone National Park. Appl. Environ. Microb. *79*, 2932–2943.

Ruepp, A., Graml, W., Santos-Martinez, M.L., Koretke, K.K., Volker, C., Mewes, H.W., Frishman, D., Stocker, S., Lupas, A.N., and Baumeister, W. (2000). The genome sequence of the thermoacidophilic scavenger *Thermoplasma acidophilum*. Nature *407*, 508–513.

San Martin-Uriz, P., Gomez, M.J., Arcas, A., Bargiela, R., and Amils, R. (2011). Draft genome sequence of the electricigen *Acidiphilium* sp. strain PM (DSM 24941). J Bacteriol. *193*, 5585–5586.

Schönknecht, G., Chen, W.-H., Ternes, C.M., Barbier, G.G., Shrestha, R.P., Stanke, M., Bräutigam, A., Baker, B.J., Banfield, J.F., Garavito, R.M., *et al.* (2013). Gene Transfer from Bacteria and Archaea Facilitated Evolution of an Extremophilic Eukaryote. Science *339*, 1207–1210.

Selkov, E., Overbeek, R., Kogan, Y., Chu, L., Vonstein, V., Holmes, D., Silver, S., Haselkorn, R., and Fonstein, M. (2000). Functional analysis of gapped microbial genomes: amino acid metabolism of *Thiobacillus ferrooxidans*. Proc. Natl. Acad. Sci. U.S.A. *97*, 3509–3514.

Servin-Garciduenas, L.E., and Martinez-Romero, E. (2014). Draft Genome Sequence of the *Sulfolobales* Archaeon AZ1, Obtained through Metagenomic Analysis of a Mexican Hot Spring. Genome Announc. *2*, e00164–14.

She, Q., Singh, R.K., Confalonieri, F., Zivanovic, Y., Allard, G., Awayez, M.J., Chan-Weiher, C.C., Clausen, I.G., Curtis, B.A., De Moors, A., *et al.* (2001). The complete genome of the crenarchaeon *Sulfolobus solfataricus* P2. Proc. Natl. Acad. Sci. U.S.A. *98*, 7835–7840.

Shemesh, M., Pasvolsky, R., Sela, N., Green, S.J., and Zakin, V. (2013). Draft Genome Sequence of *Alicyclobacillus acidoterrestris* Strain ATCC 49025. Genome Announc. *1*, e00638–13.

Talla, E., Hedrich, S., Mangenot, S., Ji, B., Johnson, D.B., Barbe, V., and Bonnefoy, V. (2014). Insights into the pathways of iron- and sulfur-oxidation, and biofilm formation from the chemolithotrophic acidophile *Acidithiobacillus ferrivorans* CF27. Res. Microbiol. *165*, 753–760.

Travisany, D., Di Genova, A., Sepulveda, A., Bobadilla-Fazzini, R.A., Parada, P., and Maass, A. (2012). Draft genome sequence of the *Sulfobacillus thermosulfidooxidans* Cutipay strain, an indigenous bacterium isolated from a naturally extreme mining environment in Northern Chile. J Bacteriol. *194*, 6327–6328.

Travisany, D., Cortes, M.P., Latorre, M., Di Genova, A., Budinich, M., Bobadilla-Fazzini, R.A., Parada, P., Gonzalez, M., and Maass, A. (2014). A new genome of *Acidithiobacillus thiooxidans* provides insights into adaptation to a bioleaching environment. Res. Microbiol. *165*, 743–752.

Urbieta, M.S., Rascovan, N., Castro, C., Revale, S., Giaveno, M.A., Vazquez, M., and Donati, E.R. (2014). Draft Genome Sequence of the Novel Thermoacidophilic Archaeon *Acidianus copahuensis* Strain ALE1, Isolated from the Copahue Volcanic Area in Neuquen, Argentina. Genome Announc. *2*, e00259–14.

Valdes, J., Pedroso, I., Quatrini, R., Dodson, R.J., Tettelin, H., Blake, R., 2nd, Eisen, J.A., and Holmes, D.S. (2008). *Acidithiobacillus ferrooxidans* metabolism: from genome sequence to industrial applications. BMC Genomics *9*, 597.

Valdes, J., Quatrini, R., Hallberg, K., Dopson, M., Valenzuela, P.D., and Holmes, D.S. (2009). Draft genome sequence of the extremely acidophilic bacterium *Acidithiobacillus caldus* ATCC 51756 reveals metabolic versatility in the genus *Acidithiobacillus*. J Bacteriol. *191*, 5877–5878.

Valdes, J., Ossandon, F., Quatrini, R., Dopson, M., and Holmes, D.S. (2011). Draft genome sequence of the extremely acidophilic biomining bacterium *Acidithiobacillus thiooxidans* ATCC 19377 provides insights into the evolution of the *Acidithiobacillus* genus. J Bacteriol. *193*, 7003–7004.

Ward, N.L., Challacombe, J.F., Janssen, P.H., Henrissat, B., Coutinho, P.M., Wu, M., Xie, G., Haft, D.H., Sait, M., Badger, J., *et al.* (2009). Three genomes from the phylum Acidobacteria provide insight into the lifestyles of these microorganisms in soils. Appl. Environ. Microb. *75*, 2046–2056.

Yelton, A.P., Comolli, L.R., Justice, N.B., Castelle, C., Denef, V.J., Thomas, B.C., and Banfield, J.F. (2013). Comparative genomics in acid mine drainage biofilm communities reveals metabolic and structural differentiation of co-occurring archaea. BMC Genomics *14*, 485.

Yin, H., Zhang, X., Liang, Y., Xiao, Y., Niu, J., and Liu, X. (2014). Draft Genome Sequence of the Extremophile *Acidithiobacillus thiooxidans* A01, Isolated from the Wastewater of a Coal Dump. Genome Announc. *2*, e00222-14.

You, X.Y., Guo, X., Zheng, H.J., Zhang, M.J., Liu, L.J., Zhu, Y.Q., Zhu, B., Wang, S.Y., Zhao, G.P., Poetsch, A., *et al.* (2011a). Unraveling the *Acidithiobacillus caldus* complete genome and its central metabolisms for carbon assimilation. J. Genet. Genomics. *38*, 243–252.

You, X.Y., Liu, C., Wang, S.Y., Jiang, C.Y., Shah, S.A., Prangishvili, D., She, Q., Liu, S.J., and Garrett, R.A. (2011b). Genomic analysis of *Acidianus hospitalis* W1 a host for studying crenarchaeal virus and plasmid life cycles. Extremophiles *15*, 487–497.

The Flexible Genome of Acidophilic Prokaryotes

12

Raquel Quatrini, Francisco J. Ossandon and Douglas Rawlings

Abstract

Over the last couple of decades there has been considerable progress in the identification and understanding of the mobile genetic elements that are exchanged between microbes in extremely acidic environments, and of the genes piggybacking on them. Numerous plasmid families, unique viruses of bizarre morphologies and life cycles, as well as plasmid-virus chimeras, have been isolated from acidophiles and characterized to varying degrees. Growing evidence provided by omic-studies have shown that the mobile elements repertoire is not restricted to plasmids and viruses, but that a plethora of integrative elements ranging from miniature inverted repeat transposable elements to large integrative conjugative elements populate the genomes of acidophilic bacteria and archaea. This chapter reviews the diversity of elements that have been found to constitute the flexible genome of acidophiles. Special emphasis is put on the knowledge generated for *Sulfolobus* (archaea) and species of the bacterial genera *Acidithiobacillus* and *Leptospirillum*. Also, recent knowledge on the strategies used by acidophiles to contain deleterious exchanges while allowing innovation, and the emerging details of the molecular biology of these systems, are discussed. Major lacunae in our understanding of the mobilome of acidophilic prokaryotes and topics for further investigations are identified.

Introduction

Bacteria and archaea have evolved efficient genome modification mechanisms that allow them to adapt rapidly and effectively to ever-changing environmental conditions and to colonize a plethora of ecological niches. Some of these mechanisms are universal (e.g. transposition), while others are significantly more frequent in the prokaryotic world (e.g. transduction, conjugation). Prokaryotic genomes are modified by intrareplicon, inter-replicon and interorganismal gene movement of pre-evolved traits, occasionally circumventing the species barrier. These mechanisms are collectively known as horizontal gene transfer (HGT).

Scientific literature on the adaptive role of HGT in prokaryotes is extremely rich; for instance, the acquisition of resistance to harmful compounds and elements or the acquisition of metabolic capabilities allowing the use of new resources are frequent events in all well-studied ecosystems. Extensive HGT between lineages that occupy similar ecological niches, which may drive radical changes of lifestyle, have also been documented. HGT seems to have contributed to the adaptation of *Thermoplasmatales* archaea to extremely acidic and hot environments. *Thermoplasma acidophilum* and *Picrophilus torridus* share approximately the same fraction of genes (65% of their proteomes) between them, as they do with *Sulfolobus solfataricus*, a very distant relative that often inhabits the same environment, whereas only 35% of their genes are present in *Pyrococcus furiosus*, a closer relative (Fütterer *et al.*, 2004). Beyond its role in prokaryotic genome innovation and evolution, HGT also influences dramatically the function of complex prokaryotic communities.

Gene flow between and within microbes is driven by a vast repertoire of Mobile Genetic Elements (MGEs). MGEs are mostly DNA-based platforms that encode enzymes and other proteins that mediate or facilitate self movement and that support different genetic cargoes. The full set of MGEs of a microorganism and the genes they carry constitute the mobilome or the flexible genomic compartment. Research in microbial MGEs has progressed rapidly over the last couple of decades. Some MGEs are mainly intrachromosomal (e.g. transposons) and others extrachromosomal or episomal (e.g. plasmids), and yet others exist in a fluctuating equilibrium between an integrated and an episomal state (e.g. temperate viruses). MGEs are classified depending on the molecular mechanism underlying their physical movement. Translocative elements, including transposons and insertion sequences, depend on transposases which catalyse 'cut and paste' or 'copy–paste' movements of DNA. Integrative elements relying on site-specific recombination of DNA mediated by a diverse family of enzymes called integrases (or recombinases), include integrons, temperate viruses and a wide diversity of chimeric elements. Other replicative and non-replicative elements which can move between microorganisms, such as plasmids and genomic islands, are reckoned as dispersive or transmissible elements and are further classified as conjugative (those that code all components needed for transfer) or mobilizable according to their transfer ability.

Thus, the mobilome is an assortment of selfish MGEs (Fig. 12.1) composed of many different types of elements, and accretion of individual MGEs (e.g. transposon) on higher-order mobile elements (e.g. a conjugative plasmid). Several chimeric elements such as transposable proviruses, integrative plasmids and mobile integrons muddle the picture considerably and favour the view of MGEs as mosaics of functional gene modules of diverse evolutionary origins.

Lack of informative classification systems and adequate tools for MGE prediction in sequenced genomes has hampered analysis of the prokaryotic mobilome. Both impediments have began to be addressed in recent years. Improved classification criteria have began to emerge for certain types of MGEs. Such is the case of proteobacterial plasmids, for which a relatively comprehensive picture of the diversity, types and mechanisms has started to emerge (Garcillán-Barcia *et al.*, 2013). Also, several MGE prediction strategies, tools and pipelines have been developed and made publicly available in recent years. Although comprehensive genomics of the prokaryotic 'mobilome' still lags behind most other genomics initiatives, sequencing efforts and comparative genomics and metagenomics studies have revealed that the different types of MGEs are found in all branches of the prokaryotic tree of life, and that they occur at varying frequencies, even in the most extreme environments. From these studies new relevant definitions have emerged, for example (i) core genome, which defines the genes present in all strains of a prokaryotic species; (ii) flexible (or dispensable) genome, which concerns genes that are present in some, but not all, strains of a species; (iii) pan genome, which entails the sum of the first two, and (iv) supergenome (or communal gene pool), which consists of all the genes encoded on MGEs and that are readily available to all permissive prokaryotes within a particular setting. This has signified a paradigm shift in that the individual gene pool is now thought to represent a vast resource for MGEs to exploit, enabling the sharing of beneficial traits among prokaryotic populations restricted only by the host ranges of the resident MGEs and the selection pressures exerted by the surrounding environment. Stable appropriation of foreign genetic material by different hosts is determined by a number of factors and is less likely as the phylogenetic distances increase. In the case of plasmids, the host-range is influenced by factors affecting their capacity of transference and replication within their new hosts. In the case of viruses, it is mainly determined by specific host attachment sites and cellular factors necessary for viral multiplication. In addition, a number of molecular mechanisms encoded by prokaryotes that counteract uptake and stabilization of foreign DNA molecules serve to further limit MGE host ranges: these include CRISPR (clustered regularly interspaced short palindromic repeats)-Cas and RM (restriction–modification) systems.

Given that the mobilome and its genetic cargo seem, in many cases, tightly linked to the environment of the host population, it is thought to vary considerably from one environment to another. In this chapter, current knowledge gathered on MGEs that favour HGT in extreme acidic econiches and the strategies used by acidophiles to contain their dispersal, are explored. Knowledge on the mobilome of acidophiles has been organized on the basis of the type of element and covers both acidophilic bacteria and archaea. Longstanding lacunae in our understanding of the of the mobilome of acidophilic prokaryotes are highlighted.

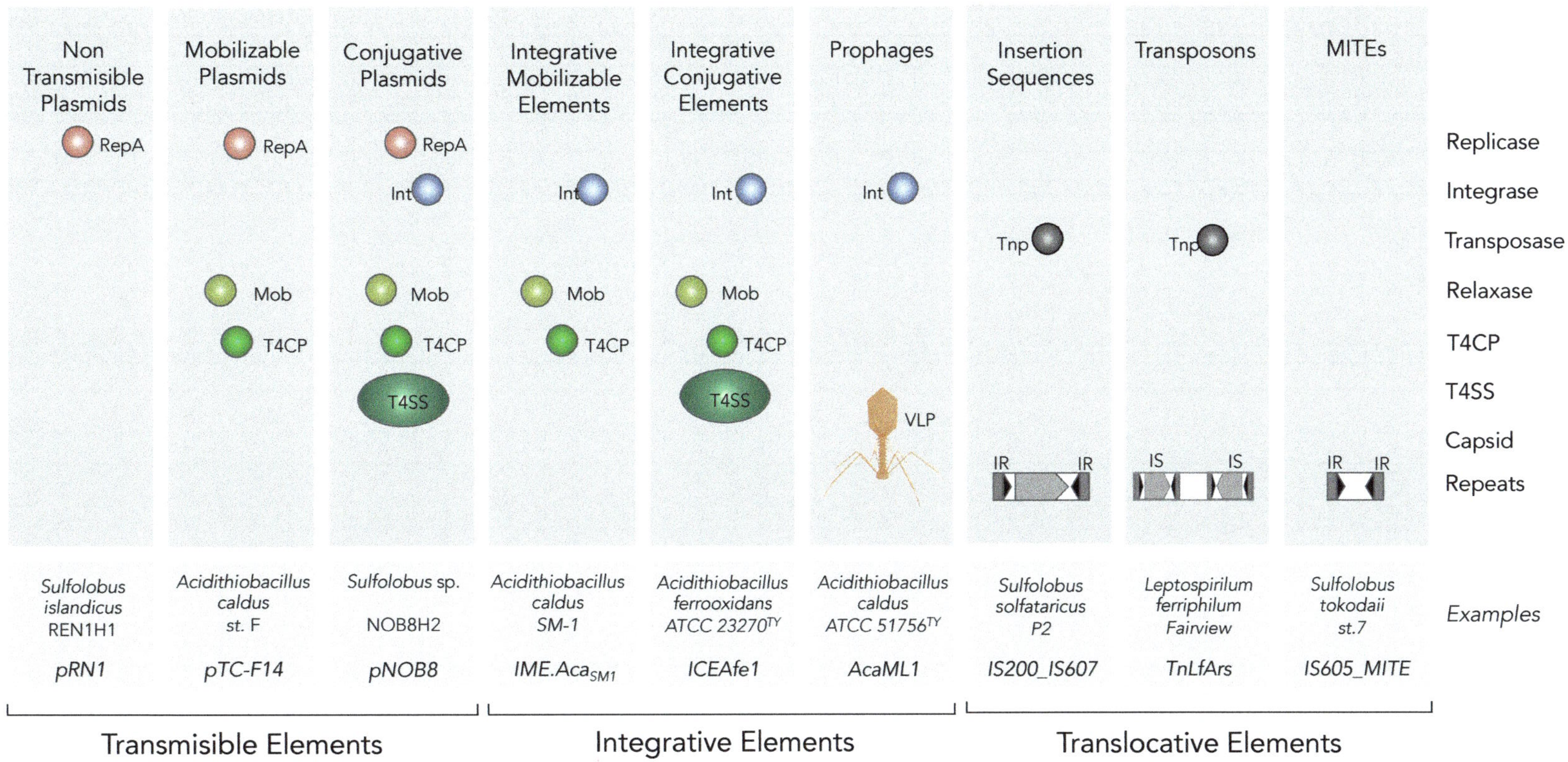

Figure 12.1 Diversity of elements in the mobilome of acidophiles.

Transmissible elements

Plasmids are self-replicating, extrachromosomal, transmissible replicons that act as key agents of change in microbial populations. They range from a few kilobase (kb) to few hundred kb replicons, termed megaplasmids. Although more than a thousand plasmids from all three domains of life and almost every conceivable environmental niche have been described, only 70 plasmids from nearly 160 acidophiles are known by means of their sequence or description (Table 12.1).

Although the genome sequences of an increasing number of plasmids from microbial acidophiles have been determined, knowledge on their biology, biochemistry and molecular mechanisms lags considerably far behind. Insights into the replication, mobilization/conjugation and regulation of some of these plasmids has been gained, yet in most cases they remain cryptic from a functional perspective, and their impact on genome plasticity remains largely unexplored. Knowledge generated for well-characterized plasmid families is covered herein, including archaeal conjugative plasmids, plasmid–virus hybrids and bacterial mobilizable and cryptic plasmids. Replicons of acidophiles, which based on size criteria fall within the category of megaplasmids (larger than 100 kb) have been omitted from this chapter, since no thorough study of these has been reported to date.

Small cryptic plasmids

Some acidophilic archaea and bacteria possess small plasmids of less than 10 kb, considered to be cryptic in nature. Examples include the high copy number crenarchaeal pRN and 'pRN-like' family plasmids, the medium copy number *Acidiphilum* plasmids pAM/pACRY/pACM and the broad host range, low copy number *Acidithiobacillus* pTF/pAca plasmids, among others recently added from sequencing projects (Table 12.1).

The research on cryptic plasmids of acidophiles has focussed mostly on the pRN plasmid family from *Sulfolobus* spp. and *Acidianus* spp. Members of this family seem to be exclusive to the crenarchaeal phylum and appear unrelated to any other known eubacterial or archaeal plasmids. Several of these plasmids have been used for the construction of *Escherichia coli–Sulfolobus* shuttle vectors (Berkner and Lipps, 2008). The pRN1 plasmid was the first crenarchaeal plasmid isolated and sequenced, and represents the prototype of this rather remarkable family (Lipps, 2009). All family members share a conserved genomic region and three highly conserved gene products which have been characterized biochemically. Recent studies have revealed that a single multifunctional replication protein (ORF904) with helicase, primase and DNA polymerase activity, reminiscent of the type of replication performed by some linear bacteriophages, is required to initiate plasmid replication (Berkner *et al.*, 2014). In marked contrast to all DNA polymerases, ORF904 is able to polymerize DNA *de novo*, i.e. without a primer. The enzyme is not a processive DNA polymerase, it does not have proofreading activity and its specific activity is rather low. Gene orthologues of *orf56* encode a fairly variable sequence-specific DNA-binding protein, designated CopG, that has a role in plasmid copy control through down-regulation of plasmid replication initiation. The *orf80* orthologues are partly annotated as *plrA* (plasmid regulatory gene A) but their function is currently unclear.

The pRN family also includes pSSVx from *Sulfolobus islandicus* REY 15/4, which is a hybrid between a pRN plasmid and a fusellovirus. The genome of pSSVx contains the three highly conserved ORFs of the plasmid family and two conserved ORFs originated from SSVs fuselloviruses, probably involved in packaging. It is believed that pSSVx replicates in the host cell as a plasmid and that upon superinfection with the helper SSV viruses it spreads as a virus satellite (Stedman *et al.*, 2003). pSSVi resembles members of the pRN plasmid family in genome organization but encodes an SSV-type integrase. Indeed, a few other pRN-type elements (pXQ1, pST1 and pST3) have been discovered in an integrated form in the chromosomes of *Sulfolobus* species, all of which encode an SSV-type integrase, likely responsible for plasmid integration into the crenarchaeal genome.

Additional small high-copy-number cryptic plasmids isolated from other *Sulfolobus* species and strains, grouped in the 'pRN-like' family, have been reported and some of them characterized. The plasmids pIT3, pTAU4, pORA1 and pTIK4 differ from the pRN family plasmids in that they encode a much less conserved or a completely different replication protein. This provides evidence that the replication apparatus of *Sulfolobus* plasmids can be quite diverse.

Table 12.1 Transmissible elements found in sequenced acidophiles

Phylum	*Species*	Strain	Accession no.	Size (kb)	No. of CDs	% G+C	Relaxase (MOB-type)	Type IV coupling protein	Conjugation system	Copy number	Host range	Name	Family
Small cryptic													
Crenarchaea	*S. islandicus*	REN1H1	NC_001771	5.4	6	37.3	–	–	–	H	B	pRN1	pRN
Crenarchaea	*S. islandicus*	REN1H1	NC_002101	7.0	6	37.4	–	–	–	H	B	pRN2	pRN
Crenarchaea	*S. islandicus*	HEN7H2	NC_004853	7.8	11	35.0	–	–	–	H	B	pHEN7	pRN
Crenarchaea	*Ad. ambivalens*	Lei 10	NC_005562	7.6	10	38.0	–	–	–	H	B	pDL10	pRN
Crenarchaea	*S. islandicus*	REY 15/4	NC_010011	5.7	8	38.7	–	–	–	H	B	pSSVx	pRN; VPH
Crenarchaea	*S. islandicus*	ARN3/6	NC_010365	7.0	7	39.5	–	–	–	H	B	pXZ1	pRN-like
Crenarchaea	*S. solfataricus*	IT3	NC_005907	5.0	6	44.0	–	–	–	H	B	pIT3	pRN-like
Crenarchaea	*S. neozealandicus*	EC	NC_006906	9.7	13	37.8	–	–	–	H	B	pORA1	pRN-like
Crenarchaea	*S. solfataricus*	P2	NC_013777	5.7	8	39.0	–	–	–	H	B	pSSVi	pRN; VPH
Euryarchaea	*P. oshimae*	DSM 9789	NC_016049	7.6	6	30.5	–	–	–	L	B	pPO1	pRN-like
Euryarchaea	*Tp. acidophilum*	H0–122	NC_008318	15.7	18	46.7	–	–	–	L	–	pTA1	
Proteobacteria	*At. ferridurans*	ATCC 33020	X52699	–	–	–	Q	–	–	L	–	pTF1	
Proteobacteria	*A. multivorum*	AIU301	NC_015181	12.1	14	59.6	Q	–	–	M	–	pACMV6	
Proteobacteria	*A. cryptum*	JF-5	NC_009472	8.8	9	61.0	F	–	–	M	–	pACRY06	
Proteobacteria	*A. multivorum*	JCM8867	NC_008691	5.2	5	57.5	Q	–	–	M	–	pAM5	
Proteobacteria	*At. caldus*	SM-1	NC_015852	9.8	10	59.8	P_A	–	–	L	B	pLAtc1	
Proteobacteria	*At. caldus*	ATCC 51756	CP005989	9.8	11	50.3	–	–	–	L	B	$pAca_{TY}$-2	
Proteobacteria	*At. ferrooxidans*	MAL4–1	NC_001520	4.1	2	62.2	P_A	–	–	L	–	pTF4.1	
Proteobacteria	*A. cryptum*	JF-5	NC_009473	5.6	5	57.5	Q	–	–	M	–	pACRY07	
Proteobacteria	*A. multivorum*	AIU301	NC_015189	5.2	8	57.6	Q	–	–	M	–	pACMV7	
Proteobacteria	*A. cryptum*	JF-5	NC_009474	4.9	5	58.7	–	–	–	M	–	pACRY08	
Proteobacteria	*A. multivorum*	AIU301	NC_015182	1.7	1	60.9	–	–	–	M	–	pACMV8	
Firmicutes	*D. acidiphilus*	SJ4	NC_018067	3.9	4	43.3	Q	–	–	–	–	pDESACI.02	
Firmicutes	*Alb. acidocaldarius*	DSM 446	NC_013208	7.9	6	54.3	–	–	–	–	–	pAACI03	

Table 12.1 Continued

Phylum	*Species*	Strain	Accession no.	Size (kb)	No. of CDs	% G+C	Relaxase (MOB-type)	Type IV coupling protein	Conjugation system	Copy number	Host range	Name	Family
Conjugative													
Crenarchaea	*S. islandicus*	EC	NC_006422	28.9	41	37.1	–	+	+	M	B	pKEF9	pKEF-like
Crenarchaea	*S. islandicus*	EC	NC_004852	24.6	38	37.3	–	+	+	M	B	pING1	pKEF-like
Crenarchaea	*S. islandicus*	EC	NC_006425	35.4	54	37.3	–	+	+	M	B	pHVE14	pKEF-like
Crenarchaea	*Sulfolobus* sp.	NOB8H2	NC_006493	41.2	52	37.3	–	+	+	M	B	pNOB8	pKEF-like
Crenarchaea	*S. islandicus*	SOG2/4	NC_010598	26.0	41	36.7	–	+	+	M	B	pSOG2	pKEF-like
Crenarchaea	*S. islandicus*	EC	NC_006423	26.2	40	36.3	–	+	+	M	B	pARN3	pARN-like
Crenarchaea	*S. islandicus*	EC	NC_006424	26.5	1	37.1	–	–	+	M	B	pARN4 (DV)	pARN-like
Crenarchaea	*S. islandicus*	SOG2/4	NC_010597	29.0	46	35.8	–	+	+	M	B	pSOG1	pARN-like
Crenarchaea	*S. islandicus*	Y.N.15.51	NC_012624	42.2	50	36.1	–	+	+	M	B	pYN01	pARN-like
Crenarchaea	*S. islandicus*	L.D.8.5	NC_013770	26.6	32	36.1	–	–	+	M	B	pLD8501	pARN-like
Crenarchaea	*S. solfataricus*	P2	NC_021914	28.0	34	39.5	–	+	+	M	B	pMGB1	pARN-like
Crenarchaea	*S. tengchongensis*	EC	NC_005969	20.4	27	41.4	–	–	+	M	B	pTC	pARN-like
Crenarchaea	*Ad. hospitalis*	W1	NC_011299	28.6	39	35.9	–	–	+	M	B	pAH1	pARN-like
Proteobacteria	*A. multivorum*	AIU301	NC_015187	65.6	69	61.9	F/Q	–	+	–	–	pACMV2	
Proteobacteria	*A. multivorum*	AIU301	NC_015179	54.2	61	61.2	MOB_F	–		–	–	pACMV3	
Proteobacteria	*A. cryptum*	JF-5	NC_009469	89.0	77	62.1	FF	+	+	–	–	pACRY03 (DV)	
Firmicutes	*Sb. thermotolerans*	Y0017	NC_016040	59.2	61	58.2	P/C	+	+ (NP)	–	–	pY0017	
Firmicutes	*Sb. thermotolerans*	L15	JN119829	–	–	–	–	–	–	–	–	pL15	

Mobilizable												
Firmicutes	*Alb. acidocaldarius*	DSM 446	NC_013206	91.7	98	54.1	–	+	–	–	–	pAACI01	
Firmicutes	*Alb. acidocaldarius*	DSM 446	NC_013207	87.3	92	55.4	–	+	–	–	–	pAACI02	
Firmicutes	*D. acidiphilus*	SJ4	NC_018066	60.4	70	39.5	–	+	–	–	–	pDESACI.01	
Proteobacteria	*A. cryptum*	JF-5	NC_009470	37.4	32	59.6	P_A	+	–	–	–	pACRY04	
Proteobacteria	*At. caldus*	MNG	NC_010600	65.2	45	57.1	Q/P_A	–	–	L	B	pTcM1	pTcM1
Proteobacteria	*At. caldus*	SM-1	NC_015854	29.7	28	58.8	Q	–	–	L	B	pLAtc3	pTcM1
Proteobacteria	*At. caldus*	ATCC 51756	CP005988	27.5	33	59.4	Q	–	–	L	B	pAca$_{TY}$.1	pTcM1
Proteobacteria	*At. caldus*	SM-1	NC_015853	14.1	13	57.7	Q	–	–	L	B	pLAtc2	
Proteobacteria	*A. cryptum*	JF-5	NC_009471	37.2	28	60.9	F	–	–	–	–	pACRY05	
Proteobacteria	*A. multivorum*	AIU301	NC_015188	40.6	44	60.0	F	–	–	–	–	pACMV4	
Proteobacteria	*A. multivorum*	AIU301	NC_015180	14.3	19	58.9	P_A	–	–	–	–	pACMV5	
Proteobacteria	*At. caldus*	F	NC_004734	14.2	16	55.7	P_A	–	–	M	B	pTC-F14.	
Proteobacteria	*At. ferrooxidans*	EC	M57717	12.2	–	–	P_A	–	–	M	B	pTF-FC2	
Nitrospira	*L. ferrooxidans*	ATCC 49879	NC_006909	28.0	33	55.2	Q	–	–	–	–	p49879.2	
Nitrospira	*L. ferrooxidans*	ATCC 49879	NC_006907	28.9	32	57.8	Q	–	–	–	–	p49879.1	
Others													
Proteobacteria	*At. ferridurans*	ATCC 33020	NC_005023	19.8	13	56.1	–	–	–	–	–	pTF5	

B, broad host range; DV, deletion variant; EC, enrichment culture; H, high plasmid copy number (>50); L, low plasmid copy number (<15); M, medium plasmid copy number (>15, <50); NP, non-pheromone conjugation system; VPH, virus–plasmid hybrid.

Unlike the situation in the crenarchaea, where a large number of genetic elements have been characterized, little is known of extrachromosomal elements in the *Thermoplasmatales*. To date, the only sequenced and characterized plasmids from this phylogenetic order are the low copy number pTA1 isolated from *Tp. acidophilum* and pPO1 isolated from *Picrophilus oshimae*. Small cryptic plasmids are missing in closely related *Thermoplasma volcanicum* and *P. torridus*.

The search for native plasmids in the acidithiobacilli has yielded a large number of cryptic plasmids ranging in size from 4 to nearly 10 kb. Comparative studies of these plasmids using restriction endonuclease digestion and southern hybridization have profiled plasmid occurrence in diverse strains from different geographical origins but have yielded little insight into the nature of the sequence similarity of this group of small independent replicons.

In 1995 Chakravarty and colleagues presented evidence for a common plasmid replicon present in several strains of *At. ferrooxidans* (Chakravarty *et al.*, 1995). They found that four strains (TFI91, TFI92, TFI85 and TFI29) contained a related 9.8 kb plasmid sharing similar restriction endonuclease sites to the prototype plasmid pTFI91. Similarity of pTFI91-family plasmids was also extended to plasmid pTfA-4 and related plasmids present in Chilean isolates of *At. ferrooxidans* obtained from copper mines several hundred kilometres apart, as well as to the larger pTF5 plasmid isolated from the type strain of *Acidithiobacillus ferridurans* (ATCC 33020) (Dominy *et al.*, 1998). This indicated that these plasmids have a very wide geographic distribution and are either positively selected for or difficult to cure. Later findings confirmed that the pTFl9I-family replicon is a unique entity not related at the nucleotide level to other known broad host range plasmids. *Acidithiobacillus caldus* sequenced strains ATCC 51756 and SM-1 also contained 9.8 kbp small cryptic plasmids. Plasmids $pAca_{TY}$-2 and pLAtc1 have been characterized bioinformatically and shown to be different from all currently sequenced acidithiobacilli plasmids and also from each other (Acuña *et al.*, 2013; You *et al.*, 2011a). Plasmid pLAtc1 encodes a mobilization system similar to that described for pRSB105, but is otherwise very different from it. In turn, plasmid $pAca_{TY}$-2 from the type strain encodes a disrupted orthologue of the replication protein RepA of other acidithiobacilli and a gene whose product is similar to the plasmid exclusion protein Exc1, known to prevent the entry of IncI-type plasmids. Restriction profile analysis has recently revealed that plasmid $pAca_{TY}$-2 corresponds to plasmid pTK1 reported in the original description of the type strain (Shelly Dean, Stellenbosch University, personal communication). Representatives of this set of broad host range plasmids or parts of them (e.g. the replication origin) have been recently used for the development of expression vectors (Zhang *et al.*, 2014) and shuttle vectors between *E. coli* and the *Acidithiobacillus* spp. (US Patent 8.163.558 B2; 2012).

Conjugative plasmids

The thermoacidophilic crenarchaeote *Sulfolobus* hosts a number of medium sized, moderate to high copy number, conjugative plasmids (Table 12.1). These appear to be present in about 3% of all isolated *Sulfolobus* strains (Prangishvili *et al.*, 1998). *Sulfolobus* conjugative plasmids have been grouped on the basis of their genomic characteristics as pKEF-like plasmids and pARN-like plasmids. Full genomic sequence comparison and phylogenetic analysis of representative genes has revealed that pKEF9, pING1, pHVE14, pSOG2 and pNOB8 are very similar to each other in gene content and organization, whereas pARN3, pARN4, pSOG1 and pTC form a related, but separate, group (Wang *et al.*, 2015). Natural deletion variants fitting this classification have also been described (e.g. pING2 del. var, of pING4; pARN4 del. var. of pARN3). They all encode a cluster of five or six core proteins that have been implicated in the conjugative process and, with few exceptions, also a non-partitioning type integrase. This integrase is known to be responsible for reversible chromosomal integration of pNOB8 plasmid in *S. tokodaii*, *S. acidocaldarius* and *Acidianus hospitalis* chromosomes. The presence of a defective conjugative plasmid-like element in the chromosome of several *Sulfolobus* species and *Metallosphaera sedula* DSM 5348, and the discovery of pAH1 in *Ad. hospitalis* W1, extends the host range of this group of conjugative plasmids to genera other than *Sulfolobus*. Being also integrative elements, these plasmids have been implicated in facilitating chromosomal DNA conjugation for some *Sulfolobus* species.

The best-characterized plasmids are the pKEF family plasmids. They range in size between 20 and

40 kb and are all medium copy number plasmids in *S. solfataricus* test strain PH1 (20–40 copies/chromosome); although slightly lower copy numbers occur in their natural hosts (Prangishvili *et al.*, 1998). Conserved genes within this family are organized in three functionally distinct genomic sections grouping the conjugation genes, the origin of replication and the proposed replication genes which resemble pRN family plasmids *repA*, *copG* and *plrA* genes. The genomic regions between these sections are more variable in size, gene content and sequence, and harbour putative recombination motifs which contribute to plasmid variability. Indeed, mixtures of variant plasmids have been detected upon propagation of these plasmids in laboratory *S. solfataricus* strains. Even so, pKEF plasmids are stably maintained in the original plasmid-carrying strains, which is consistent with them carrying orthologues of bacterial proteins involved in partitioning (ParA, ParB).

Despite the fact that conjugational exchange of plasmid DNA has been observed between *Sulfolobus* spp. (Prangishvili *et al.*, 1998), it is still unclear what the exact mechanism is, what proteins are involved, and what the origin of transfer looks like. However, one of the well-conserved regions of pKEF plasmids has provided hints in this respect. This region contains six ORFs predicted to encode the core conjugative apparatus in *Sulfolobus*, and two of these genes products (pKEF9: p12 and p01) exhibit sequence motifs and domain structures characteristic of the bacterial Type IV secretion system coupling protein (VirD4/T4CP) and relaxase (VirB4) proteins required for DNA transfer. The *Sulfolobus* T4CPs are considerably larger than bacterial VirD4, while the VirB4-like proteins are slightly smaller than their bacterial counterparts, and both proteins have C-termini and/or N-termini that are unrelated to other family members. The other four ORFs encode predicted membrane proteins with up to 10 transmembrane helix motifs, and are believed to be involved in transmembrane pore formation required for conjugational exchange of plasmid DNA, mimicking the single membrane spanning protein that facilitates self-transfer in some small plasmids of Gram-positive bacteria. Although VirB2 orthologues are absent from pKEF family plasmids, genes encoding small proteins with pilin-like hydropathy profiles (pKEF9: p05) are collinear with genes encoding VirB4-like subunits and might assemble to form the pili detected on *Sulfolobus* pKEF9 donor cells (Schleper *et al.*, 1995).

To date, limited progress has been made in characterizing the mechanisms of replication and copy number control of the conjugative plasmids of archaeal acidophiles, and very little is known about the interplay between the different elements within a host. Superconjugation immunity studies, in which a plasmid containing host is challenged with a second plasmid after successive conjugative transfers, have been performed for some of these plasmids (Prangishvili *et al.*, 1998). These experiments have revealed incompatibility of closely related plasmids and different quantitative responses on the copy number of certain conjugative plasmids. While the incompatibility of bacterial plasmids is mainly a consequence of competition between two plasmids for the same replication or partition apparatus, the mechanisms behind incompatibilities between archaeal plasmids still needs to be determined.

In the case of acidophilic bacteria, only a few of the known plasmids encode the T4SS required for conjugative self-transfer. These include the 59 kb pY0007 (and its related pL15) from *Sulfobacillus thermotolerans* (Dean and Rawlings, 2011) and two MOBF-type large plasmids, pACVM2 and pACRY03, from *Acidiphilium multivorum* and *Acidiphilium cryptum*. A deletion variant of pACRY03 lacking the coupling protein and the mating bridge genes is also found in *A. multivorum* under the pACMV3 designation. Both *Sb. thermotolerans* large, high G+C content plasmids pY0017 and pL15 encode a potential non-pheromone conjugation 14-gene operon typical of Gram-positive bacteria. Although relatively little is known about this conjugation system, it has been suggested that it might be involved in establishing physical contact between donor and recipient cells in order to facilitate conjugal transfer of the plasmids (Dean and Rawlings, 2011). None of the *Acidiphilium* conjugative plasmids have been analysed to any extent to date and conjugative transfer capacity still awaits experimental evaluation for most acidophilic bacterial conjugative plasmids.

Mobilizable plasmids

A considerably larger number of mobilizable plasmids have been identified in acidophilic bacteria, mostly in the *Proteobacteria*. Most of the reported or predicted mobilizable plasmids are fairly large elements (ranging from ~ 5 to 90 kb), encoding at least a relaxase and/or a coupling protein. Species-specific plasmid backbones accompanied by different accessory genes appear to be a recurring theme in acidophiles such as *At. ferrooxidans* (e.g. Dominy *et al.*, 1998) and *At. caldus* (e.g. van Zyl *et al.*, 2008a).

A family of mobilizable plasmids (pTcM1 family) ranging in size from approximately 30 to 70 kb has been found in a variety of *At. caldus* strains from different continents (Table 12.1). A number of these plasmids have been characterized to different degrees, including pTcM1, pTcF1 and pCSH12, pLAtc3 and $pAca_{TY}$1. These plasmids share a common region that spans 11 to 26 kb and includes the plasmid replicon (RepA-type) and other functions associated with plasmid stability and maintenance (ParA-ParB), as well as an IncQ-like mobilization relaxase of the MobA/RepB-type and a PasAB plasmid addiction system (van Zyl *et al.*, 2003, 2008a). No *oriT* site was identified within the common 26-kb region, yet pTcM1 is mobilizable. The plasmid backbone appears to be conserved with respect to replicons from microorganisms thriving in different environments. Also, a gene encoding a predicted invertase-like protein with motifs in common with the PinR family of site-specific recombinases is present in pTcM1-family plasmids and other plasmids isolated from acidophilic bacteria (pTCF14 and pTF5), the role of which still remains elusive.

Two closely related broad-host-range, mobilizable, medium copy number plasmids (10–15 plasmids per chromosome), were discovered in the acidithiobacilli (Rawlings, 2005). Plasmids pTF-FC2 (IncQ-2a) and pTC-F14 (IncQ-2b) share with other IncQ-family replicons the presence of *repB*, *repA* and *repC* genes encoding the primase, helicase and iteron-binding protein, respectively, as well as a 22-bp iteron based *oriV*. Despite their similarity, their origins of replication are compatible (Gardner *et al.*, 2004). Both plasmids lack an active partitioning system and their stability depends on the toxin–antitoxin system PasAB, which also seem to play a role in copy number control (Matcher and Rawlings, 2009; Deane and Rawlings, 2004; Gardner *et al.*, 2001), resembling CopG-mediated control of crenarchaeal pRN1-family plasmids. Both plasmids have a five-gene mobilization module encoding a MobA relaxase of the MOBP-type. While MobA and MobC appear to be essential for plasmid mobilization, the non-essential MobD and MobE only seem to affect the frequency of mobilization (van Zyl *et al.*, 2003). Even though no conjugative partner plasmids were identified at the time of isolation for pTF-FC2 and pTC-F14, both can be efficiently mobilized by RP4, an IncP conjugative plasmid (van Zyl *et al.*, 2003). The pTC-F14 IncQ replicon has been cloned into plasmid R388, which could be transferred from *E. coli* HB101 to *At. caldus*, and used to allow the replication of this shuttle vector in *At. caldus* for the construction of allelic exchange mutants (van Zyl *et al.*, 2008b).

Additional mobilizable plasmids have been isolated from *At. ferrooxidans* (pTF4.1) and *At. ferridurans* ATCC 33020^{T} (previously referred to as *At. ferrooxidans* pTF1) encoding either MobA/MobL or MobL/MobS relaxases that enable the plasmids to be mobilized at high frequency by IncP group plasmids (Drolet and Lau, 1992).

Significantly less is known about the plasmids of the leptospirilli. No plasmid from this group of acidophiles was known until 2005 when a survey in 10 strains of *L. ferrooxidans* and 6 strains of *L. ferriphilum*, identified the first two plasmids for a member of this unique group of bacteria (Table 12.1). Plasmids p49879.1 and p49879.2 were isolated from *L. ferrooxidans* ATCC 49879 (previously referred to as R3), and thereafter sequenced and experimentally characterized (Coram *et al.*, 2005). Plasmids p49879.1 and p49879.2 were found to be of similar sizes, i.e. ~30 kb, but with slightly different overall G + C mol% ratios. Approximately half of the ORFs present on the two plasmids had products related to known proteins in databases. Although no clear candidates for replication proteins could be identified and neither plasmid was capable of replication in *E. coli* or *Pseudomonas putida*, both plasmids contained four 20-bp iterons that may serve as replication protein binding sites. Other plasmid-associated functions such as a mobilization and stability systems have also been characterized: two divergently transcribed ORFs encoding MobS-like and MobL-like proteins, a MOBQ relaxase and a region of DNA with features typical of the *oriT* regions of pTF1 and the IncQ plasmids. In spite of this, when cloned in a test plasmid, neither of the candidate *oriT* could be mobilized by IncP plasmids such as RP4 in *E. coli* S17.1 (van Zyl *et al.*, 2003). Both plasmids encode proteins with high amino acid sequence identity to ParA-like ATPases responsible for active partitioning of plasmids into progeny cells upon cell division, albeit from different ParA families. This divergence may partly account for the observed plasmid compatibility of p49879.1 and p49879.2. Stability experiments showed the *par* system of p49879.2, but not that of p49879.1, to be functional in *E. coli*.

A number of accessory genes carried by the mobilizable plasmids found in acidophiles encode an eclectic collection of functions. These include energy metabolism related functions (e.g. a terminal quinol oxidase complex), cofactor biogenesis functions (e.g. haem maturation proteins), electron carriers (e.g. 4Fe–4S ferredoxin), transporters, detoxification functions and regulators (Coram *et al.*, 2005; Gardner *et al.*, 2001). However, their functionality and whether they contribute to host cell fitness remains unclear.

Integrative genetic elements

Many mobile elements that physically integrate in the chromosome (or accessory replicons) populate most prokaryotes. Although different bacteria and archaea seem to be differently prone to acquire Integrative Elements (IEs), few of them have been characterized in any detail, particularly in acidophiles. This section focuses on genetic elements of acidophiles that integrate in a site-specific fashion in the host chromosome. This group of elements includes proviruses, conjugative self-transmissible elements (known as ICEs; Integrative Conjugative Elements), mobilizable elements (known as IMEs, integrative mobilizable elements) and other elements that are no longer or were never self-transmissible but can be mobilized under certain circumstances with the help of other MGEs, generally referred to as 'genomic islands' (GIs).

These general definitions group together an overarching diversity of elements ranging in size from less than 10 kb to nearly 500 kb, sharing one or more diagnostic features. Hallmarks of IEs are functional or cryptic genes encoding integrases and small (16–50 bp) perfect or almost perfect direct repeats flanking the element. The integrases are necessary and sufficient to mediate integration and are also required for excision, although this process requires additional factors in most cases. Most of the bacterial integrases that mediate site-specific recombination between extra-chromosomal DNA elements and chromosomes are members of the tyrosine recombinase family, and most integration target sites are tRNAs or other structural RNAs (e.g. tmRNAs). Chromosomal integration in archaea differs from bacteria in that the integrase genes become split into two segments which border the integrated element, and that integration can be reversed only if the intact integrase is produced (Wang *et al.*, 2015 and references therein). These mechanistic differences may have implications for productive HGT across the boundaries of the three domains of life.

In recent years, many IEs have been identified on the basis of sequence data and comparative studies of closely related strains or species representing most of the prokaryotic taxonomic spectra. Although several of these elements may actually be degenerate, experimental evidence has began to build up that supports different aspects of their functionality.

Viruses and proviruses

Evidence collected over the past 30 years points to a rich diversity of viruses infecting acidophiles from both high and low temperature environments (Table 12.2). Most of these viruses have been associated with archaeal acidophiles including *Sulfolobus* spp., *Acidianus* spp., the ARMAN nanoarchaea and *Ferroplasma* spp. (Wang et., 2015; Dellas *et al.*, 2014; Andersson and Banfield, 2008; Allen *et al.*, 2007). These viruses have proved to be unique in terms of their morphology, which includes bottle, lemon and droplet shapes, and also in terms of their life cycles, which span integrated and silent proviruses that parasitize their host, infective and lytic forms that kill the cells they infect, and chronic infecting virions that keep their hosts viable while releasing viral progeny (Wang *et al.*, 2015; Dellas *et al.*, 2014). Tailed viruses infecting bacterial extreme acidophiles of the genus *Acidiphilium* and *Acidithiobacillus* have also been reported (Tapia *et al.*, 2012).

Viruses from acidophiles seem to be capable of integrating their genome into that of their host, establishing lysogeny and moving genes around in acidic econiches. Support for the occurrence of proviruses in acidophiles comes from genomic, metagenomic and experimental evidence. Blocks

Table 12.2 Known viruses of acidophilic prokaryotes

	Virus							
Host	Type	Genus	Name	No. of species	Genome size	Release strategies	Staus	Integrase
Sulfolobus	Circular dsDNA	*Fuselloviridae*	SSV	9	14.7–17.6	Non-lytic	Provirus	+
			SMV	1	48.7	Non-lytic		–
		Unclassified	STSV	2	75.2–76	Non-lytic	Provirus	+
		Guttaviridae	SNDV	1	20	Non-lytic		–
		Turriviridae	STIV	2	16.6–17.6	Non-lytic		–
	Linear dsDNA	*Rudiviridae*	SIRV	2	32.3–35.4	Non-lytic		–
Acidianus	Circular dsDNA	*Fuselloviridae*	ASV1	1	24.1	Non-lytic		–
		Bicaudaviridae	ATV	1	62.7	Non-lytic	Provirus	+
	Linear dsDNA	*Ampullaviridae*	ABV	1	23.8	Non-lytic		–
		Rudiviridae	SIRV	2	32.3–35.4	Non-lytic		–
		Lipothrixiviridae	AFV	6	21–41.1	Non-lytic	Provirus	–
	Linear ssDNA	*Lipothrixiviridae*	AFV3	1	40.4	Non-lytic		–
A-/E-/G-plasma	Circular dsDNA	ND	AMDV	4	>10 kb	Non-lytic	Provirus	ND
ARMAN	ND	Lemon-shaped	ND	ND	ND	Non-lytic	ND	ND
		Rod-shaped	ND	ND	ND	Non-lytic	ND	ND
Acidiphilium	Circular dsDNA	*Siphoviridae*	ΦAc1	1	102	Lytic	Provirus	ND
Acidithiobacillus	Circular dsDNA	*Myoviridae*	AcaML1	1	59.3	ND	Provirus	+
Leptospirillum	dsDNA	ND	AMDV1	1	ND	ND	Provirus	ND

ND, not determined.

of unique genes with anomalous G + C contents, of putative viral origin, have been found in the '*F. acidarmanus*' fer1 genome, and in the 'alphabet-plasma' *Thermoplasmatales* and the leptospirilli datasets from the Iron Mountain acid mine drainage (AMD) system (Andersson and Banfield, 2008; Allen *et al.*, 2007). Also fuselloviruses, bicaudaviruses and integrative plasmid–virus hybrids inserted at tRNA genes in the genomes of *Sulfolobus* spp. and *Acidianus* spp. (e.g. Guo *et al.*, 2011) have been described in the literature. In the case of bacteria, a 59-kb inducible temperate Myoviridae-like prophage located in the *srrA* tmRNA of the *At. caldus* type strain genome has been identified (Tapia *et al.*, 2012). Further research on pro/viruses of acidophiles is certain to provide greater insights into the diversity and the unique biology of viruses from acidophiles.

Genomic islands and conjugative/mobilizable elements

A number of non-viral integrated elements have been formally described in the acidithiobacilli in recent years. These include several genomic islands in *At. ferrooxidans* (Bustamante *et al.*, 2012; Orellana *et al.*, 2011) and *At. caldus* (Acuña *et al.*, 2013). Bioinformatic evidence of additional IEs has also been generated for *At. ferrivorans* (Gonzalez *et al.*, 2014) and *At. thiooxidans* (Travisany *et al.*, 2014).

Two large unique genome segments have been identified in *At. ferrooxidans* strains ATCC 23270^{T} and ATCC 53993 which account for about 16% difference in gene content between the two strains (Holmes *et al.*, 2009). The first is a ~150 kb long genomic island found exclusively in *At. ferrooxidans* ATCC 53993 that contains genes encoding mercury, arsenic and copper metal resistance determinants (Valdés *et al.*, 2010). Experimental data have shown that *At. ferrooxidans* ATCC 53993 has a much higher resistance to $CuSO_4$ (> 100 mM) than the type strain (< 25 mM), which could be possibly explained by the presence of the additional copper resistance genes in this GI (Orellana *et al.*, 2011). Another genomic island, twice as long, is present in *At. ferrooxidans* ATCC 23270^T (Holmes *et al.*, 2009; Levicán *et al.*, 2009). Its genetic organization resembles that of other well-characterized integrative conjugative elements (ICEs), and includes integration/excision, conjugative DNA processing, transfer, maintenance and regulation proteins organized in a number of discrete gene modules. Excision of the element, and expression of a complete set of genes encoding self-transfer functions under normal and DNA-damaging growth conditions, suggests this element is indeed functional (Bustamante *et al.*, 2012). As in most mobile genetic elements, approximately 60% of its ORFs encode proteins with unknown or hypothetical functions, but additional genes encoding features that may confer selective advantages to the host strain have been identified, including a exopolysaccharide biosynthesis cluster, a CRISPR-Cas system and a cluster of 36 tRNA genes covering all 20 amino acids. All these accessory functions have been found to occur in the mobilome of other sequenced acidithiobacilli (Talla *et al.*, 2014; Travisany *et al.*, 2014; Acuña *et al.*, 2013).

Comparative genomic studies of *At. caldus* ATCC 51756 and SM-1 have revealed the presence and activity of > 10 integrative mobile elements in *At. caldus* ranging in size from ~10 to 200 kb (Acuña *et al.*, 2013). Five of them encode components of type IV secretion systems (T4SS) producing P-type pili for conjugative plasmid transfer that define them as ICE, but only two of these (ICE-2 and ICE-3 type elements) have all the essential components of a functional T4SS, and one encodes at least the relaxase and the coupling protein typically present in non conjugative mobilizable plasmids (IME-1). Experimental analyses revealed the co-existence of the integrated and excised forms of all elements in a collection of *At. caldus* strains, indicating that the elements are capable of forming circular intermediates suitable for further transfer to recipient hosts. Most ICEs and the IME are widespread geographically, indicating that conserved or variant versions occur in the *At. caldus* population as active excising elements that may be transferred horizontally when needed.

High levels of genomic divergence due to differences in the mobile gene complement have also been identified in other acidophiles. Two dominant and recombining strains of *Leptospirillum* group II from the Iron Mountain AMD showing a very high similarity at the rRNA level (0.3% difference), have differences in gene content approaching 20% (Simmons *et al.*, 2008). A similar scenario has been pinpointed for *Hydrogenobaculum* sp. strains (Romano *et al.*, 2013). Integrated plasmid-like elements, virus-like elements and genomic islands encoding conjugal transfer proteins are present in representative strains of all these acidophilic genera. Several of these IEs encode copper-, silver-, arsenic-, and mercury- resistance genes, among other diverse functions, suggesting that they are relevant contributors to fitness and niche adaptation aiding in the dispersal of these and other functions.

Nevertheless, little evidence on the activity and impact on fitness of these IEs and their genetic cargo has been obtained to date. Proteomic studies performed on Iron Mountain AMD leptospirilli found that few proteins encoded by genes in the integrated mobile regions were identifiable, while many of the conjugative transfer proteins encoded by genes in a similar extrachromosomal plasmid were indeed expressed (Goltsman *et al.*, 2009). Further studies have shown that proteins associated with MGEs in the leptospirilli were significantly elevated in low developmental stage biofilms from this environment (Mueller *et al.*, 2010). In turn, studies on *Leptospirillum* group IV grown in bioreactors have shown that genes encoded in mobile elements are overrepresented in transcriptomic data sets (Goltsman *et al.*, 2013). Further work is required to understand the role and relevance of these large MGEs in adaptation and evolution of acidophiles.

Translocative elements

Translocative or transposable elements (TPEs) are components of nearly all prokaryotic genomes. They are capable of moving from one chromosome location to another, regardless of the existence of sequence homology between the element and the target site, by the action of a transposase (TPase) in a process called transposition. Transposases constitute a highly diverse group of enzymes and they are the single largest component in sequenced genomes (encompassing 1–10% of the genome) and environmental metagenomes. By transposing into transmissible and integrative elements, they can transfer between hosts and serve as sites of homologous recombination. This allows plasmids and other elements to become inserted into chromosomes and to loop out again, sometimes carrying segments of adjacent chromosomal DNA. These elements can also serve as sites of chromosomal DNA deletions, inversions and a variety of other rearrangements; therefore, they contribute greatly to genetic flexibility (or plasticity) and the entry of genetic material into the horizontal gene pool. Microbial transposable elements can be classified into two large groups: (i) autonomous elements with open reading frames that encode the products required for transposition (i.e. the TPase) together with insertion sequences (ISs) and transposons; (ii) non-autonomous elements that derive from the former by deletion of the TPase while retaining the sequences necessary for *in trans* transposition by cognate TPases, which include miniature inverted repeat transposable elements (MITEs).

Insertion sequences

Insertion sequences (ISs) are fairly abundant in acidophilic archaea and bacteria (Fig. 12.2). More than 200 copies of intact ISs of at least 25 different types were found in the *S. solfataricus* P2 genome and a comparable number in *S. tokodaii*, being this the largest number of transposable elements found in any sequenced genome to date. In the *Sulfolobales*, the types of ISs vary substantially between species, and even closely related strains can differ greatly with respect to their complement of total and active ISs (Blount and Grogan, 2005). While *S. acidocaldarius* maintains a very stable genome organization, evidence of extensive IS-catalysed rearrangements in *S. solfataricus P2* and *S. tokodaii* genomes has been described. In turn, in *S. islandicus* and *Ad. hospitalis*, the majority of potentially transposable autonomous and non-autonomous mobile elements, as well as many degenerate copies of the former, fall into a large variable region of their genomes, helping preserve overall genome organization (e.g. Guo *et al.*, 2011; You *et al.*, 2011b). In contrast to the crenarchaeota, acidophilic euryarchaeota have significantly fewer IS elements. Allen *et al.* (2007) reported less than 100 transposase insertions in the genome of the '*F. acidarmanus*' fer1 isolate and ~50 in the environmental population fer1(env) from which fer1 was derived, representing at most 10 families of ISs. Even fewer ISs are present in *Tp. volcanium* (27 ISs) and *Tp. acidophilum* (four ISs), while *P. torridus* seems to be completely devoid of them. The small size of *P. torridus* genome (1.55 Mb) and its extreme gene density have been suggested to explain its lack of ISs, yet the dramatic differences observed between these acidophilic archaeal phyla still lack an adequate explanation.

Although none of the sequenced acidophilic bacteria contains as many ISs as the *Sulofolobales*, the number and diversity of IS is also high in this group (López de Saro *et al.*, 2013). The occurrence of many IS elements has been reported in recent genome announcements and comparative genomic analyses, but descriptions on active transposable elements have only been published for a small number of acidophiles (e.g. Holmes *et al.*, 2001). Early southern hybridization experiments discovered two families of actively transposing ISs of about 1.5 kb in size (originally named IST1 and IST2) in a variety of laboratory strains and natural isolates of *At. ferrooxidans* obtained from different parts of the world. Among IS-positive strains, the number of IST2 copies varied from 15 to 25, and the number of IST1 copies, from 1 to 10 (Yates and Holmes, 1987). IST2 occurs in a small region of the genome of *At. ferrooxidans* ATCC 19859, considered to be a hot spot of transposition, and seems to exhibit two different patterns of insertion depending on the growth conditions (Cádiz *et al.*, 1994). ISTI (currently known as ISAfe1) is an ISL3 family insertion sequence from *At. ferrooxidans* ATCC 19859 (Holmes *et al.*, 2001). It is about 1.3 kbp in size, has 26-bp imperfectly paired inverted terminal repeats, 5-bp target duplications, and encodes a transposase (OrfA) and a second ORF of unknown function (OrfB). Specific changes in the position of IST1 have been associated with the formation of a large-colony variant of *At. ferrooxidans* ATCC 19859 that reversibly lost the capacity to oxidize iron and exhibited a swarming phenotype. Phenotypic switching was later shown to be correlated with the high frequency insertion and excision of ISAfe1 into, and out of, the *resB* gene which encodes a cytochrome c-type maturation protein relevant for the iron oxidation pathway (Cabrejos

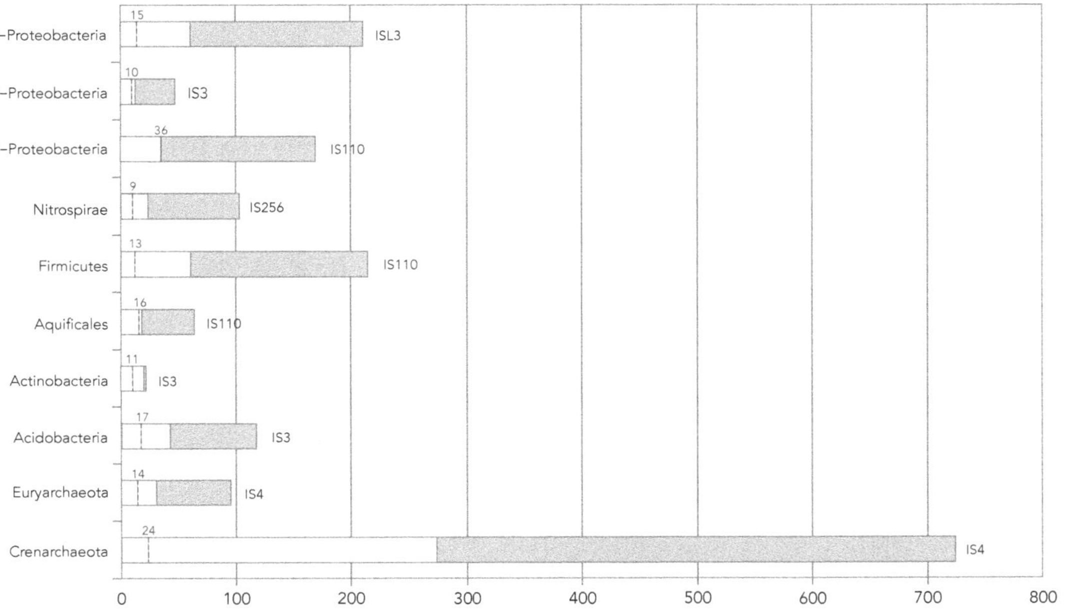

Figure 12.2 Number of insertion sequences in complete genomes of acidophilic prokaryotes. The number of predicted insertion sequences found in sequenced genomes of acidophiles, as organized by phylum, is shown in grey. The number of ISs for the most abundant family in each phylum is shown in white. The number of ISs normalized by the number of genomes in each category is indicated above the dotted lines.

et al., 1999). The cause of the swarming phenotype remains unexplained.

Recent full and partial sequencing of several strains of *Acidithiobacillus* have provided further insight into the diversity of ISs and their genomic distribution. For instance, natural isolates tend to have more ISs than long laboratory-kept type strains, and these tend to cluster within a few genomic regions, several of which have been identified as different types of genomic islands and transposition hotspots (e.g. Acuña *et al.*, 2013). The genome of an *At. caldus* strain (SM-1) isolated from a gold bioreactor in China carried many more copies of ISL3, IS4, IS5 and IS21 elements than all other known *Acidithiobacillus* genomes (You *et al.*, 2011a), and the genome of *At. thiooxidans* Licanantay strain, isolated from a copper pile in northern Chile, carried eight times more transposases that the type strain (Travisany *et al.*, 2014). Comparative genomic analysis has also provided evidence for recent proliferation of certain families; IS5 and ISL3 in the *At. caldus* (27 and 37 copies, respectively) and for the ISAs1 family in *At. ferrooxidans* ATCC 53993 (24 copies) (López de Saro *et al.*, 2013). Only a fairly small percentage of these elements (<10%) have been found to actually interrupt coding regions or to mediate gene acquisition.

A large number of ISs have also been found in *Leptospirillum* species (Mueller *et al.*, 2010), and comparable differences in the relative abundance of some IS families between closely related species and strains have also been reported for these acidophiles.

Transposons

Tn21 transposons are probably the best examples to illustrate how TPEs can be vehicles of transmission of specific phenotypic traits. Highly arsenic-resistant strains of *At. caldus* and *L. ferriphilum* isolated from the same bio-oxidation process treating a gold-bearing arsenopyrite concentrate at the Fairview mine in South Africa have been found to contain distantly related Tn21-like transposons (Tuffin *et al.*, 2005, 2006). Both the *L. ferriphilum* and *At. caldus ars* transposons (TnLfArs and TnAtcArs) were capable of transposition in *E. coli* and thus, believed to be functional in *L. ferriphilum* and *At. caldus*, respectively (Kotze *et al.*, 2006; Tuffin *et al.*, 2006). Despite their presumed functionality in their cognate acidophilic hosts, TnLfArs and

TnAtcArs were sufficiently different from each other to indicate that they were obtained independently and not to have been transferred from one bacterium to the other during their coexistence in the biooxidation tank. This suggests either that HGT might not be possible between these phylogenetically distant bacteria or that bio-oxidation tanks are not a suitable environment for HGT between them to occur.

Miniature inverted repeat transposable elements

Miniature inverted repeat transposable elements (MITEs), common in eukaryote genomes, are also present in the genomes of acidophilic archaea. Some MITEs are present in multiple copies of identical or near-identical sequence (> 140 copies in *S. solfataricus* P2), consistent with their having recently transposed within the genome (Redder *et al.*, 2001). Although they are more prevalent in *S. solfataricus* P2 than in any other sequenced archaeal genome, MITEs are common in other *Sulfolobus* species, as well as in *Tp. volcanium* and *Ad. hospitalis.* Since MITEs can constitute homologous recombination sites, they have been implicated as being responsible for many of the genomic rearrangements observed in archaea (Brügger *et al.*, 2004). Although MITEs in general show a low level of transpositional activity, they can also transpose to different genomic regions and mobilise genomic regions located between two MITEs. Since they arise from internal deletions in IS elements but retain the terminal sequences, the parent IS element can be frequently identified. Archaeal MITEs have been classified into two main types, type I MITEs, including two representatives in *Tp. volcanium* (which derive from ISE1247) and nine in *S. tokodaii* (which derive from the IS605 family), and type II MITEs present in the genomes of *S. tokodaii* and *S. islandicus* at low abundance and of *S. solfataricus,* where they constitute up to 0.6% of the genome with predicted partner IS elements ISC1048, ISC1217, ISC1058 and ISC1173. One challenge for the future will be to determine whether, and how frequently, the sequence diversity created by MITE insertions has resulted in altered gene expression or gene products in acidophiles.

Defence mechanisms

Protection against potentially deleterious effects of entry and propagation of foreign MGEs has warranted the emergence and selection of alternative defence strategies by both bacteria and archaea. Some of these strategies entail distinction of self DNA from foreign DNA by means of modification (methylation or phosphorothioation) of self DNA and degradation or restriction of non-self DNA; very much like a prokaryotic version of innate immunity. Other systems, in turn, memorize the encounters with MGEs and attack them specifically afterwards. This is the case of the CRISPR-Cas systems, which have been compared to prokaryotic adaptive immunity systems. Little is known about the RM-systems in acidophiles. On the contrary, substantial evidence on the occurrence and relevance of CRISPR-Cas systems in acidophiles has accumulated in recent years. Hints for HGT between distant prokaryotes has been obtained for both RM and CRISPR-Cas systems; interestingly, these systems act to limit HGT of the mobilome elements but are themselves part of it.

CRISPR-Cas systems

CRISPR-Cas systems are used to acquire heritable protection against invading MGEs by both bacteria and archaea. Essential for the function of these systems are the Cas proteins and the CRISPR loci. CRISPRs are made up of a variable number of repetitive DNA motifs, averaging 30 bp, separated by spacer sequences of similar size. The spacers are derived from MGEs and incorporated into active loci after exposure to, and successful destruction of, the foreign elements. Upon entrance of a MGE into the cell, the CRISPR loci precursor transcripts are processed into a number of short RNAs, called crRNAs, and displayed on Cas protein complexes. Recognition of the invading foreign DNA is achieved by an RNA-guided sequence specific mechanism through crRNAs matching the invader DNA or RNA, and subsequent degradation is exerted by Cas nucleases. Because of their inherent mechanism of action, CRISPR-Cas systems are both adaptive and heritable and thus constitute a memory of previous attacks which the cell has survived or which the population as a whole has been able to cope with.

CRISPR loci are well represented in acidophiles

and have been studied quite extensively in a couple of acidophilic archaeal and bacterial models, namely the *Sulfolobales* (reviewed by Garrett, 2015) and the leptospirilli (Tyson and Banfield, 2008). Nearly all sequenced acidophilic archaea and approximately half of acidophilic bacteria encode CRISPR loci and/or Cas proteins in their genomes, fitting general statistics derived from other econiches. By combining evidence from phylogenetic, comparative genomic and structural analyses, basic structural and functional blocks have been recognized and accommodated into a three-types based (and 12-subtypes based) classification (Makarova *et al.*, 2011). Diversity of CRISPR-Cas system types in acidophiles is summarized in Table 12.3. The most frequent types in acidophilic archaea are I-A, III-B and III-A and in bacteria I-B, I-C and III-B, with 60% of the acidophilic archaea and only 30% of acidophilic bacteria having simultaneously a type I and a type III system. Type II interference systems are absent from all currently known acidophiles. The accepted consensus is that most type I systems target dsDNA, while type III systems target RNA. Thus, acidophilic archaea seem to be better prepared to confront DNA and RNA MGEs, (e.g. archaeophages) than bacterial acidophiles. For instance, the *Sulfolobales* host subtypes I-A, I-D, III-B, III-D and III-VIII-I CRISPR-Cas. Such diversity of the CRISPR-Cas system types and spacers (>4000 in complete sequence genomes of the genus) correlate with the considerable diversity and abundance of genetic elements and viruses infecting the *Sulfolobales* and thriving in thermo-acidic econiches (Garrett *et al.*, 2015; Dellas *et al.*, 2015). Indeed, several viral families containing circular or linear dsDNA genomes as well as positive strand RNA viruses have been detected in the econiches dominated by these thermo-acidophiles. Less impressive numbers are found in other acidophiles (Table 12.3), but still more than 70% of the completely sequenced acidophiles carry CRISPRs in their genomes, suggesting that viruses and other MGEs are as common a threat in acidic econiches as in other types of environments. Archaeal acidophiles tend to have more arrays per genome (up to nine) and more repeats per array (up to 457) than bacteria. These observations have been made also in microbial communities inhabiting other types of environments. Such comparisons suggest that extremophiles frequently have more CRISPR-Cas systems and larger numbers of CRISPRs. Most acidophiles carry repeats that are highly conserved in sequence within each predicted locus, fitting well-accepted criteria for functionality. Nevertheless, with few exceptions, functionality and activity against invading MGEs of most CRISPR loci of acidophiles remain to be evaluated.

A number of studies of the CRISPR-Cas systems of acidophiles have contributed significantly to the general understanding of CRISPR biology. The *Sulfolobales* have served as a model of study for several aspects of the basic molecular and structural biology of CRISPR-Cas systems, providing seminal insights into both the adaptation and interference processes, and their regulation in type I and type III systems (for recent review see Garrett *et al.*, 2015). The CRISPR-Cas systems of the acidithiobacilli, which host rather unique loci (type U loci for Unclassified and subtype Aferro for *At. ferrooxidans* ATCC 23270^T), have provided insights into the diversity of systems yet to be discovered and the challenges associated with the classification of these systems (Makarova *et al.*, 2011). CRISPR-Cas loci of the acidithiobacilli are located within ICE elements (Acuña *et al.*, 2013; Bustamante *et al.*, 2012), lack the ubiquitous core genes involved in adaptation (*cas1* and *cas2*) and encode proteins that are highly divergent from other known and characterized Cas (Acuña *et al.*, 2013; Holmes *et al.*, 2009). Despite the fact that the number and diversification of spacer sequences suggests these loci to be active, it remains unclear whether these are self-sufficient systems or defective systems that capture and utilize pre-existing CRISPR arrays. Further study of the Aferro-subtype CRISPR-Cas systems is likely to enlarge the list of emerging alternative roles of these systems. Also, population-level genomic studies in the leptospirilli have proved CRISPRs to be amazing tools for matching viruses to their hosts by means of spacer sequence analysis (Anderson and Banfield, 2008), and for grasping the interactions and dynamics of viruses and microbial host populations (Tyson and Banfield, 2008). Further details on CRISPR analyses in the Iron Mountain AMD system are nicely covered in Chapter 13.

Restriction–modification systems

Foreign DNA is frequently cut into pieces by endonucleases encoded in the genome of the recipient cell, which recognize and cut at specific DNA

Table 12.3 Defence systems present in sequenced acidophiles

	Taxonomy	Name	No. of species	No. of strains	Genome size[a] (kb)	CRISPR–Cas System Type[b]									# CRISPR loci	# Spacers	R–M System Type[c]				# R–M loci
						IA	IB	IC	ID	IE	IF	U	IIIA	IIIB			I	II	III	IV	
Archaea	Crenarchaeota	*Acidilobus*	1	1	1.5	+	–	–	–	–	–	–	–	+	6	103	–	+	–	–	2
		Caldisphaera	1	1	1.5	–	–	–	–	–	–	–	–	–	0	0	+	+	–	–	11
		Acidianus	1	1	2.1	+	–	–	–	–	–	–	+	–	6	123	–	+	–	–	2
		Metallosphaera	2	2	1.8	+	–	–	–	–	–	–	+	–	2–5	282–373	–	+	–	–	2
		Sulfolobus	4	17	2.0–2.9	+	–	–	+	–	–	–	–	+	2–6	54–457	+	+	–	–	30
		Vulcanisaeta	2	2	2.3	+	–	–	–	–	–	–	+	+	4–9	62–120	–	+	–	–	2
	Euryarchaeota	*Ferroplasma*	1	1	1.9	–	+	–	–	–	–	–	–	–	2	20	+	+	+	–	10
		Picrophilus	1	1	1.5	–	–	–	+	–	–	–	+	–	4	116	+	+	–	+	3
		Thermoplasma	2	2	1.5–1.6	–	–	–	–	–	–	–	+	–	1–2	34–46	+	+	+	–	19
		Aciduliprofundum	2	2	1.4–1.5	+	+	–	–	–	–	–	+	–	2–2	21–23	+	+	+	–	8
Bacteria	Acidobacteria	*Acidobacterium*	*1*	*1*	4.1	–	–	+	–	–	–	–	–	–	1	23	+	+	+	–	4
		Granulicella	2	2	4.3	–	–	–	–	–	–	–	–	–	0	0	+	+	–	–	4
		Terriglobus	2	2	5.0–5.2	–	–	–	–	–	–	–	–	–	0	0	–	+	–	–	6
		Solibacter	1	1	9.9	–	–	–	–	–	–	–	–	–	0	0	–	+	–	–	1
		Koribacter	1	1	5.5	–	–	–	–	–	–	–	–	–	0	0	+	+	–	–	5
	Actinobacteria	*Acidimicrobium*	1	1	2.1	–	–	–	–	+	–	–	–	–	2	67	–	–	+	–	1
	Aquificae	*Hydrogenobaculum*	1	3	1.5	–	+	–	–	–	–	–	–	–	1–6	29–69	+	+	–	–	5
	Firmicutes	*Alicyclobacillus*	1	2	3.0–3.1	+	–	+	–	–	–	–	–	+	3–4	46–58	–	+	+	–	23
		Kyrpidia	1	1	3.4	–	+	–	–	–	–	–	–	+	4	418	ND	ND	ND	ND	ND
		Desulfosporosinus	1	1	4.9	–	–	+	–	–	–	–	–	–	1	34	+	+	–	+	7
		Sulfobacillus	1	2	3.5	–	+	–	–	–	–	–	–	–	2–2	14–68	–	+	+	–	20
	Nitrospirae	*Leptospirillum*[a]	2	2	2.4 –2.5	–	–	–	–	+	–	–	–	–	1–3	4–157	+	+	+	–	11
	Proteobacteria	*Acidiphilium*	1	1	3.7	–	–	–	–	–	–	–	–	–	0	0	–	+	–	–	8
		Acidithiobacillus	3	5	2.8–3.2	–	–	–	–	–	–	+	–	–	0–2	0–42	+	+	+	–	52
	Verrucomicrobia	*Methylacidiphilum*			2.3	–	–	+	–	–	–	–	–	+	5	19	+	+	–	+	6

Only complete genome have been considered with the exception of the leptospirilli[a].
[a]Data correspond to draft genomes derived from the Iron Mountain AMD.
[b]Data were obtained from CRISPR Database (http://http://crispr.u-psud.fr/crispr/) and CRISPI (http://crispi.genouest.org).
[c]Data were obtained from ReBase Database (http://rebase.neb.com/rebase/rebase.html).

sequences. Cognate methylases, in turn, protect the hosts DNA from the action of the cutting enzyme by chemically modifying the target sequences, thus achieving an accurate distinction between native and exogenous DNA. Overall, the RM systems limit the frequency of HGT and increase the resistance of cells to viruses. Four types of restriction systems (I–IV) are currently recognized, on the basis of subunit composition, NTP requirement and cleavage mechanism. Information stored in the ReBase database reveals that 95% of sequenced bacteria and nearly 100% of archaea encode RM systems, with type II RM systems being by far the most common. This general tendency is conserved in acidophiles (Table 12.3). Despite their widespread occurrence and high diversity, few RM systems have been characterized so far in acidophiles. These include the *Tp. acidophilus* ThaI system and SuaI and SuiI systems from *S. acidocaldarius* and *S. islandicus*, respectively. While all three systems are capable of DNA restriction, their cognate methylases await discovery and evaluation. Multiple RM-systems populate the genomes of acidophilic bacteria (e.g. > 50 in sequenced acidithiobacilli) and several of these have been shown to map to MGEs (e.g. Acuña *et al.*, 2013; van Zyl *et al.*, 2008a). Studies of RM-systems of acidophiles in general are required in order to overcome some of the barriers to genetic manipulation that many of these models impose.

Outlook

Although acidophiles can be differentiated from neutrophiles and other extremophiles in many aspects of their biology, their mobilome seems to be as diverse and untapped as that of other prokaryotes. Transmissible, integrative and translocative elements described in this chapter represent evidence of the active exchanges of genetic material that have taken place between acidophiles, and have likely had an impact on their adaptation, differentiation and/or niche expansion. Many of the genes being exchanged along with these MGEs belong to functional categories that have previously been shown to be over-represented in the mobilome of model systems and communities thriving in other econiches. In contrast, several other genes provide a clear ecological advantage under niche-specific (acidic) conditions. Several other genes encode hypothetical proteins whose function awaits further discovery. To date, very little insight into the potential advantages (e.g. gene acquisition, genome plasticity, etc.) and disadvantages (e.g. gene inactivation, genome instability, etc.) for cells carrying MGEs has been obtained for acidophiles. Also, the elucidation of the environmental conditions that enhance the proliferation and propagation or containment of these elements by defence systems, are relevant topics to be addressed to improve the understanding of the physiology and ecology of these environmentally and industrially relevant microorganisms. Many mechanistic details are likely to emerge from further studies. Both bioinformatic approaches and genetic experiments are required to uncover the dynamic interactions between the genome, the transcriptome and the mobilome. For this to be possible, larger sets of closely related genomes need to get sequenced and a more comprehensive toolbox for genetic manipulation of acidophiles needs to be developed. Both these challenging milestones have begun to be addressed in recent years and will shortly provide new opportunities to expand our knowledge and understanding of the acidophilic mobilome.

Acknowledgements

The authors gratefully acknowledge Fundación Ciencia y Vida, Universidad Andres Bello and the University of Stellenbosch for facilitating the undertaking of our studies. The authors also wish to thank FONDECYT (Grant ID 1140048 and 3140005) for financial support and numerous students that have been contributed to our research.

References

Acuña, L.G., Cárdenas, J.P., Covarrubias, P.C., Haristoy, J.J., Flores, R., Nuñez, H., Riadi, G., Shmaryahu, A., Valdés, J., Dopson, M., *et al.* (2013). Architecture and gene repertoire of the flexible genome of the extreme acidophile *Acidithiobacillus caldus*. PLoS One *8*, e78237.

Allen, E.E., Tyson, G.W., Whitaker, R.J., Detter, J.C., Richardson, P.M., and Banfield, J.F. (2007). Genome dynamics in a natural archaeal population. Proc. Natl Acad. Sci. U.S.A. *104*, 1883–1888.

Andersson, A.F., and Banfield, J.F. (2008). Virus population dynamics and acquired virus resistance in natural microbial communities. Science *320*, 1047–1050.

Berkner, S., and Lipps, G. (2008). Genetic tools for *Sulfolobus* spp.: vectors and first applications. Arch. Microbiol. *190*, 217–230.

Berkner, S., Hinojosa, M.P., Prangishvili, D., and Lipps, G. (2014). Identification of the minimal replicon and the origin of replication of the crenarchaeal plasmid pRN1. Microbiologyopen *3*, 688–701.

Blount, Z.D., and Grogan, D.W. (2005). New insertion sequences of *Sulfolobus*: functional properties and implications for genome evolution in hyperthermophilic archaea. Mol. Microbiol. *55*, 312–325.

Brügger, K., Torarinsson, E., Redder, P., Chen, L., and Garrett, R.A. (2004). Shuffling of *Sulfolobus* genomes by autonomous and non-autonomous mobile elements. Biochem. Soc. Trans. *32*, 179–183.

Bustamante, P., Covarrubias, P., Levicán, G., Katz, A., Tapia, P., Bonnefoy, V., Holmes, D.S., Quatrini, R., and Orellana, O. (2012). ICEAfe1, an actively excising integrative-conjugative element from the biomining bacterium *Acidithiobacillus ferrooxidans*. J. Mol. Microbiol. Biotech. *22*, 399–407.

Cabrejos, M.E., Zhao, H.L., Guacucano, M., Bueno, S., Levican, G., Garcia, E., Jedlicki, E., and Holmes, D.S. (1999). IST1 insertional inactivation of the *resB* gene: implications for phenotypic switching in *Thiobacillus ferrooxidans*. FEMS Microbiol. Lett. *175*, 223–229.

Cádiz, R., Gaete, L., Jedlicki, E., Yates, J., Holmes, D.S., and Orellana, O. (1994). Transposition of IST2 in *Thiobacillus ferrooxidans*. Mol. Microbiol. *12*, 165–170.

Chakravarty, L., Zupancic, T.J., Baker, B., Kittle, J.D., Fry, I.J., and Tuovinen, O.H. (1995). Characterization of the pTFI91-family replicon of *Thiobacillus ferrooxidans* plasmids. Can. J. Microbiol. *41*, 354–365.

Coram, N.J., van Zyl, L.J., and Rawlings, D.E. (2005). Isolation, sequence analysis, and comparison of two plasmids (28 and 29 kilobases) from the biomining bacterium *Leptospirillum ferrooxidans* ATCC 49879. Appl. Environ. Microbiol. *71*, 7515–7522.

Deane, S.M., and Rawlings, D.E. (2004). Plasmid evolution and interaction between the plasmid addiction stability systems of two related broad host-range IncQ-like plasmids. J. Bacteriol. *186*, 2123–2133.

Deane, S.M., and Rawlings, D.E. (2011). Two large, related, cryptic plasmids from geographically distinct isolates of *Sulfobacillus thermotolerans*. Appl. Environ. Microbiol. *77*, 8175–8180.

Dellas, N., Snyder, J.C., Bolduc, B., and Young, M.J. (2014). Archaeal Viruses: Diversity, Replication, and Structure. Annu. Rev. Virol. *1*, 399–426.

Dominy, C.N., Coram, N.J. , and Rawlings, D.E. (1998). Sequence analysis of plasmid pTF5, a 19.8-kb geographically widespread member of the *Thiobacillus ferrooxidans* pTF191-like plasmid family. Plasmid *40*, 50–57.

Dorrington, R.A., and Rawlings, D.E. (1990). Characterization of the minimum replicon of the broad-host-range plasmid pTF-FC2 and similarity between pTF-FC2 and the IncQ plasmids. J. Bacteriol. *172*, 5697–5705.

Drolet, M., and Lau, P.C.K. (1992). Mobilization protein-DNA binding and divergent transcription at the transfer origin of *the Thiobacillus ferrooxidans* pTF1 plasmid. Mol. Microbiol. *6*, 1061–1071.

Fütterer, O., Angelov, A., Liesegang, H., Gottschalk, G., Schleper, C., Schepers, B., Dock, C., Antranikian, G., and Liebl, W. (2004). Genome sequence of *Picrophilus torridus* and its implications for life around pH 0. Proc. Natl. Acad. Sci. U.S.A. *101*, 9091–9096.

Garcillán-Barcia, M.P., and de la Cruz, F. (2013). Ordering the bestiary of genetic elements transmissible by conjugation. Mob. Genet. Elements *3*, e24263.

Gardner, M.N., and Rawlings, D.E. (2004). Evolution of compatible replicons of the related IncQ-like plasmids, pTC-F14 and pTF-FC2. Microbiology *150*, 1797–1808.

Gardner, M.N., Deane, S.M., and Rawlings, D.E. (2001). Isolation of a new broad-host-range IncQ-like plasmid, pTC-F14, from the acidophilic bacterium *Acidithiobacillus caldus* and analysis of the plasmid replicon. J. Bacteriol. *183*, 3303–3309.

Garrett, R.A., Shah, S.A., Erdmann, S., Liu, G., Mousaei, M., León-Sobrino, C., Peng, W., Gudbergsdottir, S., Deng, L., Vestergaard, G., *et al.* (2015). CRISPR-Cas adaptive immune systems of the *Sulfolobales*: unravelling their complexity and diversity. Life *5*, 783–817.

Goltsman, D.S.A., Denef, V.J., Singer, S.W., VerBerkmoes, N.C., Lefsrud, M., Mueller, R.S., Dick, G.J., Sun, C.L., Wheeler, K.E., Zemla, A., *et al.* (2009). Community genomic and proteomic analyses of chemoautotrophic iron oxidizing "*Leptospirillum rubarum*" (group II) and "*Leptospirillum ferrodiazotrophum*" (group III) bacteria in acid mine drainage biofilms. Appl. Environ. Microbiol. *75*, 4599–4615.

Goltsman, D.S., Dasari, M., Thomas, B.C., Shah, M.B., VerBerkmoes, N.C., Hettich, R.L., and Banfield, J.F. (2013). New group in the *Leptospirillum* clade: cultivation-independent community genomics, proteomics, and transcriptomics of the new species "*Leptospirillum* group IV UBA BS". Appl. Environ. Microbiol. *79*, 5384–5393.

González, C., Yanquepe, M., Cárdenas, J.P., Valdés, J., Quatrini, R., Holmes, D.S., and Dopson, M. (2014). Genetic variability of psychrotolerant *Acidithiobacillus ferrivorans* revealed by (meta)genomic analysis. Res. Microbiol. *65*, 726–734.

Greve, B., Jensen, S., Brügger, K., Zillig, W., and Garrett, R.A. (2004). Genomic comparison of archaeal conjugative plasmids from *Sulfolobus*. Archaea *1*, 231–239.

Guo, L., Brügger, K., Liu, C., Shah, S.A., Zheng, H., Zhu, Y., Wang, S., Lillestøl, R.K., Chen, L., Frank, J., *et al.* (2011). Genome analyses of Icelandic strains of *Sulfolobus islandicus*: model organisms for genetic and virus–host interaction studies. J. Bacteriol. *193*,1672–1680.

Holmes, D.S., Zhao, H.L., Levican, G., Ratouchniak, J., Bonnefoy, V., Varela, P., and Jedlicki, E. (2001). ISAfe1, an ISL3 family insertion sequence from *Acidithiobacillus ferrooxidans* ATCC 19859. J. Bacteriol. *183*, 4323–4329.

Holmes, D.S., Cárdenas, J.P., Valdés, J., Quatrini, R., Esparza, M., Osorio, H., Duarte, F., Lefimil, C., and Jedlicki, E. (2009). Comparative genomics begins to unravel the ecophysiology of bioleaching. Adv. Mat. Res. *71–73*, 143–150.

Kotze, A.A., Tuffin, I.M., Deane, S.M., and Rawlings, D.E. (2006). Cloning and characterization of the chromosomal arsenic resistance genes from *Acidithiobacillus caldus* and enhanced arsenic resistance on conjugal transfer of ars genes located on transposon TnAtcArs. Microbiology *152*, 3551–3560.

Levicán, G., Katz, A., Valdés, J., Quatrini, R., Holmes, D.S., and Orellana, O. (2009). A 300 kpb genome segment, including a complete set of tRNA genes, is dispensable for *Acidithiobacillus ferrooxidans*. Adv. Mat. Res. *71–73*, 187–190.

Lipps, G. (2009). Molecular biology of the pRN1 plasmid from *Sulfolobus islandicus*. Biochem. Soc. Trans. *37*, 42–45.

López de Saro, F.J., Gómez, M.J., González-Tortuero, E., and Parro, V. (2013). The dynamic genomes of acidophiles. In Polyextremophiles: Life Under Multiple Forms of Stress, Seckbach, J., Oren, A., and Stan-Lotter, H. eds. (Springer, Dordrecht, Netherlands). pp.81–97.

Makarova, K.S., Haft, D.H., Barrangou, R., Brouns, S.J., Charpentier, E., Horvath, P., Moineau, S., Mojica, F.J., Wolf, Y.I., Yakunin, A.F., *et al.* (2011). Evolution and classification of the CRISPR-Cas systems. Nat. Rev. Microbiol. *9*, 467–477.

Matcher, G.F., and Rawlings, D.E. (2009). The effect of the location of the proteic post-segregational stability system within the replicon of plasmid pTF-FC2 on the fine regulation of plasmid replication. Plasmid *62*, 98–107.

Mueller, R.S., Denef, V.J., Kalnejais, L.H., Suttle, K.B., Thomas, B.C., Wilmes, P., Smith, R.L., Nordstrom, D.K., McCleskey, R.B., Shah, M.B., *et al.* (2010). Ecological distribution and population physiology defined by proteomics in a natural microbial community. Mol. Syst. Biol. *6*, 374.

Orellana, L.H., and Jerez, C.A. (2011). A genomic island provides *Acidithiobacillus ferrooxidans* ATCC 53993 additional copper resistance: a possible competitive advantage. Appl. Microbiol. Biotechnol. *92*, 761–767.

Prangishvili, D., Albers, S.V., Holz, I., Arnold, H.P., Stedman, K., Klein, T., Singh, H., Hiort, J., Schweier, A., Kristjansson, J.K., *et al.* (1998). Conjugation in archaea: frequent occurrence of conjugative plasmids in *Sulfolobus*. Plasmid. *40*, 190–202.

Rawlings, D.E. (2005). The evolution of pTF-FC2 and pTC-F14, two related plasmids of the IncQ family. Plasmid *53*, 137–147.

Redder, P., She, Q., and Garrett, R.A. (2001). Non-autonomous mobile elements in the Crenarchaeon *Sulfolobus solfataricus*. J. Mol. Biol. *306*, 1–6.

Romano, C., D'Imperio, S., Woyke, T., Mavromatis, K., Lasken, R., Shock, E.L., and McDermott, T.R. (2013). Comparative genomic analysis of phylogenetically closely related *Hydrogenobaculum* sp. isolates from Yellowstone National Park. Appl. Environ. Microbiol. *79*, 2932–2943.

Schleper, C., Holz, I., Janekovic, D., Murphy, J., and Zillig, W. (1995). A multicopy plasmid of the extremely thermophilic archaeon *Sulfolobus* effects its transfer to recipients by mating. J. Bacteriol. *177*, 4417–4426.

Simmons, S.L., Dibartolo, G., Denef, V.J., Goltsman, D.S., Thelen, M.P., and Banfield, J.F. (2008). Population genomic analysis of strain variation in *Leptospirillum* group II bacteria involved in acid mine drainage formation. PLoS Biol. *6*, e177.

Stedman, K.M., She, Q., Phan, H., Arnold, H.P., Holz, I., Garrett, R.A., and Zillig, W. (2003). Relationships between fuselloviruses infecting the extremely thermophilic archaeon *Sulfolobus*: SSV1 and SSV2. Res. Microbiol. *154*, 295–302.

Talla, E., Hedrich, S., Mangenot, S., Ji, B., Johnson, D.B., Barbe, V., and Bonnefoy, V. (2014). Insights into the pathways of iron- and sulfur-oxidation, and biofilm formation from the chemolithotrophic acidophile *Acidithiobacillus ferrivorans* CF27. Res. Microbiol. *165*, 753–760.

Tapia, P., Flores, F.M., Covarrubias, P.C., Acuña, L.G., Holmes, D.S., and Quatrini, R. (2012). Complete genome sequence of temperate bacteriophage AcaML1 from the extreme acidophile *Acidithiobacillus caldus* ATCC 51756. J. Virol. *86*, 12452–12453.

Travisany, D., Cortés, M.P., Latorre, M., Di Genova, A., Budinich, M., Bobadilla-Fazzini, R.A., Parada, P., González, M., and Maass, A. (2014). A new genome of *Acidithiobacillus thiooxidans* provides insights into adaptation to a bioleaching environment. Res. Microbiol. *165*, 743–752.

Tuffin, I.M., de Groot, P., Deane, S.M., and Rawlings, D.E. (2005). An unusual Tn21-like transposon containing an ars operon is present in highly arsenic-resistant strains of the biomining bacterium *Acidithiobacillus caldus*. Microbiology *151*, 3027–3039.

Tuffin, I.M., Hector, S.B., Deane, S.M., and Rawlings, D.E. (2006). Resistance determinants of a highly arsenic-resistant strain of *Leptospirillum ferriphilum* isolated from a commercial biooxidation tank. Appl. Environ. Microbiol. *72*, 2247–2253.

Tyson, G.W., and Banfield, J.F. (2008). Rapidly evolving CRISPRs implicated in acquired resistance of microorganisms to viruses. Environ. Microbiol. *10*, 200–207.

Valdés, J., Cárdenas, J.P., Quatrini, R., Esparza, M., Osorio, H., Duarte, F., Lefimil, C., Sepulveda, R., Jedlicki, E., and Holmes, D.S. (2010). Comparative genomics begins to unravel the ecophysiology of bioleaching. Hydrometallurgy *104*, 471–476.

Wang, H., Peng, N., Shah, S.A., Huang, L., and She, Q. (2015). Archaeal extrachromosomal genetic elements. Microbiol. Mol. Biol. Rev. *79*, 117–152.

Yates, J.R., and Holmes, D.S. (1987). Two families of repeated DNA sequences in *Thiobacillus ferrooxidans*. J. Bacteriol. *69*, 1861–1870.

You, X.Y., Guo, X., Zheng, H.J., Zhang, M.J., Liu, L.J., Zhu, Y.Q., Zhu, B., Wang, S.Y., Zhao, G.P., Poetsch, A., *et al.* (2011a). Unraveling the *Acidithiobacillus caldus* complete genome and its central metabolisms for carbon assimilation. J. Genet. Genomics. *38*, 243–252.

You, X.Y., Liu, C., Wang, S.Y., Jiang, C.Y., Shah, S.A., Prangishvili, D., She, Q., Liu, S.J., and Garrett, R.A. (2011b). Genomic analysis of *Acidianus hospitalis* W1 a host for studying crenarchaeal virus and plasmid life cycles. Extremophiles *15*, 487–497.

Zhang, M.J., Jiang, C.Y., You, X.Y., and Liu, S.J. (2014). Construction and application of an expression vector from the new plasmid pLAtc1 of *Acidithiobacillus caldus*. Appl. Microbiol. Biotechnol. *98*, 4083–4094.

van Zyl, L.J., Deane, S.M., and Rawlings, D.E. (2003). Analysis of the mobilization region of the broad-host-range IncQ-like plasmid pTC-F14 and its ability to interact with a related plasmid, pTF-FC2. J. Bacteriol. *185*, 6104–6111.

van Zyl, L.J., Deane, S.M., Louw, L.A., and Rawlings, D.E. (2008a). Presence of a family of plasmids (29 to 65 kilobases) with a 26-kilobase common region in different strains of the sulfur-oxidizing bacterium *Acidithiobacillus caldus*. Appl. Environ. Microbiol., *74*, 4300–4308.

van Zyl, L.J., van Munster, J.M., and Rawlings, D.E. (2008b). Construction of arsB and tetH mutants of the sulfur-oxidizing bacterium *Acidithiobacillus caldus* by marker exchange. Appl. Environ. Microbiol. *74*, 5686–5694.

Metagenomics of Acid Mine Drainage at Iron Mountain California: Expanding Our View From Individual Genes and Cultures to Entire Communities[1]

13

Brett J. Baker and Jillian F. Banfield

Abstract

The microbial world is vast and complex, which makes addressing fundamental questions related to ecology and evolution challenging. Until recently, microbial ecology relied on cultivation-based methods, which can only target a tiny faction of the diversity of microorganisms present in the environment. Therefore, microbiologists have turned to new tools to interrogate natural communities. This revolution has been fuelled by new DNA sequencing technologies that probe community composition and metabolic potential (metagenomics), and linked 'omics' approaches, such as such as metaproteomics and metatranscriptomics that provide metrics for activity. Metagenomic and metaproteomics methods were first applied to study natural microbial communities at an acid mine drainage site (Richmond Mine at Iron Mountain, California). Research conducted there demonstrated that it is possible to obtain complete genomes representing natural populations and to track their metabolic activity *in situ*. As a result dozens of novel genomes (down to the strain level) of uncultured acidophiles are now available. These genomes have provided a robust framework upon which it has been possible to address fundamental questions related to ecological processes such as colonization and to document the modes and rate of genome evolution.

Introduction

For over a hundred years, since Sergei Winogradsky began culturing microorganism from the environment (Dworkin, 2011), microbial ecology had been reliant on getting community members to grow in the laboratory. The development of PCR-based (16S rRNA gene) sequencing surveys in the late 1980s by Norman Pace and others (Pace *et al.*, 1986) resulted in a paradigm shift in microbiology. This led to realization that the vast majority of microorganisms in nature are unknown to us as they have not been cultured (Hugenholtz *et al.*, 1998). Also revolutionary was the innovation of taxonomic-specific fluorescent labels for microscopic visualization and identification of cells, now commonly referred to as FISH (fluorescence *in situ* hybridization) (DeLong *et al.*, 1989). The era of whole genome sequencing, launched in the 1990s through application of Sanger sequencing to pure cultures of single organisms, provided another approach for characterization of the microbiology of natural environments (Fleischmann *et al.*, 1995). This approach provided new insights into the capacities and evolutionary history of many organisms, but it was inherently restricted to the small subsets of organisms for which isolates existed.

Although certain organisms prominent in low-pH environments have been long studied in pure culture (Norris *et al.*, 1980, 1986; Johnson, 1998) and isolates were available for sequencing, it took many years before their genome sequences began to appear. The first isolate genomes were completed in 2000 and 2001 belonging to *Thermoplasma acidophilum* (Ruepp *et al.*, 2000), *Sulfolobus solfataricus* (She *et al.*, 2001), and *Acidithiobacillus ferrooxidans* (Selkov *et al.*, 2000). First published

1 We dedicate this chapter to the late Professor Katrina Edwards. Katrina was the first graduate student in our research group to work on the geomicrobiology of the Richmond Mine system. Through her pioneering work, she established a foundation for understanding of the interactions between microorganisms and metal sulfide minerals in acid mine drainage environments.

was a draft sequence for '*Ferroplasma acidarmanus*', isolated from the Richmond Mine (Allen *et al.*, 2007). The sequence for a *Sulfobacillus* sp. isolate did not appear until 2011 (Li *et al.*, 2011), and a *Leptospirillum ferrooxidans* in 2012 (Fuijmura *et al.*, 2012), and *At. caldus* in 2009 and 2013 (Acuña *et al.*, 2013; Valdés *et al.*, 2009).

Perhaps ironically, most of the genome sequences for acid mine drainage acidophiles were generated in another way. The initial attempt to extend the isolate sequencing method to whole community genomic DNA involved DNA extracted from a biofilm sampled within the Richmond Mine acid mine drainage system at Iron Mountain (near Redding) in northern California. This approach simultaneously recovered five partial and relatively complete genomes from bacteria and archaea (Tyson *et al.*, 2004). In recent years innovations in high-throughput sequencing technologies (e.g. 454, Illumina) have resulted in a dramatic decrease in cost and increase in the number of DNA reads that can be generated from any sample. The power of this approach has led to an explosion in the number of researchers using it, and its application across a wide variety of environment types of highly varying complexity (e.g. Baker *et al.*, 2012; Wrighton *et al.*, 2012).

Almost two decades of microbial research has been conducted at sites within the Richmond Mine. Here and elsewhere, AMD is generated by the dissolution of pyrite (FeS_2) and other sulfide minerals, a process that results in formation of acidic, iron sulfate solutions that are enriched in a variety of toxic elements (see Chapter 1). Ferric iron increases the rate of pyrite dissolution relative to the rate at which dissolution occurs following exposure to O_2. Thus, microorganisms that oxidize ferrous iron can substantially increase the rates at which sulfide minerals dissolve. Some organisms can derive energy from oxidation of iron and sulfur; these are the primary producers (autotrophs) in these subsurface communities. As it occurs within the Richmond Mine, the AMD formation process is relatively unusual in that the deposit is highly enriched in pyrite and the surrounding felsic (silica-rich) rocks have a limited buffering capability. This, in combination with evaporative concentration of AMD pools, has yielded the lowest pH values that have been documented in nature (pH −3.6; Nordstrom and Alpers, 1999).

Early microbiological studies of the Richmond Mine relied upon single-gene (16S rRNA gene) surveys (Goebel and Stackebrandt, 2001; Bond *et al.*, 2000; Edwards *et al.*, 1999) and microscopic visualization of cells labelled with FISH probes (Bond *et al.*, 2001). Prior to employing these culture-independent approaches to characterize AMD microbial communities it was widely thought that *At. ferrooxidans*, which had been cultured, were the dominant iron-oxidizers in AMD systems, though there was some evidence to the contrary (e.g. Walton and Johnson, 1992). This is often true in communities present at higher pH solutions (pH > 2.0), although *Leptospirillum* spp. are also widely recognized in these environments (Norris and Kelly, 1980; Lopez-Archilla *et al.*, 2001). However, Richmond Mine tunnels provided the ability to sample biofilms present in the subsurface directly on the pyrite surfaces and in, and on the surfaces of pools supported by pyrite accumulations. By characterizing the communities in close association with the sulfide ore deposit and not relying on culturing, it was determined that *Leptospirillum* and not *Acidithiobacillus* is the dominant iron-oxidizer in this system (Schrenk *et al.*, 1998). Soon after, a new iron-oxidizing species of Archaea, '*Fp. acidarmanus*', was found to constitute up to 85% of certain biofilm communities at this site (Edwards *et al.*, 2000).

Despite AMD microbial communities having considerably less diversity than is characteristic of most natural ecosystems, it has become apparent that numerous lineages have evolved species capable of growth in this hot, metal-rich acidic environment (Bond *et al.*, 2000). In addition to *Ferroplasma* spp., a variety of other *Thermoplasmatales*-order *Archaea* were identified in the biofilms. These have been referred to as A- through I-plasma (Druschel *et al.*, 2004). In addition to *L. ferrooxidans* and other closely related (group I) leptospirilli bacteria, which have been extensively described in other AMD systems, two other groups (II and III) of *Leptospirillum* are abundant in Richmond Mine biofilm communities (Druschel *et al.*, 2004; Bond *et al.*, 2000). These bacteria have been detected elsewhere, including other AMD-impacted environments (for example Rio Tinto in Spain). A distinct *Leptospirillum* lineage referred to as *L.* group IV was described more recently (Aliga-Goltsman *et al.*, 2013). A variety of other bacterial lineages within *Actinobacteria, Firmicutes* (namely *Sulfobacillus* spp.), and *Proteobacteria* have been detected in 16S rRNA gene libraries. Notable was the discovery of novel, deep-branching archaea that were originally referred to as 'ARMAN' (Baker *et al.*, 2006, detailed below). Also present in the environment are eukaryotes (Aliaga-Goltsman *et al.*, 2014), both fungi and protists (Baker *et al.*, 2009). However, individual communities tend to have low species richness and in many, there is dominance by a single population from each lineage. This pattern has been attributed to the limited number of energy-deriving reactions present in this system, and the extreme acidity (Baker and Banfield, 2003). These ecosystem features led us to propose, over 10 years ago, that AMD communities might be the ideal system to employ and further develop novel whole-community genomic approaches.

> The low species richness of AMD communities may permit us to largely reconstruct genomes of populations from whole community libraries... The remarkable simplicity of these habitats may permit a more fundamental understanding of how microbial communities work than is possible through study of more complex ecosystems.
>
> Baker and Banfield (2003)

Acid mine drainage at Iron Mountain, a model system for the development of culture-independent genomic (metagenomic) and functional approaches

One of the most important findings from 16S rRNA gene sequence surveys was the discovery of dozens of previously unknown lineages of microorganisms (e.g. Hugenholtz *et al.*, 1998). These included newly detected phyla, also referred to as candidate phyla. In some cases, members of groups for which there are no cultured representatives are the most abundant species in their environments. This was the case for certain new groups of archaea detected in AMD biofilms (Druschel *et al.*, 2004). Several novel approaches were developed to understand the metabolic functions of these uncultured groups, for example isotopic DNA labelling (DNA-SIP) (Dunford and Neufield, 2010). However, these approaches are relatively difficult and cannot be scaled to high-throughput to screen a variety of functions with a whole community.

In the late 1990s as these previously unknown lineage of microorganisms were being discovered, Ed DeLong's laboratory was developing a new approach involving large-insert DNA cloning, which enabled the recovery of dozen of genes from uncultured cells (Stein *et al.*, 1996). Around the same time, Jo Handelsman's laboratory was also developing a similar approach for screening of novel forms of antibiotics soil communities, and they coined the term 'metagenomics' (Handelsman *et al.*, 1998). Subsequently, this term was adopted as a descriptor for all cultivation-independent sequencing approaches (although the application of the term to 16S rRNA surveys is controversial and probably best avoided). These early whole genome-sequencing approaches, however, required the screening of the clone libraries for phylogenetic markers and other genes of interest. Therefore, only those random fragments in the library that contain these genes were analysed, and it was impossible to reconstruct near-complete or complete genomes, in a high-throughput manner.

A different approach was developed for reconstruction of near-complete genomes from uncultivated organisms (Tyson *et al.*, 2004). This involved generation of 'shotgun' (or random) genomic libraries from DNA extracted from biofilms that were sampled from the Iron Mountain site. The DNA was sheared into small fragments, and then individual fragments were cloned and sequenced from both ends using Sanger technology. The idea was to develop a high-throughput, whole-community approach, leveraging the fact that the AMD system is a less complex (less species rich) environment than ocean water or soils. A total of 76 million DNA base pairs (after trimming the reads for quality) were sequenced and assembled in this first study, which resulted in the near-complete reconstruction of genomes for several abundant AMD community members. Notably, this included genomes for the first organisms from the *Nitrospirae* phylum (Tyson *et al.*, 2004). Analyses of this initial metagenomic dataset inspired the development of new computational applications for dealing with natural microbial communities in more complex systems (e.g. Wrighton *et al.*, 2014; Castelle *et al.*, 2013).

All of the computational tools for genomic analyses were originally developed for reconstructing individual genomes of isolated microorganisms (i.e. the datasets were clonal). They were not designed reconstruct multiple genomes from multiple populations. Natural microbial communities are composed of mixture of different species present and within each of the species there are multiple variants or strains (Simmons *et al.*, 2008; Allen *et al.*, 2007; Tyson *et al.*, 2004). These species and closely related strain variants are all present at different abundance levels. The abundance of each organism type is reflected in the number of reads that derive from those community members in the genomic libraries. Because this representation is often calculated by mapping reads to genomes, the abundance metric is referred to as coverage.

Genomic assembly is the reconstruction of individual sequencing reads into longer fragments (referred to as 'contigs'), and ultimately, into near-complete or complete genomes (sometimes possible via careful manual assembly curation). Assembly of the data is crucial because it provides better identification of gene functions (annotation), gene order (operon structure), and genomic context than can be made based on individual reads (Fig. 13.1). Being able to identify all the genes present in the uncultured community members enables the prediction of metabolic pathways, and their potential ecological roles and interactions. The assembly process can generate composite genomes that reflect the mixture of sequences of the strain variants (Eppley *et al.*, 2007). These provide information for evolutionary analyses.

The larger contigs often contain phylogenetic markers for taxonomic identification, and they can often be linked to other contigs by identifying pairs of sequencing reads that derived from a single DNA

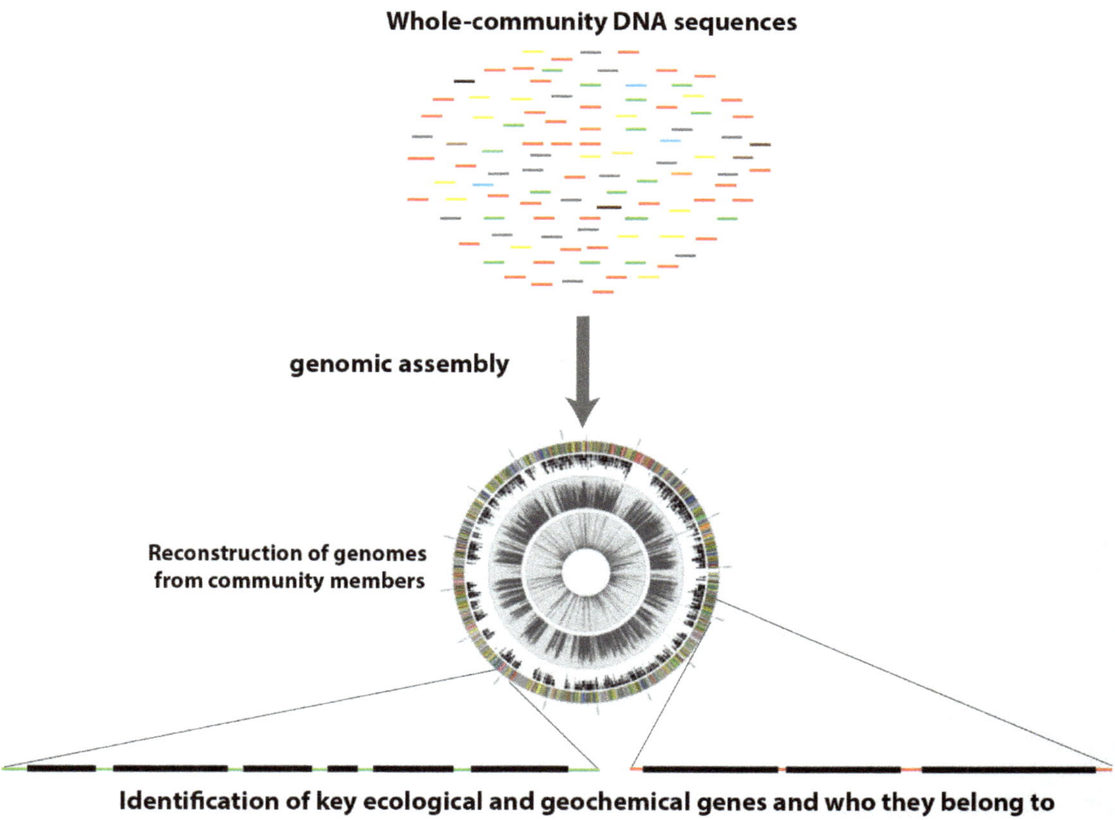

Figure 13.1 Depiction of the assembly of a genomic dataset to generate the genome for one community member from a shotgun metagenomic library. Notice that by assembling the genome it is possible to identify which member of the community have which genes, and groups of genes (e.g. operon structure). This is not possible by analysing unassembled reads.

fragment ('mate-pairs') yet are placed at the ends of different contigs. It is often possible to join contigs based on a mate-paired reads, inserting Ns in the unsequenced region of the original DNA fragment (this is called scaffolding, and the resulting sequences are termed scaffolds). For well-sampled genomes, many scaffolding gaps can be closed through manual curation).

Most of the contigs present in genomic assemblies are short (< 20,000 nucleotides), and do not contain phylogenetic markers needed for taxonomic identification. The process of determining which genome these contigs belong to, regardless of the presence/absence of phylogenetically informative genes, is called 'binning'. The first attempts at binning involved clustering based on the combination of GC content (%GC) and coverage of the scaffolds (Tyson *et al.*, 2004). However, the resolution of GC/coverage binning is limited because many organisms can have genomes with similar GC content and are present at similar abundance levels. This motivated us to develop other methods. The basic requirement is a characteristic of all assembled genome fragments that is common to fragments of the same genome and different for different genomes. GC content is such a characteristic, but more dimensions of resolution are needed. Codon use is more effective tool, but better still is the tetranucleotide signatures of the DNA (Teeling *et al.*, 2004). Using a sliding window and counting the number of all of the possible tetramers (256, but only 136 are unique four-nucleotide sequences) across the contig sequences provides a fingerprint that is often effective for contigs larger than a few thousand base pairs. These signatures can be effectively clustered and visualized using emergent self-organizing maps (tetra-ESOM; Dick *et al.*, 2010). Sequencing of samples from multiple sites within the Richmond Mine and tetra-ESOM binning enabled us to reconstruct genomes of nearly all of the lineages that had been identified using 16S rRNA gene surveys, several novel *Thermoplasmatales* archaea (Yelton *et al.*, 2013), *Ferroplasma* strain variants (Allen *et al.*, 2007; Eppley *et al.*, 2007), *Actinobacteria* (unpublished), *Leptospirillum* groups (II (Tyson *et al.*, 2014; Denef *et al.*, 2012; Aliaga-Goltsman *et al.*, 2009), III (Aliaga-Goltsman *et al.*, 2009; Tyson *et al.*, 2005) and IV (Aliaga-Goltsman *et al.*, 2013), and *Sulfobacillus* spp. (Justice *et al.*, 2014).

One of the surprisingly findings from whole-community genome sequencing was the discovery of lineages that had been overlooked in the PCR-based gene surveys. From the first metagenomic study new groups of archaea were detected that commonly used PCR primers did not amplify; these were named ARMAN (Acidophilic Richmond Mine Archaeal Nanoorganisms) groups (I and 2) (Baker *et al.*, 2006). Microscopic visualization of these groups in the biofilms revealed that they are considerably smaller than other community members, in fact that close to the predicted lower limit of cell size (Comolli *et al.*, 2009). This guided a new metagenomic approach in which filtration was used to enrich for these tiny cells prior to sequencing. This enabled the first near-complete sampling of the genomes of these organisms (ARMAN groups 4 and 5; Baker *et al.*, 2010). The ARMAN groups have since been identified in AMD sites around the world (Ziegler *et al.*, 2013; Bohorquez *et al.*, 2012; Amaral-Zettler *et al.*, 2011; Ziegler *et al.*, 2009; Juottonen *et al.*, 2008; Sunna and Bergquist, 2003) and have been detected in other environments (Murakami *et al.*, 2011; Juottonen *et al.*, 2008). These organisms have been implicated in carbon cycling, specifically in the degradation of dying (sunken) biofilms (Justice *et al.*, 2012).

Metagenomics enables the identification of physiological traits that can guide cultivation attempts. One of the first genomes to be reconstructed from the first metagenomic assemblies belonged to an uncultured group of *Leptospirillum* (group III) (Tyson *et al.*, 2004). These bacteria had first been detected in Richmond Mine biofilms (Bond *et al.*, 2000). Based on the observation that this was the only organism (known at the time) whose genome encoded genes for nitrogen fixation, a cultivation medium was designed that resulted in isolation (Tyson *et al.*, 2005). Subsequently, partial genomic information for another *Leptospirillum* was obtained that is present at much lower abundance levels in some biofilms and that also can fix N_2. This organism, referred to as *Leptospirillum* Group IV, has not yet been isolated.

In addition to providing clues to culturing new microbial groups, metagenomics is powerful for a variety of community-level genomic analyses. Specifically, it enables the reconstruction of metabolic pathways of different species, thus ecological roles can be partitioned. Furthermore, it is possible to characterize fine-scale genomic variability in natural microbial populations and to determine how these evolutionary processes are important in community functioning (discussed in detail below) (Denef *et al.*, 2009; Lo *et al.*, 2007). A key finding from such analyses, described in more detail below, was the importance of homologous recombination in shaping the genomes of both bacteria and archaea (Denef *et al.*, 2009; 2012; Eppley *et al.*, 2007; Tyson *et al.*, 2004). Another significant result was the reconstruction of genomes of viruses of archaea and bacteria (phage). Because these entities were sampled in community context, it was possible to link them to their hosts by using the matches between their sequences and the spacer sequence in the CRISPR locus on their CRISPR-Cas systems (Anderson and Banfield, 2008). This approach is generally applicable in metagenomic studies, and represents an important component because viruses/phage exert critical controls on carbon turnover (via cell lysis) and transfer of genetic material.

Harnessing proteomics to interrogate the activity of microbial populations in the wild

> Related gene expression studies should enable development of AMD ecosystem models that resolve the contributions of individual species at the molecular (pathway) level, describe and quantify flows of materials and energy...
>
> Baker and Banfield (2003)

Cataloguing metabolic potential through metagenomics has become commonplace for microbial ecologists working on a variety of habitats. However, this approach only provides characterization of the community at the DNA level, and does not measure the organism investments in specific functions in the environment. Furthermore, it is impossible to know if predicted genes actually encode for proteins (hypothetical genes). A particular metabolic process may be associated with several organisms in the community, and it can be difficult to determine which of these organisms are most active and ecologically important. To address the disconnect between gene prediction and protein presence, tandem mass-spectrometry-based proteomic methods were employed to evaluate the protein inventories of organisms in Richmond Mine microbial communities (Ram *et al.*, 2005). This was the first proteomic characterization of natural microbial community. The study revealed that many proteins of unknown function were among the most abundant, suggesting that these are important. Notable was a highly expressed operon of proteins of unknown function (Ram *et al.*, 2005); this is now identified as the Cas operon that is central to the CRISPR-Cas microbial immune system (Marraffini *et al.*, 2008; Barrangou *et al.*, 2007). Also, by identifying proteins derived from hypothetical genes made it possible to re-annotate them as proteins of unknown function.

Being able to analyse AMD microbial communities at the genomic- and proteomic-levels proved to be extremely powerful, and several key processes were identified. Perhaps most notably, several new proteins involved in iron oxidation were identified, including a novel c-type cytochrome (cyt579) (Ram *et al.*, 2005). Given that iron oxidation is the primary process that links microbial activity to AMD generation, this finding is central to understanding of this important environmental problem (and it is foundational to understanding of the ecology of acidic environments where *Leptospirillum* spp. dominate).

A community succession occurs during the development of biofilms that float on the surfaces of AMD pools and stream margins (Wilmes *et al.*, 2009). Early colonizers consist of autotrophic iron-oxidizers *Leptospirillum* group II, with *Leptospirillum* group III (nitrogen fixers) at lower abundance. As the biofilms age and thicken, a diverse collection of Archaea and eukaryotes appear and the *Leptospirillum* group II genotype shifts. Characterization of the proteomes of microorganisms in communities from several sites in the Richmond Mine made it possible to evaluate metabolic activity as a function of the variety of biofilm types. In one study (Mueller *et al.*, 2010), the proteomes of 28 different communities were analysed and changes in physiologies of the organisms were correlated to changes in geochemistry and biofilm development. This study established an effective strategy to understand the relationship between ecology and microbial metabolism of a whole community in nature. A more in depth comparison of proteins from biofilms at different stages of development revealed that early colonizing organisms have responses to abiotic stresses to growth in the AMD solutions (Mueller *et al.*, 2011). Also, organisms in early stages biofilms were more invested in proteins involved in cell division, protein synthesis, and utilization of 1- and 2-carbon compounds. Then, as the community matures, resources are in shorter supply and SOS responses inhibit cell division in *Leptospirillum* group II. In the more mature biofilms there is repression of translation and an increase in synthesis of complex carbohydrates and amino acids.

Strain-level variability of natural populations, linking ecology and evolution through space and time

The flow of ecological and evolutionary processes are intertwined, as one is constantly affecting the other. The short generation time of microorganisms makes them ideal for studying this interplay, as during relatively short time periods new strains (sub-species) appear within natural communities. Until recently, the understanding of these evolutionary processes was mostly limited to cultivation-based studies. This is one of the truly powerful advantages of metagenomic datasets, they enable the analysis of within population heterogeneity, genome-wide, and for many coexisting organisms (Eppley *et al.*, 2007; Tyson *et al.*, 2005).

Inspection of the first metagenomic AMD dataset revealed extensive fine-scale genomic variability within natural populations. This variability exists in part because closely related individuals within the community are distinguished by single nucleotide polymorphisms (SNPs) (or single base pair changes) in their sequences. In addition, genomic islands, often associated with phage or plasmid insertions, distinguish otherwise genomically very similar individuals (Allen *et al.*, 2007). Remarkably, however, large blocks (tens to hundreds of kilobases) of DNA are also being exchanged between closely related *Leptospirillum* populations, a process that is described as homologous recombination (Simmons *et al.*, 2008; Lo *et al.*, 2007). An indication that these exchanges occurred relatively recently was the observation that the donor and recipient genotypes were both present in the Richmond Mine communities. By studying, in detail, the genomic changes in *Leptospirillum* (group II) in samples collected at various sites in the mine over a period of 9 years it was possible to show that these evolutionary changes occurred over the past few decades (Denef and Banfield, 2012). These are relatively recent genetic exchanges that appear to have conferred a strong selective advantage (in one case, the region encodes the CRISPR-Cas locus central to defence against phage), as a sweep (population bottleneck) appears to have reduced to low abundance or eliminated the pre-existing genotypes from that site. The proliferation of certain variants in the face of changes in the environment is the first step of speciation, and is the fundamental link between ecology and evolution. Analyses of proteomic data from Iron Mountain led to the realization that this variability can be detect in proteins of the communities as well (Lo *et al.*, 2007; Ram *et al.*, 2005). Detailed examination of the gene expression of these recombinant blocks in *Leptospirillum* group II revealed that genes involved in motility, signal transduction, and transport were more highly represented than housekeeping genes (Denef *et al.*, 2009).

Using 'omic' approaches, it is possible to track evolutionary processes in natural communities that result in the ability of species to occupy distinct ecological niches (Wilmes *et al.*, 2010). As noted above, biofilms thicken with age (Wilmes *et al.*, 2009). By examining the proteomes of genotypic variants using proteomics applied to a variety of communities from sites with different physical and geochemical attributes, it was possible to detect differences in physiology as a function of environment type (Denef *et al.*, 2010). It was estimated that roughly 15% of the genes in the dominant bacteria were involved in niche differentiation between early- and late-stage communities. Interestingly, the activity of lower abundance community members (namely *Archaea* and *Leptospirillum* group III) was restricted by environment conditions, mainly pH, temperature, and ionic strength of the water (Mueller *et al.*, 2010). However, higher abundance populations were not effected by similar physical conditions in the ecosystem. In fact, shifts in physiologies of the abundant community members depended on the activity of these lower abundance populations, thus confirming that intra-species interactions are extremely important in overall community functioning. The ability to track

Table 13.1 List of genomes obtained from the Richmond Mine at Iron Mountain, CA

Organism	Lineage	strain	Size	No. of contigs	No. of genes	Reference
Bacteria						
Leptospirillum ferriphilum Group II – 5way-CG	*Nitrospira*	CGL2	2.63 Mbp	103	2670	Tyson *et al.* (2014)
Leptospirillum rubarum Group II	*Nitrospira*	UBAL2	2.66 Mbp	10	2629	Aliaga-Goltsman *et al.* (2009)
Leptospirillum sp. Group II – C75 Meter	*Nitrospira*	C75L2	2.59 Mbp	86	2532	Denef *et al.* (2012)
Leptospirillum ferrodiazotrophum Group III	*Nitrospira*	UBAL3	2.8 Mbp	43	2654	Tyson *et al.* (2005)
Leptospirillum sp. Group IV	*Nitrospira*	UBABSL4	1.52 Mbp	296	1936	Aliaga-Goltsman *et al.* (2013)
Sulfobacillus acidophilus IM3	*Firmicutes*	AMDSBA3	3.68 Mbp	128	3650	Justice *et al.* (2014)
Sulfobacillus benefaciens IM1	*Firmicutes*	AMDSBA1	4.56 Mbp	155	4629	Justice *et al.* (2014)
Sulfobacillus benefaciens IM4	*Firmicutes*	AMDSBA4	4.11 Mbp	179	4254	Justice *et al.* (2014)
Sulfobacillus sp. IM6	*Firmicutes*	AMDSBA2	3.07 Mbp	417	3296	Justice *et al.* (2014)
Sulfobacillus thermosulfidooxidans IM5	*Firmicutes*	AMDSBA5	3.65 Mbp	161	3871	Justice *et al.* (2014)
Sulfobacillus Unknown Bin	*Firmicutes*	AMDSBAU	4.65 Mbp	773	5832	Justice *et al.* (2014)
Sulfobacillus sp. type 1	*Firmicutes*	FWFSBA1	n.a.	216	2715	Unpublished
Sulfobacillus sp. type 3	*Firmicutes*	FWFSBA3	n.a.	91	2790	Unpublished
Unassigned firmicute	*Firmicutes*	UNLFIR	n.a.	544	1409	Unpublished
Actinobacterium	*Actinobacteria*	UNLACT1	n.a.	719	2590	Unpublished
Actinobacterium	*Actinobacteria*	UNLACT2	n.a.	711	1770	Unpublished
Archaea						
G-plasma	*Thermoplasmatales*	UNLGPL	1.83 Mbp	21	1923	Yelton *et al.* (2011)
I-plasma	*Thermoplasmatales*	UNLIPL	1.70 Mbp	8	1711	Yelton *et al.* (2011)
A-plasma	*Thermoplasmatales*	S4JAPL	1.99 Mbp	118	2280	Yelton *et al.* (2011)
A-plasma	*Thermoplasmatales*	S4JAP2	n.a.	169	597	Unpublished
C-plasma	*Thermoplasmatales*	S4JCPL	1.58 Mbp	125	1720	Yelton *et al.* (2013)
D-plasma	*Thermoplasmatales*	FWFDPL	n.a.	112	1812	Yelton *et al.* (2013)
E-plasma	*Thermoplasmatales*	UNLEPL	1.66 Mbp	8	1676	Yelton *et al.* (2011)
Ferroplasma acidarmanus	*Thermoplasmatales*	ISOLFER1	1.94 Mbp	1	1955	Allen *et al.* (2007)
Ferroplasma acidarmanus Group I	*Thermoplasmatales*	CGFER1	n.a.	1	1742	Allen *et al.* (2007)
Ferroplasma acidarmanus Group 2	*Thermoplasmatales*	CGFER2	2.49 Mbp	177	2736	Eppley *et al.* (2007)
ARMAN-1	Micrarchaeota	S4JAR1	867.98 kbp	66	998	Unpublished
Micrarchaeum acidiphilum ARMAN-2	Micrarchaeota	UNLAR2	1.01 Mbp	1	1006	Baker *et al.* (2010)
Parvarchaeum acidiphilum ARMAN-4	Parvarchaeota	UNLAR4	800 kbp	44	935	Baker *et al.* (2010)
Parvarchaeum acidophilus ARMAN-5	Parvarchaeota	UNLAR5	921 kbp	73	936	Baker *et al.* (2010)
Viruses						
Unassigned Virus	Virus	UNLVIR	n.a.	52	180	Unpublished

Table 13.1 Continued

Organism	Lineage	strain	Size	No. of contigs	No. of genes	Reference
VirusAMDV	Virus	AMDVIR	n.a.	102	518	Andersson and Banfield (2008)
VirusAMDV1	Virus	AMDV1	n.a.	10	50	Andersson and Banfield (2008)
VirusAMDV2	Virus	AMDV2	14 kbp	1	6	Andersson and Banfield (2008)
VirusAMDV2s	Virus	AMDV2s	n.a.	16	24	Andersson and Banfield (2008)
VirusAMDV3	Virus	AMDV3	n.a.	47	227	Andersson and Banfield (2008)
VirusAMDV4	Virus	AMDV4	n.a.	22	153	Andersson and Banfield (2008)
VirusAMDV5	Virus	AMDV5	n.a.	2	12	Andersson and Banfield (2008)

variations in genomic and proteomic responses among multiple coexisting microorganisms in nature is unprecedented, and was possible in this AMD system because samples were collected and genomically characterized over a large time span.

Conclusions: Iron Mountain, a reflection of nearly two decades of work on fascinating microbial world

Prior to the development and implementation of metagenomics and proteomics to study microbial biofilm communities that grow in the Richmond Mine AMD environment, understanding the ecology of low-pH systems was restricted to single molecule and cultivation-based approaches. A massive sequencing effort from 2001 to 2011, advances in computational approaches (especially binning), and significant investments in manual genome curation resulted in the recovery of genomes for all the common organisms in the Richmond Mine biofilms, *L. ferriphilum*, '*L. rubarum*', *L. ferrodiazotrophum*, actinobacteria, *Ferroplasma* (types I and II), A, C, D, E, G, I-plasmas, ARMAN groups 2, 4 and 5 (Dick *et al.*, 2009) (Table 13.1). These genomic datasets are now available via, http://ggkbase.berkeley.edu/amd/organisms. Having genomes for nearly all the populations throughout this ecosystem, and the ability to track proteomic profiles, and even link to metabolites (Fischer *et al.*, 2012) *in situ*, enabled fundamental biological questions to be addressed. The comprehensive catalogue of metabolic pathways within the community provides a map of how niches are partitioned in the mine. For example, among the dominant populations at Iron Mountain, the ability to oxidize iron is limited to *Leptospirillum* and *Ferroplasma* spp. (Denef *et al.*, 2010) and *Archaea* species are primarily involved in organic carbon degradation of sunken biofilms (Justice *et al.*, 2012).

For many years, it remained uncertain whether the genome-resolved metagenomic, (meta)proteomic and (meta)metabolomic methods developed to study of the Richmond Mine biofilms could be scaled to achieve comparable results in more complex systems. It is now clear, however, that this is indeed the case. Thus, as the current era of Richmond Mine studies draws to a close, it can be concluded that the main accomplishments (from the effort of numerous researchers) were the recognition of the multiple values afforded by shotgun sequence information (from population biology to extremophile physiology and phage-host dynamics) and innumerable methods that are now finding application across a wide variety of environment types. The 'omic'-based approach of studying Iron Mountain has laid the foundation by which other microbial communities can be studied.

Acknowledgements

We would also like to thank the many people who worked on the Iron Mountain AMD project over the years including; Eric Allen, Daniela Aliaga Goltsman, Anders Andersson, Chris Belnap, Luis Comolli, Vincent Denef, John Eppley, Gregory Dick, Genevieve DiBartolo, Greg Druschel, Katrina Edwards, Judith Flanagan, Tom Gihring, Brett Goebel, Lauren Hauser, Robert Hettich, Phil Hugenholtz, Nicholas Justice, Linda Kalnejais, Ian Lo, Annika Mosier, Ryan Mueller, Rachna Ram, Matt Schrenk, Chris Miller, Chongle Pan, Christine Sun, Sheri Simmons, Steve Singer, Susan Spaulding, Brian Thomas, Michael Thelen, Robert Hettich, Phil Bond, Steve Smriga, Gene Tyson, Nathan VerBerkmoes, Richard Webb, Paul Wilmes, Rachel Whitaker, and Alexis Yelton. We also thank the US Department of Energy, Office of Biological and Environmental Research, Office of Science for the majority of funding that supported this work.

References

Acuña, L.G., Cardenas, J.P., Covarrubias, P.C., Haristoy, J.J., Flores, R., Nuñez, H., Riadi, G., Shmaryahu, A., Valdes, J., Dopson, M., *et al.* (2013). Architecture and Gene Repertoire of the Flexible Genome of the Extreme Acidophile *Acidithiobacillus caldus*. PLoS One *8*, e78237.

Allen, E.E., Tyson, G.W., Whitaker, R.J., Detter, J.C., Richardson, P.M., and Banfield, J.F. (2007). Genome dynamics in a natural archaeal population. Proc. Natl. Acad. Sci. U.S.A. *104*, 1883–1888.

Aliaga Goltsman, D.S., Denef, V.J., Singer, S.W., VerBerkmoes, N.C., Lefsrud, M., Mueller, R., Dick, G.J., Sun, C., Wheeler, K., Zemla, A., *et al.* (2009). Community genomic and proteomic analysis of chemoautotrophic, iron-oxidizing "*Leptospirillum rubarum*" (Group II) and *Leptospirillum ferrodiazotrophum* (Group III). Appl. Environ. Microbiol. *75*, 4599–4615.

Aliaga Goltsman, D.S., Dasari, M., Thomas, B.C., Shah, M.B., VerBerkmoes, N.C., Hettich, R.L., and Banfield, J.F. (2013). A new group in the *Leptospirillum* clade, cultivation-independent community genomics, proteomics and transcriptomics of the new species *Leptospirillum* Group IV UBA BS. Appl. Environ. Microbiol. *79*, 5384–5393.

Aliaga Goltsman, D.S., Comolli, L., Thomas, B.C., and Banfield, J.F. (2015). Community transcriptomics reveals unexpected high microbial diversity in acidophilic biofilm communities. ISME J. *9*, 1014–1023.

Amaral-Zettler, L.A., Zettler, E.R., Theroux, S.M., Palacios, C., Aguilera, A., and Amils, R. (2011). Microbial community structure across the tree of life in the extreme Río Tinto. ISME J. *5*, 42–50.

Andersson, A., and Banfield, J.F. (2008). Virus population dynamics and acquired virus resistance in natural microbial communities. Science *230*, 1047–1050.

Barrangou, R., Fremaux, C., Devaeu, H., Richards, M., Boyaval, P., Moineau, S., Romero, D.A., and Horvath, P. (2007). CRISPR provides acquired resistance against viruses in prokaryotes. Science *315*, 1709–1712.

Baker, B.J., and Banfield, J.F. (2003). Microbial communities in acid mine drainage. FEMS Microbial. Ecol. *44*, 1–14.

Baker, B.J., Tyson, G.W., Webb, R.I., Flanagan, J., Hugenholtz, P., Allen, P., Allen, E.E., and Banfield, J.F., (2006). Lineages of acidophilic archaea revealed by community genomic analysis. Science *314*, 1933–1935.

Baker, B.J., Tyson, G.W., Goosherst, L., and Banfield, J.F. (2009). Insights into the diversity eukaryotes in acid mine drainage biofilm communities. Appl. Environ. Microbiol. *75*, 2192–2199.

Baker, B.J., Comolli, L., Dick, G., Hauser, L., Hyatt, D., Dill, B.D., Land, M., VerBerkmoes, N.C., Hettich, R.L., and Banfield, J.F. (2010). Enigmatic, ultra-small uncultivated Archaea, Proc. Natl. Acad. Sci. U.S.A. *107*, 8806 – 8811.

Baker, B.J., Lesniewski, R.A., and Dick, G.J. (2012). Genome-enabled transcriptomics reveals archaeal populations that drive nitrification in a deep-sea hydrothermal plume. ISME J. *6*, 2269–2279.

Bohorquez, L.C., Delgado-Serrano, L., López, G., Osorio-Forero, C., Klepac-Ceraj, V., Kolter, R., Junca, H., Baena, S., and Zambrano, M.M. (2012). In-depth characterization via complementing culture-independent approaches of the microbial community in an acidic hot spring of the Colombian Andes. Microb. Ecol. *63*, 103–115.

Bond, P.L., and Banfield, J.F. (2001). Design and performance of rRNA targeted oligonucleotide probes for in situ detection and phylogenetic identification of microorganisms inhabiting acid mine drainage environments. Microbial. Ecol. *41*, 149–161.

Bond, P.L., Smriga, S.P., and Banfield, J.F. (2000). Phylogeny of microorganisms populating a thick, subaerial, predominantly lithotrophic biofilm at an extreme acid mine drainage site. Appl. Environ. Microbiol. *66*, 3842–3849.

Castelle, C.J., Hug, L.A., Wrighton, K.C., Thomas, B.C., Williams, K.H., Wu, D., Tringe, S.G., Singer, S.W., Eisen, J.A., and Banfield, J.F. (2013). Extraordinary phylogenetic diversity and metabolic versatility in aquifer sediment. Nat. Comm. *4*, 2120.

Comolli, L., Baker, B.J., Downing, K.H., Siergerist, C.E., and Banfield, J.F. (2009). Three-dimensional analysis of the structure and ecology of a novel, ultra-small archaeon. ISME J. *3*, 159–167.

DeLong, E.F., Wickham, G.S. , and Pace, N.R. . (1989). Phylogenetic stains, ribosomal RNA-based probes for the identification of single cells. Science *243*, 1360–1363.

Denef, V.J., Kalnejais, L.H., Mueller, R.S., Wilmes, P., Baker, B.J., Thomas, B.C., VerBerkmoes, N.C., Hettich, R.L., and Banfield, J.F. (2009). Proteogenomic basis for ecological divergence of closely related bacteria in natural acidophilic microbial communities. Proc. Natl. Acad. Sci. U.S.A. *107*, 2383–2390.

Denef, V.J., and Banfield, J.F. (2012). In situ evolutionary rate measurements show ecological success of recently emerged bacterial hybrids. Science *336*, 462–466.

Dick, G.J., Andersson, A., Baker, B.J., Simmons, S.S., Thomas, B.C., Yelton, A.P., and Banfield, J.F. (2009).

Community-wide analysis of microbial genome sequence signatures. Gen. Biol. *10*, R85.

Druschel, G.K., Baker, B.J., Gihring, T.H., and Banfield, J.F. (2004). Acid mine drainage biogeochemistry at Iron Mountain, California. Geochem. Trans. *5*, 13–32.

Dunford, E.A., and Neufield, J.D. (2010). DNA stable-isotope probing (DNA-SIP). J. Vis. Exp. *42*, e2027.

Dworkin, M. (2011). Sergei Winogradsky, a founder of modern microbiology and the first microbial ecologist. FEMS Micrbiol. Ecol. *36*, 364–379.

Edwards, K.J., Goebel, B.M., Rodgers, T.M., Shrenk, M.O., Gihring, T.M., Cardona, M.M., Hu, B., McGuire, M.M., Hamers, R.J., Pace, N.R., *et al.* (1999). Geomicrobiology of Pyrite (FeS2) Dissolution, Case study at Iron Mountain, Calfornia. Geomicrobiol. J. *16*, 155–179.

Edwards, K.J., Bond, P.L., Gihring, T.M., and Banfield, J.F. (2000). An archaeal iron-oxidizing extreme acidophile important in acid mine drainage. Science *287*, 1796–1799.

Eppley, J.M., Tyson, G.W., Getz, W.M., and Banfield, J.F. (2007). Genetic exchange across a species boundary in the archaeal genus *Ferroplasma*. Genetics *177*, 407–416.

Fischer, C.R., Wilmes, P., Bowen, B.P., Northen, T.R., and Banfield, J.F. (2012). Dueterium-exchange metabolimics identifies *N*-methyl *lyso* phophatidylethanolamines as abundant lipids in acidophilic mixed microbial communities. Metabolomics *8*, 566–578.

Fleischmann, R.D., Adam, M.D., White, O., Clayton, R.A., Kirkness, E.F., Kerlavage, A.R., Bult, C.J., Tomb, J.F., Dougherty, B.A., Merrick, J.M., *et al.* (1995). Whole-genome random sequence and assembly of *Haemophilus influenza* Rd. Science *269*, 496–512.

Fuijmura, R., Sato, Y., Nishizawa, T., Oshima, K., Kim, S.W., Hattori, M., Kamijo, T., and Ohta, H. (2012). Complete genome sequence of *Leptospirillum ferrooxidans* strain C2–3, isolated from a fresh volcanic ash deposit on the island of Miyake, Japan. J. Bacteriol. *194*, 4122–4123.

Goebel, B.M., and Stackebrandt, E. (1994). Cultural and phylogenetic analysis of mixed microbial populations found in natural and commercial bioleaching environments. Appl. Environ. Microbiol. *60*, 1614–1621.

Handelsman, J., Rondon, M.R., Brady, S.F., Clardy, J., and Goodman, R.M. (1998). Molecular biological access to the chemistry of unknown soil microbes, A new frontier for natural products. Chem. Biol. *5*, R245–R249.

Hugenholtz, P., Pitulle, C., Hershberger, K.L., and Pace, N.R. (1998). Novel division level bacterial diversity in a Yellowstone hot spring. J. Bacteriol. *180*, 366–376.

Johnson, G.B. (1998). Biodiversity and ecology of acidophilic microorganisms. FEMS Microbiol. Ecol. *27*, 307–317.

Juottonen, H., Tuittila, E.S., Juutinen, S., Fritze, H., and Yrjala, K. (2008). Seasonality of rDNA- and rRNA-derived archaeal communities and methanogenic potential in a boreal mire. ISME J. *2*, 1157–1168.

Justice, N.B., Pan, C., Mueller, R., Spaulding, S.E., Shah, V., Sun, C.L., Yelton, A.P., Miller, C.S., Thomas, B.C., Shah, M., *et al.* (2012). Heterotrophic archaea contribute to carbon cycling in low-pH, suboxic biofilm communities. Appl. Environ. Microbiol. *78*, 8321–8330.

Justice, N.B., Zhou, L., Wang, Y., Spaulding, S.E., Mosier, A.C., Hettich, R.L., Pan, C., and Banfield, J.F. (2014). ^{15}N- and ^{2}H proteomic stable isotope probing link nitrogen flow to archaeal heterotrophic activity. Environ. Microbiol. *16*, 3224–3237.

Li, B., Chen, Y., Liu, Q., Hu, S., and Chen, X. (2011). Complete genome analysis of *Sulfobacillus acidphilus* strain TPY, isolated from a hydrothermal vent in the Pacific Ocean. J. Bacteriol. *193*, 5555–5556.

Lo, I., Denef, V.D., VerBerkmoes N.C., Shah, M.B., Goltsman, D., DiBartolo, G., Tyson, G.W., Allen, E.E., Ram, R.J., Detter, J.C., *et al.* F. (2007). Strain-resolved community proteomics reveals recombining genomes of acidophilic bacteria. Nature *446*, 537–541.

Lopez-Archilla, A.I., Marin, I., and Amils, R. (2001). Microbial community composition and ecology of an acidic aquatic environment, the Tinto River, Spain. Microb. Ecol. *41*, 20–35.

Marraffini, L.A., and Sontheimer, E.J. (2008). CRISPR interference limits horizontal gene transfer in *Staphylococci* by targeting DNA. Science *322*, 1843–1845.

Mueller, R., Denef, V.J., Kalnejais, L., Suttle, B., Thomas, B.C., Wilmes, P., Smith, R., Nordstrom, D.K., McCleskey, B., Shah, M., *et al.* (2010). Ecological distribution and population physiology defined by proteomics in a natural microbial community. Mol. Syst. Biol. *6*, 4819–4828.

Mueller, R.S., Dill, B.D., Pan, C., Belnap, C.P., Thomas, B.C., VerBerkmoes, N.C., Hettich, R.L., and Banfield, J.F. (2011). Proteome changes in the initial bacterial colonist during ecological succession in an acid mine drainage biofilm community. Environ. Microbiol. *13*, 2279–2292.

Murakami, S., Fujishima, K., Tomita, M., and Kanai, A. (2011). Metatranscriptomic analysis of microbes in an ocean-frong deep subsurface hot spring reveals novel small RNAs and type-specific tRNA degradation. Appl. Environ. Microbiol. *78*, 1015–1022.

Nordstrom, D.K., and Alpers, C.N. (1999). Negative pH, efflorescent mineralogy, and consequences for environmental restoration at the Iron Mountain Superfund site, California. Proc. Natl. Acad. Sci. U.S.A. *96*, 3455–3462.

Norris, P.R., and Kelly, D.P. (1980). Dissolution of pyrite (FeS_2) by pure and mixed cultures of some acidophilic bacteria. FEMS Microbiol. Lett. *4*, 143–146.

Norris, P.R., Marsh, R.M., and Lindstrom, E.B. (1986). Growth of mesophilic and thermophilic acidophilic bacteria on sulfur and tetrathionate. Biotechnol. Bioeng. *8*, 318–329.

Pace, N.R., Stahl, D.A. , Lane, D.J., and Olsen, G.J. . (1986). The analysis of natural microbial populations by ribosomal RNA sequences. Adv. Microb. Ecol. *9*, 1–55.

Ram, R.J., VerBerkmoes, N., Thelen, M.P., Tyson, G.W., Baker, B.J., Shah, M., Blake, R.C. II, Hettich, R., and Banfield, J.F. (2005). Community proteomics of a natural microbial biofilm. Science *308*, 1915–1920.

Ruepp, A., Grami, W., Santos-Marinez, M.L., Koretke, K.K., Volker, C., Mewes, H.W., Frischman, D., Stocker, S., Lupas, A.N., and Baumeister, W. (2000). The genome sequence of the thermoacidophilic scavenger *Thermoplasma acidophilum*. Nature *407*, 508–513.

Schrenk, M.O., Edwards, K.J., Goodman, R.M., Hamers, R.J., and Banfield, J.F. (1998). Distribution of *Thiobacillus ferrooxidans* and *Leptospirillum ferrooxidans*, implications for generation of acid mine drainage. Science *279*, 1519–1522.

Selkov, E., Overbeek, R., Kogen, Y., Chu, L., Vonstein, V., Holmes, D., Silver, S., Haselkorn, R., and Fonstain, M. (2000). Functional analysis of gapped microbial genomes: amino acid metabolism of *Thiobacillus ferrooxidans*. Proc. Natl. Acad. Sci. U.S.A. *97*, 3509–3514.

She, Q., Singh, R.K., Confalonieri, F., Zivanovic, Y., Allard, G., Awayez, M.J., Chan-Weiher, C.C.-Y., Clausen, G., Curtis, B., De Moors, A.D., *et al.* (2001). The complete genome of the crenarchaeon *Sulfolobus solfoatarícus* P2. Proc. Natl. Acad. Sci. U.S.A. *98*, 7835–7840.

Simmons, S.L., DiBartolo, G., Denef, V.J., Aliaga Goltsman, D., Thelen, M.P., and Banfield, J.F. (2008). Population genomic analysis of strain variation in *Leptospirillum* Group II bacteria involved in acid mine drainage formation. PLoS Biol. *6*, e177, 1427–1442.

Stein, J.L., Marsh, T.L., Wu, K.Y., Shizuya, H., and DeLong, E.F. (1996). Characterization of uncultivated prokaryotes: isolation and analysis of a 40-kilobase-pair genome fragment from a planktonic marine archaeon. J. Bacteriol. *178*, 591–599.

Sunna, A., and Bergquist, P.L. (2003). A gene encoding a novel extremely thermostable 1,4-beta-xylanase isolated directly from an environmental DNA sample. Extremophiles *7*, 63–70.

Teeling, H., Waidmann, J., Lombardot, T., Bauer, M., and Glockner, F.O. (2004). TETRA, a web-service and a stand-alone program for the analysis and comparison of tetranucleotide usage patterns in DNA sequences. BMC Bioinformatics *5*, 163.

Tyson, G.W., Chapman, J., Hugenholtz, P., Allen, E.E., Ram, R.J., Richardson, P.M., Solovyev, V.V., Rubin, E.M., Rokhsar, D.S., and Banfield, J.F. (2004). Insights into community structure and metabolism by reconstruction of microbial genomes from the environment. Nature *428*, 37–43.

Tyson, G.W., Lo, I., Baker, B.J., Allen, E.E., Hugenholtz, P., and Banfield, J.F. (2005). Genome-directed isolation of the key nitrogen fixer, *Leptospirillum ferrodiazotrophum* sp. nov., from an acidophilic microbial community. Appl. Environ. Microbiol. *71*, 6319–6324.

Valdés, J., Quatrini, R., Hallberg, K., Dopson, M., Valenzuela, P.D., and Holmes, D.S. (2009). Draft genome sequence of the extremely acidophilic bacterium *Acidithiobacillus caldus* ATCC 51756 reveals metabolic versatility in the genus *Acidithiobacillus*. J. Bacteriol. *191*, 5877–5878.

Walton, K.C., and Johnson, D.B. (1992). Microbiological and chemical characteristics of an acidic stream draining a disused copper mine. Environ. Pollut. *76*, 169–175.

Wilmes, P., Remis, J.P., Hwang, M., Auer, M., Thelen, M.P., and Banfield, J.F. (2009). Natural acidophilic biofilm communities reflect distinct organismal and functional organization. ISME J. *3*, 266–270.

Wilmes, P., Benjamin, P., Bowen, B.P., Brian, C., Thomas, B.C., Mueller, R.S., Denef, V.J., VerBerkmoes, N.C., Hettich, R.L., Northen, T.R., *et al.* (2010). Metabolome-proteome differentiation coupled to microbial divergence. Mbio *1*, e00246.

Wrighton, K.C., Thomas, B.C., Sharon, I., Miller, C.S., Castelle, C.J., VerBerkmoes, N.C., Wilkins, M.J., Hettich, R.L., Lipton, M.S., Williams, K.H., *et al.* (2012). Fermentation, hydrogen, and sulfur metabolism in multiple uncultivated bacterial phyla. Science *337*, 1661–1665.

Wrighton, K.C., Castelle, C.J., Wilkins, M.J., Hug, L.A., Sharon, I., Thomas, B.C., Handley, K.M., Mullin, S.W., Nicora, C.D., Singh, A., *et al.* (2014). Metabolic interdependencies between phylogenetically novel fermenters and repiratory organisms in an unconfined aquifer. ISME J. *8*, 1452–1463.

Yelton, A.P., Thomas, B.C., Simmons, S.L., Wilmes, P., Zemla, A., Thelen, M.P., Justice, N., and Banfield, J.F. (2011). A semi-quantitative, synteny-based method to improve functional predictions for hypothetical and poorly annotated bacterial and archaeal genes. PLoS Comp. Biol. *7*, e1002230.

Yelton, A.P., Comolli, L.R., Justice, N.B., Castelle, C., Denef, V.J., Thomas, B.C., and Banfield, J.F. (2013). Comparative genomics in acid mine drainage biofilm communities reveals metabolic and structural differentiation of co-occurring archaea. BMC Genomics *14*, 485.

Ziegler, S., Ackermann, S., Majzian, J., Asskamp, M., and Eusterhues, K. (2009). Matrix composition and community structure analysis of a novel bacteria pyrite leaching community. Environ. Microbiol. *11*, 2329–2338.

Ziegler, S., Dolch, K., Geiger, K., Krause, S., Asskamp, M., and Eusterhues, K. (2013). Oxygen-dependent niche formation of a pyrite-dependent acidophilic consortium built by archaea and bacteria. ISME J. *7*, 1725–1737.

Proteomics of Acidophilic Prokaryotes

14

Francisco Remonsellez, Fernando A. Pagliai, Claudio Navarro, Rodrigo Almárcegui and Carlos A. Jerez

Abstract

Proteomics, metaproteomics, genomics and metagenomics, together with metabolomics, have been widely used in recent years to study the global regulatory responses of all kinds of cells. By this approach it may be possible to explore the new properties of microorganisms that arise from the interplay of genes, proteins, other macromolecules, small molecules, and the environment. Proteomics offers direct information of the dynamic protein expression in tissue or whole cells, giving us a global analysis of the cell proteins and their cellular behaviour. In this chapter, recent advances in standard and quantitative proteomics applied to individual acidophilic bacteria and archaea are reviewed. Specifically, adaptation responses of acidophiles to extremely high metal concentrations, different minerals or solid substrates and possible applications of some of these findings are illustrated.

Importance of proteomics in the study of acidophilic microorganisms

Genomics, metagenomics, proteomics and metaproteomics, together with metabolomics have been widely used in recent years to study the global regulatory responses of all kinds of cells, including extreme acidophiles. Through this systemic approach it has become possible to explore the new properties of microorganisms that arise from the interplay of genes, proteins, other macromolecules, small molecules, and the environment (Buckley, 2004). Analysis of the genomic sequences currently available for a large number of acidophilic microorganisms has begun to unravel the molecular adaptations of these microorganisms to their environment and the interactions they establish with other members of the acidophilic microbial communities. In turn, on-going efforts to study the proteomics of acidophiles have started to answer many fundamental and applied questions on their biology, such as: What is the nature of bacteria–mineral interactions? Are proteins exported to the external medium during bacteria–mineral interactions? Do these secreted proteins alter the behaviour of other members of the community? What protein complexes are involved in mineral oxidative reactions? What proteins confer the extremely high metal resistances of many acidophilic microorganisms?

Only two reviews on the proteomics of extreme acidophiles have been published in the last decade (VerBerkmoes *et al.*, 2009; Jerez, 2008). An interesting and useful review on the most recent proteomic methodologies used in the study of other extremophilic microorganisms is also available (Burg *et al.*, 2011). In the present chapter we review some of the most recent proteomic studies of both acidophilic bacteria and archaea and how these studies have helped advance the understanding of the microbial processes driven by them. In addition, the ways in which proteomics may impact the application of acidophiles to improve industrial bioleaching of minerals is also to be discussed.

Basically, proteomics enable the identification of differentially expressed proteins by dissimilar cell types or similar cell types exposed to different environmental conditions. In general, total proteins are extracted from the sample of interest and are denatured and digested to obtain peptides. Alternatively, if the samples are too complex, sub-proteomes such as the proteins secreted to the medium (secretome) or the different subcellular fractions can be obtained separately and digested to peptides independently. The resulting peptides are then analysed by liquid chromatography–electrospray ionization–tandem mass spectrometry (LC-ESI-MS/MS) or by matrix-assisted laser desorption/ionization mass spectrometry (MALDI-MS/MS). Aspects of the preparation of protein samples for proteomic analysis from extremophiles and introductory principles of mass spectrometry (MS), protein fractionation and processing procedures to suit the capabilities of different individual mass spectrometers have been covered elsewhere (Van Oudenhove and Devreese, 2013; Burg *et al.*, 2011).

Proteomics of acidophilic bacteria

Standard 2D-PAGE proteomics

This section illustrates, through a few examples, how standard two-dimensional polyacrylamide gel electrophoresis (2-D PAGE) proteomics has begun to contribute knowledge on the adaptation of acidophilic bacteria to their harsh environment (Table 14.1). The standard 2D-PAGE in combination with MS is a widely used technology for comparative bacterial proteomics. It allows separation of over 1000 protein spots in a single run and its high reproducibility is valuable for multiple sample comparisons (Fig. 14.1). In addition, it directly correlates the changes observed at the peptide level to individual protein isoforms. Several studies using 2D-PAGE to assess changes in protein expression of *Acidithiobacillus ferrooxidans* under different growth conditions have been conducted. Proteins changing their levels under conditions such as heat shock, pH stress or phosphate limitation have been reported (Jerez, 2008 and references therein). The use of MS to identify 2D-PAGE spots has greatly improved the procedure and is still widely used with great advantage. For example, it has aided in the identification of proteins involved in the differential response of *At. ferrooxidans* after short exposure to the copper sulphides containing iron bornite (Cu_5FeS_4) or chalcopyrite ($CuFeS_2$). The response to bornite involved thirteen proteins

Table 14.1 Extremely acidophilic chemolithotrophic bacteria studied by proteomics

Microorganism	Methods	Targets	Source or Reference
Acidithiobacillus ferrooxidans	2D-PAGE/MS ICPL	Copper resistance determinants	Almárcegui *et al.* (2014a,b)
	SDS-PAGE MS	Biofilm formation mechanisms on pyrite	Vera *et al.* (2013a)
	SDS-PAGE MS	Biochemistry of change from elemental sulfur to ferrous iron	Kucera *et al.* (2013)
	SDS-PAGE MS	Response to high potassium concentration	Ouyang *et al.* (2013)
	SDS-PAGE MS	Biochemistry of S°-coupled iron reduction	Osorio *et al.* (2013)
	SDS-PAGE MS	Mechanisms contributing to pH homeostasis	Mangold *et al.* (2013a)
	2D-PAGE MS	Metabolic response in different minerals	Felício *et al.* (2011)
	2D-PAGE MS	Identification of extracellular proteins in elemental sulfur and ferrous sulfate	Zhang *et al.* (2008)
	2D-PAGE MS	Biochemistry of the sulfur metabolism	Valenzuela *et al.* (2008)
	2D-NEPHGE/SDS-PAGE MS	Identification of periplasmatic proteins	Chi *et al.* (2007)
	2D-PAGE MS	Biochemistry of growth under phosphate starvation	He *et al.* (2005)
	2D-PAGE MS	Characterization of secreted proteins or secretome	Pagliai and Jerez (2009)
Acidithiobacillus thiooxidans	2D-PAGE MS	Characterization of secreted proteins or secretome	Bobadilla Fazzini *et al.* (2009, 2011)
Acidithiobacillus caldus	2D-PAGE MS	Biochemistry of growth on elemental sulfur and tetrathionate	Mangold *et al.* (2011)
Leptospirillum ferriphilum	2D-PAGE MS	Response to arsenic stress	Li *et al.* (2010)

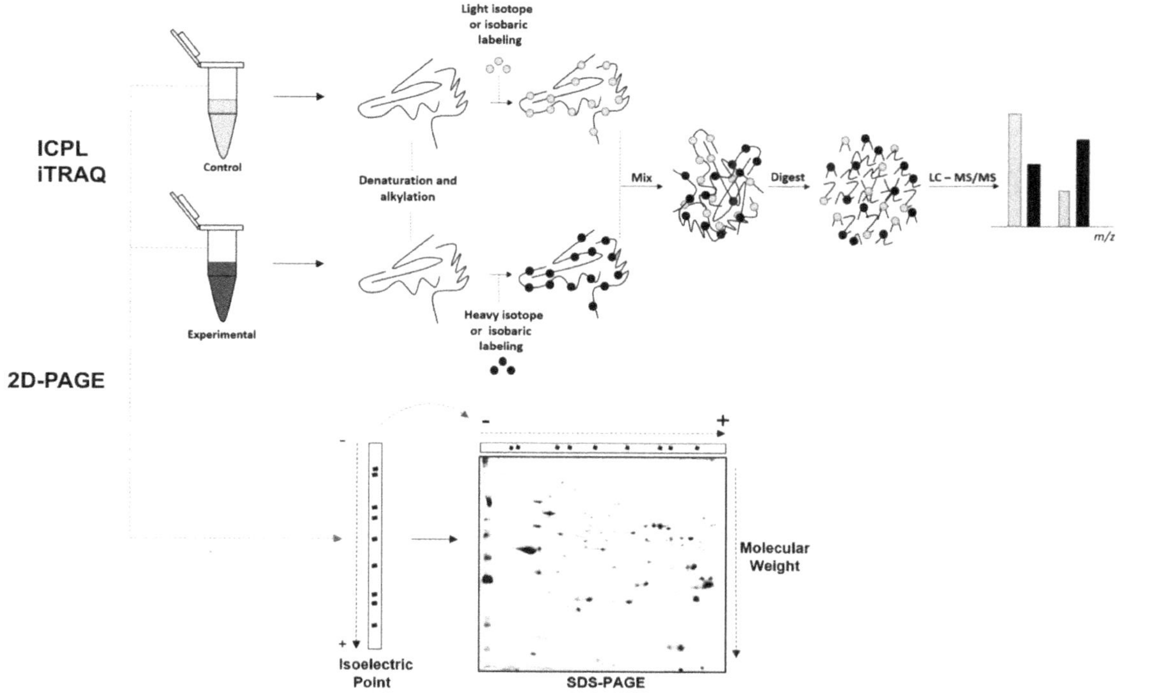

Figure 14.1 Two methods to quantify proteins in cells grown under different conditions. In quantitative proteomics studies (upper panel), after labelling with light and heavy stable isotopes (ICPL) or isobaric tags (iTRAQ), proteins of control and experimental samples are combined. The peptides obtained are analysed by LC-MS/MS and quantified based on the intensity ratio of labelled peptide pairs. In 2D-PAGE (lower panel), proteins are separated by their isoelectric point (isoelectric focusing or IEF) (first dimension) followed by mass by means of SDS-PAGE (second dimension) and proteins resolved are identified by MS/MS.

related to energy metabolism, detoxification and protein synthesis whereas the same response was not observed for chalcopyrite (Felício *et al.*, 2011). A likely explanation for the distinct metabolic responses observed in *At. ferrooxidans* is that these two minerals have very different crystalline structures and susceptibilities to bacterial or chemical attack, chalcopyrite being described as the most refractory mineral.

Two-dimensional gel electrophoresis has also been used in quantitative proteomic analysis during adaptation of *At. ferrooxidans* grown in elemental sulfur to ferrous iron. In this experiment, 39 proteins from total cell lysates changed their expression levels (Kucera *et al.*, 2013). Some of the proteins that were up- or down-regulated, were related to iron or sulfur metabolism, among several other functions. This preliminary study indicated that substrate adaptation is a complex process entailing a wide regulatory network and that further studies are required to elucidate the physiological changes experienced by *At. ferrooxidans* during substrate switching in both environmental and biotechnological contexts.

During growth of acidophiles in bioleaching operations the pregnant metal-containing solutions are subjected to solvent extraction to recover the solubilized metals and the spent leach liquors are recirculated through the bioleaching process to save water. This generates high osmotic potentials due to elevated concentrations of various soluble anions and cations present in these liquors. In this context, the response of *At. ferrooxidans* to high potassium concentrations was tested using two-dimensional electrophoresis coupled with matrix-assisted laser desorption/ionization time-of-flight tandem mass spectrometer (MALDI-TOF–TOF). Eight proteins were differentially expressed in the presence of high potassium concentration. These included some up-regulated proteins such as GDP-mannose 4,6-dehydratase (GMD), ribose 5-phosphate isomerase A (RpiA) and ribose-phosphate pyrophosphokinase. It was postulated that these proteins involved in the generation of sugar precursors are important to produce glycocalyx. Congruently, thickening of the glycocalyx layer was observed by transmission electron microscopy in *At. ferrooxidans* cells cultivated under high potassium concentration (Ouyang *et al.*, 2013). The authors concluded that this thick glycocalyx layer would protect the cells against the high potassium concentration stress. Some of the down-regulated proteins identified were the phosphoglycolate phosphatase and the ATP synthase F1 beta and delta subunits. Since the ATP synthase is directly linked to transport of potassium into the cell, it is possible that the microorganism reduces the levels of this protein complex to offset the high potassium concentration in the culture medium.

Also, the mechanisms contributing to pH homeostasis have been analysed in *Acidithiobacillus caldus* grown at different pH values by means of 2D-PAGE. Differential expression of transcription factors such as the alternative sigma factor RpoS and an ArsR family transcriptional regulator, among many other proteins, suggested a tight transcriptional control of pH-induced genes. Still, further studies are required to understand in detail the role of these and other differentially expressed proteins during growth under different pH conditions. Enzymes involved in sulfur metabolism have also been found to be up-regulated at pH 1.1, suggesting either that cells required more energy for maintenance, or that the increased metabolic activity is part of a specific acid stress response built to export intracellular protons via proton pumps (Mangold *et al.*, 2013a).

The basis of S°-coupled iron reduction in *At. ferrooxidans* has also been approached from a proteomic perspective. *At. ferrooxidans* protein profiles were analysed during aerobic growth with S° as electron source and O_2 as electron acceptor or under anaerobic conditions with Fe^{3+} as electron acceptor (Osorio *et al.*, 2013). Differential protein levels were observed in each condition. The results of this study suggest models for the coupling of S° oxidation to Fe^{3+} reduction, indicating that Fe^{3+} reduction by *At. ferrooxidans* grown anaerobically using S° as electron donor is mediated, at least in part, by an indirect mechanism involving H_2S.

High-throughput proteomics

Fourier transform ion cyclotron resonance mass spectrometer (FT-ICR MS) is well suited for the analysis of differential protein expression due to the high mass accuracy determination and high resolution, as well as for its inherent wide dynamic range. For example attomole-levels of peptides can be detected. FT-ICR MS is also quite useful in analysing phosphorylation states, O- or N-linked glycosylation, and sulfating of proteins.

This highly sophisticated technology has been applied in the study of the protein complement of a modest number of acidophiles and also in acidophilic microbial communities.

For instance, total proteins extracted from *At. ferrrooxidans* grown on the surface of a sulfur prill were analysed by FT-ICR MS and compared with those seen in ferrous iron or thiosulfate. In this study, the highest scores obtained were for peptides derived from proteins identified as TrbB and TrbJ, amongst others. These proteomic results were in agreement with RT-PCR, quantitative RT-PCR and DNA macroarray results (Chi *et al.*, 2007; Valenzuela *et al.*, 2006).

Recently, a shotgun proteomic study of an *At. ferrooxidans* ATCC 23270 biofilm at early stages of formation on a pyrite surface was conducted (Vera *et al.*, 2013). Planktonic and sessile subpopulations of cells were separated from the mineral and their proteins extracted. A total of 1319 proteins were detected in both samples. Sixty-two of these proteins showed increased levels and 25 were decreased in the biofilm subpopulation. Results of this study indicated that at early steps of biofilm formation, following adherence to the mineral surface, *At. ferrooxidans* undergoes several changes. For example, proteins related to stress responses, respiratory chains and fatty acid synthesis were found to decrease in biofilm cells. On the other hand, some transcription factors, and proteins involved in cell wall or EPS biosynthesis, sulphur assimilation and glutathione metabolism among others, were found to be increased in the cells present in the biofilm. Of all these changes, the biosynthesis of extracellular polymeric substances appeared to be crucial (Vera *et al.*, 2013).

The term 'metaproteomics' was first proposed for the large-scale characterization of the entire protein complement of environmental microbiota (Wilmes and Bond, 2004). These authors applied 2D-PAGE and downstream analysis to a mixed community of prokaryotic microorganisms related to activated sludge after enhanced biological phosphorus removal, extracting and purifying from it the entire proteome to map it. Highly expressed proteins were then identified by MS. High throughput MS has since been used to study the community proteomics of a natural acid mine drainage microbial biofilm (Ram *et al.*, 2005). These authors were able to detect 2033 proteins from the five most abundant species in the biofilm, including 48% of the predicted proteins from the dominant biofilm organism, *Leptospirillum* group II. An abundant novel protein with an absorption at 579 nm characteristic of *Leptospirillum* cytochrome type was identified. It was called Cyt_{579} and this protein had both kinetic and thermodynamic properties consistent with a central role in iron oxidation in *Leptospirillum* group II. Its high abundance, extracellular localization and enzymatic activity implies that Cyt_{579} is the primary iron oxidant in the electron transport chain of this bacterium (Ram *et al.*, 2005). This novel approach together with functional metagenomics has offered an integrated way of studying microbial communities to establish the role of each of the members and how they change under different conditions (Denef *et al.*, 2010) (see Chapter 13). These potent approaches may be successfully applied to communities such as those usually found in copper or other metals industrial biomining operations.

Quantitative proteomics

Although 2D-PAGE has a high resolving power and can be quantitative, it cannot be used for high-throughput proteome analysis, mainly due to problems with automation and with the detection rate of low abundant membrane proteins (Schmidt *et al.*, 2005). Several methods besides 2D-PAGE have been used to quantify the differences in proteomes of cells grown under different conditions (Tebbe *et al.*, 2005; Schmidt *et al.*, 2005). Comparison of MS, or MS/MS, signals using isotopic labelling of proteins are among the preferred methods (Gevaert *et al.*, 2008). The iTRAQ method, where digested peptide samples can be independently labelled, allows the comparison of several proteomes in a single experiment (Wiese *et al.*, 2007).

Another simple and reliable method is the isotope-coded protein labelling (ICPL) proteomics, a powerful technique that combines the labelling of proteins with stable isotopes with mass spectroscopy for the identification and quantification of complex protein mixtures (Paradela *et al.*, 2010). In this case, the proteomes derived from two experimental conditions are differentially labelled with a light isotope (^{12}C) and a heavy isotope (^{13}C). The ICPL label is incorporated in all free amino groups (e.g. lysine and the free protein N-terminus). Samples are then mixed and separated by electrophoresis and enzymatically digested to identify and quantify, using liquid chromatography-tandem mass spectrometry (LC-MS/MS), peptides exhibiting changes in their levels as a result of the different experimental conditions. Relative quantification is achieved by comparison of the relative abundance (regarded as either peak intensity or area) of the differentially labelled peptides (Tebbe *et al.*, 2005; Paradela *et al.*, 2010) (Fig. 14.1). The method is highly accurate and reproducible quantification of proteins with high sequence coverage, indispensable for the detection of post-translational modifications and protein isoforms, has been achieved (Schmidt *et al.*, 2005).

Acidic solutions where acidophiles thrive are characterized by concentrations of base and transition metals that are toxic to most living organisms. As might be expected, microorganisms that grow in mineral-rich environments are, in most cases, remarkably tolerant to a wide range of metal ions, and should possess robust metal resistance mechanisms (Dopson *et al.*, 2014; Orell *et al.*, 2012). Recently, the ICPL procedure has been used to study *At. ferrooxidans* response to high copper concentrations (Almárcegui *et al.*, 2014a,b). Key elements involved in *At. ferrooxidans* copper-resistance include a wide and redundant repertoire of known determinants (Orellana and Jerez, 2011), novel copper-chaperones (Almárcegui *et al.*, 2014a,b; Navarro *et al.*, 2009), an abundant reserve of inorganic polyphosphate (polyP) and a robust defensive response against oxidative stress (Orell *et al.*, 2010, 2012 and references therein). Conventional copper homeostasis determinants used by *At. ferrooxidans* and other bacteria (Rensing and Grass, 2002) include copper transporting ATPases, related to a system of the RND gene family involved in the extraction of copper from the cell, as well as genes coding for periplasmic copper chaperones. These proteins were found to be overexpressed in *At. ferrooxidans* ATCC 23270 when this bacterium was adapted to 5–25 mM copper(II). In addition, when expressed in *Escherichia coli* cells, several of these proteins conferred a greater resistance to copper compared to wild-type cells, supporting their functionality (Navarro *et al.*, 2009). On the other hand, *At. ferrooxidans* strain ATCC 53993 has been reported to be much more resistant to copper than strain ATCC 23270. Although both strains contain the same copper resistance determinants encoded in their genomes, the difference is apparently due to the presence of a genomic island exclusive of strain ATCC 53993, which contains several genes coding an additional predicted copper-ATPase and a Cus system (Orellana and Jerez, 2011).

Using ICPL qProteomics the copper-responsive protein complement of iron-grown *A. ferrooxidans* ATCC 23270 cells was recently evaluated (Almárcegui *et al.*, 2014a). Up-regulation of RND-type Cus systems and different RND-type efflux pumps was reported, together with the overexpression of several genes involved in histidine and cysteine synthesis. Under these conditions, down-regulation of the major outer membrane porin and some ionic transporters was observed, suggesting a general decrease in the influx of the metal into the cell (Almárcegui *et al.*, 2014a). Similar results were observed in sulfur-grown *At. ferrooxidans* ATCC 23270 cells (Almárcegui *et al.*, 2014b). In addition, new putative copper resistance determinants in *At. ferrooxidans* grown in elementary sulfur were identified via quantitative proteomic approaches. Although many of the protein levels changed in similar ways in cells grown in sulfur or ferrous iron, the copper response of sulfur-grown cells appeared to be modest compared with that of iron-grown bacteria. Furthermore, the overexpression of rusticyanin in sulfur-grown cells subjected to copper (Almárcegui *et al.*, 2014b) supports the idea that this blue-copper protein has a role in *At. ferrooxidans* copper resistance, as previously suggested by Felício *et al.*, 2011 and references therein. Fig. 14.2 summarizes a working model for copper resistance in *At. ferrooxidans* derived mostly from proteomic studies (Almárcegui *et al.*, 2014a,b).

Proteomics in acidophilic *Archaea*

The best-known and most widely studied acidophilic *Archaea* are members of the order Sulfolobales. These extremely thermophilic sulfur- and iron(II)-oxidizing crenarchaeotes group the genera *Sulfolobus* (*S.*), *Acidianus* (*Ad.*), *Metallosphaera* (*M.*), *Sulfurisphaera* (*Ss.*) and *Sulforococcus* (*Sc.*) (see Chapter 6). Within the euryarchaeotes (family *Ferroplasmaceae*), acidophilic iron (II)-oxidizing species, *Ferroplasma* (*Fp.*) *acidiphilum* and '*Fp. acidarmanus*' have been characterized (see Chapter 6). A new genus, *Acidiplasma*, of the family *Ferroplasmaceae* has been described as containing two species, *Acidiplasma* (*Ap.*) *aeolicum* and *Ap. cupricumulans*, isolated from the hydrothermal pool located on Vulcano Island (Italy) and from chalcocite/copper-containing heaps (Myanmar), respectively. Extremely thermoacidophilic archaea have received considerable attention not only for the uniqueness of their biology, but also because these organisms may play an important role, not yet fully understood, in the leaching of refractory sulfide minerals (Watling, 2006).

While proteomic studies of *Archaea* are simplified by their typically unicellular structure and relatively small genome size (0.49–5.75 Mbp), *Archaea* are also distinguished by a number of unusual properties that can complicate MS-based analysis of their proteome. In particular, most archaea models are poly-extremophilic, able to withstand amazingly harsh conditions such as high temperatures, saturating salt and/or extreme pH. Thus, many of the proteins of *Archaea* are relatively resistant to the standard chemical and thermal treatments used to denature proteins for separation by 2D-PAGE or trypsin digest. MS-based identification of archaeal proteins is also complicated by our current inability to correctly identify ribosomal binding sites and promoter consensus sequences in *Archaea*. This results in miss-prediction of coding sequences during genome annotation and can compromise the database used for MS-based searches (Tebbe *et al.*, 2005). Because archaeal proteomic studies remain at an early stage, some authors suggest to couple traditional genetic and biochemical approaches with advanced technologies, such as genomics and proteomics, which will provide an avenue for scientists to discover new aspects related to the molecular physiology of

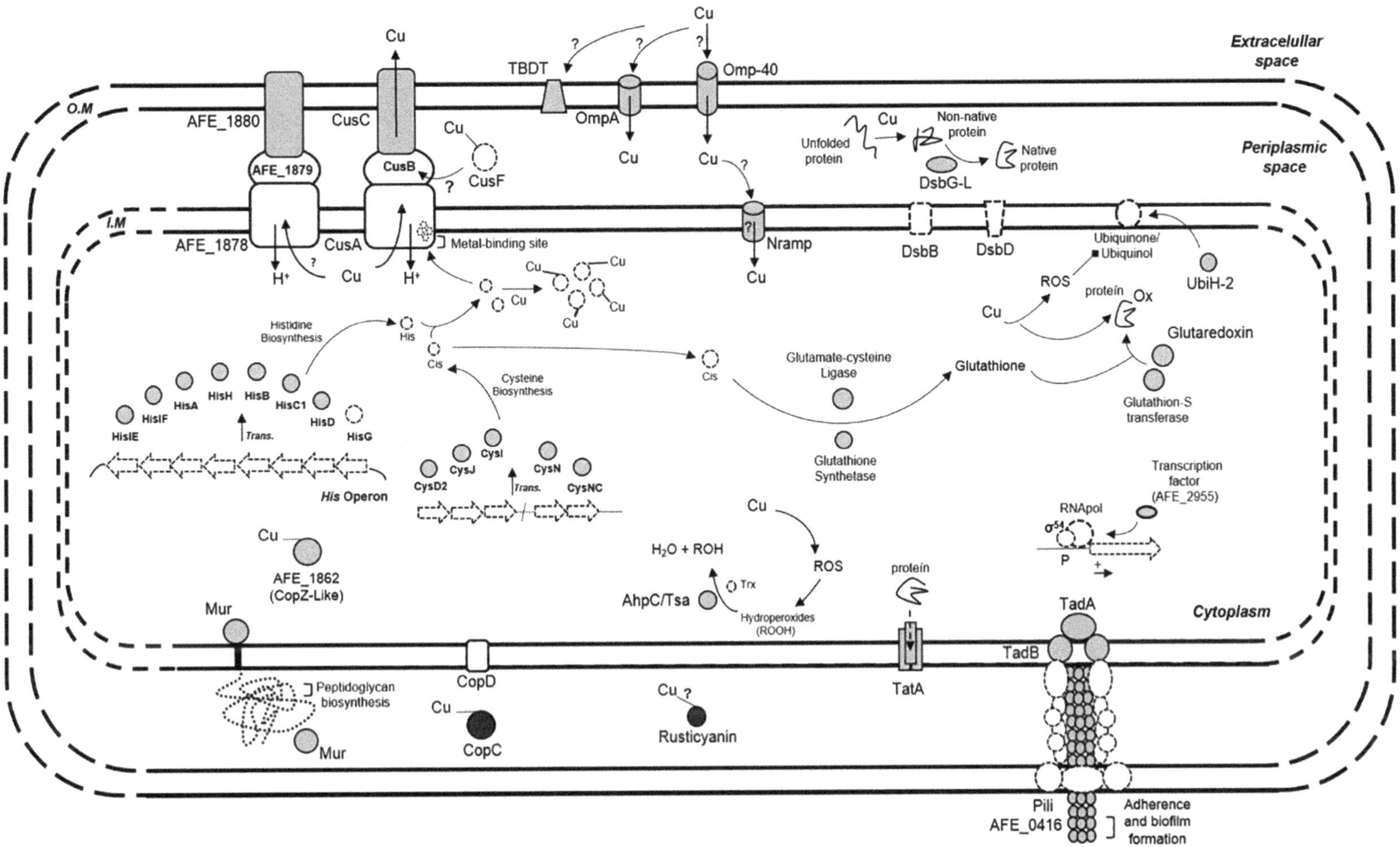

Figure 14.2 Model of the *At. ferrooxidans* copper response. The scheme shows several of the proteins that change their levels in *At. ferrooxidans* when it is adapted to grow in the presence of copper. Putative proteins shaded in light grey were identified by means of ICPL quantitative proteomic studies (Almárcegui *et al*., 2014a,b) and proteins shaded in black were identified by means of 2D-PAGE and FT-ICR MS (Chi *et al*., 2007). Those in white have been identified in *A. ferrooxidans* by transcriptional studies (Almárcegui *et al*., 2014a,b; Orellana and Jerez, 2011; Navarro *et al*, 2009). Elements of the response that have not been experimentally determined are depicted in dotted lines. The location and order in which proteins are illustrated is arbitrary.

Archaea (Maupin-Furlow *et al.*, 2012). A summary of the proteomic studies performed on acidophilic *Archaea* is shown in Table 14.2.

Standard proteomics in acidophilic *Archaea*

Standard proteomics has been the main tool used to characterize proteins differentially expressed in '*Fp. acidarmanus*', *Fp. acidiphilum*, *M. sedula*, *S. metallicus* and *S. solfataricus*. Most of proteomic work performed in acidophilic *Archaea* has focused on the understanding of relevant biochemical pathways, the identification of genetic determinants involved in heavy metals resistance, the study of the mechanisms of biofilm formation and of the functions of extracellular proteins (Table 14.2). Specifically, *S. solfataricus* has been used as a model to study unique or previously unknown properties, to uncover the existence of novel metabolic pathways, to determine protein secretion and to provide a better understanding of evolution and adaptation of extremophile microorganisms (e.g. Ellen *et al.*, 2010). Technological advances in the field of proteomics, such as the combination of peptide separation approaches, and improvements in peptide fractionation efficiency and proteome coverage (including soluble, insoluble and secreted proteins) have emerged from studies of acidophilic archaea, mostly due to the physico-chemical properties of their hydrophobic proteins and to the difficulties in resolving proteins with extreme pIs (e.g. Chong and Wright, 2005).

Heavy metals resistance mechanisms have been investigated in '*Fp. acidarmanus*' cells exposed to copper, arsenic and zinc using proteomic approaches (Mangold *et al.*, 2013b; Baker-Austin *et al.*, 2005, 2007). Cells of this microorganism exposed to copper revealed the induction of stress proteins associated with protein folding and DNA repair (including RadA, thermosome and DnaK homologues), suggesting that '*Fp. acidarmanus*' uses multiple mechanisms for resistance to high levels of copper (Baker-Austin *et al.*, 2005). Studies of '*Fp. acidarmanus*' revealed also enhanced expression of proteins associated with protein refolding (including the thermosome Group II HSP60 family chaperonin and HSP70 DnaK type heat shock proteins) when cells were exposed to arsenite (Baker-Austin *et al.*, 2007). Neither molecular chaperones nor zinc transporters were found to be differentially expressed when '*Fp. acidarmanus*' cells were exposed to zinc (Mangold *et al.*, 2013b). This might be explained, at least partially, by the unavailability of zinc as a free

Table 14.2. *Archaea* present in extreme acidic environments studied by proteomics

Microorganism	Methods	Targets	Source or reference
Sulfolobus metallicus	2D-PAGE MS	Copper and cadmium resistance mechanisms	Orell *et al.* (2013)
	2D-Blue Native PAGE/SDS-PAGE MS	Ferrous iron oxidation mechanisms	Bathe and Norris (2007)
Metallosphaera sedula	2D-PAGE MS	Copper resistance mechanisms	Maezato *et al.* (2012)
'*Ferroplasma acidarmanus*'	2D-PAGE MS	Zinc tolerance	Mangold *et al.* (2013b)
	2D-PAGE MS	Biofilm formation mechanisms	Baker-Austin *et al.* (2010)
	2D-PAGE MS	Arsenic resistance mechanisms	Baker-Austin *et al.* (2007)
	2D-PAGE MS	Copper resistance mechanisms	Baker-Austin *et al.* (2005)
'*Ferroplasma acidarmanus*' and *Ferroplasma acidiphilum*	2D-PAGE MS	Iron homeostasis mechanisms	Potrykus *et al.* (2011)
	2D-PAGE MS	Biochemistry of iron oxidation and chemo-organotrophic growth	Dopson *et al.* (2005)
Ferroplasma types I and II	High-throughput MS	Functions of microbial biofilms in AMD	Ram *et al.* (2005)
Ferroplasma types I-, A-, G- and E-plasma	Quantitative MS	Biofilms communities in laboratory AMD	Belnap *et al.* (2010)
Ferroplasma types I and II, and G-plasma	Quantitative MS	Effect of pH conditions	Belnap *et al.* (2011)

ion under the conditions tested. Several proteins involved in carbohydrate metabolism/energy production were down-regulated during zinc exposure. In turn, a number of proteins with unknown functions were up-regulated under the same conditions. In addition, Maezato *et al.* (2012) reported the specific relationship between metal resistance and litho-autotrophy during biotransformation of chalcopyrite by *M. sedula* mutant strains by using qRT-PCR and 2D-PAGE. These authors identified induction of proteins related with carbon monoxide dehydrogenase, AAA family ATPase, Glu/Leu/Phe/Val dehydrogenase, pyruvate flavodoxin/ferredoxin oxidoreductase and lactate/malate dehydrogenase, when *M. sedula* mutant (CuR1) cells resistant to heavy metals were exposed to copper (Maezato *et al.*, 2012). Recently, proteins involved in copper and cadmium resistance in *S. metallicus* were identified. In this work, 23 proteins were found to be down-regulated after copper treatment. This kind of response was also seen when the microorganism was exposed to cadmium (Orell *et al.*, 2013) and when '*Fp. acidarmanus*' was challenged by either arsenite or copper (Baker-Austin *et al.*, 2005, 2007). In addition, *S. metallicus* cells revealed the up-regulation of 30 proteins after copper treatment. Most of these proteins are related to the production and conversion of energy, the biosynthesis of amino acids and to stress responses. Ten of these proteins were also up-regulated in *S. metallicus* grown in the presence of cadmium, suggesting that at least in part a common response to these two heavy metals exists in this archaea. Within proteins up-regulated in *S. metallicus* cells exposed to copper, a putative ATP synthase subunit B was detected (Orell *et al.*, 2013). When cells are subjected to some stressing conditions such as the presence of heavy metals, a greater cellular demand for energy has been reported to occur (Baker-Austin *et al.*, 2005).

Two other proteins identified through proteomic studies are the ferredoxin oxidoreductase and the alcohol dehydrogenase, which have been generally involved in electron transport chains and use NAD^+ as an electron acceptor. The levels of ferredoxin oxidoreductase were also up-regulated in response to cadmium (Orell *et al.*, 2013). Several previous reports have suggested that oxidoreductases provide oxidative protection in response to heavy metals in both bacteria and archaea. Some proteins involved in oxidative damage repair, such as NADH dependent oxidases and thioredoxin reductases, were expressed in cells of '*Fp. acidarmanus*' exposed to arsenite (Baker-Austin *et al.*, 2007). The expression of this group of proteins has also been observed when the same microorganism was exposed to copper (Baker-Austin *et al.*, 2005). Enzymes related to biosynthesis of amino acids, such as phosphoglycerate dehydrogenase and a glutamate dehydrogenase, were expressed in *S. metallicus* in response to copper or cadmium (Orell *et al.*, 2013) and in '*Fp. acidarmanus*' cells exposed to arsenite (Baker-Austin *et al.*, 2007). Proteins involved in the general stress response, such as one subunit of the HSP60 chaperonin and one subunit of the proteosome, were detected in *S. metallicus* cells exposed to copper or cadmium (Orell *et al.*, 2013).

Different growth conditions, such as chemomixotrophic, chemo-organotrophic and iron limitation growth, have been analysed in '*Fp. acidarmanus*' and *Fp. acidophilum* (Potrykus *et al.*, 2011; Dopson *et al.*, 2005). Proteins with high sequence similarity to proteins having suggested functions in central metabolic pathways such as glycolysis, gluconeogenesis, tricarboxylic acid (TCA) and fatty acid metabolism were up-regulated in '*Fp. acidarmanus*' under chemo-organotrophic conditions in comparison to the chemomixotrophic condition. These results reflect a higher energy demand to be derived from the organic carbon during chemo-organotrophic growth (Dopson *et al.*, 2005). Under iron-limitation conditions, a putative isochorismatase, presumably involved in siderophore biosynthesis, was up-regulated in '*Fp. acidarmanus*' cells, and the SufBCD system was up-regulated in *Ferroplasma* genus (Potrykus *et al.*, 2011). On the other hand, changes in protein expression under different growth conditions have also been analysed in *S. metallicus*. Blue native PAGE of membrane proteins from pyrite-grown cells showed the presence of a complex related with FoxA that was absent in the membrane proteins from sulfur-grown cells. The expression of this protein was supported by using transcriptional analysis of the *foxA* gene. Other genes of this *fox* system may be involved also in oxidation of ferrous iron by *S. metallicus* (Bathe and Norris, 2007).

Finally, differential protein expression patterns between biofilm and planktonic populations have been analysed in '*Fp. acidarmanus*'. The mature biofilms in this microorganism showed the induction of six enzymes associated with anaerobic growth, including a subunit of pyruvate ferredoxin oxidoreductase, an electron transfer flavoprotein, a malate dehydrogenase, a thioredoxin reductase, a RNase exoribonuclease and a hypothetical protein. The results of this work suggest that the dominating phenotype in the mature biofilm was associated with anaerobic modes of growth (Baker-Austin *et al.*, 2010).

High-throughput and quantitative mass spectrometry

High throughput proteomic analysis has been used to demonstrate the existence of functionally disrupted genes in the archaeal model *S. solfataricus*. The presence of disrupted genes (DGs) is usually considered as an evidence of non-functional pseudogenes. Nevertheless, the role of pseudogenes, which are more common in eukaryotic genomes, is still poorly understood in the *Archaea* domain. Interestingly, functional pseudogenes have the potential for becoming new genes useful for the survival and adaptation of organisms to particular environmental changes. Using a combination of RT-PCR with shotgun proteomics Cobucci-Ponzano and colleagues demonstrated that interrupted genes are expressed *in vivo* in *S. solfataricus* (Cobucci-Ponzano *et al.*, 2010). For instance, the ORFs encoding a disrupted orthologue of Kae1/Bud32, an essential complex in yeast and humans implicated in transcription and telomere maintenance, were expressed independently in this microorganism. In addition, the universal translation regulator SUI-1 was expressed by programmed −1 frameshifting. This is the first proteomic study on interrupted genes performed to date and it demonstrates that DGs should be taken into account in the annotation of the sequenced genomes and in protein predictions.

Although diverse techniques have been applied for quantitative proteomics in *Archaea*, these studies have mostly focused on cytosolic proteins or whole cell lysates. In recent years, quantitative proteomic approaches have been used in model *Sulfolobus* spp. to determine membrane proteins and to initiate comprehensive studies on archaeal biofilms (Pham *et al.*, 2010). Analysis of the membrane of the archaeon *S. solfataricus* P2 using the combination of 3 different iTRAQ experiments enabled the detection of 395 proteins, 373 of which had predicted membrane properties, improving our understanding of the membrane-associated proteosome of microorganisms in *Archaea* domain. On the other hand, a combination of transcriptomic and proteomic approaches revealed that biofilm-forming lifestyle is strain specific. *Sulfolobus* spp. cells grown as biofilms have been shown to have altered expression of proteins involved predominantly in energy production and conversion, adaptation to environmental changes, stress, substrate transport/binding activities and regulatory events (Koerdt *et al.*, 2011). Genes and proteins involved in energy metabolism were highly altered in *Sulfolobus* spp. biofilm-associated population. These changes included mainly genes related to respiratory complexes, tricarboxylic acid cycle (TCA) enzymes and V-ATPases subunits. Most of these genes showed down-regulation in biofilm-associated cells, suggesting a decreased metabolic activity in this cell population in comparison qirh the planktonic counterparts (Koerdt *et al.*, 2011).

High-throughput MS has also been used in the metaproteomic study of the biofilm community present in a natural acid mine drainage (Ram *et al.*, 2005). This biofilm was dominated by *Leptospirillum* group II, but it also contained *Leptospirillum* group III, *Sulfobacillus* and *Archaea* related to '*Fp. acidarmanus*' and 'G-plasma'. Ram and colleagues were able to detect 84 *Ferroplasma* type I, 99 *Ferroplasma* type II and 122 'G-plasma' proteins. In addition, they found 30 proteins on unassigned archaeal scaffolds. Despite the much lower degree of sampling of *Ferroplasma* type II genome, proteins possibly involved in creating biofilm architecture were detected.

A proteomic quantitative analysis of acid mine drainage microbial communities at a laboratory scale using biofilms reactors revealed an abundance of stress response proteins compared with natural acidophilic communities (Belnap *et al.*, 2010). Biofilm communities in this study were dominated by *Leptospirillum* group II, with *Leptospirillum* group III present at lower abundance. In addition, many low abundance members including several archaeal species such as *Ferroplasma* types I and II and 'G-plasma', were detected (Belnap *et al.*, 2010). The majority of proteins for low abundance community members, such as *Leptospirillum* group III and 'G-plasma', were related to essential functions such as energy conversion, nucleotide metabolism, transcription, translation and coenzyme metabolism. Several chaperone proteins belonging to *Ferroplasma* type I and 'A-plasma' were detected, and were ~ 4–20 times more abundant in the laboratory community compared with the natural communities, suggesting high stress levels for these organisms. In addition, 'G-plasma' and 'E-plasma' proteins involved in fatty acid metabolism such as acetyl CoA acetyltransferase, acyl-CoA synthetase and enoyl-CoA hydratase were detected in the laboratory system (Belnap *et al.*, 2010).

The response of these acidophilic microbial communities to different pH conditions (pH values of 1.45 or 0.85) has also been analysed by quantitative proteomics (Belnap *et al.*, 2011). Comparison of specific protein abundances indicates that low-pH conditions promote the growth of certain *Archaea* in *in vitro* biofilms. These authors were able to detect more than 2000 proteins, of which ~ 300 were from *Ferroplasma* spp. type I and type II, ~ 280 from the Thermoplasmatales 'G-plasma', and considerably fewer from 'A-plasma', 'E-plasma' and 'I-plasma', reflecting the activity of the acidophilic archaea in the Richmond system. A high number of proteins from G-plasma were identified in the low-pH condition. These included ribosomal proteins, enzymes involved in energy conversion reactions, as well as carbohydrate transport proteins. A similar trend was observed for *Ferroplasma* type II, with several hypothetical proteins from this microorganism being among the most abundant proteins detected. In contrast to both G-plasma and *Ferroplasma* type II, *Ferroplasma* type I showed abundant proteins related to amino acid and multidrug transporters, ribosomal proteins and peroxiredoxin at high pH, and proteins related to energy conversion pathways at low pH. These results confirm the importance of pH and related geochemical factors in fine-tuning acidophilic microbial community structure and function at the species and strain level, and demonstrate the broad utility of proteomics in laboratory community studies (Belnap *et al.*, 2011; see Chapter 13).

Possible industrial applications derived from proteomics

Secretome analysis and effects on bioleaching

During dissolution of ores, some *At. ferrooxidans* cells adhere to the solid surface of the mineral via hydrophobic and hydrophilic interactions (Arredondo *et al.*, 1994; Devasia *et al.*, 1993; Ohmura *et al.*, 1993). Elemental sulfur is generated as an intermediate product of the bioleaching process and can accumulate in the surface of the mineral (Rohwerder *et al.*, 2003). Some acidophilic bacteria capable of oxidizing this elemental sulfur also adhere to the mineral surface. The combined action of secreted proteins, exopolysaccharides, and other extracellular products may be critical in the initial attachment of bacteria to the minerals (Rohwerder *et al.*, 2003; Amaro *et al.*, 1993).

The identification of the proteins secreted by *At. ferrooxidans* grown in elemental sulfur has been considered of great significance for the understanding of the mechanisms that allow the bacterium to obtain energy from the minerals and to survive in its extreme environment. One of the first attempts to recover extracellular proteins from bioleaching bacteria compared *At. ferrooxidans* ATCC 23270 grown in elemental sulfur and on ferrous sulfate (Zhang *et al.*, 2008). Extracellular proteins were obtained by incubating cell pellets in acid at 60°C and further separated by 2D electrophoresis. Some abundant spots present in the sulfur-grown preparation were identified by MALDI-TOF/TOF. Most of the identified spots corresponded to hypothetical proteins, with the exception of a conjugal transfer protein, a pilus protein, two lipoproteins, a polysaccharide deacetylase and a Ser/Thr phosphatase. Noteworthy, the method performed by Zhang *et al.* (2008) would only identify proteins released from the outer membrane after acid treatment.

In another approach, the extracellular proteins of *At. ferrooxidans* were obtained from the acidic culture supernatant of cells grown in elemental sulfur (Pagliai and Jerez, 2009). Initially, the culture supernatant of cells grown in elemental sulfur was used as the source of extracellular proteins, since greater amounts of proteins were obtained in the presence of this substrate compared to those from pyrite- or chalcopyrite-grown *At. ferrooxidans* cells.

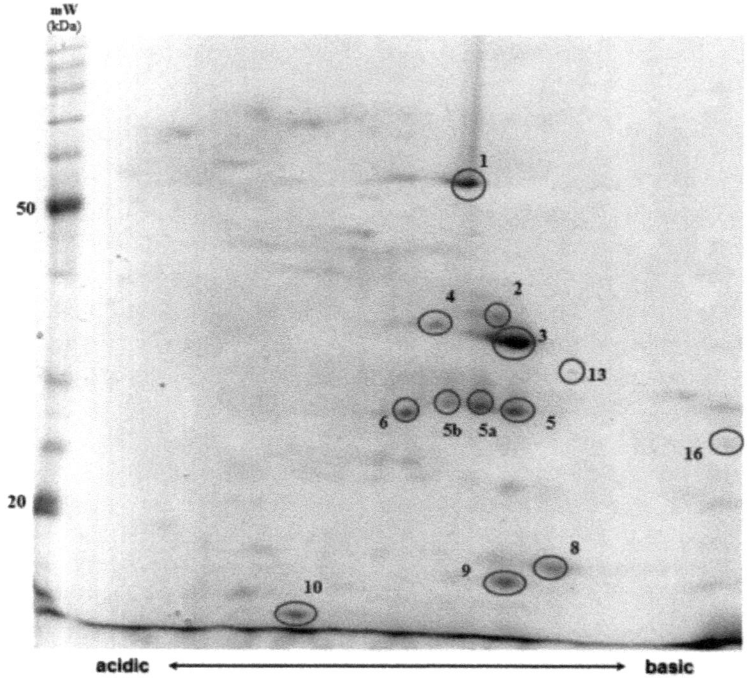

Figure 14.3 2D-NEPGHE of the extracellular proteins of *At. ferrooxidans* grown in elemental sulfur. *At. ferrooxidans* ATCC 23270 was grown in elemental sulphur. The excised spots subjected to MS sequencing are circled.

The extracellular proteins of *At. ferrooxidans* were separated by 2D-NEPHGE (Fig. 14.3) and the protein spots were visualized using Blue Silver staining. The most intense spots were quantified and excised from the gel, followed by the determination of their amino acid sequence via mass spectrometry (ESI-QTOF). Many of the proteins identified in the extracellular fraction showed basic isoelectric points and molecular weights under 50 kDa. This profile of small and basic proteins correlated with the proteins found in the periplasm of *At. ferrooxidans* (Chi *et al.*, 2007). This would be expected in part, since most secreted proteins, after crossing the inner membrane via Sec translocase, can be found in the periplasmic space during their transit to the external medium. However, some of the highly represented periplasmic proteins were not found in the extracellular fraction. Likewise, the elongation factor Tu, one of the major cytoplasmic proteins, could not be detected via Western blot in the extracellular fraction of *At. ferrooxidans*. These results suggest that the presence of many of the proteins identified in the extracellular medium is not the result of either cell lysis or periplasm leaking, but most likely the action of the bacterium secretion machinery. The major functional roles of *At. ferrooxidans* secreted proteins were solute-binding and transport proteins, cell envelope, energy metabolism and folding proteins. For instance, spot 1 was described as an outer membrane-associated tetrathionate hydrolase (TTH) (Kanao *et al.*, 2007). It has been suggested that the THH would be the major protein responsible of tetrathionate consumption in the extracellular media of *At. ferrooxidans* grown in oxygen-limited and oxygen–sufficient conditions. This extracellular TTH activity is related to the formation of sulfur globules, which may be involved in the adaptation of this bacterium to an anaerobic lifestyle (Beard *et al.*, 2011).

However, the function of most of these secreted proteins in the extracellular space remains to be elucidated. A very interesting preliminary result showed that secretome extracts from *A. ferrooxidans* or *At. thiooxidans* when added to fresh *At. ferrooxidans* cell cultures growing in elemental sulfur greatly shortens the lag period of growth. On the contrary, a control secretome extract from

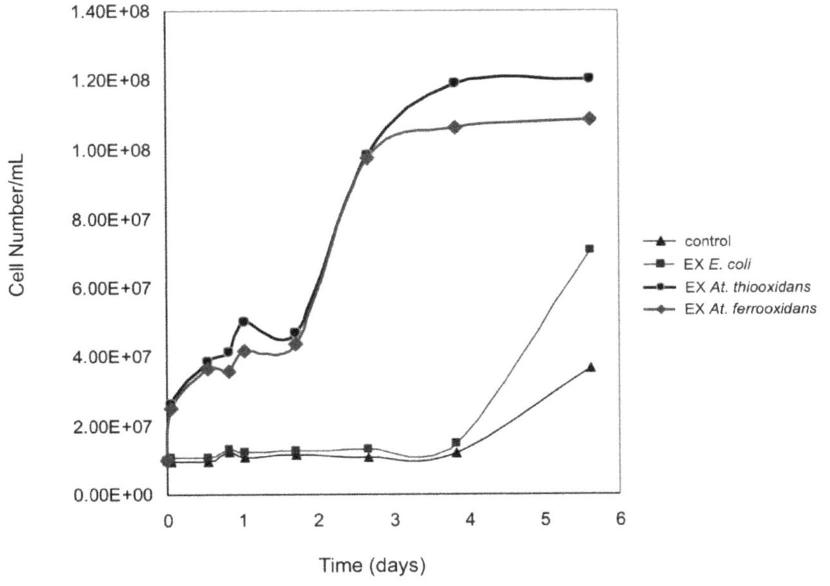

Figure 14.4 Effect of adding secretome proteins to the culture medium on the growth of *At. ferrooxidans*. *At. ferrooxidans* ATCC 23270 was grown in elemental sulfur and the indicated secretome (EX) was added at the beginning of growth.

E. coli did not produce this effect (INAPI patent application, 2009) (Fig. 14.4).

In this regard, the extracellular proteins of a mixed culture of *At. thiooxidans* DSM 17318 and *At. ferrooxidans* DSM 16786 grown in elemental sulfur were obtained by concentrating the culture supernatants and separating the concentrated extracellular proteins by 2D-PAGE. The most intense spots were cut from the gel and their amino acid sequence identified by LC-MS/MS. A putative lipoprotein (named Licanantase) was found as the major component of the *At. thiooxidans* pure and mixed culture's secretome and this preparation stimulated copper bioleaching rate when added to cells grown in chalcopyrite (Bobadilla-Fazinni and Parada, 2009). A second major hit in this secretome was a porin with high amino acid similarity to the Omp40/TTH previously described in *At. ferrooxidans* (Beard *et al.*, 2011; Guiliani and Jerez, 2000).

The role of individual proteins from the *At. ferrooxidans* secretome (Pagliai and Jerez, 2009) on the growth of biomining microorganisms is currently under study since it is expected that these extracts or purified proteins from the secretomes enhance not only the growth of biomining microorganisms on copper ores but also increase metal recovery. In this context, it has been claimed that a lipoprotein secreted by *At. thiooxidans* significantly enhances the rate of chalcopyrite dissolution (Bobadilla-Fazinni *et al.*, 2011). It will be of great interest to find out whether secretomes from moderate thermophiles and hyperthermophiles show similar effects on growth and mineral dissolution. If that were the case, the secretome proteins would constitute an excellent improvement for metal extraction by using hyperthermophiles grown in reactors. The microorganisms used in these processes should be highly resistant to copper and other metals (Dopson and Holmes, 2014; McCarty *et al.*, 2014; Jerez, 2013; Orellana and Jerez, 2011).

Conclusions

Proteomic methods together with the other omic procedures have greatly facilitated our understanding of some of the mechanisms extreme acidophiles use to adapt to their environment, both at the organismal and communal level. Proteomics has been used to identify the localization of proteins within the cells, to understand protein complexes and networks, to assess co-translational and post-translational modification of proteins, to compare growth conditions and to investigate the molecular physiology of acidophilic microbial communities.

Compared to other ecosystems, acidophilic communities have a relatively low diversity, with only a few dominant taxa. This has allowed the use of metaproteomics and proteogenomics approaches to resolve in great detail organization and function of acidophilic consortia. Nevertheless, full understanding of biology and ecology of these microbial communities awaits further methodological and technological improvements to obtain, handle and quantitate protein samples, as well as more powerful bioinformatic tools to handle the increasing amounts of data obtained. Although instruments and procedures used in proteomics have greatly improved in recent years, further advances will be needed to have ultrasensitive and extremely accurate mass spectrometers in order to identify and measure the wide range of proteins present in the cells in varying abundances.

Another important challenge required to complement and confirm proteomic information is the development of efficient genetic tools to transform and generate knockout mutations in acidophiles. This would help to assign function to many hypothetical and unknown proteins changing their expression levels under different growth conditions as detected by proteomic analysis. Additionally, this knowledge could have an impact on the manipulation of acidophiles of biotechnological interest for further application.

Acknowledgements

Our work was supported by grants FONDECYT 1110214 and in part by FONDECYT 1150791.

References

Almárcegui, R.J., Navarro, C.A., Paradela, A., Albar, J.P., von Bernath, D., and Jerez, C.A. (2014a). New copper resistance determinants in the extremophile *Acidithiobacillus ferrooxidans*: A quantitative proteomic analysis. J. Proteome Res. *13*, 946–960.

Almárcegui, R.J., Navarro, C.A., Paradela, A., Albar, J.P., von Bernath, D., and Jerez, C.A. (2014b). Response to copper of *Acidithiobacillus ferrooxidans* ATCC 23270 grown in elemental sulfur. Res. Microbiol. *165*, 761–772.

Amaro, A.M., Seeger, M., Arredondo, R., Moreno, M., and Jerez, C.A. (1993). The growth conditions affect *Thiobacillus ferrooxidans* attachment to solids. Biohydrometall. Technol. II. p. 577–585.

Arredondo, R., García, A., and Jerez, C.A. (1994). Partial removal of lipopolysaccharide from *Thiobacillus ferrooxidans* affects its adhesion to solids. Appl. Environ. Microbiol. *60*, 2846–2851.

Baker-Austin, C., Dopson, M., Wexler, M., Sawers, R.G., and Bond, P.L. (2005). Molecular insight into extreme copper resistance in extremophilic archaeon *'Ferroplasma acidarmanus'* Fer1. Microbiology *151*, 2637–2646.

Baker-Austin, C., Dopson, M., Wexler, M., Sawers, R.G., Stemmler, A., Rosen, B.P., and Bond, P.L. (2007). Extreme arsenic resistance by the acidophilic archaeon *'Ferroplasma acidarmanus'* Fer1. Microbiology *11*, 425–434.

Baker-Austin, C., Potrykus, J., Wexler, M., Bond, P.L., and Dopson, M. (2010). Biofilm development in the extremely acidophilic archaeon *'Ferroplasma acidarmanus'* Fer1. Extremophiles *14*, 485–491.

Bathe, S., and Norris, P.R. (2007). Ferrous iron- and sulfur-induced genes in *Sulfolobus metallicus*. Appl. Environ. Microbiol. 73, 2491–2497.

Beard, S., Paradela, A., Albar, J.P., and Jerez, C.A. (2011). Growth of *Acidithiobacillus ferrooxidans* ATCC 23270 in thiosulfate under oxygen-limiting conditions generates extracellular sulfur globules by means of a secreted tetrathionate hydrolase. Front. Microbiol. 2, 79.

Belnap, C.P., Pan, C., VerBerkmoes, N.C., Power, M.E., Samatova, N.F., Carver, R.L., Hettich, R.L., and Banfield, J.F. (2010). Cultivation and quantitative proteomic analyses of acidophilic microbial communities. ISME J. *4*, 520–530.

Belnap, C.P., Pan, C., Denef, V.J., Samatova, N.F., Hettich, R.L., and Banfield, J.F. (2011). Quantitative proteomic analyses of the response of acidophilic microbial communities to different pH conditions. ISME J. *5*, 1152–1161.

Bobadilla-Fazzini, R., and Parada, P. (2009). Analysis of sulfur metasecretome in mixed cultures of *Acidithiobacillus thiooxidans* and *Acidithiobacillus ferrooxidans* Adv. Mat. Res. *71–73*, 151–154.

Bobadilla Fazzini, R., Levican, G., and Parada, P. (2011). *Acidithiobacillus thiooxidans* secretome containing a newly described lipoprotein Licanantase enhances chalcopyrite bioleaching rate. Appl. Microbiol. Biotechnol. *89*, 771–780.

Buckley, M.R. (2004). Systems microbiology: beyond microbial genomics. In A Report of the American Academy of Microbiology, Buckley, M.R., ed. (American Academy of Microbiology, Washington, DC, USA), pp. 1–15.

Burg, D., Ng, C., Ting, L., and Cavicchioli, R. (2011). Proteomics of extremophiles. Environ. Microbiol. *13*, 1934–1955.

Chi, A., Valenzuela, L., Beard, S., Mackey, A.J., Shabanowitz, J., Hunt, D.F., and Jerez, C.A. (2007). Periplasmic proteins of the extremophile *Acidithiobacillus ferrooxidans*: a high throughput proteomics analysis. Mol. Cell. Proteomics. *6*, 2239–2251.

Chong, P.K., and Wright, P.C. (2005). Identification and characterization of the *Sulfolobus solfataricus* P2 proteome. J. Proteome. Res. *4*, 1789–1798.

Cobucci-Ponzano, B., Guzzini, L., Benelli, D., Londei, P., Perrodou, E., Lecompte, O., Tran, D., Sun, J., Wei, J., Mathur, E.J., Rossi, M., and Moracci, M. (2010). Functional characterization and high-throughput proteomic analysis of interrupted genes in the archaeon *Sulfolobus solfataricus*. J. Proteome. Res. 9, 2496–2507.

Denef, V.J., Kalnejais, L.H., Mueller, R.S., Wilmes, P., Baker, B.J., Thomas, B.C., VerBerkmoes, N.C., Hettich,

R.L., and Banfield, J.F. (2010). Proteogenomic basis for ecological divergence of closely related bacteria in natural acidophilic microbial communities. Proc. Natl. Acad. Sci. U.S.A. *107*, 2383–2390.

Devasia, P., Natarajan, K.A., Sathyanarayana, D.N., and Rao, G.R. (1993). Surface chemistry of *Thiobacillus ferrooxidans* relevant to adhesion on mineral surfaces. Appl. Environ. Microbiol. *59*, 4051–4055.

Dopson, M., and Holmes, D.S. (2014). Metal resistance in acidophilic microorganisms and its significance for biotechnologies. Appl. Microbiol. Biotechnol. *98*, 8133–8144.

Dopson, M., Baker-Austin, C., and Bond, P.L. (2005). Analysis of differential protein expression during growth states of *Ferroplasma* strains and insights into electron transport for iron oxidation. Microbiology *151*, 4127–4137.

Ellen, A.F., Albers, S.-V., and Driessen, A.J.M. (2010). Comparative study of the extracellular proteome of *Sulfolobus* species reveals limited secretion. Extremophiles *14*, 87–98.

Felício, A.P., Oliveira, E., Odena, M.A., Garcia, O., Bertolini, M.C., Caldas, L.F., Mariscal, L.M., and Marques, M.T. (2011). Differential proteomic analysis of *Acidithiobacillus ferrooxidans* cells maintained in contact with bornite or chalcopyrite: Proteins involved with the early bacterial response. Process Biochem. *46*, 770–776.

Gevaert, K., Impens, F., Ghesquière, B., Van Damme, P., Lambrechts, A., and Vandekerckhove, J. (2008). Stable isotopic labeling in proteomics. Proteomics *8*, 4873–4885.

Guiliani, N., and Jerez, C.A. (2000). Molecular cloning, sequencing and expression of Omp40, the gene coding for the major outer membrane protein from the acidophilic *Thiobacillus ferrooxidans*. Appl. Environ. Microbiol. *66*, 2318–2324.

He, Z., Zhong, H., Hu, Y., Xiao, S., Liu, J., Xu, J., and Li, G. (2005). Analysis of differential-expressed proteins of *Acidithiobacillus ferrooxidans* grown under phosphate starvation. J. Biochem. Mol. Biol. *38*, 545–549.

Jerez, C.A. (2008). The use of genomics, proteomics and other OMIC technologies for the global understanding of biomining microorganisms. Hydrometallurgy *94*, 162–169.

Jerez, C.A. (2013). The use of extremophilic microorganisms in industrial recovery of metals. In Extremophiles: Sustainable Resources and Biotechnological Implications, O. Songh, ed. (John Wiley & Sons, Inc., Hoboken, NJ, USA), pp. 319–334.

Kanao, T., Kamimura, K., and Sugio, T. (2007). Identification of a gene encoding a tetrathionate hydrolase in *Acidithiobacillus ferrooxidans*. J. Biotechnol. *132*, 16–22.

Koerdt, A., Orell, A., Pham, T.K., Mukherjee, J., Wlodkowski, A., Karunakaran, E., Biggs, C.A., Wright, P.C., and Albers, S.-V. (2011). Macromolecular fingerprinting of *Sulfolobus* species in biofilm: A transcriptomic and proteomic approach combined with spectroscopic analysis. J. Proteome Res. *10*, 4105–4119.

Kucera, J., Bouchal, P., Lochman, J., Potesil, D., Janiczek, O., Zdrahal, Z., and Mandl, M. (2013). Ferrous iron oxidation by sulfur-oxidizing *Acidithiobacillus ferrooxidans* and analysis of the process at the levels of transcription and protein synthesis. Antonie Van Leeuwenhoek *103*, 905–919.

Li, B., Lin, J., Mi, S., and Lin, J. (2010). Arsenic resistance operon structure in *Leptospirillum ferriphilum* and proteomic response to arsenic stress. Bioresour. Technol. *101*, 9811–9814.

McCarthy, S., Chenbing, A., Wheaton, G., Tevatia, R., Eckrich, V., Kelly, R., and Blum, P. (2014). Role of an Archaeal PitA transporter in the copper and arsenic resistance of *Metallosphaera sedula*, an extreme thermoacidophile. J. Bacteriol. *196*, 3562–3570.

Maezato, Y., Johnson, T., McCarthy, S., Dana, K., and Blum, P. (2012). Metal resistance and lithoautotrophy in the extreme thermoacidophile *Metallosphaera sedula*. J. Bacteriol. *194*, 6856–6863.

Mangold, S., Valdés, J., Holmes, D.S., and Dopson, M. (2011). Sulfur metabolism in the extreme acidophile *Acidithiobacillus caldus*. Front. Microbiol. *2*, 17.

Mangold, S., Rao Jonna, V., and Dopson, M. (2013a). Response of *Acidithiobacillus caldus* toward suboptimal pH conditions. Extremophiles *17*, 689–696.

Mangold, S., Potrykus, J., Bjorn, E., Lovgren, L., and Dopson, M. (2013b). Extreme zinc tolerance in acidophilic microorganisms from the bacterial and archaeal domains. Extremophiles *17*, 75–85.

Maupin-Furlow, J.A., Humbard, M.A., and Kirkland, P.A. (2012). Extreme challenges and advances in archaeal proteomics. Curr. Opin. Microbiol. *15*, 351–356.

Navarro, C.A., Orellana, L.H., Mauriaca, C., and Jerez, C.A. (2009). Transcriptional and functional studies of *Acidithiobacillus ferrooxidans* genes related to survival in the presence of copper. Appl. Environ. Microbiol. *75*, 6102–6109.

Ohmura, N., Kitamura, K., and Saiki, H. (1993). Selective adhesion of *Thiobacillus ferrooxidans* to pyrite. Appl. Environ. Microbiol. *59*, 4044–4050.

Orell, A., Navarro, C.A., Arancibia, F., Mobarec, J.C., and Jerez, C.A. (2010). Life in blue: copper resistance mechanisms of Bacteria and Archaea used in industrial biomining of minerals. Biotechnol. Adv. *28*, 839–848.

Orell, A., Navarro, C.A., Rivero, M., Aguilar, J.S., and Jerez, C.A. (2012). Inorganic polyphosphates in extremophiles and their possible functions. Extremophiles. *23*, 573–583.

Orell, A., Remonsellez, F., Arancibia, R., and Jerez, C.A. (2013). Molecular characterization of copper and cadmium resistance determinants in the biomining thermoacidophilic archaeon *Sulfolobus metallicus*. Archaea *2013*, 289236.

Orellana, L.H., and Jerez, C.A. (2011). A genomic island provides *Acidithiobacillus ferrooxidans* ATCC 53993 additional copper resistance: a possible competitive advantage. Appl. Microbiol. Biotechnol. *92*, 761–767.

Osorio, H., Mangold, S., Denis, Y., Ñancucheo, I., Esparza, M., Johnson, D.B., Bonnefoy, V., Dopson, M., and Holmes, D.S. (2013). Anaerobic sulfur metabolism coupled to dissimilatory iron reduction in the extremophile *Acidithiobacillus ferrooxidans*. Appl. Environ. Microbiol. *79*, 2172–2181.

Ouyang, J., Guo, W., Li, B., Gu, L., Zhang, H., and Chen, X. (2013). Proteomic analysis of differential protein expression in *Acidithiobacillus ferrooxidans* cultivated in high potassium concentration. Microbiol. Res. *168*, 455–460.

Pagliai, F.A., and Jerez, C.A. (2009). The secretome of the extremophilic *Acidithiobacillus ferrooxidans* ATCC 23270. Adv. Mat. Res. *71–73*, 183–186.

Paradela, A., Marcilla, M., Navajas, R., Ferreira, L., Ramos-Fernandez, A., Fernández, M., Mariscotti, J.F., García-del Portillo, F., and Albar, J.P. (2010). Evaluation of isotope-coded protein labeling (ICPL) in the quantitative analysis of complex proteomes. Talanta *80*, 1496–1502.

Pham, T.K., Sierocinski, P., Van der Oost, J., and Wright, P.C. (2010). Quantitative proteomic analysis of *Sulfolobus solfataricus* membrane proteins. J. Proteome Res. *9*, 1165–1172.

Potrykus, J., rao Jonna, V., and Dopson, M. (2011). Iron homeostasis and responses to iron limitation in extreme acidophiles from the *Ferroplasma* genus. Proteomics. *11*, 52–63.

Ram, R.J., VerBerkmoes, N.C., Thelen, M.P., Tyson, G.W., Baker, B.J., Blake, R.C., Shah, M., Hettich, R.L., and Banfield, J.F. (2005). Community proteomics of a natural microbial biofilm. Science *308*, 1915–1920.

Rensing, C., and Grass, G. (2002). *Escherichia coli* mechanisms of copper homeostasis in a changing environment. FEMS Microbiol. Rev. *27*, 197–213.

Rohwerder, T., Gehrke, T., Kinzler, K., and Sand, W. (2003). Bioleaching review part A: progress in bioleaching: fundamentals and mechanisms of bacterial metal sulfide oxidation. Appl. Microbiol. Biotechnol. *63*, 239–248.

Schmidt, A., Kellermann, J., and Lottspeich, F. (2005). A novel strategy for quantitative proteomics using isotope-coded protein labels. Proteomics. *5*, 4–15.

Tebbe, A., Klein, C., Bisle, B., Siedler, F., Scheffer, B., Garcia-Rizo, C., Wolfertz, J., Hickmann, V., Pfeiffer, F., and Oesterhelt, D. (2005). Analysis of the cytosolic proteome of *Halobacterium salinarum* and its implication for genome annotation. Proteomics. *5*, 168–179.

Valenzuela, L., Chi, A., Beard, S., Orell, A., Guiliani, N., Shabanowitz, J., Hunt, D.F., and Jerez, C.A. (2006). Genomics, metagenomics and proteomics in biomining microorganisms. Biotechnol. Adv. *24*, 197–211.

Valenzuela, L., Chi, A., Beard, S., Shabanowitz, J., Hunt, D.F., and Jerez, C.A. (2008). Differential-expression proteomics for the study of sulfur metabolism in the chemolithoautotrophic *Acidithiobacillus ferrooxidans*. In Microbial Sulfur Metabolism, Dahl, C., and Friedrich, C.G., eds (Springer-Verlag, Berlin, Heidelberg, Germany), pp. 77–86.

Van Oudenhove, L., and Devreese, B. (2013). A review on recent developments in mass spectrometry instrumentation and quantitative tools advancing bacterial proteomics. Appl. Microbiol. Biotechnol. *97*, 4749–4762.

Vera, M., Krok, B., Bellenberg, S., Sand, W., and Poetsch, A. (2013). Shotgun proteomics study of early biofilm formation process of *Acidithiobacillus ferrooxidans* ATCC 23270 on pyrite. Proteomics *13*, 1133–1144.

VerBerkmoes, N.C., Denef, V.J., Hettich, R.L., and Banfield, J.F. (2009). Functional analysis of natural microbial consortia using community proteomics. Nat. Rev. Microbiol. *7*, 196–205.

Watling, H.R. (2006). The bioleaching of sulphide minerals with emphasis on copper sulphides-a review. Hydrometallurgy. *84*, 81–108.

Wiese, S., Reidegeld, K.A., Meyer, H.E., and Warscheid, B. (2007). Protein labeling by iTRAQ: a new tool for quantitative mass spectrometry in proteome research. Proteomics *7*, 340–350.

Wilmes, P., and Bond, P.L. (2004). The application of two-dimensional polyacrylamide gel electrophoresis and downstream analyses to a mixed community of prokaryotic microorganisms. Env. Microbiol. *6*, 911–920.

Zhang, C., Zhang, R., Xia, J., Zhang, Q., and Nie, Z. (2008). Sulfur activation-related extracellular proteins of *Acidithiobacillus ferrooxidans*. Trans. Nonferrous. Met. Soc. China. *18*, 1398–1402.

Metabolomic Approaches to the Study of Acidophiles

15

Patricio Martínez and Pilar Parada

Abstract

After the rapid development of genome sequencing tools in recent decades, the holistic study of biological systems has been gaining strength. Additionally, new technologies of high-throughput analysis such as transcriptomics, proteomics and metabolomics, have helped to integrate different types of data favouring a more comprehensive understanding of cells physiology and ecology. Metabolomics, in particular, has allowed the study of the complexities of cellular metabolism, facilitating the overall analysis of metabolic pathways and the understanding of the biochemistry associated. Metabolic biomarkers derived from metabolomic studies, that allow monitoring cell behaviour, have become important tools in cellular physiology studies and in biotechnological applications.

In acidophiles, advanced scientific studies have provided a better understanding of their physiology and utilization in the biotechnological industry. However, much remains to be investigated, and many of these microorganisms are still unknown or have not been successfully isolated. The use of high-throughput 'omics' tools, such as metabolomics, has contributed to integrated data analysis of their behaviour and complexity, giving a new view of the acidophiles system biology. In this chapter, we review the use of metabolomics as an advanced platform for the study of acidophile biology and its contribution and prospects for biotechnological applications.

Introduction to metabolomics

Metabolomics is one of the youngest omics sciences and comprises the global study of cell metabolism using high-throughput techniques for the identification and quantification of metabolites. This discipline has contributed with an integral understanding of metabolic pathways and their variations in different organisms, has enabled the discovery and detection of new diseases through the identification of new metabolite biomarkers and has generated new data relevant to the study of systems biology.

From a historical perspective, one of the most important milestones in the development of this discipline were the studies by the Nobel Prize in Physics, Joseph John Thomson, on the basic foundations of mass spectrometry (Thomson, 1912). Following his discoveries, modern techniques of mass spectrometry (MS) were revised by Arthur Jeffrey Dempster and Francis William Aston in 1918 and 1919, respectively. Their work laid the foundation for modern mass spectrometry, contributing to significant improvements in accuracy and sensitivity of the MS equipment, which were key to the implementation of high resolution techniques in metabolomics and other disciplines. In parallel, the development of other detection techniques such as nuclear magnetic resonance (NMR) in the 1930s, and molecular separation by chromatography during 1930s and 1940s, allowed to approach the analysis of complex samples identifying multiple metabolites.

The first metabolomic study ever published, though named differently at the time, was a study of urine and breath human samples by Linus Pauling and Arthur Robinson in 1971, reporting that the chemical constituents of the samples were loaded with useful information that could be correlated with health and the treatment of disease (Pauling *et al.*, 1971). After 27 years of analytical performances and metabolic studies, in 1998 the term of metabolomics was finally coined by Stephen Oliver and colleagues (Oliver *et al.*, 1998) while performing functional studies in yeasts. In 2005, the first publicly available metabolomic database was released (METLIN), helping with the characterization of metabolites and providing reliable data from mass spectrometry. Currently this database contains over 240,000 metabolites and over 62,000 high resolution mass spectra.

In particular, metabolomics applied to the study of microorganisms has increased over the past 20 years due to the growing number of sequenced genomes, which has greatly facilitated the acquisition and interpretation of data. This is also the case of acidophiles. The present chapter gives a special look into the progress of metabolomics studies of acidophiles and their contributions to the understanding of the biology of this group of extremophiles.

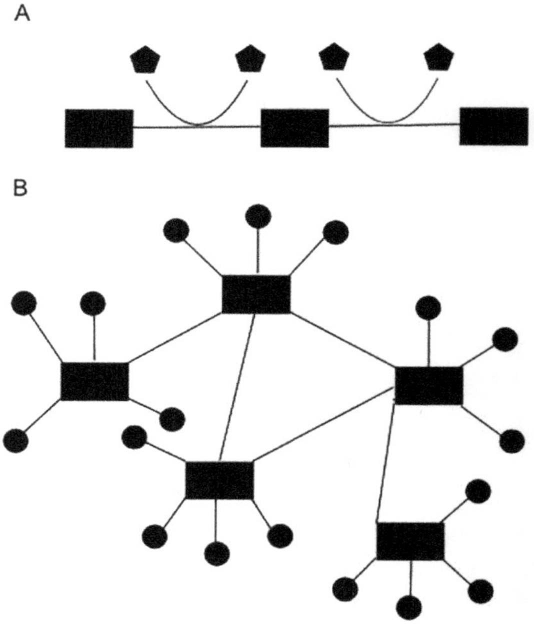

Figure 15.1 General scheme for metabolic studies. (A) Traditional 'linear' view of a metabolic pathway. (B) Integrated 'metabolic network', where groups of metabolites respond to fluxes and interact with other groups in complex ways. Modified from Ryan and Robards (2006).

The study of cellular metabolism in the omics era

The study of cellular metabolism is not a new discipline. Even if metabolism is reckoned to be an integrated process involving a large number of interconnected pathways and reactions, its study has been mostly pathway- and enzyme-centric. Analyses of metabolites have long focused on a relatively small group of substances, usually related to energy producing pathways such as glycolysis and cellular respiration, and have been carried out using 'linear' analytical approaches (Fig. 15.1A). The modern view, however, favours the study of metabolism using holistic approaches (Krastanov, 2010; Ryan and Robards, 2006) which facilitate the understanding of the metabolic network inside the cell (Fig. 15.1B).

Recent advances in the so-called 'omic' disciplines (genomics, transcriptomics, proteomics, and metabolomics) and in data integration using bioinformatics tools has allowed the reconstruction of complete metabolic networks in different microorganisms and improved our understanding of metabolic functionality, as well as the effects of genetic mutations in cell metabolism (e.g. Edwards and Palsson, 2000). This trend has strongly promoted the development and improvement of techniques suitable for high-throughput proteomic and metabolomic analyses, such as gas chromatography (GC), liquid chromatography (LC) and capillary electrophoresis (CE) for molecular separation and MS and NMR for detection or identification of molecules. Thanks to these techniques the metabolism of an organism can now be studied globally through targeted or untargeted metabolomics (Roberts *et al.*, 2012; Vinayavekhin and Saghatelian, 2010). While the former requires the use of standards, the latter identifies metabolites based on mass ranges without the use of standards, and has the potential to detect uncharacterized metabolic components. Metabolomics can relate the state of a cell under certain environmental conditions to unknown metabolic pathways composed of unidentified proteins. Furthermore, metabolomic studies have made it possible to measure both intracellular and extracellular metabolites related to an ever growing list of processes such as ion homeostasis, redox state, nutrition cycle, energy, and intercellular communication, among others. In short, the capability to measure metabolite pools allow us to understand the dynamics of gene expression, complete pathways and biological processes, facilitating our understanding of microbial behaviour and adaptation. Additionally, metabolomics facilitates the identification of biomarkers that allow correlation of metabolic cellular responses to particular conditions, and thus has become an extremely useful tool for biological diagnostics. Thanks to this development, many metabolic models produced by genomic and metagenomic projects have begun to be validated through direct detection of the metabolites involved. These important advances in the study of cellular metabolism have favoured the incorporation of metabolomics in the 'omics' platforms of various scientific disciplines, providing valuable input with respect to the complexity and integration of a systems biology approach.

Difficulties and limitations of metabolomics studies

Despite being a robust platform, metabolomics is not free from the complications of emerging methodologies. The physicochemical properties of the metabolites, coupled with the complexity of the biological samples, make it impossible to use standard analytical methods and raising the need for newly developed protocols.

High-throughput metabolome analysis is significantly more complex than other well known high-throughput omics approaches (e.g. proteomics). In the analysis of biochemical target compounds, genomics, transcriptomics and proteomics involve only four nucleotides or 22 amino acids, respectively. Moreover, the chemical similarity of these compounds facilitates both data analysis and interpretation. In contrast, metabolomics involves greater numbers of highly diverse chemical compounds (for example, hydrophilic, hydrophobic, and volatile groups of hundreds of compounds in a single sample). Because of this, the selection of the methods used for identification must be made with extreme care and with a clear focus on targets. Another issue that is very important in quantitative analysis is the matrix effect, which can disturb a sample preparation and affect metabolite extraction, alter the quantity of certain metabolites and mask other metabolites, generating inconsistency in data interpretation. To avoid these problems, the use of internal standards as references for evaluating extraction protocols is very important. Mixes of standard metabolites are also being used to identify and quantify targeted metabolites. For untargeted metabolomic analysis, the use of stable isotopes should be considered, as it can be very difficult to correlate biological significance with identified molecules and filter out the large number of artefacts that can be generated with this type of highly sensitive technology (Burton *et al.*, 2008). Finally, it is essential to emphasize that qualitative and quantitative analysis must be carried out under stringent standards and that rigorous statistical analysis design need to be considered to ensure reproducibility and significance of the data collected.

Current knowledge of the metabolism of acidophiles

Genomic sequencing and data analysis have provided interesting insights into several aspects of the metabolism of extreme acidophiles (Chapter 11). Valuable information has also been derived from gene and protein expression analysis using 'omic' disciplines such as microarray analysis (Quatrini *et al.*, 2009) and proteomic approaches (Chapter 14) in model acidophiles, as well as in communities of coexisting microorganisms (Chapter 13). In recent years a number of metabolomics studies of acidophiles have been performed to understand these microorganisms, their relationship with the environment, their functionality and ecology when living in community and their use in biotechnological applications. These studies have pointed out the importance of certain molecules in the metabolic performance of microorganisms when forming biofilm versus being in a planktonic condition and how this is related to their ability to endure metal-rich environments in both AMD or bioleaching operations (Martínez *et al.*, 2013, Mosier *et al.*, 2013; Halter *et al.*, 2012; Fischer *et al.*, 2011). Relevant aspects emerging form snapshots of the entire physiology of selected acidophiles obtained under different conditions (e.g. biofilms, changes in the energy source, etc.) are addressed bellow.

Biological behaviours in need of metabolomics answers

Many aspects of the metabolism of acidophiles could be target of metabolomic analysis. Some of the most compelling form an applied perspective are discussed below.

The oxidizing capability of some acidophiles is the driving force of biotechnological applications such as the bioleaching of metal sulfides for valuable metal recovery (Chapter 16), and thus significant research efforts have focused on understanding two principal functionalities: sulfur and iron oxidation. Mechanisms of sulfur and iron oxidation have been extensively studied in the *Acidithiobacillus* and valuable information has been obtained regarding the genes, enzymes and pathways involved in these processes. Little is known, however, about the role of *Acidithiobacillus* metabolites in mineral dissolution and how these affect the efficiency of metal recovery. Even less is known about metabolites from other acidophiles such as *Leptospirillum*, *Sulfobacillus*, and *Ferroplasma*.

Back in 2001, Hallberg and Johnson (2001) defined two types of mineral leaching mechanisms, referred to as 'non-contact' and 'contact' leaching. In non-contact leaching planktonic microorganisms oxidize ferrous iron to ferric iron, which acts as a strong chemical oxidant attacking the mineral surface. In contact leaching, the microorganisms are attached to the surface of the sulfide minerals and embedded in a matrix formed by extracellular polymeric substances (EPS) and other metabolites and proteins, forming a biofilm. In this attached state, the EPS generates an interface between the microorganisms and the ore, where the electrochemical processes that solubilize the mineral are facilitated. It is important to understand that both mechanisms likely involve several different microorganisms, making it even more challenging to understand how they interact and contribute to the leaching process.

In non-contact leaching mediated by iron-oxidizing microorganisms, oxidation of ferrous ion to ferric ion depends directly on environmental conditions (Rawlings, 2005). Conditions that are unfavourable and affect cell viability (e.g. the presence of toxic elements) will affect oxidation of ferrous ion, resulting in lower metal recovery (Sand and Gehrke, 2006). The network of metabolic reactions being affected by these unfavourable conditions could be interrogated to determine which metabolites are undergoing major changes and to understand what the major physiological implications are. This type of analysis has been performed in different organisms, but not yet in acidophiles. For example, Booth *et al.* (2011) studied the effects of copper-mediated stress on metabolism in *Pseudomonas fluorescens* in the planktonic state versus biofilms. Observations gathered through metabolomic analyses pointed interesting differences in the metabolic response between the two groups analysed. The response of planktonic microorganisms was characterized by changes in the TCA cycle, glycolysis, and metabolism of pyruvate, nicotinate, and nicotinamide, while the response of cells in biofilms was characterized mostly by changes in the exopolysaccharide composition. This supports the existence of a protective response in biofilm cells with high metal resistance and a reactive response in planktonic cells with low metal resistance profiles.

Mineral adherence and biofilm formation play a major role in contact leaching of minerals (Rohwerder *et al.*, 2003). Microorganisms that produce EPS (composed of polysaccharides, proteins and lipids) have an active role in mineral dissolution (Sand and Gehrke, 2006). Gene expression studies in *Acidithiobacillus ferrooxidans* have demonstrated differential regulation of EPS synthesis gene operons when the cells were grown in either iron or sulfur (Barreto *et al.*, 2005). Metabolomic studies could help identify the metabolite precursors used in the synthesis of EPS and to better understand the functional response of microorganisms faced with different mineral substrates.

Biofilm dynamics is a phenomenon that has been studied extensively in various microorganisms, helping to identify molecular signal modulators involved in a process called quorum sensing (Riedel *et al.*, 2001). In acidophiles, quorum sensing has been observed to regulate EPS expression and biofilm formation (González *et al.*, 2013). Using tandem MS (MS/MS), Ruiz *et al.* (2012) identified a messenger molecule in *At. ferrooxidans* (c-di-GMP) that was strongly related to adherence and biofilm formation through regulation of EPS. Using qualitative and quantitative analysis, these authors showed that c-di-GMP is primarily produced by cells adhering to the substrate and forming a biofilm. While the literature does provide interesting examples in other microorganisms where the use of metabolomics tools has provided a comprehensive understanding of the biofilm phenomenon and its implications (Takahashi *et al.*, 2012; Zhang and

Powers, 2012), there are only very few metabolomics studies in acidophiles touching upon this subject.

Advances in metabolomics of acidophiles

Richmond acid mine drainage biofilms

Recently, Mosier *et al.* (2013) conducted an interesting study of the metabolites associated with adaptation of microorganisms to acidic and metal-enriched conditions. A remarkable feature in this study is the analysis of microorganisms under natural conditions, where it is feasible to study members of an ecosystem that may not necessarily be cultured in the laboratory, an approach analogous to metagenomic studies. Although the complexity of acidic ecosystems and the technical difficulty of metabolite extraction from the environmental matrix complicates data analysis, the authors used tools that facilitate data collection in order to confirm the physiological significance of observed metabolites. These tools included stable isotopes, untargeted metabolomic analysis and MS. Samples of natural biofilm were obtained from acid mine drainage (Richmond Mine, Iron Mountain, California), categorized according to stage of development of the biofilm, and cultured in the laboratory in the presence of ^{15}N-labelled ammonium sulfate to facilitate identification of metabolites and the number of nitrogen atoms present. The microbial composition of the biofilm was determined by fluorescence *in situ* hybridization, which identified mostly members of *Leptospirillum* (groups II and III), archaea and (in lesser amounts) *Sulfobacillus*. Metabolites were extracted and analysed by LC–coupled MS, which detected 3500 ions with different retention times and mass/charge values. The use of blank and control conditions without ^{15}N eliminated the inherent noise of this extremely sensitive technology. Among these ions, 241 were released, confirming that 80% respond to criteria of specific technical modifications (e.g. adduct ion fragmentation). After manual depuration, 56 metabolites were identified and associated with chemical formulas.

The authors then used MS/MS to disrupt all 56 candidates and compare the molecular disruption profiles and the chemical formulas previously obtained against known databases (MetaCyc, KEGG, MassBank, and MET-LIN). The results indicated that more than 90% of the metabolites tested were unknown and did not match MS/MS spectra found in the databases. It is important to note that these databases are associated with commercial standards that are estimated to represent only half of the existing biological metabolites. Of the metabolites identified, the authors highlighted three that are likely to play significant roles in the biofilm analysed: phosphatidylethanolamine lipids, taurine, and hydroxyectoine. The phosphatidylethanolamine lipids detected (and also some derivatives) had been reported in previous studies of acidophilic microbial communities (Fischer *et al.*, 2011), and were associated with *Leptospirillum* group II (*L. ferriphilum*) based on correlations of lipid and protein abundance. An interesting role was suggested for this type of molecule in the biofilm: a potential mechanism of resistance to toxic cations released from the mineral environment by other members of the community. This type of lipid metabolite has also been found in eukaryotes, such as the acidophilic fungus *Acidomyces richmondensis*, an organism found in abundance in acid mine drainage. Genetic studies of this organism demonstrated the presence of eight genes involved in synthesis of phosphatidylethanolamine lipids, suggesting the production of such molecules in biofilms associated with bacteria and fungi.

Another metabolite found by Moiser *et al.* (2013) was taurine. This molecule has been widely studied and reported to have multiple functions in different organisms. It is utilized as the sole source of carbon, nitrogen and sulfur in some organisms, and as an osmotic regulator in others. This is the likely role in microorganisms present in extremely acidic environments and exposed to high concentrations of soluble ions. The authors analysed the genome sequences of the microorganisms known to be present in the environmental samples in search for genes related to taurine synthesis. They found that neither leptospirilli, sulfobacilli, nor *Ferroplasma* spp. had the genes required to synthesize taurine using known metabolic pathways, but that *Sulfobacillus* possessed transporters for taurine uptake into the cell in rapid response to unfavourable environmental conditions. Moreover, the fungus *A. richmondensis* could be responsible for taurine synthesis and degradation in biofilms, as the genes involved in taurine synthesis were found in its genome. It is noteworthy that the authors were unable to identify taurine in pure cultures of this organism by MS, possibly owing to unfavourable conditions for taurine synthesis in laboratory or a potential cooperative role.

A third metabolite detected in this study is hydroxyectoine, present in natural biofilms as well as laboratory-grown enriched biofilms. This molecule may act as a compatible solute in the adaptation to hyperosmotic stress, and can also provide resistance to high temperatures, dehydration and freezing. The genome of *Leptospirillum rubarum* has the genetic components for synthesis of this metabolite (Goltsman *et al.*, 2009) and a similar metabolite, ectoine, was also detected in this study (not confirmed by MS/MS). Previous proteomic analysis (Mueller *et al.*, 2011) detected proteins related to hydroxyectoine synthesis, especially in the early stages of biofilm formation, corroborating this observation. Other members of the biofilm community also have genes for ectoine synthesis, including *Ferroplasma* spp. and *Sulfobacillus* spp.

These studies contribute to our understanding of the metabolic dynamics of acidophiles exposed to extreme conditions. Some osmoprotectant metabolites are synthesized in greater quantities in the early stages of colonization of a mineral surface, supporting the hypothesis that microorganisms are more exposed to high ion concentrations in their planktonic state than in biofilms. This type of metabolites, generated in response to adverse conditions in an ecosystem are good candidates for use as biomarkers in industrial processes, as indicators of cell activity in biomining applications. Monitoring of these biomarkers can serve to alert unfavourable physiological conditions and suggest changes to the operating conditions.

Carnoulès acid mine drainage

Another interesting metabolomic study was conducted by Halter *et al.* (2012) in *Euglena mutabilis*, a protist found in acid mine drainages. This photosynthetic microorganism is believed to have an important role in the production of bioavailable organic matter for other organisms in oligotrophic ecosystems (Ñancucheo and Johnson, 2012). In order to understand their overall metabolism, the authors took samples of *E. mutabilis* from an abandoned mine in southern France (Carnoulès). Metabolomic analyses of *in situ* samples of cell fractions and of liquid fractions from surrounding waters were performed. In parallel, *in vitro* experiments were performed using isolated microorganism grown in the laboratory. In this case, metabolites were extracted from cell fractions and from culture medium, providing an endo- and exo-metabolome, respectively. Samples were analysed by GC–coupled MS using a set of standard metabolites, and the obtained spectra were compared to identify metabolites (targeted metabolomics). A greater amount of metabolites was detected in lab-isolated samples than in *in situ* samples, which could be explained by the small amount of biological material obtained from *in situ* sampling. From a total of 57 metabolites identified in *E. mutabilis* by the *in situ* analysis, 22 were also found in the surrounding water, suggesting that they were either taken up or synthesized and secreted by the cells. In *in vitro* samples, a total of 78 metabolites were identified; in the cell fraction, only three of the 57 metabolites identified in the *in situ* condition were not found. These results indicate that *E. mutabilis* is able to synthesize under laboratory conditions most of the metabolites detected *in situ*. To identify which metabolites were secreted by this microorganism, the authors compared metabolites detected in the endo- and exo-metabolomes after 7 days of incubation, grouping them in three categories. The first group comprised metabolites that accumulated only in the cell during incubation, which were mainly fatty acids. The second group comprised cell metabolites that disappeared during incubation but did not accumulate in the supernatant, and these were sugars and organic acids. Remarkably, some metabolites that were present in high concentration inside the cells (such as fatty acids and some organic acids) were not detected in supernatants, ruling out cell lysis. The third group comprised metabolites that accumulated in the external environment, and these included amino acids, sugars, and other metabolites, such as polyamines putrescine and spermidine. From changes in the concentration estimated during the incubation phase, fatty acids showed the highest rate of accumulation within the cell (6-fold), and putrescine had the highest rate of excretion (16-fold). Regarding secreted metabolites, from the 26 metabolites identified *in situ* in the surrounding water, the authors detected

15 that also were secreted under laboratory conditions. These included sugars, amino acids, and other compounds such as glycerol and putrescine. An estimate of the concentration of some of these metabolites indicated that they correspond to about half of the total organic matter present in the water from the sampling site. This observation emphasizes the possible role *E. mirabilis* may have in the supply of organic matter to the ecosystem, though the authors did not address the physiological role that some of these secreted metabolites (e.g. putrescine) may have within the microorganism. Many of the metabolites secreted by *E. mirabilis* may be used as energy sources, as well as nitrogen sources, to support growth of other microorganisms present in this carbon and nitrogen-poor ecosystem, as is also the case for members of genus *Sulfobacillus*, *Ferroplasma*, *Acidithiobacillus* and *Leptospirillum*, among others (Ñancucheo and Johnson, 2010). These observations are a clear contribution to the understanding of the physiology of acidophiles in communities where mutualism is often the key to survival.

Biomining model microorganisms

A metabolomic study of acidophilic bacteria isolated from biomining environments in Chile, *At. ferrooxidans* strain Wenelen and *At. thiooxidans* strain Licanantay, has been described (Martínez *et al.*, 2013). Both strains have been extensively studied and are recognized to possess improved oxidizing activity compared with other known microbes from the same species (Ohata, 2010; Sugio, 2009; Morales and Badilla, 2010), an aspect that has allowed their use in optimized copper bioleaching processes.

This metabolomic study employed both targeted and untargeted analytical strategies for analysis and detection of metabolites, using capillary electrophoresis and MS for separation and identification, respectively. The metabolomic analysis was performed after exposing microorganisms to different energy sources, such as iron, sulfur and the copper sulfide mineral chalcopyrite ($CuFeS_2$). The two latter substrates correspond to insoluble energy sources, which allowed analysis of metabolism in planktonic cells, as well as in sessile cell components of biofilms. Samples were collected at different stages of growth, and intracellular and secreted metabolites were analysed. The overall objective of this work was to extend existing functional knowledge on these microorganisms, based on the search for active metabolic pathways that may be key to cell physiology under different growth conditions. In contrast to the previously described studies, all analyses were performed in the laboratory under controlled conditions; however, the use of chalcopyrite as an energy source made the experimental conditions comparable to natural conditions. Of all 3143 unidentified metabolites detected by untargeted metabolomics in both bacteria, fewer were detectable in cells attached to the substrate in contrast with free cells. This can be related to impurities dragged from the substrate during the extraction, interfering in the ionization of the molecules. This phenomenon is characteristic of the analysis of complex samples, as previously mentioned, and is thus important to be considered during data interpretation. The analyses revealed clear differences between metabolic profiles of *At. ferrooxidans* and *At. thiooxidans* exposed to different growth conditions, indicating that their metabolism changed substantially while adapting to each particular condition. Similarly, there were differences between metabolic profiles of attached and free microorganisms, a phenomenon previously discussed in other microorganisms (Booth *et al.*, 2011).

Metabolomic analysis using standards for metabolite identification (targeted metabolomics) provided additional interesting results on the metabolic pathways that may play a major role in adaptation of these organisms to the different substrates mentioned above. Synthetic pathways for polyamine, glutathione, and amino acids and metabolites involved in energy processes were highlighted and analysed in greater detail in both acidophiles. The results indicated that glutathione synthesis routes and some of its precursors, such as glutamate, were overexpressed in both microorganisms when grown in the presence of sulfur, a phenomenon not observed when organisms were grown in the presence of ferrous iron or chalcopyrite. This observation is consistent with the previously described function of reduced glutathione (GSH) in sulfur oxidation, where it plays a catalytic role in elemental sulfur activation (Rohwerder and Sand, 2003). Significant differences were also found in amino acid pool synthesized under various growth conditions in each microorganism.

In *At. thiooxidans* strain Licanantay, the amino acid differences observed between sulfur and chalcopyrite conditions could be explained by differential synthesis of proteins required for adaptation to a new energy source and specific detoxification mechanisms under each condition. A similar phenomenon is found in the *At. ferrooxidans* strain Wenelen grown in ferrous iron versus chalcopyrite. While this comparative analysis seems useful, future integration with proteomic data would allow a more thorough investigation, leading to a more comprehensive conclusion. An interesting observation was the abundance of certain amino acids (such as aspartate and glutamate) in the extracellular space. This has also been reported in other organisms that secrete these amino acids to produce polymers (i.e. polyglutamate and polyaspartate) related to biofilm synthesis (Morikawa *et al.*, 2006). However, Martínez *et al.* (2013) reported that such external polymeric structures were not detected surrounding the target microorganisms. Another possibility is that secretion of these amino acids is related to an intracellular accumulation phenomena or to disturbances in the cytoplasmic membranes, as reported for other microorganisms (Hoischen and Kramer, 1990). Further investigation is needed to determine the physiological significance of this observation and any potential applications.

Another relevant observation of this study was the detection of abundant intracellular sugars in *At. thiooxidans* strain Licanantay when grown in sulfur. These sugars are linked to the formation of large structures, such as EPS, involved in biofilm synthesis, and may be correlated with previously described genomic studies (Barreto *et al.*, 2005). Metabolites related to energy metabolism pathways (e.g. synthesis of NADP, AMP, ADP) were produced in greater abundance by *At. thiooxidans* strain Licanantay than in *At. ferrooxidans* strain Wenelen when both were grown in sulfur. This is consistent with the improved efficiency of *At. thiooxidans* for processing of sulfur.

The possibility of discovering new and important metabolic pathways in acidophilic microorganisms using metabolomics and 'omics' data integration increases the power of these tools in understanding the physiology of these organisms. Analysis of metabolomes from *At. ferrooxidans* and *At. thiooxidans* under different growth conditions (Martínez *et al.*, 2013), identified metabolites related to the polyamine synthesis pathway, suggesting a role in adherence and biofilm formation. Polyamines are polycationic compounds present in the intracellular space of all cells (Tabor and Tabor, 1985) involved in a variety of biological responses such as cellular proliferation, differentiation, biofilm formation (Igarashi and Kashiwagi, 1999), protein synthesis (Friedman and Oshima, 1989), and DNA synthesis and stabilization (Terui *et al.*, 2005). Different kinds of polyamines have been reported, the most common being putrescine, spermidine and spermine. The study analysed the polyamine pathway in both strains, and found that levels of *S*-adenosyl-L-methionine, a spermidine synthesis precursor, were below the detection limit in early growth stages but elevated in exponential and stationary phases under all growth conditions (sulfur, ferrous iron and chalcopyrite). Nevertheless, spermidine, which was also present in bacterial supernatants, was detected at levels five- to six-fold higher with respect to *S*-adenosyl-L-methionine during all growth phases, indicating a tendency for accumulation during growth. Interestingly, intracellular levels of spermidine in *At. ferrooxidans* strain Wenelen are approximately one-third higher in iron-containing medium than in sulfur-containing medium, probably because spermidine accumulates in the intracellular space owing to insufficient secretion. In contrast, data from cells growing in sulfur-containing medium with solid surface stimuli, showed reduced intracellular accumulation and the presence of extracellular spermidine, probably owing to secretion. For *At. thiooxidans* strain Licanantay, it was expected that both solid energy sources (sulfur and chalcopyrite) would enhance biofilm production and reveal similar tendencies with respect to spermidine secretion and concentration. Surprisingly, secretion was observed only under sulfur growth conditions. The reason for this phenomenon is not clear, but may be related to the efficient of use of this energy source.

Two additional intermediates in the spermidine biosynthesis pathway, arginine and ornithine, showed opposite behaviour: arginine production increased, whereas ornithine decreased over time. Common polyamine synthesis routes have been described in bacteria, with ornithine and arginine used as precursors (Fig. 15.2). Additionally, a variety of new polyamines and synthesis routes have been described, mainly in thermophiles from the *Archaea*

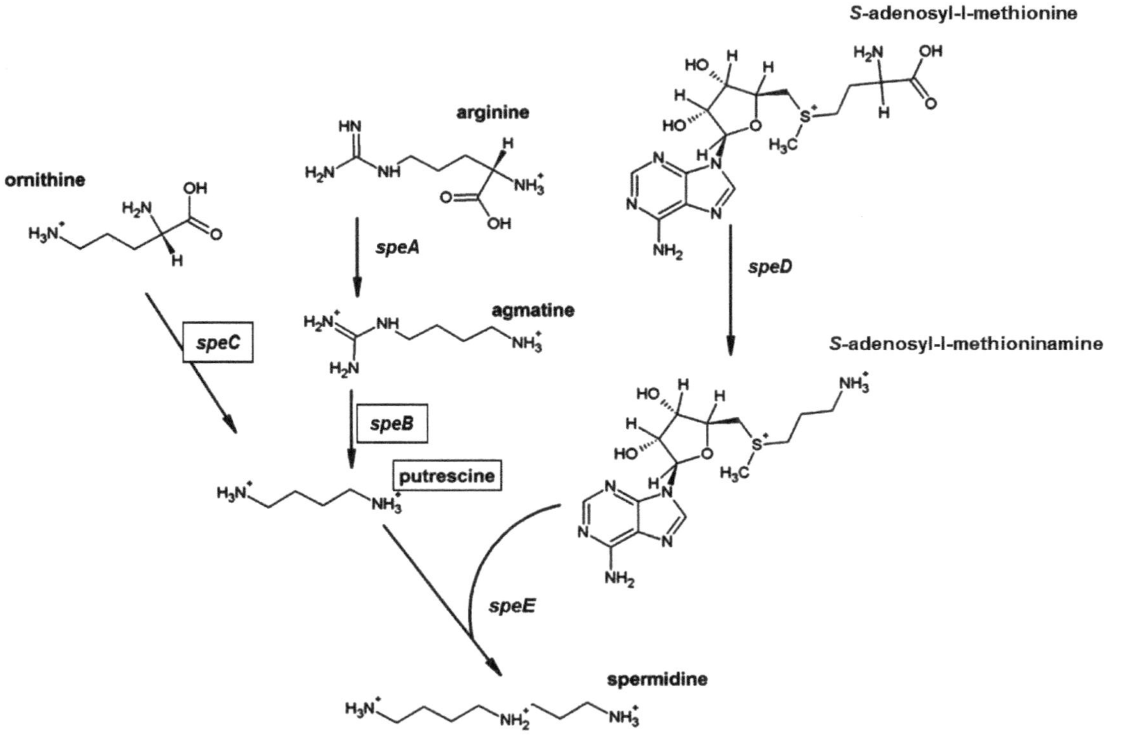

Figure 15.2 General spermidine synthesis pathway. Spermidine is produced from putrescine and *S*-adenosyl-L-methioninamine. Ornithine and arginine are precursors of putrescine. *speA*, arginine decarboxylase; *speB*, agmatine ureohydrolase; *speC*, ornithine decarboxylase; *speD*, *S*-adenosyl-methionine decarboxylase; *speE*, spermidine synthase. Squares highlight results of metabolomic analysis of *At. ferrooxidans* and *At. thiooxidans*: putrescine was not detected in either strain, and *speB* and *speC* genes were not present in either genome. Modified from Martínez *et al*. (2013).

and *Bacteria* domains (Ohnuma *et al.*, 2005; Terui *et al.*, 2005). The metabolic findings extracted from this work suggest the possible existence of an alternative route for polyamine synthesis, different from the canonical pathway illustrated in Fig. 15.2. The common bacterial spermidine intermediate, putrescine, was not detected in any condition, suggesting that spermidine synthesis might be analogous to routes described in other extremophiles (Ohnuma *et al.*, 2005). However, an MS-based search of some intermediates reported in these alternative routes was not successful in either organism, suggesting the presence of new molecules not previously described for this metabolic pathway in this kind of bacteria or similar acidophiles.

To complement these observations using an integrated approach, genomic and transcriptomic tools were also employed. First, a search for genes related to polyamine synthesis identified genes for arginine decarboxylase (*speA*), *S*-adenosyl-L-methionine decarboxylase (*speD*), and polyamine agmatine aminopropyl transferase or spermidine synthase (*speE*) in the genomes of both organisms (Fig. 15.2). Interestingly, the gene for agmatine ureohydrolase (*speB*) was not found, consistent with the absence of putrescine in metabolite identification. The gene for ornithine decarboxylase (*speC*) was also absent, again suggesting that there could be an alternative pathway for spermidine synthesis that has not been reported in these bacteria, and that probably uses a route through agmatine. Next, *spe* expression levels were analysed in *At. ferrooxidans* strain Wenelen under various growth conditions to identify differences that could be related to polyamine behaviour extracted from metabolomic data. In general, most *spe* genes analysed were overexpressed when the strain was faced with solid energy sources (sulfur and chalcopyrite) compared with soluble energy sources (ferrous iron and tetrathionate). This observation suggests a possible relationship between the spermidine pathway and chemotaxis, attachment, or biofilm formation in this type of bacteria.

The observation that spermidine is secreted under sulphur growth conditions suggests that it could be acting as a communication molecule for bacterial adhesion and biofilm formation on insoluble energy sources. There are different reports that connect biofilm synthesis and polyamines in different bacteria and their potential role in cell-to-cell signalling (Karatan *et al.*, 2005; Sturgill and Rather, 2004). As mentioned previously, biofilm formation in *At. ferrooxidans* is closely related to production of extracellular components, such as EPS, and changes in the EPS composition and expression of biosynthesis genes according to the energy substrate have been previously reported (Barreto *et al.*, 2005; Gehrke *et al.*, 1998). This appears to be consistent with the abundant spermidine found in *At. ferrooxidans* strain Wenelen and *At. thiooxidans* strain Licanantay supernatants, particularly when grown on sulfur. The changes in EPS composition appear to be related to attachment mechanisms used by microorganisms in different energy sources. At this level, spermidine may be regulating sulfur attachment and biofilm formation mechanisms. Despite this, the possibility remains that spermidine in the supernatant could be linked to a detoxification response. Genomic analysis shows that canonical *pot* genes related to polyamine transporters are not present in *At. ferrooxidans.* Nevertheless, some functional domains present in other proteins are candidates for these transporters. These data suggest the existence of a new secretion and/or incorporation model for spermidine in *At. ferrooxidans* that has not yet been identified.

This metabolomic analysis, integrated with other omics, has allowed us to conclude that the presence of spermidine has a functional significance for these acidophiles, and to determine that its regulation is related to specific growth stages and the presence of sulfur. Moreover, the consistent presence of spermidine in supernatants of sulfide solid substrate conditions has allowed the proposal that this metabolite can be used as a biomarker to follow sulfur-oxidizers activity (Martínez, 2012).

These studies provide clear examples of how metabolomics can be used to further understand the role of key metabolic pathways in acidophiles and to derive applied knowledge to improve industrial operations using acidophiles.

Future challenges to be addressed using metabolomics of acidophiles

Although metabolomic data are not simple to interpret, the difficulties to genetically modify most acidophiles have made 'omics' data integration an attractive approach for studying the biology of this group of extremophiles. To enhance and facilitate 'omics' integration, improved procedures for metabolite extraction and increased efficiency of metabolites detection methods are needed. Understanding the role that metabolites have in these ecosystems is important for an efficient use of the biological and biotechnological potential of these microorganisms. In this respect, understanding the role of microbe-specific metabolites can contribute to the discovery of biomarkers related to microbial activity, with which to monitor industrial applications. For example, even in our days, bioleaching of minerals in heaps and dumps, are evaluated mainly by metallurgical parameters, like pH and redox potential. Some attempts have been reported, where composition of the microbial communities inhabiting industrial bioleaching systems is assessed using molecular tools like quantitative PCR (qPCR), DGGE (Denaturing Gradient Gel Electrophoresis) or DNA microarrays (Parada *et al.*, 2013; Badilla *et al.*, 2008; Moreno *et al.*, 2007). Nevertheless, in all these cases the information generated is redrawn from liquid samples circulating through the heaps or dumps. Little information has been obtained from the actual mineral samples inside the heaps, principally to avoid the potential perturbation to the irrigation systems during drilling. As described by Vera *et al.* (2013), microbial composition in sessile and planktonic cells can be substantially different, thus information on the microbial composition inside the heaps and dumps can give crucial data with respect to the functionality of the communities present and their bioleaching efficiency. A biomarker detected at the effluent of a heap or dump, representing specific microbial activity can also be a very valuable tool helping productive decisions at on-site operations.

Conclusions

In this chapter, we have addressed the state of the art of metabolomics in the study of acidophiles, with a special focus on AMD and biomining microorganisms owing to their importance in environmental contamination and industrial processes such as bioleaching. From the studies reviewed, it is apparent that in most cases certain metabolites have a direct or indirect relationship with structural phenomena such as biofilm formation. This should not be a surprise since one of the strategies adopted by these microorganisms to tolerate and adapt to the extremely adverse ecosystem has been to develop the ability to group and synthesize structures that generate a micro-environment conducive to their survival. This has also promoted certain evolutionary phenomena of cell cooperation, such as mutualism, where the presence of other metabolites, mostly secreted, are part of the energetic homeostasis allowing cell populations in the ecosystem to be preserved. In the generation of microbial communities, communication between members is crucial, and signalling metabolites play an important role in these systems (Chapter 10). Furthermore, osmoprotective metabolites also seem to play an important role in protecting microorganisms from ionic stress conditions, especially in the early stages of proliferation and adhesion to substrates, prior to biofilm formation.

The application of metabolomics in the study of the acidophiles is very recent, and thus much remains to be elucidated and developed. Owing to the complexity of the matrix analysed, improved methods of extraction and detection are imperative in order to obtain reliable data. This type of analysis allows the study of complex biological systems, providing information that, once integrated, facilitates understanding of the physiology of individual acidophiles or of the whole community.

Finally, from a biotechnological point of view, metabolomics can enhance our understanding of the mineral bioprocessing phenomena when microbes are exposed to different environments, can provide useful biomarkers discovery for identification and therefore can enhance the operation and control of industrial processes.

Acknowledgements

The authors thank BioSigma S.A. for financial support in the development of the metabolomics platform, and CORFO funding in the project 'Application of metabolomics to mining to improve industrial processes associated with bioleaching of mineral resources', 2009–5855.

References

Badilla, R., Maass, A., Parada, P., Aravena, A., Gonzalez, M., Martínez, S., Ehrenfeld, K., and Moreno, P. (2008). DNA fragments array from biomining microorganisms and method for detection of them. U.S. Patent No. 20080207462.

Barreto, M., Jedlicki, E., and Holmes, D.S. (2005). Identification of a gene cluster for the formation of extracellular polysaccharide precursors in the chemolithoautotroph *Acidithiobacillus ferrooxidans*. Appl. Environ. Microbiol. *71*, 2902–2909.

Booth, S.C., Workentine, M.L., Wen, J., Shaykhutdinov, R., Vogel, H.J., Ceri, H., Turner, R.J., and Weljie, A.M. (2011). Differences in metabolism between the biofilm and planktonic response to metal stress. J. Proteome Res. *10*, 3190–3199.

Burton, L., Ivosev, G., Tate, S., Impey, G., Wingate, J., and Bonner, R. (2008). Instrumental and experimental effects in LC-MS-based metabolomics. J. Chromatogr. B Anal. Technol. Biomed. Life Sci. *871*, 227–235.

Edwards, J.S., and Palsson, B.O. (2000). The *Escherichia coli* MG1655 in silico metabolic genotype: its definition, characteristics, and capabilities. Proc. Natl. Acad. Sci. U.S.A. *97*, 5528–5533.

Fischer, C.R., Wilmes, P., Bowen, B.P., Northen, T.R., and Banfield, J.F. (2011). Deuterium-exchange metabolomics identifies N-methyl lyso phosphatidylethanolamines as abundant lipids in acidophilic mixed microbial communities. Metabolomics *8*, 566–578.

Friedman, S.M., and Oshima, T. (1989). Polyamines of sulfur-dependent archaebacteria and their role in protein synthesis. J. Biochem. *105*, 1030–1033.

Gehrke, T., Telegdi, J., Thierry, D., and Sand, W. (1998). Importance of Extracellular Polymeric Substances from *Thiobacillus ferrooxidans* for Bioleaching. Appl. Environ. Microbiol. *64*, 2743–2747.

Goltsman, D.S., Denef, V.J., Singer, S.W., VerBerkmoes, N.C., Lefsrud, M., Mueller, R.S., Dick, G.J., Sun, C.L., Wheeler, K.E., Zemla, A., *et al.* (2009). Community genomic and proteomic analyses of chemoautotrophic iron- oxidizing "*Leptospirillum rubarum*" (group II) and "*Leptospirillum fer- rodiazotrophum*" (group III) bacteria in acid mine drainage biofilms. Appl. Environ. Microbiol. *75*, 4599–4615.

González, A., Bellenberg, S., Mamani, S., Ruiz, L., Echeverría, A., Soulère, L., Doutheau, A., Demergasso, C., Sand, W., Queneau, Y., *et al.* (2013). AHL signaling molecules with a large acyl chain enhance biofilm formation on sulfur and metal sulfides by the bioleaching bacterium *Acidithiobacillus ferrooxidans*. Appl. Microbiol. Biotechnol. *97*, 3729–3737.

Hallberg, K.B., and Johnson, D.B. (2001). Biodiversity of acidophilic prokaryotes. Adv. Appl. Microbiol. *49*, 37–84.

Halter, D., Goulhen-Chollet, F., Gallien, S., Casiot, C., Hamelin, J., Gilard, F., Heintz, D., Schaeffer, C., Carapito,

C., Van Dorsselaer, A., *et al.* (2012). *In situ* proteo-metabolomics reveals metabolite secretion by the acid mine drainage bio-indicator, *Euglena mutabilis*. ISME J. *6*, 1391–1402.

Hoischen, C., and Krämer, R. (1990). Membrane alteration is necessary but not sufficient for effective glutamate secretion in *Corynebacterium glutamicum*. J. Bacteriol. *172*, 3409–3416.

Igarashi, K., and Kashiwagi, K. (1999). Polyamine transport in bacteria and yeast. Biochem. J. *344*, 633–642.

Karatan, E., Duncan, T.R., and Watnick, P.I. (2005). NspS, a predicted polyamine sensor, mediates activation of *Vibrio cholerae* biofilm formation by norspermidine. J. Bacteriol. *187*, 7434–7443.

Krastanov, A. (2010). Metabolomics – The state of art biotechnology. Biotechnol. Equip. *24*, 1537–1543.

Martínez, P., and Parada, P. (2012). Spermidine biomarker for the detection of sulfoxidizing activity. PCT Patent No. 2013080157.

Martínez, P., Gálvez, S., Ohtsuka, N., Budinich, M., Cortés, M.P., Serpell, C., Nakahigashi, K., Hirayama, A., Tomita, M., Soga, T., *et al.* (2013). Metabolomic study of Chilean biomining bacteria *Acidithiobacillus ferrooxidans* strain Wenelen and *Acidithiobacillus thiooxidans* strain Licanantay. Metabolomics 9, 247–257.

Morales, P., and Badilla, R. (2010). Process To Increase The Bioleaching Speed Of Ores Or Concentrates Of Sulfide Metal Species, By Means Of Continuous Inoculation With Leaching Solution That Contains Isolated Microorganisms, With Or Without Presence Of Native Microorganisms. U.S. Patent No. 7837760.

Moreno, P., Aravena, A., and Parada, P. (2007). Method and array for detection and identification of microorangisms present in a sample using the genomic regions coding for different tRNA synthetases. U.S. Patent No. 20070134700.

Morikawa, M., Kagihiro, S., Haruki, M., Takano, K., Branda, S., Kolter, R., and Kanaya, S. (2006). Biofilm formation by a Bacillus subtilis strain that produces gamma-polyglutamate. Microbiology *152*, 2801–2807.

Mosier, A., Justice, N., Bowen, B., Baran, R., Thomas, B., Northern, T., and Banfield, J. (2013). Metabolites associated with adaptation of microorganisms to an acidophilic, metal-rich environment identified by stable-isotope-enabled metabolomics. MBio *4*, e00484–12.

Mueller, R.S., Dill, B.D., Pan, C., Belnap, C.P., Thomas, B.C., VerBerkmoes, N.C., Hettich, R.L., and Banfield, J.F. (2011). Proteome changes in the initial bacterial colonist during ecological succession in an acid mine drainage biofilm community. Environ. Microbiol. *13*, 2279–2292.

Ñancucheo, I., and Johnson, D.B. (2010). Production of glycolic acid by chemolithotrophic iron- and sulfur-oxidizing bacteria and its role in delineating and sustaining acidophilic sulfide mineral-oxidizing consortia. Appl. Environ. Microbiol. *76*, 461–467.

Ñancucheo, I., and Johnson, D.B. (2012). Acidophilic algae isolated from mine-impacted environments and their roles in sustaining heterotrophic acidophiles. Front. Microbiol. *3*, 235.

Ohata, A., Manabe, M., and Parada, P. (2010). Sulfur-oxidizing bacteria and their use in bioleaching processes for sulfured copper minerals. U.S. Patent No. 7700343.

Ohnuma, M., Terui, Y., Tamakoshi, M., Mitome, H., Niitsu, M., Samejima, K., Kawashima, E., and Oshima, T. (2005). N1-aminopropylagmatine, a new polyamine produced as a key intermediate in polyamine biosynthesis of an extreme thermophile, *Thermus thermophilus*. J. Biol. Chem. *280*, 30073–30082.

Oliver, S.G., Winson, M.K., Kell, D.B., and Baganz, F. (1998). Systematic functional analysis of the yeast genome. Trends Biotechnol. *16*, 373–378.

Parada, P., Ehrenfeld, N., Pacheco, I., Maass, A., Aravena, A., Gonzalez, M., and Martinez, S. (2013). Method for the identification and quantification of microorganisms useful in biomining processes. U.S. Patent No. 8492093.

Pauling, L., Robinson, A.B., Teranishi, R., and Cary, P. (1971). Quantitative Analysis of Urine Vapor and Breath by Gas-Liquid Partition Chromatography. Proc. Natl. Acad. Sci. U.S.A. *68*, 2374–2376.

Quatrini, R., Appia-Ayme, C., Denis, Y., Jedlicki, E., Holmes, D.S., and Bonnefoy, V. (2009). Extending the models for iron and sulfur oxidation in the extreme acidophile *Acidithiobacillus ferrooxidans*. BMC Genomics *10*, 394.

Rawlings, D.E. (2005). Characteristics and adaptability of iron- and sulfur-oxidizing microorganisms used for the recovery of metals from minerals and their concentrates. Microb. Cell Fact. *4*, 13.

Riedel, K., Hentzer, M., Geisenberger, O., Huber, B., Steidle, A., Wu, H., Høiby, N., Givskov, M., Molin, S., and Eberl, L. (2001). N-acylhomoserine-lactone-mediated communication between *Pseudomonas aeruginosa* and *Burkholderia cepacia* in mixed biofilms. Microbiology *147*, 3249–3262.

Roberts, L.D., Souza, A.L., Gerszten, R.E., and Clish, C.B. (2012). Targeted metabolomics. In Current Protocols in Molecular Biology, Ausubel, F.M., Brent, R., Kingston, R.E., Moore, D.D., Seidman, J.G., Smith, J.A., Struhl, K., Coen, D.M., Gardner, A.F., Sadreyev, R.I., *et al.*, eds. (John Wiley & Sons, Inc., Hoboken, NJ), Chapter 30, Unit 30.2.1–24.

Rohwerder, T., and Sand, W. (2003). The sulfane sulfur of persulfides is the actual substrate of the sulfur-oxidizing enzymes from *Acidithiobacillus* and *Acidiphilium* spp. Microbiology *149*, 1699–1710.

Rohwerder, T., Gehrke, T., Kinzler, K., and Sand, W. (2003). Bioleaching review part A: progress in bioleaching: fundamentals and mechanisms of bacterial metal sulfide oxidation. Appl. Microbiol. Biotechnol. *63*, 239–248.

Ruiz, L.M., Castro, M., Barriga, A., Jerez, C.A., and Guiliani, N. (2012). The extremophile *Acidithiobacillus ferrooxidans* possesses a c-di-GMP signalling pathway that could play a significant role during bioleaching of minerals. Lett. Appl. Microbiol. *54*, 133–139.

Ryan, D., and Robards, K. (2006). Metabolomics: the greatest omics of them all? Anal. Chem. *78*, 7954–7958.

Sand, W., and Gehrke, T. (2006). Extracellular polymeric substances mediate bioleaching/biocorrosion via interfacial processes involving iron(III) ions and acidophilic bacteria. Res. Microbiol. *157*, 49–56.

Sturgill, G., and Rather, P.N. (2004). Evidence that putrescine acts as an extracellular signal required for swarming in *Proteus mirabilis*. Mol. Microbiol. *51*, 437–446.

Sugio, T., Miura, A., Parada, P., and Badilla, R. (2009). Bacteria strain wenelen DSM 16786, use of said bacteria for leaching of ores or concentrates containing metallic sulfide mineral species and leaching processes based on the use of said bacteria or mixtures that contain said bacteria. U.S. Patent No. 7601530.

Tabor, C.W., and Tabor, H. (1985). Polyamines in microorganisms. Microbiol Rev. *49*, 81–99.

Takahashi, N., Washio, J., and Mayanagi, G. (2012). Metabolomic approach to oral biofilm characterization-A future direction of biofilm research. J. Oral Biosci. *54*, 138–143.

Terui, Y., Ohnuma, M., Hiraga, K., Kawashima, E., and Oshima, T. (2005). Stabilization of nucleic acids by unusual polyamines produced by an extreme thermophile, *Thermus thermophilus*. Biochem. J. *388*, 427–433.

Thomson, J.J. (1912). Further experiments on positive rays, Philosophical Magazine Series 6, vol. 24, no. 140, pp. 209–253.

Vera, M., Schippers, A., and Sand, W. (2013). Progress in bioleaching: fundamentals and mechanisms of bacterial metal sulfide oxidation--part A. Appl Microbiol Biotechnol. *97*, 7529–7541.

Vinayavekhin, N., and Saghatelian, A. (2010). Untargeted metabolomics. In Current Protocols in Molecular Biology, Ausubel, F.M., Brent, R., Kingston, R.E., Moore, D.D., Seidman, J.G., Smith, J.A., Struhl, K., Coen, D.M., Gardner, A.F., Sadreyev, R.I., et al., eds. (John Wiley & Sons, Inc., Hoboken, NJ), Chapter 30, Unit 30.1.1–24.

Zhang, B., and Powers, R. (2012). Analysis of bacterial biofilms using NMR-based metabolomics. Future Med. Chem. *4*, 1273–1306.

Part VI

Applications and Outlook

Biotechnologies that Utilize Acidophiles 16

Susan T.L. Harrison

Abstract

The bioleaching of mineral sulfides provides the most prevalent anthropogenic application of chemolithotrophic acidophiles. The same reactions, occurring unconstrained in the environment, are also the cause of acid mine drainage (AMD), a major issue with respect to water pollution. In this chapter, the role of these microorganisms in processes for metal recovery through tank, heap, dump and *in situ* leaching systems is described. Their potential in gold and copper recovery is demonstrated through discussion of existing processes and operations. The potential to expand the relevance of these technologies to other metals, low grade and complex mineral ores and concentrates is explored. Further, the importance of developing processes for the environmentally responsible processing of wastes from both mining operations and end-of-life products is raised as are the challenges for ensuring process efficiency while minimizing the environmental footprint.

Introduction

The ubiquity of acidophilic chemolithotrophic microorganisms in nature and our awareness of the results of their activities (even if not originally aware of the microorganisms themselves when first harnessing them) has resulted in a broad variety of biotechnological uses of these microorganisms developing over time, including application in biomining. The most relevant biotechnology driven by acidophiles, both in terms of global impact and usage of this group of microorganisms, is bioprocessing (bioleaching and bio-oxidation) of metal-bearing ores. Currently, application of acidophiles in bioleaching and bio-oxidation spans across the extraction of a range of metals, predominantly from sulfidic minerals, finding application to both niche high grade ores as well as increasingly important low grade mineral reserves. Interest is growing in the integration of these biotechnologies with conventional technologies to augment extraction and the recovery of multiple metal products. Furthermore, potentially contaminant waste materials, including both waste rock as well as end-of-life metal-containing materials such as the growing accumulation of 'e-wastes' show potential for processing for metal recovery using these biotechnologies. Additionally, metal recovery from liquid process effluents or acid mine drainage can be facilitated through microbial metabolism (Johnson, 2014a). While considering the many desirable applications for extraction of value and maximizing resource productivity, it is necessary to be mindful of the role of the same chemistry in generation of acid mine drainage and associated deportment of metals from these waste rocks, and to use knowledge of biohydrometallurgy in preventing pollution of water bodies.

The biotechnologies for metal extraction, forming a subset of hydrometallurgy, compete with pyrometallurgical (roasting and smelting) and hydrometallurgical processes such as pressure oxidation. The extractive biotechnologies include bioleaching, a microbially mediated process in which the metal of interest is solubilized, and bio-oxidation, in which the minerals occluding the metals of interest are solubilized in a pre-treatment step. Jointly, these are termed biomining technologies (Rawlings and Johnson, 2007). Specific advantages of the biomining approaches are two-fold:

- the ability to handle niche minerals such as those bearing arsenic or complex polymetallics that are not well suited to pyrometallurgical treatments (Johnson, 2014b; Neale *et al.*, 2011).
- the ability to treat low grade ores through enabling processing without the need for the energy-intensive preparation of a concentrate of finely divided material.

The latter is achieved through heap bioleaching and carries a large energy saving through reduced comminution energy costs. The potential of *in situ* bioleaching could further reduce energy requirements, and associated carbon footprint, through reducing the blasting and mining required (Batterham, 2011) as well as the transporting of the ore and removal and disposal of the waste rock (Johnson, 2015). Further to these, biohydrometallurgy has potential for application at the smaller scale than some of its competitive technologies, particularly where the grade is sufficiently high (Neale *et al.*, 2011). For instance, heap bioleaching projects that run at less than 50,000 tonnes per annum (tpa) copper are characterized by higher grades than projects run at greater than 50,000 tpa; in the latter copper grades as low as 0.3 to 0.2% are tolerated. Similarly, the mineral type and grade impact the choice of process, as illustrated in Fig. 16.1.

Each of these bioprocesses can readily accommodate the recovery of both the major metal of interest and other metals present (Johnson, 2014b) through either selective recovery following both or all metals entering the solution or a two stage dissolution process (Johnson, 2014a; Mwase *et al.*, 2012; Logan *et al.*, 2007). Cited benefits of the biomining technologies include their operation at lower temperatures and pressures than alternative pyro- and hydrometallurgical options, reduced chemical addition with respect to chemical leaching approaches and reduced energy footprint of the extraction stage (Johnson, 2014b). Challenges include the longer reaction times and perceived issues centred on the robustness and reliability of the biomining technologies (Johnson, 2014b) although the robustness of these reactions in natural systems suggests this perception requires challenging.

In this chapter, the major biotechnological approaches used for metal extraction using acidophiles are overviewed. Specific classes of application are then considered, starting with those currently in operation and moving towards those under development. In closing the chapter, some challenges that require ongoing development in our understanding for process optimization are highlighted.

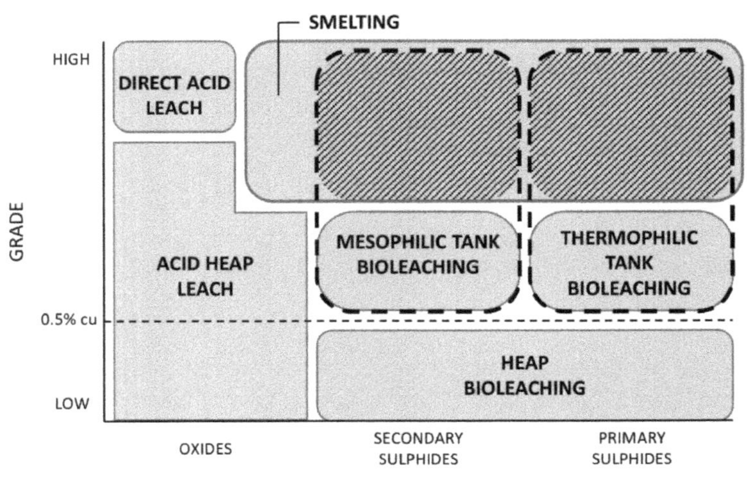

Figure 16.1 The selection of technology for metal extraction from a resource is influenced by the type of ore/mineral and its grade. Furthermore the presence of impurities, such as arsenic, can remove options such as smelting, as demonstrated by the hashed areas (adapted from Clark *et al.*, 2006).

Biotechnologies for metal extraction using acidophiles

The chemistry of mineral bioleaching

Metal extraction through bioleaching is chiefly mediated by the microorganisms serving to supply the lixiviant for a chemical leach process, e.g. ferric iron, inorganic acid such as sulfuric acid (H_2SO_4), organic acids such as citric acid and oxalic acid, or cyanide. Following colonization and the establishment of an active microbial community, these lixiviants are provided continuously and in close proximity to the mineral reaction site. In each of these cases, the actual leaching reaction is a chemical reaction.

While the use of microbially produced organic acids has been investigated for the recovery of metals from laterites, and microbially produced cyanide for the recovery of gold and platinum group metals (PGMs) from concentrates and electronic waste (Pham and Ting, 2009), the commercial relevance of these processes has not yet been demonstrated. Demonstrated biotechnologies for metal extraction at the commercial scale are restricted to the recovery of metals from mineral sulfides using acidophiles or the pre-treatment of mineral ores where the metal of interest is made available through removal of mineral sulfides. This chapter focuses on these latter approaches for leaching mineral sulfides, using protons and ferric iron.

Acid insoluble minerals, such as pyrite (FeS_2) and molybdenite (MoS_2), are solublized by ferric iron through six one electron steps to release thiosulfate (Schippers and Sand, 1999). This is oxidized biologically to sulfate or rearranges to elemental sulfur as in the following equations:

$$FeS_2 + 6Fe^{3+} + 3H_2O \rightarrow S_2O_3^{2-} + 7Fe^{2+} + 6H^+$$

$$8S_2O_3^{2-} \rightarrow S_8 + 8SO_3^{2-}$$

$$S_2O_3^{2-} + 2O_2 + H_2O \rightarrow 2\,SO_4^{2-} + 2H^+$$

$$S_2O_3^{2-} + 8Fe^{3+} + 5H_2O \rightarrow 2SO_4^{2-} + 8Fe^{2+} + 10H^+$$

Acid-soluble minerals such as galena (PbS), arsenopyrite (FeAsS), sphalerite (ZnS), alabandite (MnS) and chalcopyrite ($CuFeS_2$) are solubilized through the combined action of protons and ferric iron, with the release of hydrogen sulfide (H_2S) (Schippers and Sand 1999). The latter undergoes spontaneous rearrangement to polysulfide, chemical oxidation to elemental sulfur and biological oxidation to sulfate, as shown in the following equations:

$$ZnS + Fe^{3+} + H^+ \rightarrow Zn^{2+} + 0.5H_2S_n + 7Fe^{2+}\ (n \geq 2)$$

$$H_2S_n + 2Fe^{3+} + 5H_2O \rightarrow 0.25S_8 + 2\,Fe^{2+} + 2H^+$$

$$0.25S_8 + 3O_2 + 2H_2O \rightarrow 2SO_4^{2-} + 4H^+$$

A key consideration in mineral bioleaching is the mechanism by which the contacting of the leach agent and the mineral occurs. In bioleaching, this also requires consideration of the proximity or contacting of the microorganisms and mineral surface. It is further influenced by the liberation of the mineral in the ore. Leaching efficiency is enhanced by forming a finely divided concentrate for bioleaching through crushing, milling and concentration; however, this has associated resource implications, especially energy, demanding a minimum grade of ore for its implementation (Fig. 16.1). Alternative approaches allow contacting of leach agent and the microbial community generating it with crushed whole ore, run-of-mine ore, waste rock or urban waste (electronic or e-waste). Each of these is characterized by a different contacting system, extending from the intimate contacting in the tank bioleach system, through vat leaching, heap leaching to dump leaching. These are described below.

Tank bioleaching

Tank leaching has been developed as a competitive technology for the extraction of metals from milled mineral concentrates such as flotation concentrates, competing with smelting and pressure oxidation (Rawlings and Johnson, 2007). The finely divided concentrate (typically less than 100 μm) is processed as a slurry in a series of continuous stirred tank reactors under well mixed conditions. Excellent contacting between mineral, leach agent and microorganism is maintained, while active aeration ensures the supply of the essential reactant, oxygen. The process is temperature controlled and the finely divided mineral ensures good liberation of the mineral surface. Tank bioleaching is carried out at solids loadings in the range 12–20% (m/v), depending on the microbial consortia used. Typically overall retention times of four to seven days are required (van Aswegen *et al.*, 2007), preventing large inventory hold-up. Critical considerations are the energy input required for comminution, agitation and aeration and the materials of construction of the tank system (Brierley and Brierley, 2001). Tank leaching is described in detail later in section entitled 'Tank Bioleaching'.

Heap bioleaching

In heap leaching, very large beds (often several kilometres long and more than 6 m high), of typically low grade ores are built on an impervious pad. Microorganisms are introduced into these beds or the natural community associated with the ore is nurtured to colonize the ore bed. The ore beds may be aerated through forced aeration from below or constructed to maximize natural convention of air through the heap to provide the oxygen and carbon dioxide requirements. The heap is irrigated from above with the aim of creating an unsaturated bed with optimal conditions for both microbial generation of the leach agents and mineral leaching. In the active heap, leach agents, ferric iron and protons, are generated through ferrous iron oxidation and sulfur oxidation. These contribute to mineral leaching. The free flowing liquid percolating through the heap transports the extracted metal out of the heap where it is removed from solution, typically by solvent extraction, and the solution returned to the heap (Logan *et al.*, 2007; Petersen and Dixon, 2007). Owing to the heterogeneous nature of the ore bed, some variability in conditions within the heap occur. The key factors for control of the efficiency of the heap bioleach include the manner of packing of the ore bed, the sulfide content of the ore bed, the aeration achieved and the flow rate of the irrigating liquor; the latter two impacting heat flow and thus the temperature within the bed (Petersen and Dixon, 2007). In addition, the microbiology and gangue mineralogy impact the process. Heaps may be constructed with run-of-mine (ROM) ores, crushed ores (particle sizes of the order of 15–25 mm) or crushed and agglomerated ores. Heap bioleaching is further discussed in section entitled 'Demonstrations of heap bioleaching'.

Dump bioleaching

Dump leaching has been used extensively for the recovery of base metals (mostly copper) from low grade waste rock materials for centuries (Johnson, 2014b; Gilbertson, 2000). The Kennecott Copper operation, extracting copper from a sub-marginal grade of ROM ore, was initiated in the 1950s (Brierley and Brierley, 2001). These dump leaches are typically characterized by large particle size (e.g. ~150 mm), although very variable, and high stacking heights, from 18 m to as much as 100 m, leached by irrigation on the disposal site using mesophilic microorganisms which occur naturally (Johnson, 2014b) or are added by irrigation to accelerate the dump leach. Examples of such operations are Rio Tinto (Spain), producing 8000 tpa copper from a dump process over the period 1950 to 1980; Cananea (Mexico), producing 9000 tpa copper over a similar period (Gentina and Acevedo, 1985); and Duval (USA) producing 2500 tpa copper from material containing 0.15–0.20% (Gentina and Acevedo, 2013). In dump leaching, irrigation of the dump occurs using very simple and low cost approach. While productivities and recoveries are low, dump leaching recovers metals that are otherwise wasted using low investment and operating costs, so rendering it economical.

In situ bioleaching

The extension of the principles of dump leaching to *in situ* leaching has great promise for future development (Johnson, 2015; Gentina and Acevedo, 2013; Batterham, 2011). With *in situ* leaching, the rock is not removed from its buried location, avoiding drilling, blasting, excavation and haulage. Instead, it is fractured by design and leached in place, allowing considerable energy savings and so reduced costs of metal recovery of 10- to 20-fold relative to conventional processing of low grade ores (Johnson, 2015). Some examples of potential operations are currently under investigation. While care must be taken to control fluid flow and percolation precisely following fracture such that it does not enter the ground water, a substantial energy requirement and environmental burden associated with the mining, waste handling and disposal legacy may be prevented by using deep *in situ* biomining.

Vat bioleaching

Vat leaching is not currently widely used for bioleaching but is an area of interest for further development. In vat leaching, the coarse ore (~8 mm) or e-waste is contacted with the leach agent in a submerged tank or vat (Jena *et al.*, 2009). The material for leaching is not fully suspended and the residence time of the leach agent can be varied. Counter-current contacting of the ore and leach agent is preferred as it allows fresh leach agent to contact the most leached materials. Study of the potential of vat leaching is motivated by the need for good contacting between mineral and leach agent while minimizing energy input. The ability to meet the oxygen requirements will form a key challenge in application of this system to bioleaching.

Extraction and recovery of metals using bioreductive dissolution processes

The processes described to date in this chapter focus on the use of iron and sulfur oxidation to facilitate metal dissolution, restricting their application to mineral sulfides. Recent work through collaboration of du Plessis and Johnson's laboratory has led to the conceptual Ferredox process in which oxidized iron ores such as goethite (FeO-OH) and asbolane [$Mn_3O_3(OH)_6$] are subjected to microbially catalysed bioreductive dissolution with associated release of metals into solution (Johnson, 2014b; du Plessis *et al.*, 2011). This conceptual process is of substantial interest as some 70% of the nickel and cobalt reserves are found trapped in laterites, not as discrete minerals making their recovery difficult.

In the Ferredox flowsheet development, anoxic bioreductive dissolution of limonite laterites using bacteria such as *At. ferroxidans*, *At. ferridurans* and *At. ferrovorans* has been demonstrated, resulting in a ferrous iron-rich solution containing dissolved nickel or cobalt. The dissolution reaction uses elemental sulfur as reductant as shown below:

$$S^0 + 6FeO(OH) + 10H^+ \rightarrow SO_4^{\ 2-} + 6Fe^{2+} + 8H_2O$$

Target metals, such as nickel and cobalt, can be recovered from solution by selective precipitation. Thereafter, iron oxidation to ferric iron results in its removal by precipitation as schwertmannite or jarosite. The remaining Fe^{3+} in solution is used for reductive acid generation for oxidation of the elemental sulfur. The conceptual flowsheet of the Ferredox process with laboratory scale validation for nickel recovery from limonite laterites is in place (du Plessis *et al.*, 2011). Significant potential for application to other oxidized iron ores is expected. Further the bioreductive dissolution may have a role to play in conventional bioleaching for the control of iron precipitation in heap processes or removal of elemental sulfur in gold recovery (Johnson, 2014b).

Recovery of metals from solution

Recovery of base metals from solution is most typically carried out by solvent extraction and electrowinning (SX-EW), resembling other hydrometallurgical approaches. However, this is recognized to contribute substantially to the environmental footprint of the bioleaching of base metals. Investigation of alternate approaches, such as the precipitation of the metal salts demonstrated on the Talvivaara operation (Riekkola-Vanhanen., 2013), is required. Bioreactors containing populations of iron-oxidizing and sulfate-reducing acidophilic bacteria have been used to remove soluble metals selectively from AMD and mine process waters, thereby facilitating their recovery and recycling (e.g. Hedrich and Johnson, 2014; Bijmans *et al.*, 2008, 2010).

Demonstrations of heap bioleaching

Heap leaching of secondary copper sulfides

The leaching of low-grade secondary sulfide ores containing copper minerals such as chalcocite, covellite and enargite resemble approaches used for acid leaching of transition metal oxides. In these heaps, percolation of an acid lixiviant is maintained to ensure effective microbial colonization for the ongoing regeneration of the leach agents, ferric iron and protons, for the leaching of the secondary sulfides. These heaps may be stacked with ROM material or agglomerated ore following crushing to $D_{90} < 15$ mm (du Plessis *et al.*, 2007). Particle size distribution is recognized to play a key role in both mineral liberation and heap permeability, the latter affecting fluid flow, channelling and air distribution (Bouffard and West-Sells, 2009) and which is particularly affected by the relative amounts of fine particles.

Heap aeration favours the leaching rates both

through the effective supply of O_2 (required for oxidation of iron and sulfur) and CO_2 (required for bacterial growth) to the microorganisms, enhancing their activity (Bryan *et al.*, 2012; Petersen *et al.*, 2010). Furthermore, by controlling liquid and air flows, elevated temperatures can be achieved, shortening leaching times (Dew *et al.*, 2011; Petersen and Dixon, 2007; Harvey and Bath, 2007). This was demonstrated at the Quebrada Blanca mine in northern Chile, situated at an elevation of 4000 m OSL, where both low temperatures and low partial pressures of oxygen present challenges to biomining. Here 17,300 tons of low grade ore were processed each day, crushed to < 9 mm, agglomerated, and stacked in aerated heaps of 6–6.5 m (Brierley and Brierley, 2001). Some 80% copper recovery was achieved from the resource over 500 days at 15–18°C whereas the inventory time has been reduced to 300–360 days at 22–25°C (Domic, 2007), demonstrating the robustness of the technology.

The commissioning of 11 heap bioleach processes between 1980 and 1996 and the commissioning of nine further operations in the ensuing ten years demonstrates the relevance of biomining in such situations. These are listed, with data on the operations, in Table 16.1. They are concentrated in South America, particularly Chile, and Australia, with expansion expected in China and Africa.

High temperature heaps for chalcopyrite leaching

With growing demand and decreasing grade, an increasing fraction of the primary copper ores currently exploited contains less than 1% copper, with much of this present as chalcopyrite (du Plessis *et al.*, 2007). It is widely documented that primary sulfides, such as chalcopyrite, leach poorly under mesophilic conditions owing to passivation of the mineral surface, requiring temperatures in excess of 50°C, and preferably 55–80°C (Dew *et al.*, 1999, 2011; du Plessis *et al.*, 2007; Clark *et al.*, 2006). The generation of the heat within the heap to enable chalcopyrite extraction results from the leaching of the secondary sulfides present and, particularly, pyrite on commissioning of the heap and its first stage of operation. The resource requires a sufficient accessible sulfide content to reach the requisite temperature. The exothermic reactions resulting in these temperatures are predominantly the oxidation reactions shown below (du Plessis *et al.*, 2007):

$$Fe^{2+} + 0.25O_2 + H^+ \rightarrow Fe^{3+} + 0.5H_2O$$
$$(\Delta H_0^{\,j} = -97.3\,kJ/mol)$$

$$S^0 + 1.5O_2 + H_2O \rightarrow SO_4^{\,2-} + 2H^+$$
$$(\Delta H_0^{\,j} = -606\,kJ/mol)$$

The heats of reaction show that sulfur oxidation is of prime importance, highlighting the need for active sulfur-oxidizing acidophiles in the microbial community.

These heaps are prepared using crushed agglomerated ore (e.g. 80% < 15 mm diameter), stacked on an impermeable pad. One to two metres of overburden are used at the base of the heap to embed the drainage system for collection of pregnant leach liquor (PLS, irrigation fluid draining through the heap and now laden with soluble metal) and aeration lines (du Plessis *et al.*, 2007). The heap is irrigated at a typical rate of 4 to 18 l/(m^2 h) with raffinate (du Plessis *et al.*, 2007). The PLS collected at the base of the heap is transported to ponds from where it proceeds to solvent extraction and electrowinning for copper recovery. The barren raffinate resulting is used to irrigate the heap.

As mineral heaps are typically initiated at ambient temperature, it is necessary for a succession of microorganisms to display sufficient activity through the heating period of the heap, requiring a dynamic microbial community, as illustrated in Table 16.2. This community is required to colonize the heap on start-up or be added following increase of temperature. While initial studies of succession have been presented (Africa *et al.*, 2013; Tupikina *et al.*, 2013; Bromfield *et al.*, 2011; Dew *et al.*, 2011), further understanding is essential. The microbial community is dominantly associated with the ore, at concentrations of two to three orders of magnitude higher than in the PLS (Govender *et al.*, 2013). The microbial reactions are thus dominantly occurring in the vicinity of the ore and heat generation depends largely on the sulfide content and active microbial community. This provides a major challenge for the monitoring of the heaps, as the PLS does not reflect the microbial community on the ore (Govender *et al.*, 2013; Tupikina *et al.*, 2013) nor potentially the physicochemical conditions at the ore surface.

Table 16.1 Commercial operations for the heap bioleaching of copper

Mine (owners)	Ore type	Ore processed (tpd)	Ore, reserves and grade	Cu production (tpa)	Dates of operation
Lo Aguirre, Santiago region, Chile[a,b,c] (Sociedad Minera Pudahuel Ltda)	Oxides/ chalcocite	16,000	12×10^6 ton at 1.5% Cu	14,000–15,000	1980–2001
Gunpowder's Mammoth Mine, Australia[c,d] (Aditya Birda Minerals)	*In situ*				1991–2013, then under C&M
Mt Leyshon, Australia[a,c](formerly Normandy Poseidon)	chalcocite	1370	0.15%	750	1992 till 1997
Cerro Colorado,[a,b,c] 3200 m elevation (BHP Billiton)	Chalcocite, covellite	16,000	80×10^6 ton at 1.4%, currently <1% Cu	130,000 (80–85% extraction in 450–500 days)	1993 ongoing
Girilambone, Australia[a,c] (Straits Resources)	Chalcocite, bornite	2000	2.4% Cu	14,000	1993–2003
Ivan Zar, Chile[a,b,c] (Compania Minera Milpro)	Oxide and sulfide, inclusing atacamite and chalcocite-type sulfide	1500	5×10^6 ton at 2.5%	12,000	1994 ongoing
Quebrada Blanca, Chile,[a,b,c,e] 4000 m elevation (Teck Cominco)	Chalcocite	17,300	85×10^6 ton at 1.4% Cu and 45×10^6 ton at 0.5% Cu at start. Grades in 2012–1 3 of 0.85% and 0.45%, respectively	75,000 (design); 80,000 (actual); (80% extraction in 500 days at 15–18°C; in 350 days at 22–25°C)	1994 ongoing
Punta del Cobre, Chile	Oxides/sulfides		10×10^6 ton at 1.7% Cu	7000–8000	1994 ongoing
Carmen de Andacollo, Chile[a,b,c]	Chalcocite	10,000	32×10^6 ton at 0.58% Cu	21,000	1996 ongoing
Dos Amigos, Chile[a,b,c]	Chalcocite	3000		5000 <1999; 10,000 >1999 (80% Cu recovered in 300–350 days)	1996 ongoing
Cerro Verde, Peru[a,c]	Oxide/sulfide	15,000		54,200	1996 ongoing
Phoenix deposit, Cyprus[b]	Oxide, sulfide		9.1×10^6 ton at 0.78% Cu; 5.9×10^6 ton at 0.31% Cu	8000	1996 ongoing
Zaldivar, Chile[a,b,c] (Barrick Gold Corp.)	Chalcocite	~20,000		150,000 (particle size 150–200 μm; 365 days)	1995 ongoing
Nifty, Australia[b] (Straits Resources)	Oxides, chalcocite	5000	1.2%	16,000	1998 ongoing
S&K Copper project, Monywa, Myanmar[a,c]	Chalcocite	15,000–18,000	126×10^6 ton at 0.5% Cu	40,000	1998 ongoing
Lomas Bayas, Chile (XSTRATA)	Oxide/sulfide	36,000	41×10^6 ton at 0.4% Cu	60,000	1998 ongoing
Escondida, Chile[a,b,c,f] (BHP Billiton)	Oxide/sulfide		1.5×10^9 ton at 0.3–0.7% Cu	200,000	1999 ongoing
Morenci, Arizona, USA[b] (Freeport McMoran)	Chalcocite, pyrite	75,000, 25,000	3.45×10^9 ton at 0.28% Cu	380,000	2001 ongoing

Table 16.1 Continued

Mine (owners)	Ore type	Ore processed (tpd)	Ore, reserves and grade	Cu production (tpa)	Dates of operation
Whim Creek and Mons Cupri, Australia[b]	Oxides, sulfides		900×10^3 ton at 1.1% Cu; 6×10^6 ton at 0.8% Cu	17,000	2006 ongoing
Jinchuan Copper, China[b,g] (Zijin Mining Group Ltd	Chalcocite, covellite, enargite		240×10^6 ton at 0.63% Cu	10,000	2006–2009
Sarcheshmeh Copper Complex, Iran[h,i] (Nicico)	Porphyry pyrite and chalcopyrite, chalcocite		1200×106 ton at 0.7% Cu	170,000	2006 ongoing

[a]Domic (2007).
[b]Watling (2006).
[c]Brierley and Brierley (2001).
[d]Aditya Birda Minerals website http://www.adityabirlaminerals.com.au/
[e]Teck Cominco website: http://www.teck.com/generic.aspx?PAGE=Teck+site%2fDiversified+Mining+Pages%2fCopper+Pages%2fQuebrada+Blanca&portalName=tc
[f]Clark *et al.* (2006).
[g]Abhilash *et al.* (2015).
[h]Nicico website http://www.nicico.com/
[i]Pradhan *et al.* (2008).

The mining company BHP Billiton demonstrated the performance of the chalcopyrite heap in both a demonstration heap operated at Escondida Mine in Chile and using a simulation column 6 m in height and 1 m in diameter, controlled such that the temperature profile in the ore bed was determined by the mineral sulfide oxidation and associated transport functions. The 6 m column was operated at Mintek (Randburg, South Africa) on acid-agglomerated low grade ore (0.3 to 0.7% Cu of which 85% was chalcopyrite and 4.7% pyrite) crushed to passing $D_{90} < 13$ mm and 12% passing 1 mm (Dew *et al.*, 2011). Following inoculation, the column was irrigated at $46\,l/(m^2{\cdot}h)$ and temperature programmed to heat from 15°C to 30°C over the first 30 days, then rapidly to above 50°C by day 40 with transition to thermophilic conditions (65°C) by day 60 and a maximum temperature of 71°C on day 85. A dynamic microbial community transition from mesophilic through moderately thermophilic to extreme thermophilic conditions was supported with appropriate iron- and sulfur-oxidizers dominating each phase (see Table 16.2). Operating under these conditions, a copper recovery of 75% and pyrite solubilization of 79% was achieved within 283 days (Dew *et al.*, 2011). This study demonstrated the excellent potential of high temperature heaps for the recovery of chalcopyrite from low grade reserves. However, a commercial heap of this nature has not yet been commissioned.

Extracting other base metal sulfides

Many base metals are bioleached with ferric iron and acid, generated as leach agents by iron- and sulfur-oxidizing organisms. These include cobalt, gallium, molybdenum, zinc, nickel, lead and tin (Watling, 2008; Brierley and Brierley, 2001). The tolerance of the microbial species to soluble transition metals varies between species and even strains of the same species, and can be a limiting factor (Nurmi *et al.*, 2009; Cabrerra *et al.*, 2005). Many acidophiles have far greater baseline tolerance to many (cationic) transition metals than neutrophilic prokaryotes, and can often develop enhanced tolerance, for example, adaptation of *At. ferrooxidans* to 50 g/l Ni has been demonstrated (Watling, 2008).

Nickel extraction through heap leaching has been explored at pilot scale at, amongst others, Radio Hill (Western Australia) from pentlandite, Talvivaara (Sotkama, Finland) from a complex polymetallic ore containing pentlandite and by the Jinchuan Group Ltd (Gansu Province, China) treating pyrrhotite, pentlandite and chalcopyrite. In all cases, nickel recoveries exceeding 80% were reported. Further, bioleaching of zinc-bearing

Table 16.2 Succession within the microbial community on the heating of the heap subsequent to exothermic oxidation of mineral sulfides. Adapted from van Hille *et al*. (2013) and Dew *et al*. (2011)

Temperature (°C)	Succession of typical bioleaching microorganisms with time and increasing temperature				
65–80					*Sulfolobus metallicus* *Metallosphaera hakonensis*
50–65				*Sulfobacillus* spp. *Sulfolobus* *Ac. cupricumulans* *Acidithiomicrobium* *Acidimicrobium* *Sulfobacillus thermosulfidooxidans* *Metallosphaera hakonensis*	
45–50			*Sulfobacillus thermosulfidooxidans* *Ferroplasma* spp. *Acidiplasma cupricumulans* *Leptospirillum ferriphilum* *Acidithiobacillus caldus* *Acidithiomicrobium* *Acidimicrobium*		
30–45		*Acidithiobacillus ferrooxidans* *Acidithiobacillus thiooxidans* *Acidithiobacillus caldus* *Ferroplasma acidiphilum* *Leptospirillum ferriphilum* *Leptospirillum ferrooxidans* *Sulfobacillus thermosulfidooxidans*			
15–30	*Acidithiobacillus ferrivorans* *Acidithiobacillus thiooxidans* *Leptospirillum ferrooxidans* *Ferroplasma acidiphilum*				
Time of leach®					

sphalerites have been reported in Namaqualand, South Africa.

The Talvivaara process proceeded to full scale operation in 2008, with first product delivered in 2009. This was achieved through a two-stage heap leach of crushed (D_{80} < 8 mm) and agglomerated ore, followed by metal recovery by selective precipitation (Riekkola-Vanhanen, 2013).

Heap bioleaching for gold recovery

To recover gold from low grade refractory sulfidic ores, bio-oxidation of the sulfidic (pyrite/arsenopyrite) fraction can be employed to ensure accessibility of the gold to extraction with cyanide or alternative lixiviants. Oxidation of sulfide minerals also reduces the cyanide consumption as a consequence of avoided thiocyanate formation. The best-known process utilizing heap bioleaching for gold recovery is the commercial BioPRO™, implemented at Newmont's Gold Quarry mine in Nevada, USA, in 1999 (Logan *et al.*, 2007) and motivated by increasing gold-bearing mineral sulfides in an oxide deposit. Refractory sulfidic ores of sufficient gold grade sulfide-sulfur above 0.2 were crushed (80% < 19 mm diameter) and stacked on a bio-oxidation heap, following mixing with liquid from the irrigation ponds to neutralize the carbonate, condition and wet the ore and introduce the microbial inoculum evenly throughout the ore. The heaps were stacked to 10–16 m in height with a single pad holding some 810,000 tons of ore and operated with aeration. An irrigation solution obtained from the bio-solubilization pond was recirculated to the heap through drip emitters at 4 l·m^2·h. Continuous initial irrigation was later modified to intermittent cycles of one to two weeks. The overall cycle time for the bio-oxidation phase was some 435 days with an average leach period of 160 days (range 130–300 days). An additional 90 days' preparation was required for crushing and loading, 60 days after bio-oxidation for draining and 90 days for offloading, prior to the 30 day site turn-around.

While the microorganisms originally enriched through test-work were identified as *At. ferrooxidans*, *Leptospirillum* spp. and *Sulfobacillus* spp., greater diversity was observed in the full-scale heap, corresponding to the large temperature range recorded (25 to 80°C). Therefore the inoculation ponds were supplemented with extremely thermophilic archaea, including *Sulfolobus*, *Metallosphaera* and *Acidianus* spp. Following the bio-oxidation heap, the ore was transported to the CIL process and cyanidation for gold recovery. Over the 6 years of operation, the average ore grade treated was 1.6% sulfide-S and 2.64 g/ton gold. A gold recovery of 53.6% was achieved (Logan *et al.*, 2007).

Current challenges and associated developments in heap bioleaching

In heap leaching, a number of key challenges are faced for further optimization of these processes. A key consideration is the reduction of the time required for effective leaching and improvement of the extent of extraction achieved. The core challenges are associated with achieving an active microbial community with rapid generation of the leach agents and production of heat and those associated with mineral leaching through effective contacting of leach agent and mineral surface.

In the first category, increasing understanding of the microbial consortium is sought, with identification of the role of key players within the consortium. The succession of micro-organisms with changing operational conditions within the heap, such as ion concentration, CO_2 availability and, especially, temperature requires understanding for process optimization. In addition, microbe–mineral interactions and colonization of mineral ore surfaces is of interest (Africa *et al.*, 2013; Bromfield *et al.*, 2011; Govender *et al.*, 2013; Tupikina *et al.*, 2013; Demergasso *et al.*, 2010). The supply of O_2 and CO_2 impacts on microbial growth (Bryan *et al.*, 2012; Petersen *et al.*, 2010) as does the presence of inhibitory substances (Nurmi *et al.*, 2009; Cabrera *et al.*, 2005). These relationships require clarification. Contact between the leach agent and mineral is strongly affected by the distribution of fluid within the ore bed (Fagan *et al.*, 2013, 2014; Ilankoon *et al.*, 2013), ore bed permeability (Fagan *et al.*, 2014), depth of leaching and liberation of the mineral surfaces (S.T.L. Harrison, unpublished data).

Further to this, the integrated modelling of the sub-processes of these complex systems will provide substantial value through improved prediction. While major strides have been made with the construction of the simulations by researchers such as Dixon, Petersen and Truelove, it is the integration of insightful experimental data with these modelling efforts that will provide the next level of understanding of heap leaching.

New processes under development

Bioleaching of base metals preceding cyanide leaching

As bioleaching of pyrite from gold-bearing pyritic ores or concentrates serves as a pre-treatment ahead of cyanide leaching for gold extraction, so the bioleaching of base metal sulfides from low grade platinum concentrates has been demonstrated to have potential in the improvement of the leaching of the PGMs. Low grade platinum concentrates are formed as a by-product of the flotation circuit concentrating ores such as on the Upper Grade 2 (UG2) and Platreef circuits in South Africa. The impurities in these concentrates, such as chromite, pyrrhotite and various base metal sulfides increase the cost of smelting and its environmental burden while decreasing its efficiency. Mwase *et al.* (2012) coated a low grade platinum concentrate onto pebbles, according to Geobiotics LLC's proprietary GEOCOAT© process (Harvey *et al.*, 2002) prior to exposing the material to column leaching at 65°C using a mixed mesophilic and thermophilic microbial inoculum with aeration and irrigation with a solution containing 2 g/l Fe and 10 g/l H_2SO_4 at 5 l·m^2·h. After 30 days, 52% of the copper, 95% of the nickel and 85% of the cobalt present had been leached. Following an alkaline wash, the material was again coated onto pebbles and subsequent cyanide leaching (0.15 M NaCN) for 21 days at 23°C resulted in recovery of 20%, 87% and 46% of the platinum, palladium and rhodium present respectively. Based on the experimental proof of concept, the flow sheet shown in Fig. 16.2 has been proposed to facilitate the pre-treatment of low-grade PGM concentrates to recover base metals prior to recovery of PGMs by cyanidation and CIL.

Tank bioleaching

Tank leaching of gold-bearing ores

Bioleaching technology for mineral sulfides was initially developed through the processing of refractory gold ores in which the gold is occluded in either pyrite or arsenopyrite minerals. Through the bioleaching of the pyrite and arsenopyrite into solution, the gold is made available for subsequent leaching through, for example, cyanide leaching. This technology was developed over the period from the mid-1980s to mid-1990s, with three dominant technologies emerging: BIOX™ developed in South Africa through Gencor Ltd, BHP Billiton, Goldfields and now Biomin Ltd as ownership of the technology was passed on; Bacox™ also developed in South Africa through Mintek (originally in association with Anglo American Corporation); and BacTec developed in Australia (Brierley and Brierley, 2001; Clark *et al.*, 2006). Table 16.3 lists the plants that have been and are currently operated using this technology worldwide. It is evident that the BIOX™ technology supplied by Biomin Ltd remains dominant. To date the cumulative gold extracted via BIOX™ technology is 657,110 kg (http://www.biomin.co.za/biox/).

Typically the BIOX™ process is operated with a feed particle size distribution of 80% < 75 μm diameter. The smaller the grind size, the more rapid is the sulfide oxidation rate (van Aswegen *et al.*, 2007). However, at very fine grind sizes, the formation of reactive oxygen species may be increased (Jones *et al.*, 2012) with a negative impact on the microbial activity (Nemati and Harrison, 2000). A 20% solids loading is fed to the reactor train. Maximizing the solids loading enhances the overall system productivity; however, it also increases the demand for O_2 supply and associated heat removal which are a function of the sulfur grade of the concentrate and often the constraining factors in the operation (Bailey and Hansford, 1994). Typically, the reactor train is configured with three primary reactors in parallel, followed by three reactors in series. The overall residence time in the bioleach reactors is four to six days, divided evenly between the primary reactors and the downstream reactors (van Aswegen *et al.*, 2007). Mineralogy, sulfide content and sulfide oxidation rates influence residence time.

The tank bioleaching of these refractory gold-bearing sulfidic concentrates is operated as a mesophilic process, specified at 42 to 45°C (van Aswegen *et al.*, 2007), although higher operating temperatures up to 48°C are routinely reported (van Hille *et al.*, 2013) owing to the exothermic nature of the process and potential for process perturbations. While the typical microbial consortium for BIOX™ was considered to be *At. ferrooxidans*, '*At. thiooxidans*' (later identified as *At. caldus*) and

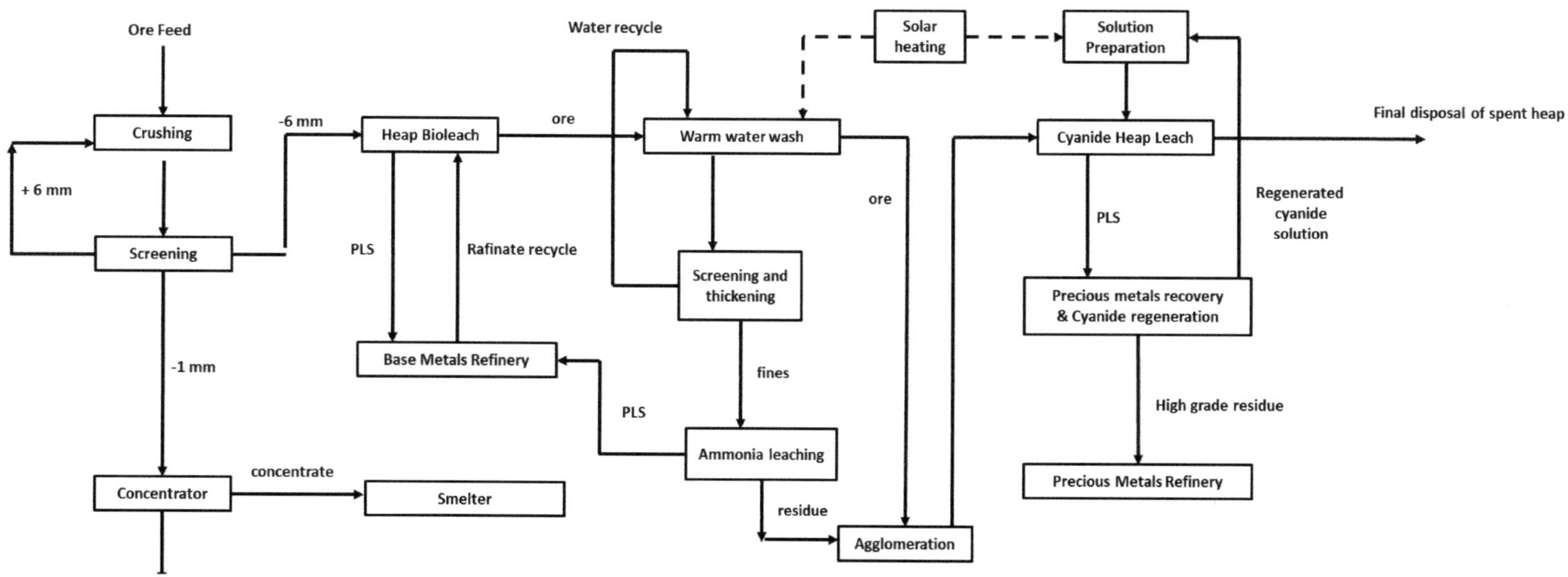

Figure 16.2 Flowsheet for two-stage metal extraction process for recovery of base metals prior to PGMs from low grade concentrates (supplied by James Mwase in 2014).

Table 16.3 Tank bioleaching plants for the recovery of gold from refractory ores

Mine (owners)	Country	Capacity (tpd)	Reactor size	Technology	Dates of operation
Fairview (Pan African Resources)[a,b,c]	South Africa	62	340	BIOX™	1986 ongoing
VaalReefs (Anglo American Corporation) [d]	South Africa	Pilot at 1 tpd, Demo at 20 tpd	6 and 12 m^3 25 and 125 m^3	Bacox™	1986
São Bento (AngloGold Ashanti)[a,b,c]	Brazil	150	550	BIOX™	1990, currently under care and maintenance
Harbour Lights[a,b,c]	Australia	40	160	BIOX™	1991–1994
Wiluna (Apex Minerals NL)[a,b,c]	Australia	158	480	BIOX™	1993 ongoing
Youanmi[c]	Australia	120		BacTech	1994–1998
Ashanti Sansu (AngloGold Ashanti)[a,b,c]	Ghana	960	900	BIOX™	1994 ongoing
Coricancha (Nyrstar)[a,b,c]	Peru	60	262	BIOX™	1998 ongoing
Tasmania Mine, formerly Beaconfield Gold Mine[c]	Tasmania	70		Bacox™	1999 ongoing
Amantatau[b]	Uzbekistan	2740			2002
Stratoni[b]	Greece	2700			2004
Shandong Tiancheng Biotechnology Gold Industry, Laizhou[d]	China	100		Bacox™	2001 ongoing
Fosterville[a,b] (Aurico)	Australia	211	900	BIOX™	2005 ongoing
Suzdal (Severstal)[a]	Kazakhstan	520	650	BIOX™	2005 ongoing
Bogoso (Golden Star Resources)[a]	Ghana	820	1500	BIOX™	2007 ongoing
Jinfeng (Eldorado Gold)[a]	China	790	1000	BIOX™	2007 ongoing
Kokpatas (Navoi Mining and Metallurgical Combinat[a]	Uzbekistan	2138	900	BIOX™	2008 ongoing
Agnes (Galaxy Gold)[a]	South Africa	20	75	BIOX™	2010 ongoing

[a]http://www.biomin.co.za/biox
[b]Clark *et al*. (2006).
[c]Brierley and Brierley (2001).
[d]http://www.mintek.co.za/technical-divisions/biotechnology-bio/

Leptospirillum ferriphilum (van Aswegen *et al.*, 2007; Coram and Rawlings, 2002), it is increasingly recognized that this consortium is more varied and dynamic (van Hille *et al.*, 2013). These species are dominant in the stock culture and culture in pilot reactors. Recently, the microbial communities in many commercial BIOX™ plants were shown to be dominant in *At. caldus* and a variety of archaea including *Acidiplasma cupricumulans*, *Ferroplasma acidiphilum* and *Thermoplasma*-like species (van Hille *et al.*, 2013). Controlled studies have demonstrated the shifts in community structure to be mediated through temperature, pH (typically maintained in the range 1.2 to 1.8) and, to a smaller extent, solids loading (van Hille *et al.*, 2013). The mechanisms of these shifts are an area of current investigation.

Oxygen supply to the tank is typically set to maintain a dissolved oxygen concentration of 1–2 mg/l, to ensure that the critical dissolved O_2 concentration is exceeded. Achieving adequate O_2 supply is a key design parameter (Bailey and Hansford, 1994). Where carbonate is present in the concentrate at a level of 2% or greater, sufficient CO_2 supply for the growth of the autotrophic organisms results. In the absence of sufficient carbonate, enrichment of the air supply with CO_2 is required and affects relative growth rates of microbial species present (Bryan *et al.*, 2012).

Following sufficient sulfide oxidation, the BIOX™ solid product is thickened, washed, neutralized to pH 7–8 and processed further through cyanidation for gold recovery. Key challenges in this process are of both a microbiological, mineralogical and engineering nature. A good understanding of the gangue chemistry and associated ions in solution, together with their effect on the microbial community, is required. Similarly, the impact of toxins entering the process on microbial activity must be understood and controlled. Examples include flotation reagents, grease and oil, cyanide and thiocyanate, water treatment chemicals. The provision of adequate O_2 transfer, CO_2 transfer and cooling require efficient mass and energy transfer systems. Agitator design is key to both and to ensure sufficiently homogeneous suspension of the concentrate (Harrison *et al.*, 2012).

Tank leaching for base metal sulfides

Following the implementation of tank bioleaching of refractory gold-bearing concentrates as a pretreatment for gold recovery, the technology has been extended and applied to the leaching of base metals from sulfidic mineral concentrates. Leaching of copper, nickel, cobalt, zinc, lead, gallium and molybdenum has been demonstrated (Clark *et al.*, 2006; Brierley and Brierley, 2001). Operations, at pilot, demonstration or commercial scale, demonstrating the tank leaching technology for base metals are listed in Table 16.4. BioNIC™ was developed by BHP Billiton for nickel recovery from sulfidic concentrates via mesophilic tank leaching followed by solvent extraction or ion exchange (Clark *et al.*, 2006). This was followed by development of similar technology for copper, BioCOP™, initially using mesophilic conditions and thereafter thermophilic conditions (Batty and Rorke, 2006). These have been demonstrated at pilot scale of 20,000 tpa (Table 16.3) through a joint venture (Alliance Copper) of BHP-Billiton and Codelco (Chile). While the thermophilic process out-performed the mesophilic process, it required special attention to the supply of O_2 owing to the much reduced solubility of O_2 at elevated temperature, requiring the provision of an O_2 plant. However, O_2 utilization was very efficient at 80% (Clark *et al.*, 2006).

The only commercial scale operation for tank bioleaching of base metals to date is the recovery of cobalt from a cobaltiferous pyrite concentrate, of which some 0.9–1.1 million tons of fine-grain tailings waste (< 35 μm) was stockpiled over the period 1956 to 1997 at the Kilembe Mine, Uganda, operated by the Kasese Cobalt Company (Morin and d'Hugues, 2007; Durance *et al.*, 2004; Brierley and Brierley, 2001). The concentrate contained ~80% pyrite with 1.38% ionic cobalt distributed through it, which was liberated by dissolution of the pyrite fraction. The tank leach technology was developed by BRGM over the period 1988 to 1998 whereafter the operation commenced at Kasese, Uganda. While designed to operate at a solids loading of 20% and 42°C, typical operating temperatures of 46 to 50°C were found. Iron-oxidizers and sulfur-oxidizers were present (Viera *et al.*, 2007), comprising a consortium of five to seven microorganisms Including *L. ferriphilum*, *At. caldus*, *Sulfobacillus benefaciens* and *Acidiplasma* sp.. Metal tolerance of this consortium was reported as 35 g/l ferric iron and 5 g/l cobalt.

The agitated bioleach tanks used were the largest built to date on commissioning in 1998, at 1350 m^3 each. Oxygen was supplied at 1750 kg O_2/h to meet the O_2 uptake rate of 1350–1400 mg/(l·h) at a 50% efficiency (Morin and d'Hugues, 2007). Limestone addition controlled pH and augmented CO_2 supply for microbial growth. At an overall residence time of 6 days, cobalt recovery of 80 to 92% reported (Morin and d'Hugues, 2007, Brierley and Brierley, 2001), of which some 60% occurred in the primary tanks. Recycle of the largest size fraction of the concentrate aided cobalt leaching efficiency through reducing cobalt loss in the solids residue (Durance *et al.*, 2004) and increased the microbial population, activity and community diversity present in the tanks. Subsequent to bioleaching, ferric iron was removed by precipitation prior to selective recovery of copper precipitate, cobalt cathode and nickel precipitate (Morin and d'Hugues, 2007). The plant operation was suspended in 2002–03 when the cobalt price decreased to \$6–8/lb, below the financial breakeven point. Production was re-commenced

Table 16.4 Operations demonstrating tank bioleaching of base metals from sulfidic mineral concentrates

Operation (owners)	Country	Mineral	Size of operation (tpd)	Technology	Time period
BHP Billiton[a]	South Africa	Pentlandite	0.3	BioNIC™	Early 1990s
Mt Lyell, Copper Mines of Australia[b]	Tasmania	Chalcopyrite	Pilot	Moderately thermophilic tank bioleaching, followed by solvent extraction – electrowinning	Pre 1998
Yabula Nickel Refinery[a]	Queensland, Australia	Nickel sulfide	Pilot	Ni recovery of 92% and cobalt recovery of 87% at pulp density of 17.5% and 7-day residence time	Pre 1999
Chuquicamata (Alliance Copper)[c,d]		Chuquicamata concentrate = mix of chalcocite and chalcopyrite Mansa Mina concentrate, containing enargite Cu_3AsS_4	55	Mesophilic at 20–50°C, leading to 25 g/l Cu in solution; BioCOP™: thermophilic at 55–80°C, leading to 35 g/l Cu in solution	1997 2003
Aguablanca[d]	Spain		pilot	Mintek through BioMinE	
Kasese Mining Company at Kilembe Mine[b,d]	Uganda	Cobaltiferous pyrite concentrate: 1.38% Co, 41% S	241 tpd; three 1350 m³ 1° reactors		1998
Mintek-Bactec-Peñoles plant, Monterrey[d]	Mexico	Complex polymetallic containing Cu, Zn, Pb sulfides	2.2	Mintek – BacTec	2001

[a]Gericke *et al.* (2014).
[b]Brierley and Brierley (2001).
[c]Clark *et al.* (2006).
[d]Neale *et al.* (2011).

in 2003–04 on increase of the cobalt price to $20/lb (Morin and d'Hugues, 2007) and operated until depletion of the stockpile.

Two other base metal tank leaching systems are currently being evaluated for commercial-scale production. In Iran, the tank leaching of a finely milled copper concentrate at Babak is under consideration using technology developed in conjunction with Canadian consultants (Jochen Petersen, personal communication). While an inoculated process is being considered, the role of the microorganisms in the presence of the fine grind should be questioned (Nemati and Harrison, 2000). In Finland, Mondo Minerals BV are investigating the recovery of nickel and cobalt from a flotation by-product from their talc business. These by-products contain 7.6–15.2% Ni, 0.3–0.7% Co, 0.9–2.2% As and 24.3–34.0% sulfide (Gericke *et al.*, 2014). Using a tank bioleaching process at 45°C and 15 to 17.5% solids loading at a D_{80} of 20 μm over five stages, nickel and cobalt recovery in excess of 95% is reported at the mini-pilot plant scale (Gericke *et al.*, 2014). The nickel recovery is proportional to the sulfide extraction. The iron and arsenic are removed as a stable precipitate and the nickel and cobalt recovered as a mixed hydroxide product (Gericke *et al.*, 2014). Commercialization of this process is under consideration.

Tank leaching of base metal sulfide concentrates does not yet typically outcompete pyrometallurgical processing where smelter capacity is available and the concentrate is well suited to smelting. However, tank bioleaching may be favourable where SS-EW capacity is available and long-distance transportation to a smelter is required. Tank bioleaching of base metal sulfides is competitive where smelting is problematic, e.g. polymetallic sulfides concentrates with significant arsenic, antimony or bismuth content which incur penalties on smelting (Johnson 2014b; Neale *et al.*, 2011; du Plessis *et al.*, 2007; Brierley and Brierley, 2001) (Fig. 16.1).

Reprocessing of mine wastes and urban wastes

As accessible primary metal ores increasingly become lower grade, re-processing of mine wastes from previous mining operations, which may contain greater concentrations of base and other metals, becomes more viable and appealing. This is especially true as mining and comminution costs are avoided. For example, in South Africa, substantial re-processing of tailings dumps and deposits is currently under way (Johan Wagner, personal communication). Bioleaching approaches will be favourable for a number of deposits, with the added advantage that re-processing sulfidic mine wastes reduces the legacy of acid mine drainage (AMD), such as demonstrated in the case of the Kasese cobalt biomining operation (see 'Tank leaching of base metal sulfides'). Hence an excellent opportunity exists to superimpose value recovery from these waste materials through bioleaching with the simultaneous generation of benign tailings through appropriate processing (Kazadi Mbamba *et al.*, 2012; Hesketh *et al.*, 2010). Attainment of such benign tailings requires excellent understanding of acid generating capacity and potential for metal deportment.

In addition to mine wastes, many urban wastes such as e-wastes, have richer metal grades than virgin ores, especially copper. Potential for recovery of these metals by (bio)hydrometallurgy holds much promise for the future (Bryan *et al.*, 2015), leading to new applications for the acidophiles.

Closing comments

The importance of acidophilic bioleaching for the extraction and recovery of metals from both virgin (primary) ores and waste materials, including mining wastes and urban wastes is governed by the ability of these technologies to operate under specific conditions, not well served by alternative technologies. Of particular importance are the following:

- potential for *in situ* extraction of metals from deep-buried ores;
- potential to extract metals from low-grade ores without the need to liberate sufficiently to form a concentrate;
- potential to process polymetallic ores or concentrates, allowing for the recovery of multiple metal products;
- potential to process mineral concentrates that contain elements not tolerated in smelter,, e.g. arsenic, antimony and bismuth.

Further to this, the relative environmental impact of the process technologies requires comparison. Biotechnologies have several potential advantages over energy expensive comminution and pyrometallurgy. However, refinement of the metal recovery step, currently most commonly carried by SX-EW, has great potential to reduce the environmental footprint of these processes and should form a focal point of future research.

Current disadvantages cited for the biotechnological approach to metal extraction include the long-time requirement of the heap leaching processes (Johnson, 2014b); perceived issues with respect to process robustness and reliability (Johnson, 2014b; Brierley, 2010) as well as the energy requirement of comminution for preparation of the mineral concentration for intensified processes (Johnson, 2014b). Further, appropriate water management is essential. For very low grade resources, the slow recovery rate is tolerated as this represents value from waste. For resources of higher value, hold-up times for these inventories can be improved by enhanced liberation, improved solid liquid contacting controlled through fluid flow and improved homogeneity of the packed heaps (as, for example in the GeoCoat™ process; Harvey and Bath, 2007), enhanced colonization reducing start-up times and efficient microbial succession in response to the heterogeneity of pH, temperature and leachate chemistry across space and time. This includes improved knowledge of engineering heaps and of their integrated modelling (Brierley, 2010). As the technology becomes increasingly established e.g. tank leaching technology and heap leaching of secondary sulfides, effective communication of process robustness is required. In addressing the energy of liberation, improved understanding of the liberation of minerals for leaching is required such as is provided through the study of the depth of leachability or new approaches to achieving such access to minerals, e.g. fracturing of ores.

One of the major advantages of using acidophiles to extract metals, when compared to other hydro- and pyrometallurgical processing, is the lower temperature and/or pressure required, contributing to lower carbon footprints and a reduced environmental burden. This may be a major factor influencing the growth of biomining technologies in future years.

References

Abhilash, Pandey, B.D., and Natarajan, K.A. (2015). Microbiology for Minerals, Metals, Materials and the Environment (CRC Press, Boca Raton, FL, USA).

Africa, C.-J., van Hille, R.P., Sand, W., and Harrison, S.T.L. (2013). Investigation and *in situ* visualisation of thermophilic microbial interfacial interactions with mineral sulfides in a heap simulated environment. Miner. Eng. *48*, 100–107.

Bailey, A.D., and Hansford, G.S. (1994). Oxygen mass transfer limitation of batch bio-oxidation at high solids concentration. Miner. Eng. *7*, 293–303.

Batterham, R. (2011). In situ leaching – how far in the future? In Biohydrometallurgy: Biotech Key to Unlock Mineral Resource Value, Qui, G., Jiang, T., Qin, W., Liu, X., Yang, Y., and Wang, H., eds. (Central South University Press, Changsa, China), pp. 21–22.

Batty, J.D., and Rorke, G.V. (2006). Development and commercial demonstration of the BioCOP™ thermophile process. Hydrometallurgy *83*, 83–89.

Bouffard, S.C., and West-Sells, P.G. (2009). Hydrodynamic behavior of heap leach piles: Influence of testing scale and material properties. Hydrometallurgy *98*, 136–142.

Brierley, J.A., and Brierley, C.L. (2001). Present and future commercial applications of biohydrometallurgy. Hydrometallurgy *59*, 233–239.

Brierley, C.L. (2010). Biohydrometallurgical prospects. Hydrometallurgy *104*, 324–328.

Bromfield, L., Africa, C.-J., Harrison, S.T.L., and van Hille, R.P. (2011). The effect of temperature and culture history on the attachment of *Metallosphaera hakonensis* to mineral sulfides with application to heap bioleaching. Miner. Eng. *24*, 1157–1165.

Bryan, C.G., van Wyk, N., Fraser, M.K., Dew, D., Rautenbach, G.F., and Harrison, S.T.L., (2012). The effect of CO_2 availability on the growth, iron oxidation and CO_2-fixation rates of pure cultures of *Leptospirillum ferriphilum* and *Acidithiobacillus ferrooxidans*. Biotechnol. Bioeng. *109*, 1693–1703.

Bryan, C.G., Watkin, E.L., McCredden, T.J., Wong, Z.R., Harrison, S.T.L., and Kaksonen, A.H. (2015). The use of pyrite as a source of lixiviant in the bioleaching of electronic waste. Hydrometallurgy *152*, 33–43.

Bijmans, M.F., Dopson, M., Ennin, F., Lens, P.N., and Buisman, C.J. (2008). Effect of sulfide removal on sulfate reduction at pH 5 in a hydrogen fed gas-lift bioreactor. J. Microbiol. Biotechnol. *18*, 1809–1818.

Bijmans, M.F., de Vries, E., Yang, C.H., Buisman, C.J., Lens, P.N., and Dopson, M. (2010). Sulfate reduction at pH 4.0 for treatment of process and wastewaters. Biotechnol. Prog., *26*, 1029–1037.

Cabrera, G., Gómez, J.M., and Cantero, D. (2005). Influence of heavy metals on growth and ferrous sulfate oxidation by *Acidithiobacillus ferrooxidans* in pure and mixed cultures. Process Biochem. *40*, 2683–2687.

Clark, M.E., Batty, J.D., van Buuren, C.B., Dew, D.W., and Eamon, M.A. (2006). Biotechnology in minerals processing: technological breakthroughs creating value. Hydrometallurgy *83*, 3–9.

Coram, N.J., and Rawlings, D.E. (2002). Molecular relationship between two groups of the genus *Leptospirillum* and the finding that *Leptospirillum ferriphilum* sp. Nov. dominates South Afrcican commercial bio-oxidation tanks that operate at 40°C. Appl. Environ. Microb. *68*, 838–845.

Demergasso, C., Galleguillos, F., Soto, P., Seron, M., and Iturriaga, V. (2010). Microbial succession during a heap bioleaching cycle of low grade copper sulfides: does this knowledge mean a real input for industrial process design and control? Hydrometallurgy *104*, 382–390.

Dew, D.W., van Buuren, C.B., McEwan, K., and Bowker, C. (1999). Bioleaching of base metal sulfide concentrates: a comparison of mesophilic and thermophilic bacterial cultures. Process Metallurgy *9*, 229–238.

Dew, D., Rautenbach, G.F., van Hille, R.P., Davis-Belmar, C.S., Harvey, I.J., and Truelove, J.S. (2011). High temperature heap leaching of chalcopyrite: method of evaluation and process model validation. In Proceedings of SAIMM Symposium: Percolation Leaching: the status globally and in Southern Africa (Misty Hills, Muldersdrift, South Africa), pp. 201–220.

Domic, E.M. (2007). A review of the development and current status of copper bioleaching operations in Chile: 25 years of successful commercial implementation. In Biomining, Rawlings, D.E., and Johnson, D.B., eds (Springer-Verlag, Berlin, Germany), pp. 81–95.

du Plessis, C.A., Batty, J.D., and Dew, D.W. (2007). Commercial applications of thermophilic bioleaching. In Biomining, Rawlings, D.E., and Johnson, D.B., eds (Springer-Verlag, Berlin, Germany), pp. 57–80.

du Plessis, C.A., Slabbert, W., Hallberg, K.B., and Johnson, D.B. (2011). Ferredox: a biohydrometallurgical processing concept for limonitic nickel laterites. Hydrometallurgy *109*, 221–229.

Durance, M.V., Brochot, S., and Mugabi, M. (2004). Material balance approach for parameter determination in bioleaching process. In Computer Applications in Biotechnology, Pons, M.-N., and van Impe, J., eds (Elsevier IFAC Publications, Nancy, France), pp. 121–126.

Fagan, M.A., Sederman, A.J., Harrison, S.T.L., and Johns, M.L. (2013). Phase distribution identification in the column leaching of low grade ores using MRI. Miner. Eng. *48*, 94–99.

Fagan, M.A., Ngoma, I.E., Chiume, R.A., Minnaar, S., Sederman, A.J., Johns, M.J., and Harrison, S.T.L. (2014). MRI and gravimetric studies of hydrology in drip irrigated heaps and its effect on the propagation of bioleaching microorganisms. Hydrometallurgy *150*, 210–221.

Gentina, J.C., and Acevedo, F. (1985). Microbial ore leaching in developing countries. Trends Biotechnol. *3*, 86–89.

Gentina, J.C., and Acevedo, F. (2013). Application of bioleaching to copper mining in Chile. Electron. J. Biotechno. *16*, 12.

Gericke, M., Pawlik, C., Dew, D.W., van Aswegen, P., Barnett, S.C.C., and Mikkola, T. (2014). Bioleaching of nickel-cobalt-arsenic containing flotation by products of the talc industry. Presented at: MEI Biohydrometallurgy 2014 (Falmouth, UK).

Gilbertson, B.P. (2000). Creating value through innovation: biotechnology in mining. Miner. Process. Extr. Metall. *109*, 61–67.

Govender, E., Bryan, C.G., and Harrison, S.T.L. (2013). Quantification of growth and colonisation of low grade sulfidic ores by acidiphilic chemoautotrophs using a novel experimental system. Miner. Eng. *48*, 108–115.

Harrison, S.T.L., Stevenson, R., and Cilliers, J.J. (2012). Assessing solids concentration homogeneity in Rushton-agitated slurry reactors using electrical resistance tomography (ERT). Chem. Eng. Sci. *71*, 392–399.

Harvey, T.J., Holder, N., and Stanek, T. (2002). Thermophilic bioleaching of chalcopyrite with GEOCOAT© process. Presented at: Alta 2002 Nickel/Cobalt 8 – Copper 7 Conference (Perth, Australia).

Harvey, T.J., and Bath, M. (2007). The GeoBiotics GEOCOAT™ technology – progress and challenges. In Biomining, Rawlings, D.E., and Johnson, D.B., eds (Springer-Verlag, Berlin, Germany), pp. 97–112.

Hedrich, S., and Johnson, D.B. (2014). Remediation and selective recovery of metals from acidic mine waters using novel modular bioreactors. Environ. Sci. Technol. *48*, 12206–12212.

Hesketh, A.H., Broadhurst, J.L., and Harrison, S.T.L. (2010). Mitigating the generation of acid mine drainage from copper sulfide tailings impoundments in perpetuity: a case study for an integrated management strategy. Miner. Eng. *23*, 225–229.

Ilankoon, I.M.S.K., Cole, K.E., and Neethling, S.J. (2013). Measuring hydrodynamic dispersion coefficients in unsaturated packed beds: Comparison of PEPT with conventional tracer tests. Chem. Eng. Sci. *89*, 152–157.

Jena, P.K., Rath, M., and Mishra, C.S.K. (2009). Metal and mineral recovery through bioleaching. In Biotechnology Applications, Mishra, C.S.K., and Champagne, P., eds (I.K. International Publishing House Pvt. Ltd., New Delhi, India), pp. 309–319.

Johnson, D.B. (2014a). Recent developments in microbiological approaches for securing mine wastes and for recovering metals from mine wastes. Minerals *4*, 279–292.

Johnson, D.B. (2014b). Biomining – biotechnologies for extracting and recovering metals from ores and wastes. Curr. Opin. Biotech. *30*, 24–31.

Johnson, D.B. (2015). Biomining goes underground. Nat. Geosci. *8*, 165–166.

Jones, G.C., van Hille, R.P., and Harrison, S.T.L. (2012). Reactive oxygen species generated in the presence of fine pyrite particles and their implication in thermophilic minerals bioleaching. Appl. Microbiol. Biot., *97*, 2735–2742.

Kazadi Mbamba, C., Harrison, S.T.L., Franzidis, J.-P., and Broadhurst, J.L. (2012). Mitigating acid rock drainage risks while recovering low-sulfur coal from ultrafine colliery wastes using froth flotation. Miner. Eng. *29*, 13–21.

Logan, T.C., Seal, T., and Brierley, J.A. (2007). Whole-ore heap bio-oxidation of sulfidic gold-bearing ores. In Biomining, Rawlings, D.E., and Johnson, D.B., eds (Springer-Verlag, Berlin, Germany), pp. 113–139.

Mwase, J.M., Petersen, J., and Eksteen, J.J. (2012). A conceptual flowsheet for heap leaching of platinum group metals (PGMs) from a low-grade ore concentrate. Hydrometallurgy *111–112*, 129–135.

Morin, D.H.R., and D'Hugues, P. (2007). Bioleaching of a cobalt-containing pyrite in stirred reactors: a case study from laboratory scale to industrial application. In Biomining, Rawlings, D.E., and Johnson, D.B., eds (Springer-Verlag, Berlin, Germany), pp. 35–54.

Neale, J.W., Gericke, M., and Ramcharan, K. (2011). The application of bioleaching to base metal sulfides in southern Africa: prospects and opportunities. In Proceedings of 6th Southern African Base Metals Conference (SAIMM) (Phalaborwa, South Africa), pp. 367–388. Available online: www.saimm.co.za/Conferences/BM2011/367-Neale.pdf

Nemati, M., and Harrison, S.T.L. (2000). Effect of solid loading on thermophilic bioleaching of sulfide minerals. J. Chem. Technol. Biotechnol. *75*, 1–7.

Nurmi, P., Özkaya, B., Kaksonen, A.H., Tuovinen, O.H., and Puhakka, J.A. (2009). Inhibition kinetics of iron oxidation by *Leptospirillum ferriphilum* in the presence of ferric, nickel and zinc ions. Hydrometallurgy *97*, 137–145.

Petersen, J., and Dixon, D.G. (2007). Modelling and optimisation of heap bioleach processes. In Biomining, Rawlings, D.E., and Johnson, D.B., eds (Springer-Verlag, Berlin, Germany), pp. 153–176.

Petersen, J., Minnaar, S.H., and du Plessis, C.A. (2010). Carbon dioxide consumption during the bioleaching of a copper ore in a large isothermal column. Hydrometallurgy *104*, 356–362.

Pham, V., and Ting, Y.P. (2009). Gold bioleaching of electronic waste by cyanogenic bacteria and its enhancement with bio-oxidation. Adv. Mat. Res. *71*, 661–664.

Pradhan, N., Nathsarma, K.C., Srinivasa Rao. K., Sukla, L.B., and Mishra, B.K. (2008). Heap bioleaching of chalcopyrite: a review. Miner. Eng. *21*, 355–365.

Rawlings, D.E., and Johnson, D.B. (2007). The microbiology of biomining: development and optimization of mineral-oxidizing microbial consortia. Microbiology *153*, 315–324.

Riekkola-Vanhanen, M. (2013). Talvivaara mining company – from project to a mine. Miner. Eng. *48*, 2–9.

Schippers, A., and Sand, W. (1999). Bacterial leaching of metal sulfides proceeds by two indirect mechanisms via thiosulfate or via polysulfides and sulfur. Appl. Environ. Microbiol. *65*, 319–321.

Tupikina, O.V., Minnaar, S.H., Van Hille, R.P., Van Wyk, N., Rautenbach, G.F., Dew, D., and Harrison, S.T.L. (2013). Determining the effect of acid stress on the persistence and growth of thermophilic microbial species after mesophilic colonisation of low grade ore in a heap leach environment. Miner. Eng. *53*, 152–159.

van Aswegen, P.C., van Niekerk, J., and Olivier, W. (2007). The BIOX™ process for the treatment of refractory gold concentrates. In Biomining, Rawlings, D.E., and Johnson, D.B., eds (Springer-Verlag, Berlin, Germany), pp. 1–34.

van Hille, R., van Wyk, N., Froneman, T., and Harrison, S.T.L. (2013). Dynamic evolution of the microbial community in BIOX leaching tanks. Adv. Mat. Res. *85*, 331–334.

Viera, M., Pogliani, C., and Donati, E. (2007). Recovery of zinc, nickel, cobaly and other metals by bioleaching. In Microbial Processing of Metal Sulfides, Donati, E.R., and Sand, W., eds (Springer; Dordrecht, the Netherlands), pp. 102–120.

Watling, H. (2006). The bioleaching of sulfide minerals with emphasis on copper sulfides – a review. Hydrometallurgy *84*, 81–108.

Watling, H. (2008). The bioleaching of nickel-copper sulfides. Hydrometallurgy *91*, 70–88.

Acidophiles and Astrobiology

17

Ricardo Amils and David Fernández-Remolar

Abstract

The *NASA Astrobiology* roadmap underlines the importance of extreme environments and the microorganisms that live in them to evaluate the possible existence of life outside of planet Earth. Acidophiles are of special astrobiological interest because the environments in which they thrive are generated by the metabolism of chemolithotrophic microorganisms that obtain energy from inorganic mineral substrates, a property that places them among the best candidates for a successful primitive energy conservation system. A thorough geomicrobiological characterization of an extreme acidic environment, the Río Tinto basin, has revealed the importance of the iron cycle in the water column of the river and detected a higher level of metabolic diversity in the sediments and subsurface of the Iberian Pyrite Belt (IPB), due to the presence of micro-niches with environmental conditions different from the acidic waters with high redox potentials existing in the river. The identification of iron oxides and iron sulfates on the surface of Mars, similar to those produced in the Tinto basin by the metabolic activity of chemolithotrophic microorganisms, has given Río Tinto the status of a geochemical and mineralogical terrestrial analogue of Mars. The argument that Mars' environmental conditions are not suitable for methanogenesis could be challenged by the methane production observed in the sediments and the subsurface of the IPB.

Introduction

The exploration of extreme environments has led to the discovery of numerous habitats that were considered uninhabitable until relatively recently, as a consequence of our anthropocentric concept of life. Subsequently, interest in the ecology of extreme environments has grown exponentially for different reasons – some fundamental and related with the search for the limits of life, and others more pragmatic focussed on the biotechnological potential of extremophiles and their enzymes (e.g. enzymatic catalysis and biohydrometallurgy; see Chapter 16).

Extremophiles have also played a fundamental role in the development of astrobiology. As stated in the *NASA Astrobiology Roadmap* (http://astrobiology.arc.nasa.gov/roadmap/), one of the main goals of astrobiology is to characterize (i) extreme habitats, (ii) the organisms developing on them and (iii) the mechanisms used to deal with the existing extreme conditions in these habitats, to evaluate their possible existence in other locations outside of the 'blue planet'. The Viking mission, considered the first astrobiological mission devoted to search for signs of life on Mars, concluded that life had little chance of developing there due to the extreme conditions detected on its surface (Margulis *et al.*, 1979). In the last forty years important advances in microbiology have challenged this rather pessimistic point of view. Thanks to research on extremophiles we know now that life is extremely robust and able to adapt to diverse conditions, and this knowledge has increased our chances of finding life in other parts of the universe.

This chapter presents a review of the astrobiological interest of acidophiles since, in contrast to other extremophiles which adapt to diverse geophysical and geochemical constraints imposed on them by extraneous factors (temperature, water activity, radiation, pressure, etc.), acidophiles thrive in the extreme conditions generated by their own chemolithotrophic metabolism, mainly centred on the iron and sulfur cycles.

Rio Tinto

The Río Tinto (Fig. 17.1) is a natural acidic river located in the south-west of Spain. It rises in Peña de Hierro (Iron Mountain) at the heart of Iberian Pyrite belt (IPB) and flows into the Atlantic Ocean at Huelva. The IPB is a geological area 250 km long and approximately 60 km wide that forms the northern-most zone of the Variscan Iberian Massif. The IPB is one of the largest sulfide deposits in the world. Massive bodies of iron and copper sulfides, as well as minor quantities of lead and zinc sulfides, constitute the main mineral ores (Leistel *et al.*, 1998). Its formation took place during the Hercynian orogenesis by hydrothermalism (Lescuyer *et al.*, 1998). An important characteristic of the Tinto ecosystem is its constant acidic pH, which is a direct consequence of the strong buffering capacity of ferric iron. When the river is diluted by tributaries or rain, hydrolysis of ferric iron generates protons which maintain the pH of the water column constant:

$$Fe^{3+} + 3\,H_2O \leftrightarrow Fe(OH)_3 + 3H^+$$

The reaction is reversible. When intense evaporation occurs during the hot summers, protons are consumed to dissolve ferric hydroxide precipitates. Owing to this buffering capacity of ferric iron a pH around 2.3 remains constant all along the river course, requiring only sufficient ferric iron and ferric hydroxide in the system. Owing to its size and easy access Río Tinto is considered an excellent model for the study of the microbial ecology of extreme acidic environments.

Microbial ecology of the Tinto basin

The combined use of conventional and molecular microbial ecology methodologies has led to the identification of the most abundant prokaryotic microorganisms of the Tinto basin (e.g. García-Moyano *et al.*, 2012; González-Toril *et al.*, 2003; López-Archilla *et al.*, 2001). Eighty per cent of the water column diversity corresponds to microorganisms belonging to only three bacterial genera – *Acidithiobacillus, Leptospirillum* and *Acidiphilium*

Figure 17.1 View of Río Tinto at Berrocal.

– all conspicuous members of the iron cycle (González-Toril *et al.*, 2003). *At. ferrooxidans* can both oxidize ferrous iron aerobically and reduce ferric iron in anaerobic conditions (e.g. Malki *et al.*, 2006). All *Leptospirillum* spp. isolated from Río Tinto are aerobic iron-oxidizers. All known species of *Acidiphilium* can couple the oxidation of organic compounds to the reduction of ferric iron under oxygen-limited conditions, though they do not grow by iron respiration under strictly anoxic conditions (e.g. Malki *et al.*, 2008). Although other iron-oxidizers (*'Ferrovum'*, *Ferrimicrobium* and *Ferroplasma* spp.) or iron-reducers (*Ferrimicrobium*, *Metallibacterium* spp. and *Acidobacterium* spp.) have been identified in the Tinto ecosystem (e.g. García-Moyano *et al.*, 2012; González-Toril *et al.*, 2003), their relatively low numbers detected by fluorescence *in situ* hybridization (FISH) suggest that they play a minor role in the iron cycle, at least in the water column of the Tinto.

Concerning the sulfur cycle, only *At. ferrooxidans* is found in significant numbers in the water column. This bacterium can oxidize both ferrous iron and reduced sulfur compounds. Sulfate-reducing microorganisms have also been detected in the sediments in some locations along the river, thus a complete sulfur cycle is also in operation in the ecosystem (e.g. García-Moyano *et al.*, 2012; Sánchez-Andrea *et al.*, 2011).

The characterization of the anoxic sediments from acidic environments had been neglected in the past, with few exceptions. Recently, a comparative analysis of the sediments and the water column of different samples along the physico-chemical gradient of Río Tinto have been performed (García-Moyano *et al.*, 2012). This study showed a significantly higher level of biodiversity in the sediments when compared to their water column counterparts, and the existence of a diversity gradient consequence of the geochemical gradient generated along the course of the river. Nearly all the microorganisms identified in this study were, in one way or another, related to the iron cycle. Most were previously detected and/or isolated in AMD sites (e.g. González-Toril *et al.*, 2010; Hallberg and Johnson, 2001) or biohydrometallurgical operations (Rawlings, 2005). Nonetheless, some bacteria, such as members of *Actinobacteria*, *Firmicutes*, *Acidobacteria*, *Planctomycetes* and *Chloroflexi*, were identified in the Tinto basin (García-Moyano *et al.*, 2012).

An in depth analysis of two anoxic sediments from the Tinto basin has recently shown major phylogenetic differences among sample sites (Sánchez-Andrea *et al.*, 2011). In one of the sediments, JL Dam, the most numerous group of bacteria belonged to the *Firmicutes* (56.6%), followed by the *Acidobacteria* (27.3%) and the Deltaproteobacteria (11.6%). In the SN Dam, the *Proteobacteria* phylum was the most represented, followed by the *Actinobacteria* (20.4%). Organisms of the *Firmicutes* and *Acidobacteria* phyla were present in relatively low percentages. This study also showed a differential pattern of distribution in the sediments at different depths. In the upper layers of JL Dam, sequences belonging to the *Acidobacteria* were the most abundant, while in the intermediate layer *Firmicutes* were mainly detected, and in the deepest layers extremely anaerobic organisms, including sulfate-reducing bacteria of the *Desulfosporosinus* and *Desulfurella* genera, were identified. In contrast, in the upper layer of the SN Dam sequences belonging to the *Acidithiobacillaceae* family were the most abundant, though in the deepest layers, organisms related to the spore-forming sulfate-reducing bacteria *Desulfosporosinus* were detected. It seems that in the strict anoxic conditions detected in the lower part of the sediments of the Tinto basin sulfate reduction is a recurrent functional activity, consequence of the high concentration of sulfate present in the system.

It is often assumed that toxicity caused by high metal concentrations in acidic habitats limits eukaryotic growth and diversity (Gross, 2000). However, colourful biofilms covering large surfaces of the Tinto basin are common features of acidic environments (see Chapter 7) and it has been estimated that eukaryotic algae account for over 60% of the biomass in the Tinto basin (López-Archilla *et al.*, 2001). An important number of eukaryotic species thriving in Río Tinto are photosynthetic. Among them, chlorophytes related to different genera such as *Chlamydomonas*, *Dunaliella*, *Chlorella*, as well as *Euglena* are the dominant eukaryotic microorganisms present in the river (see Chapter 7; Amaral-Zettler *et al.*, 2010). These species are known for their high metal tolerance. Filamentous algae, represented by the genera *Zygnemopsis* and *Klebsormidium* have been also found. The most acidic part of the river is inhabited by a eukaryotic community dominated by two species related to the genera *Dunaliella* (*Chlorophita*) and *Cyanidium*

(*Rhodophyta*). Pennate diatoms are also present in the river, forming large brown biofilms (see Chapter 7).

In addition to photosynthetic species, heterotrophic protists are also widely distributed along the river. The mixotrophic flagellates are mainly represented by members of the *Bodo* and *Ochromonas* genera. The dominant ciliate taxa belong to the order *Hypotrichida*. Amoeba are frequently found feeding on large diatoms, even in the most acidic part of the river. Heliozoa seem to be characteristic top predators of the benthic food chain in the Tinto ecosystem (Amaral-Zettler *et al.*, 2002). The only animal found in the river is a species of bdelloid rotifer related to the genus *Rotifera* (Amaral-Zetller *et al.*, 2002).

Among decomposers, fungi are the most diverse, and both unicellular and filamentous forms are present (e.g. López-Archilla *et al.*, 2001). A recent characterization of Río Tinto basin samples from different stations along the river (pH range between 1.5 and 2.9) have rendered more than three hundred and fifty fungal isolates, which have been identified by sequence analysis of the internal transcribed spacer (ITS) regions. This sequence analysis revealed the *Ascomycetes* as the most abundant phylum. *Basidiomycetes* and *Zygomycetes* accounted for less than 2% of the sequenced isolates. Of the *Ascomycetes*, 52% clustered within the *Eurotiomycetes* class, while 27% grouped with the *Dothideomycetes* and 17% with the *Sordariomycetes*. The rest of the members of this phylum clustered with the *Helotiales* (2%), *Mucorales* (1%) and less than 1% with the *Agaricomycetes*. The *Eurotiomycetes* and *Sordariomycetes* isolates showed the highest resistance to toxic heavy metals, which were much higher than the concentrations detected in the river, while members of the *Dothideomycetes* showed a level of resistance to concentrations similar to those existing in the water column (M. Oggerin, unpublished).

Sedimentary geomicrobiology of the Rio Tinto

Most of the biomass of the Tinto basin is located on the riverbed and the surface of the rocks, forming dense biofilms composed mainly of filamentous algae and fungi in which prokaryotic microorganisms are trapped. Significant iron mineral precipitation occurs on the negatively charged surface of these biofilms, generating iron precipitates, which grow following the hydrological cycles and consolidate as iron-rich deposits occupying different elevations above the present river in the form of fluvial terraces (Fig. 17.2; Fernández-Remolar *et al.*, 2005).

Seasonal evaporation of river water drives precipitation of hydronium jarosite and schwertmannite, while copiapite, coquimbite, gypsum and other sulfate minerals generate efflorescences from variable source waters which are brought to the surface by capillary action (Fernández-Remolar *et al.*, 2005). During the wet season, hydrolysis of sulfate salts added to the effect of iron hydrolysis result in the precipitation of amorphous iron oxy-hydroxides to nanophase goethite. The oldest terraces show increasing goethite crystallization and replacement of goethite with hematite through time. Organic matter does not preserve well in the Río Tinto sediments, but bio-signatures imparted to sedimentary rocks as macroscopic textures of coated microbial streamers, surface blisters formed by biogenic gas, and microfossils preserved as cast and moulds in iron oxides, help to shape strategies for astrobiological investigations in extant or future exploration missions (e.g. Amils *et al.*, 2007; Fernández-Remolar *et al.*, 2005). Interestingly, the specific biomineralization of hydronium jarosite by a filamentous fungi isolated from the Tinto basin, *Purpureocillium lilacinum*, in non-permissive ionic condition has been recently described (Oggerin *et al.*, 2013). Siderate ($FeCO_3$), often found in the modern sediments of the river (Fernández-Remolar *et al.*, 2012), has also been obtained in cultures of the acidophilic heterotrophic iron reducer *Acidiphilium* in acidic conditions (e.g. Sánchez-Roman *et al.*, 2014). These results strongly suggest that the presence of biological nucleation sites can modify the expected mineral precipitation schemes offered by the bulk physico-chemical conditions in which they grow.

The recent detection of protein fragments and other organic molecules in the ancient terraces of the Rio Tinto basin (Preston *et al.*, 2011) has provided evidence that the acidic and ferruginous environment promotes the preservation of molecules bearing information of the producing organism that inhabited this extreme environment over previous millennia.

Figure 17.2. Río Tinto terraces. (A and B) Young terraces. (C and D) Intermediate terraces. (E and F) Old terraces.

Until recently, it had been generally accepted that the extreme conditions found in Río Tinto were the direct consequence of the mining activities performed in the area during the last 5000 years (e.g. Davis Jr *et al.*, 2000). New geological, geophysical and hydrogeological information supports the hypothesis that this is not the case, and that mining has merely contributed to the extreme conditions in the river. The recharge area of the Peña de Hierro aquifer has been located northwest of the pit lake at a depth ranging from −100 to −400 m (Gómez-Ortiz *et al.*, 2014). The groundwater moves southwards along the fracture network, and when it reaches the remnants of massive and/or stockwork sulfide bodies, located at −500 m in depth, the water interact with the mineral substrate facilitating the metabolism of chemolithotrophic microorganisms and generating acidic fluids (Gómez-Ortiz *et al.*, 2014; Fernández-Remolar *et al.*, 2008). Groundwater is eventually pumped along open strike-slip normal faults to reach the surface, where it sources the acidic springs that feed the headwaters of Río Tinto. The generation of acidic water occurs naturally through the oxidation of subsurface sulfidic bodies. Therefore, mining is not necessarily the cause of the characteristic low pH and high concentration of metals found in the river. This hypothesis is strongly supported by the sedimentary record of the ancient terrace deposits along the Tinto basin, which largely predate the oldest mining activity in the area (Fernández-Remolar *et al.*, 2005). The oldest terrace, containing finely laminated and massive ironstones, has been dated 2.1 million years Ma (Fernández-Remolar *et al.*, 2008). Furthermore, the gossan deposits in the area, representing the remaining materials of the *in situ* oxidation of the massive and stockwork sulfide deposits, have been dated as older than 6 Ma (Moreno *et al.*, 2003).

Considering the geomicrobiological characteristics of the Tinto ecosystem, it has been postulated that the river is predominantly under the control of iron (e.g. Amils *et al.*, 2007). Iron is the main product of pyrite bioleaching and iron-bearing minerals like chalcopyrite, both of which are present in high concentrations in the IPB (see Chapter 1). The activities of iron-oxidizing microorganisms are responsible for both the solubilization of sulfidic minerals and the corresponding high concentrations of iron, sulfate and protons detected in the water column.

Iron has different properties, which give to the Tinto ecosystem an interesting astrobiological perspective. Iron is not only a source of energy as a chemolithotrophic electron donor in its reduced form, but can also be used as an electron acceptor for anaerobic respiration in its oxidized form. In addition, as already mentioned, ferric iron is responsible for controlling the pH of the ecosystem. Even though the reaction is reversible, the dilution factor by tributaries is stronger than evaporation, so an important part of the iron remains precipitated along the course of the river, giving rise to iron bioformations. Accordingly, the concentration of soluble iron decreases gradually from the origin to the mouth of the river. Furthermore, soluble ferric iron readily absorbs harmful UV radiation, protecting the organisms growing in its waters (e.g. Gómez *et al.*, 2010).

The iron-controlled scenario seems reasonable for the chemolithotrophic prokaryotes detected in the Tinto ecosystem. However, given that eukaryotic diversity in the Río Tinto basin is much greater than the prokaryotic diversity (see Chapter 7), and that most of the primary production of the system derives from the activity of photosynthetic protists, the question arises as whether there is an advantage for eukaryotes to develop in an extreme acidic environment with high concentration of toxic heavy metals. One possible answer might be related with the limited availability of iron at neutral pH.

Although iron is an extremely important element for life it can be a limiting factor for growth at neutral and alkaline pH (e.g. Boyd *et al.*, 2000). Organisms have developed very specific and elaborate mechanisms to take up iron anywhere they can find it (e.g. Braun and Killmann, 1999), even though iron is the second most abundant metal (after aluminium) in the lithosphere. In an oxidizing atmosphere at neutral and alkaline pH, soluble iron is rapidly oxidized into insoluble compounds, which are incorporated into anaerobic sediments where sulfate-reducing microorganisms may further transform them into pyrite, an even less reactive iron mineral at neutral pH. The geological recycling of these sediments and the microbiology associated with the iron cycle are the only ways to reintroduce this critical element into the biosphere. The possible advantage for the eukaryotes thriving in the extreme conditions of Río Tinto is an unlimited iron supply provided by the chemolithotrophs growing in the rich iron sulfides of the IPB (Gómez *et al.*, 2003). The availability of iron and other heavy metals is considered so important that a model for an anoxic Proterozoic ocean deprived of iron and other heavy metals as a consequence of an intense sulfate-reducing activity has been proposed (Anbar and Knoll, 2002).

Subsurface geomicrobiology of the Iberian Pyrite Belt

It is clear that the main characteristics of the Tinto basin are not the product of industrial contamination but a consequence of the existence of an underground bioreactor in which the massive sulfidic minerals of the IPB are the main energy source, and the river the exhaust pipe releasing the products of the metabolic reactions occurring in the subsurface. To test this hypothesis two drilling projects, the MARTE (2003–2006) and the IPBSL

(2011–2015) projects, have been designed to intersect ground waters interacting with the mineral ore to provide evidence of subsurface microbial activities.

The main goal of the MARTE (Mars Astrobiology Research and Technology Experiment) project, a collaborative effort between NASA and the *Centro de Astrobiología* in Madrid, Spain, was the search for subsurface microbial activity within the IPB. The selected study site was Peña de Hierro (Fig. 17.3) on the north flank of the Río Tinto anticline. The hydrothermal activity in the area is recorded as complex-massive sulfide lenses or stockwork veins of pyrite and quartz, which occur at the upper part of the IPB volcanic sequence (Leistel *et al.*, 1998). Faults intersect the Early Carboniferous volcanic tuff-hosted pyrite bodies, which are the primary energy source for the chemolithotrophic microorganisms detected in the Tinto basin.

Three boreholes, BH1, BH4 and BH8, were continuously cored by rotary diamond-bit drilling using a wireline system that produced 60 mm diameter cores within a plastic liner. Water was used as drilling fluid to refrigerate the bit. Sodium bromide was used as a chemical tracer for controlling contamination introduced during the drilling. Upon retrieval, cores were flushed with N_2, sealed and transported to a nearby laboratory for geomicrobiological analysis. Samples were prepared aseptically in anaerobic conditions using an anaerobic chamber.

After drilling, the wells were completed by installing PVC casings set in clean gravel packing. Underground sampling for water and gas aquifer analysis was done by the installation of multilevel diffusion samplers (MLDS) at different depths. Anion and metal concentrations and dissolved gases were determined by ion and gas chromatography.

The groundwater entering the ore body at Peña de Hierro was characterized by analysing springs upslope. The water from these springs was aerobic, with a neutral pH and a low ionic strength. The environment within the ore body was sampled by drilling boreholes BH4 and BH8. These wells cored around 165 m of pyrite stockwork. The water table was encountered at nearly 90 m below the surface.

Rock leachate analyses were performed to detect drilling contamination and to estimate resources available to microorganisms from the solid phase. Sulfate was abundant and a good indicator of the degree of oxidation of the sulfides. Surprisingly, nitrite and nitrate were present at concentrations higher than 100 ppm in many samples. Both ferrous

Figure 17.3 View of Peña de Hierro Pit Lake.

and ferric iron could be leached from powdered ore samples (e.g. Amils *et al.*, 2011).

Borehole fluids from the MLDS were analysed as a proxy for formation fluids. The measured composite pH was ca. 3.5 and remained acidic for the two sampling years after drilling. The dissolved ferric to ferrous iron ratio ranged from 0.3 to 4.3. Sulfate concentrations were relatively constant and lower than in rock leachates. Dissolved methane was detected in many of the MLDS samples, indicating active methanogenic activity within the ore body.

Dissolved H_2 concentration averaged 25 ppm, except in the zone within the massive pyrites, just below the water table, where concentrations ranged from 100 to 1000 ppm. A similar pattern was observed in borehole BH8 (e.g. Amils *et al.*, 2011).

Microorganisms were detected in different uncontaminated samples using both culture-dependent and culture-independent methods. Using enrichment cultures, aerobic chemolitho-autotrophs, mainly pyrite- and iron-oxidizers, anaerobic thiosulfate oxidizers using nitrate as electron acceptor, methanogens and sulfate reducers (SRB) were enriched from several samples. Samples from both boreholes were analysed with an antibody microarray (LDChip200) and an oligonucleotide hybridization microarray which give positive signals for Gram-positive bacteria, sulfur and metal oxidizers, sulfate reducers and methanogenic archaea. Hydrogenotrophic and denitrifying bacteria were also identified by 16S rRNA cloning and sequencing. Using fluorescence *in situ* hybridization (CARD-FISH) it was possible to prove the presence of active microorganisms in different uncontaminated samples (Puente-Sánchez *et al.*, 2014).

Electric resistivity tomography (ERT) and time-domain electromagnetic sounding (TDEM) have been used in the IPBSL (Iberian Pyrite Belt Subsurface Life Detection) project to further investigate the subsurface geomicrobiology of the IPB, by detecting the most probable subsurface areas hosting microbial activity in deep regions of the basement, those in which water intersect sulfidic ores. From the analysis of the geophysical data two wells, BH10 and BH11, were drilled (Fig. 17.4) with depths of 620 and 340 m, respectively (Gómez-Ortiz *et al.*, 2014), using conditions analogous to those described for the MARTE drilling project. In addition to the geological core login in the drilling site, selected samples were obtained for mineralogical (XRD), elemental analysis (ICP-MS), and stable isotopic and petrographic analysis. Results from mineralogical analysis showed the presence of pyrite and its alteration products, including hematite and magnetite, in both boreholes. The elemental analysis of the leachates from the core samples showed the presence of iron and other metals at different depths. The stable isotopic analysis of pyrites showed important ^{34}S fractionation at different depths, which is a clear indication of the activities of sulfate-reducing prokaryotes. Also, fractionation in ^{13}C was observed in samples from both boreholes, again providing a clear bio-signature of microbial activity at these depths (Fernández-Remolar *et al.*, 2012).

Rock leachates were analysed by ion chromatography to determine the concentration of water soluble anions. Chromatograms indicated the presence of reduced organic anions, including acetate, and oxidized inorganic anions such as nitrate and sulfate. Proteins and sugars were also detected at different depths, indicating the presence of extant or recent microbiological activities. Samples along the BH10 borehole were analysed with the immunosensor LDCChip300, an antibody microarray containing 300 antibodies with diverse and complementary specificity. Positive antigen–antibody reactions were detected against methanogenic archaea and SRB, which agreed with the results obtained by enrichment cultures. H_2, CO_2 and CH_4 have been detected by gas chromatography of different core samples from both boreholes.

DNA and RNA have been successfully extracted from different BH10 and BH11 samples. Most of them rendered positive PCR amplifications of bacterial and archaeal 16S rRNA genes, Samples from the two boreholes were fixed for CARD-FISH analysis. The results obtained have shown positive hybridizations signals for both *Bacteria* and *Archaea* at different depths. The microorganisms observed appear to form colonies attached to mineral particles, thus it is unlikely that they correspond to contamination by drilling fluid, especially because all selected samples had the detectable background concentrations of bromide. Anaerobic enrichment cultures were prepared in an anaerobic chamber using mineral salt medium with the addition of different electron donors and electron acceptors. The

Figure 17.4 Borehole BH11 drilling site.

following microbial groups have been detected after more than one year of incubation using samples from both boreholes: methanogens, methanotrophs, sulfate-reducers, iron-oxidizers, acetogens and denitrifiers. Using all the available data different hot spots have been detected, which are being used to select sampling points for metagenomic and retro-transcriptomic analysis.

The results obtained in the MARTE and the IPBSL drilling projects have allowed the following conclusions to be made: (i) as groundwater enters the volcanogenic-hosted massive sulfide system (VHMS), biological and abiotic processes are activated; (ii) electron donors available for microbial metabolism include ferrous iron, metal sulfides and H_2, probably generated by water/rock interaction; (iii) electron acceptors include nitrate, sulfate, ferric iron and CO_2; (iv) these electron donors and acceptors support a community of different microbial metabolisms; (v) oxidants driving the system appear to be supplied by the rock matrix, in contrast to conventional ARD models; (vi) only mobilization of these sources by ground-water appears to be required to promote microbial metabolisms. These observations confirmed the hypothesis that microorganisms are active in the subsurface of the IPB and are responsible for the characteristic extreme conditions detected in the Tinto basin. Besides providing deep understanding of the geomicrobiology of the Rio Tinto, these results are also of great astrobiological interest.

The Rio Tinto: a model of astrobiological interest

Recent observations of abundant layered sulfate minerals at different locations on Mars suggest a history of an aqueous, acidic, sulfate-rich environment (Ehldmann and Mustard, 2012; Farrand *et al.*, 2009; Milliken *et al.*, 2008; Klingelhöfer *et al.*, 2005; Squyres *et al.*, 2004) that might have originated from the weathering of sulfide-rich minerals (e.g. Fairén *et al.*, 2004).

In early 2004, the NASA rover *Opportunity* from the MER mission began its exploration on the Martian surface at a region called Meridiani Planum. The scientific motivation for the selection of Meridiani Planum was the identification by orbital observation in previous missions of regionally distributed hematite, inferred to have formed under aqueous conditions on the early Martian surface. Several hypothesis for hematite deposition were proposed prior to landing (e.g. Christensen *et al.*, 2004), but the observations made by the *Opportunity* decisively tipped the balance towards models that invoke pervasive chemical weathering of basalts and subsequent formation of hematite-rich spherules within sulfate rich-sediments (e.g. Squyres *et al.*, 2004). Jarosite, a basic ferric iron sulfate mineral identified by Mössbauer spectroscopy (Klingelhöfer *et al.*, 2005) placed particular constrains on the paleo-environmental interpretation of Meridiani outcrop rocks, as this mineral is considered to precipitate under acidic conditions (Bigham *et al.*, 1996).

The Río Tinto as a geochemical and mineralogical terrestrial Mars analogue provides an interesting perspective for the interpretation of Martian data for two reasons. First, both iron oxides and ferric sulfates form at Río Tinto under well characterized physico-chemical and biological conditions (Amils *et al.*, 2007; Fernández-Remolar *et al.*, 2005; González-Toril *et al.*, 2003). Second, the modern drainage, where depositional processes can be observed in action, is complementary by a historical record of deposition preserved as diagenetically stabilized sedimentary rock in terraces at different levels above the river (e.g. Fernández-Remolar *et al.*, 2005). The combination of ancient and modern deposits facilitates comparison with Meridiani Planum and other iron minerals regions like Valles Marineris, Mawrth Vallis and Syrtis Major (Ehlmann and Mustard, 2012; Farrand, 2009; Milliken *et al.*, 2008; Klingerhöfer *et al.*, 2005), where depositional and diagenetic processes must be inferred from ancient sedimentary rocks.

The mineralogical composition and sedimentary geomicrobiology reported for the Tinto basin are of use to address different questions of interest for space exploration missions in general, and for the Mars exploration in particular: (i) Which biological, chemical and physical processes left an interpretable record in Río Tinto rocks?; (ii) How different processes (dissolution, re-precipitation, lithification, diagenesis) modified the initial mineralogical and chemical composition of iron-bearing precipitates; (iii) What are the consequences of these processes for the retention of environmental or biological proxy records in Río Tinto rocks? and (iv) What proportion of the informative proxy record is expected to be captured using the suite of instruments currently used in Mars exploration missions (MER, MSL) or to be used in future space explorations missions (Amils *et al.*, 2007; Fernández-Remolar *et al.*, 2005)?

The Río Tinto and Meridiani depositional systems have both similarities and differences. The most obvious difference concerns physical setting. Whereas the Meridiani rocks accumulated via Aeolian and aqueous processes in an arid environment (Grotzinger *et al.*, 2005), Río Tinto sediments formed in seasonally arid stream beds. Moreover, the Tinto basin precipitates owe their genesis to the oxidation of hydrothermally emplaced pyrite ores. While this has been also proposed as a source of sulfuric acid on Mars, it is not the only possible source of acidic fluids that interact with the basaltic rocks in the Meridiani region (Knoll *et al.*, 2005).

The suite of minerals identified in Río Tinto rocks and sediments is quite similar to those identified in Meridiani outcrops. Mössbauer spectra of the outcrop at Merdiani reveal an assemblage that includes jarosite (Fig. 17.5), hematite and an unidentified iron bearing phase with octahedrally coordinated Fe^{3+}, probably an hydrolysis product of jarosite. The Río Tinto Mössbauer spectra are very similar to those from Meridiani (Fernández-Remolar *et al.*, 2005) providing a useful geochemical and mineralogical analogue for studying the processes in play when Meridian rocks formed. Río Tinto rocks also document the effects of diagenesis, providing a doubly useful comparison to Meridiani sediments. Like the Río Tinto rocks, Martian outcrops on the Meridiani plain contain hematite, but unlike Río Tinto terraces, Meridiani outcrops remain sulfate-rich, including ferric sulfates that do not persist into the rock record at Río Tinto. This suggests that mineral formation and diagenesis occurred under conditions of strong water limitation (e.g. Grotzinger *et al.*, 2005)

As noted in various other chapters, many physiologically and phylogenetically diverse microorganisms can thrive in acidic and strongly oxidizing environments which are inferred to be broadly similar to those at Meridiani and other iron mineral rich regions, when its sulfate-rich sedimentary successions were deposited (Ehdmann and

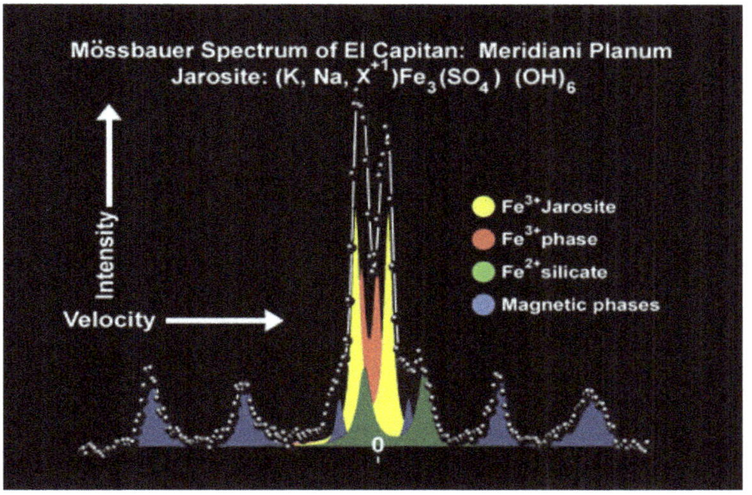

Figure 17.5 Mössbauer spectra from El Capitán rock at Meridiani Planum (NASA).

Mustard, 2012; Farrand *et al.*, 2009; Milliken *et al.*, 2008). Thus, the Río Tinto is relevant to framing our biological expectations of Mars exploration. Despite the fact that organic materials do not preserve well in Río Tinto sediments, diagenetically stabilized ironstones in the river terraces preserve macroscopic and microscopic bio-signatures in iron oxide precipitates (Fernández-Remolar *et al.*, 2005). Río Tinto sedimentary deposits record aspects of the physical, chemical, and biological environment of the regional ecosystem, and these indicators persist through diagenesis to provide a geochemical and geomicrobiological chronicle of Río Tinto processes through time. By inference, Martian outcrops carry a similar potential to preserve a record of life (if it ever existed) and the environment. To improve the characterization of Mars, the rover *Curiosity* from the MSL mission (Fig. 17.6) has been dotted with an XRD for a better mineral identification, and a mass spectrometer to provide isotopic data of the Martian minerals.

The results obtained so far in the characterization of the IPB allow suggesting that these Martian systems could support subsurface life, even if surface conditions preclude it. In addition, it has been noted that subsurface microbial metabolism coupled with sulfide weathering can produce large amounts of methane, which has been proposed as an atmospheric indicator of extant life on Mars (Webster *et al.*, 2014; Mumma *et al.*, 2009; Formisano *et al.*, 2004).

Although methane can be generated abiotically, more than 80% of Earth's methane is biologically produced as a final product of the degradation of organic matter by methanogenic archaea. Methanogens are generally found in habitats that share two physico-chemical properties: reduced redox potentials and circumneutral pH, with currently only few exceptions (e.g. Taconi *et al.*, 2008). These conditions are quite different of the extreme acidic and oxidative conditions existing in the Rio Tinto.

After the detection of methane in the borehole fluids of the MARTE project and in the Martian atmosphere, a systematic survey for methanogenic activity was initiated in the anoxic sediments of the Tinto basin (Sanz *et al.*, 2011). The first site in which methane production was detected was Campo de Galdierias. Sediments from this site showed specific positions with negative redox potentials, while in the surrounding sediments the redox potential values were highly positive and similar to those measured in the river water column. Microcosms were established using reduced sediments from this site spiked with different methanogenic substrates. The greatest methane production was observed in microcosms spiked with methanol. In all cases the production of methane was associated with a decrease in redox potentials to negative values and an increase of pH (Sanz *et al.*, 2011).

A second site, JL Dam, was selected to have access to deeper and more accessible sediments. Cores from this site showed well-defined black

Figure 17.6 Suite of instruments in Curiosity's mobile arm (NASA).

bands between the otherwise reddish-brown sediments. Black bands were associated with negative (reduced) redox potentials and higher pH values compared to the positive (oxidized) redox potentials and acidic pH of the adjacent red and brown layers. DNA from the black bands was extracted, amplified and sequences corresponding to *Methanosaeta concilii* were obtained (Sanz *et al.*, 2011). To further explore the methanogenic diversity of the cores, enrichment cultures were designed using different substrates. The highest CH_4 production occurred in the presence of lactate–methanol–sucrose mixture. Only *M. concilii* was detected in this microcosm, suggesting that this is the predominant methanogen in environments exposed to organic substrates. *Methanobacterium bryantii* and *Methanosarcina barkeri* were identified in cultures enriched with H_2 or methanol, respectively (Sanz *et al.*, 2011).

As mentioned previously, the bulk environmental conditions at Rio Tinto, especially with respect to pH and redox potential, are far from the conditions known to allow the growth of methanogenic archaea. This apparent contradiction can be resolved at the micro-site level, as the generation of micro-niches might lead to the proliferation of microorganisms with different requirements from those found in the harsh bulk conditions existing in the water column. These micro-niches could be readily generated in a semi-solid matrix, like sediments, or in a solid matrix inside a rock.

The presence of methanogens in an environment containing large concentrations of oxidized iron and sulfate has important astrobiological implications since it could be a scenario for the biological production of the atmospheric methane detected on Mars (Webster *et al.*, 2014; Mumma *et al.*, 2009; Formisano *et al.*, 2004). The argument that Mars' environmental conditions are not suitable for methanogenesis could be challenged by the methane production observed in the sediments of Río Tinto or the subsurface of the IPB. Considering the short lifetime of the Mars atmospheric methane (Mumma *et al.*, 2009), there is a possibility that extant methanogens are currently active on the subsurface of the red planet. As mentioned above, the *Curiosity* rover exploring crater Gale is equipped with a mass spectrometer capable to measure the carbon fractionation signal of the Martian methane, which might allow to clarify its biological or abiotic origin.

Although the history of iron in the Earth biosphere is still an open question, it has been suggested that the Tinto ecosystem, as well as other iron-rich acidic environments are relics of an ancient (Archaean) iron world (Gómez *et al.*, 2003), probably operating in other planetary systems, e.g. Mars (e.g. Amils *et al.*, 2011; Fernández-Remolar *et al.*, 2005). Obviously the actual conditions in which the Tinto ecosystem operates are different from those that may have prevailed during the Archaean eon or might currently prevail on Mars, but the properties of the microorganisms isolated and characterized so far in this environment allow us to predict their performance in these systems.

Liquid water is an absolute requirement for all known life forms. While liquid water is abundant in the Tinto basin, both on the surface and underground, due to environmental constraints, water appears only as solid or vapour phases on the current Mars surface. Although the current orbital technology could reveal the presence of liquid water in the subsurface of Mars, so far there is only indirect evidence of widespread subterranean ice (Boynton *et al.*, 2002), its quantification in the polar water-ice (Bibring *et al.*, 2005) and the characterization of brines from the polar region (Smith *et al.*, 2009). However, images from Mars, as well as spectral data provided by different instruments in orbit and on the surface of the planet, give support to distinctive episodes of water release on the planet's surface in the past (e.g. Bibring *et al.*, 2005; Heldmann *et al.*, 2005; Poulet *et al.*, 2005; Squyres *et al.*, 2004; Fairén *et al.*, 2003; Malin and Edgett, 2003; Parker *et al.*, 1993). Climatic studies of the early atmospheric evolution of Mars (Carr, 1999) indicate that during the Noachian, the atmospheric pressure was high enough to sustain substantial amounts of liquid water on its surface, explaining the above mentioned water-related features.

Although there is only a remote possibility that the Martian hematite formations are the product of chemolitho-autotrophy, the microbial diversity found in the Tinto basin, with metabolisms compatible with the conditions prevailing on Mars (Fig. 17.7), allows speculation that microorganisms may have grown or may still be growing in places were mineral and liquid water converge. Obviously Río Tinto is not the only acidic environment of astrobiological interest. In recent years diverse extreme acidic environments have been identified as terrestrial analogues of Mars, such as acid rock drainage in the South Shetland Island in Antarctica (Dold *et al.*, 2013), the acidic seasonally dry lakes of Kalgoorly in Australia (Mann, 1983) and cold acid drainage systems in the Canadian Artic (Lacelle and Léveillé, 2010; Michel and Everdingen, 1987). Most of these environments have been analysed from a geological and mineralogical point of view. A thorough geomicrobiological characterization of these sites and others to be explored will eventually complement the information already obtained in the IPB. From a Martian perspective it will be very convenient to design a drilling mission to gain information on the existence of redox gradients of possible use by chemolithotrophic microorganisms. The characteristics exhibited by acidophilic microorganisms make them suitable for its past or extant development in the red planet and possibly elsewhere in the universe.

Conclusions

Forty years ago, the Viking mission launched the first astrobiological mission devoted to the search for signs of life on Mars and concluded that life had little chance of developing there due to the extreme conditions detected on its surface. Since then, important advances in microbiology have challenged this somewhat pessimistic viewpoint. Today thanks to research on extremophilic microorganisms, we know that life is extremely robust and able to adapt to different extreme conditions, which increases the chances of finding it elsewhere in the universe. Acidic habitats are of special interest for astrobiology because they are the only natural extreme environment generated by the metabolic activity of chemolitho-autotrophic

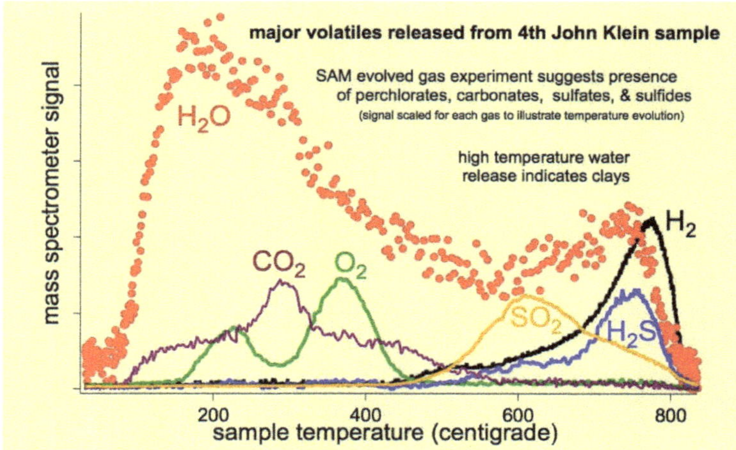

Figure 17.7 Major volatiles released from a Martian rock sample analysed by the Curiosity mass spectrometer (NASA).

microorganisms known to date. The characterization of the extremely acidic Río Tinto basin has answered some basic questions, such as the origin of the extreme conditions of the habitat and the identification and isolation of the microorganisms responsible for these conditions, but has also posed some new ones, all of astrobiological interest, such as the paradox of the existence of a high level of diversity of complex eukaryotic microorganisms or the need to explain detected functional activities in non-permissive environmental conditions. The discovery of some Noachian iron lithological units on Mars similar to those produced biologically in the Tinto basin gave Río Tinto the status of geochemical and mineralogical terrestrial analogue of Mars, through which attempts to better understand the geomicrobiological processes that may have driven the generation of iron oxides and sulfates on Mars. A thorough geomicrobiological characterization of other acidic sites should produce complementary information to better understand these extreme environments and to further evaluate the possible existence of these metabolic activities on Mars or other planetary systems.

Acknowledgement

The authors thank the ERC contract 250350-IPBSL for financial support.

References

Amaral-Zettler, L.A., Gómez, F., Zettler, E., Keenan, B.G., Amils, R., and Sogin, M.L. (2002). Eukaryotic diversity in Spain's River of Fire. Nature *417*, 137.

Amaral-Zettler, L., Zettler, E.R., Theroux, S.M., Palacios, C., Aguilera, A., and Amils, R. (20010) Microbial community structure across the tree of life in the extreme Río Tinto. ISME J. 5, 42–50.

Amils, R., González-Toril, E., Fernández-Remolar, D., Gómez, F., Aguilera, A., Rodríguez, N., Malki, M., García-Moyano, A., González-Fairén, A., de la Fuente, V. *et al.* (2007). Extreme environments as Mars terrestrial analogs: the Río Tinto case. Planet. Space Sci. 55, 370–381.

Amils, R., González-Toril, E., Aguilera, A., Rodríguez, N., Fernández-Remolar, D., Gómez, F., García-Moyano, A., Malki, M., Oggerin, M., Sánchez-Andrea, I., *et al.* (2011). From Río Tinto to Mars: the terrestrial and extraterrestrial ecology of acidophiles. In Advances in Applied Microbiology, Vol 77, Askin, A.I., Gadd, G.M., and Sariasini, S., eds. (Academic Press, Burlongton, VT, USA), pp. 41–70.

Anbar, A.D., and Knoll, A.H. (2002). Proterozoic ocean chemistry and evolution: a bioinorganic bridge. Science *297*, 1137–1142.

Bibring, J.P., Langevin, Y., Gendrin, A., Gondet, B., Poulet, F., Berthé, M., Soufflot, A., Arvidson, R., Mangold, N., Mustard, J., *et al.* (2005). Mars surface diversity as revealed by the OMEGA/Mars Express observations. Science *307*, 1576–1581.

Bigham, J.M., Schwertmann, U., Traina, S.J., Winland, R.L., and Wolf, M. (1996). Schwertmannite and the chemical modelling of iron in acid sulphate waters. Geochim. Cosmochim. Acta *60*, 211–221.

Boyd, P.W., Watson, A.J., Law, C.S., Abraham, E.R., Trull, T., Murdoch, R., Bakker, D.C., Bowie, A.R., Buesseler, K.O., Chang, H., *et al.* (2000). A mesoscale phytoplankton bloom in the polar Southern Ocean stimulated by iron fertilization. Nature *407*, 695–702.

Boyton, W.V., Feldman, W.C., Squires, S.W., Prettyman, T.H., Brückner, J., Evans, L.G., Reedy, R.C., Starr, R., Arnold, J.R., Drake, D.M., *et al.* (2002). Distribution of hydrogen in the near surface of Mars: evidence for subsurface ice deposits. Science *297*, 81–85.

Braun, V., and Killmann, H. (1999). Bacterial solution to the iron supply problems. Trends Biochem. Sci. *24* 104–109.

Carr, M.H. (1999). Retention of an atmosphere on early Mras. J. Geophys. Res. *104*, 21897–21909.

Christensen, P.R., and Ruff, S.W. (2004). Formation of the hematite-bearing unit in Meridiani Planum: evidence for deposition in standing water. J. Geophys. Res. Planets *109*, E08003.

Davis, R.A. Jr, Welty, A.T., Borrego, J., Morales, J.A., Pendon, J.G., and Ryan, J.G. (2000). Río Tinto estuary (Spain): 5000 years of pollution. Environ. Geol. *39*,1107–1116.

Dold, B., González-Toril, E., Aguilera, A., López-Pamo, E., Cisternas, M.E., Bucchi, F., and Amils, R. (2013). Acid rock drainage and rock weathering in Antartica: important sources for iron cycling in the Southern Ocean. Environ. Sci. Technol. *47*, 6129–6136.

Ehdmann, B.L., and Mustard, J.F. (2012). An in-situ record of major environmental transitions on early Mars at Northeast Syrtis Major. Geophys. Res. Lett. *39*, L11292.

Farrand, W.H., Glotch, T.D., Rice, J.W. Jr., Hurowitz, J.A., and Swayze, G.A. (2009). Discovery of jarosite within Mawrth Vallis region of Mars: Implications for the geologicalhistory of the region. Icarus *204*, 478–488.

Fairén, A.G., Dohm, J.M., Baker, V.R., de Pablo, M.A., Ruiz, J., Ferris, J., and Anderson, R. (2003). Episodic flood inundations of the northern plains of Mars. Icarus *165*, 53–67.

Fairén, A.G., Fernández-Remolar, D., Dohm, J.M., Baker, V.R., and Amils, R. (2004). Inhibition of carbonate synthesis in acidic oceans from Mars. Nature, *431*, 423–426.

Fernández-Remolar, D.C., Morris, R.V., Gruener, J.E., Amils, R., and Knoll, A.H. (2005). The Rio Tinto Basin, Spain: mineralogy, sedimentary geobiology and implications for interpretation of outcrop rocks of meridiani Planum, Mars. Earth Planet. Sci. Lett. *240*, 149–167.

Fernández-Remolar, D., Gómez, F., Prieto-Ballesteros, O., Schelble, R.T., Rodríguez, N., and Amils, R. (2008). Some ecological mechanisms to generate habitability in planetary subsurface areas by chemolithotrophic communities: the Río Tinto subsurface ecosystem as a model system. Astrobiology *8*, 157–173.

Fernández-Remolar, D.C., Preston, L.J., Sánchez-Román, M., Izawa, M.R.M., Huang, L., Southam, G., Banerjee, N.R., Osinski, G.R., Flemming, R., Gómez-Ortíz, D., *et al.* (2012). Carbonate precipitation under bulk acidic conditions as a potential biosignature for searching life on Mars. Earth Planet. Sci. Lett. *351*, 13–26.

Formisano, V., Atreya, S., Encrenaz, T., Ignatiev, N., and Giuranna, M. (2004). Detection of methane in the atmosphere of Mars. Science *306*, 1758–1761.

García Moyano, A., González-Toril, E., Aguilera, A., and Amils, R. (2012). Comparative microbial ecology study of the sediments and the water column of the Río Tinto, an extreme acidic environment. FEMS Microbiol. Ecol. *81*, 303–314.

Gómez, F., Fernández-Remolar, D., González-Toril, E., and Amils, R. (2003). The Tinto River, an extreme Gaian environment. In Gaia 2000, Margulis, L., Miller, J., Boston, P., Schneider, S., and Crist, C., eds (MIT Press, Boston, MA, USA), pp. 321–333.

Gómez, F., Mateo-Martí, E., Prieto-Ballesteros, O., Martín-Gago, J., and Amils, R. (2010). Protection of chemolithotrophic bacteria exposed to Mars environmental conditions. Icarus *209*, 482–487.

Gómez-Ortiz, D., Fernández-Remolar, D., Granda, A., Quesada, C., Granda, T., Prieto-Ballesteros, O., Molina, A., and Amils, R. (2014). Identification of the subsurface sulfide bodies responsible for acidity in Río Tinto source, water. Spain Earth Plant. Sci. Lett. *391*, 36–41.

González-Toril, E., Llobet-Brosa, E., Casamayor, E.O., Amann, R., and Amils, R. (2003). Microbial ecology of an extreme acidic environment, the Tinto River. Appl. Environ. Microbiol. *69*, 4853–4865.

González-Toril E., Aguilera, A., Rodríguez, N., Fernández-Remolar, D., Gómez, F., Díaz, E., García-Moyano, A., Sanz, J.L., and Amils, R. (2010). Microbial ecology of Río Tinto, a natural extreme acidic environment. Hydrometallurgy *104*, 329–333.

Gross, W. (2000). Ecophysiology of algae living in highly acidic environments. Hydrobiology *433*, 31–37.

Grotzinger, J.P., Arvidson, R.E., Bell, J.F. III, Calvin, W., Clark, B.C., Fike, D.A., Golombek, M., Greeley, R., Haldemann, A., Herkenhoff, K.E., *et al.* (2005). Stratigraphy, sedimentology and depositional environment of the Burns formation, Meridiani Planum, Mars. Earth Planet. Sci. Lett. *240*, 11–72.

Hallberg, K.B., and Johnson, D.B. (2001). Biodiversity of acidophilic prokaryotes. Adv. Appl. Microbiol. *49*, 37–84.

Heldmann, J.L., Toon, O.B., Pollard, W.H., Mellon, M.T., Pitlick, J., McKay, C.P., and Andersen, D.T. (2005). Formation of Martian gullies by the action of liquid water flowing under current Martian environmental conditions. J. Geophys. Res. *110*.

Klingelhöfer, G., Morris, R.V., Bernhardt, B., Schröder, C., Rodionov, D.S., de Souza, P.A. Jr, Yen, A., Gellert, R., Evlanov, E.N., Zubkov, B., *et al.* (2005). Jarosite and hematite at *Meridiani Planum* from the Mössbauer spectrometer on the Opportunity rover. Science *306*, 1740–1745.

Knoll, A.H., Carr, M., Clark, B., Des Marais, D.J., Farmer, J.D., Fisher, W.W., Grotzinger, J.P., McLennan, S.M., Malin, M., Schröder, C., *et al.* (2005). An astrobiological perspective on Meridiani Planum. Earth Planet. Sci. Lett. *240*, 179–189.

Lacelle, D., and Lévillé, R. (2010). Acid drainage generation and associated Ca-Fe-SO_4 minerals in a periglacial environment, Eagle Plains, Northen Yukon, Canada: a potential analogue for low-temperature sulfate formation on Mars. Planet. Space Sci. *5*, 509–521.

Leistel, J.M., Marcoux, E., Theiblemont, D., Quesada, C., Sánchez, A., Almodóvar, G.R., Pascual, E., and Saez, R. (1998). The volcanic-hosted massive sulphide deposists of the Iberian Pyrite Belt. Mineralium Deposita *33*, 2–30.

Lescuyer, J.L., Leistel, J.M., Mrcoux, E., Milési, J.P., and Thiéblemont, D. (1998). Late Devonian-Early Carboniferous peak sulphide mineralization in the Western Hercynides. Mineralium Deposita 33, 208–220.

López-Archilla, A.I., Marín, I., and Amils, R. (2001). Microbial community composition and ecology of an acidic aquatic environment: the Tinto River, Spain. Microbiol. Ecol. *41*, 20–35.

Malin, M.C., and Edgett, K.S. (2003). Evidence for persistent flow and aqueous sedimentation on early Mars. Science *302*, 1931–1934.

Malki, M., González-Toril, E., Sanz, J.L., Gómez, F., Rodríguez, N., and Amils, R. (2006). Hydrometallurgy *83*, 223–228.

Malki, M., De Lacey, A.L., Rodríguez, N., Amils, R., and Fernández, V.M. (2008). Preferential use of an anode as an electron acceptor by an acidophilic bacterium in the presence of oxygen. Appl. Environ. Microbiol. *74*, 4472–4476.

Mann, A.W. (1983). Hydrogeochemistry and weathering in the Yilgarn Block, Western Australuia-ferrolysis and heavy metals in continental brines. Geochim. Coscochim. Acta *47*, 181–190.

Margulis, L., Mazur, P., Barghoorn, E.S., Halvorson, H.O., Jukes, T.H., and Kaplan, I.R. (1979). The Viking Mission: implications for life on Mars. J. Mol. Evol. *14*, 223–232.

Milliken, R.E., Swayze, G.A., Arvidson, R.E., Bishop, J.L., Clark, R.N., Ehlmann, B.L., Green, R.O., Grotzinger, J.P., Morris, R.V., Murchie, S.L., *et al.* (2008). Opaline silica in young deposits on Mars. Geology *36*, 847–850.

Michel, F.A., and Everdingen, R.O. (1987). Formation of jarosite deposits on Cretaceous shales in the Fort Norman area, Northwest Territories. Can. Mineral. *25*, 221–226.

Moreno, C., Capitán, M.A., Doyle, M., Nieto, J.M., Ruiz, F., and Sáez, R. (2003). Edad mínima del gossan de Las Cruces: implicaciones sobre la edad del inicio de los ecosistemas extremos en la Faja Pirítica Ibérica. Geogaceta *33*, 67–70.

Mumma, M.J., Villanueva, G.L., Novak, R.E., Hewagama, T., Bonev, B.P., DiSanti, M.A., Mandell, A., and Smith, M.D. (2009). Strong release of methane on Mars in Northen Summer 2003. Science *323*, 1041–1045.

Oggerin, M., Tornos, F., Rodríguez, N., del Moral, C., Sánchez-Román, M., and Amils, R. (2013). Specific jarosite biomineralization by *Purpureocillium lilacinum*, an acidophilic fungi isolated from Río Tinto. Environ. Microbiol. *15*, 2228–2237.

Parker, T.J., Gorsline, D.S., Saunders, R.S., Pieri, D.C., and Schneeberger, D.M. (1993). Coastal geomorphology of the martian northern plains. J. Geophys. Res. *98*, 11061–11078.

Puente-Sánchez, F., Moreno-Paz, M., Rivas, L.A., Cruz-Gil, P., García-Villadangos, M., Gómez, M.J., Postigo, M., Garrido, P., González-Toril, E., Briones, C., *et al.* (2014). Deep subsurface sulfate reduction and methanogenesis in the Iberian Pyrite Belt revealed through geochemistry and molecular biomarkers. Geobiology *12*, 34–47.

Poulet, F., Bibring, J.P., Mustard, J.F., Gendrin, A., Mangold, N., Langevin, Y., Arvidson, R.E., Gondet, B., Gómez, C., and the OMEGA Team (2005). Phyllosilicates on Mars and implications for early martian climate. Nature *438*, 623–627.

Preston, L., Shuster, J., Fernández-Remolar, D., Banerjee, N., Osinski, G.R., and Southam, G. (2011). The preservation and degradation of filamentous bacteria and biomolecules within iron oxide deposits at Rio Tinto, Spain. Geobiology *9*, 233–249.

Rawlings, D.E. (2005). Characteristics and adaptability of iron- and sulfur-oxidizing microorganisms used for the recovery of metals from minerals and their concentrates. Microb. Cell Fact. *4*, 13–28.

Sánchez-Andrea, I., Rodríguez, N., Amils, R., and Sanz, J.L. (2011). Microbial diversity in anaerobic sediments at Río Tinto, a naturally acidic environment with a high heavy metal content. Appl. Environ. Microbiol. *77*, 6085–6093.

Sánchez-Román, M., Fernández-Remolar, D., Amils, R., Sánchez-Navas, A., Schmid, T., San Martíon-Uriz, P., Rodríguez, N., and McKenzie, J.A. (2014). Microbial mediated formation of Fe-carbonate minerals under extreme acidic conditions. Sci. Rep. *4*, 4757.

Sanz, J.L., Rodríguez, N., Díaz, E., and Amils, R. (2011). Methanogenesis in the sediments of Río Tinto, an extreme acidic environment. Environ. Microbiol. *13*, 2336–2341.

Smith, P.H., Tamppari, L.K., Arvidson, R.E., Bass, D., Blaney, D., Boyton, W.V., Carswell, A., Catling, D.C., Calk, B.C., Duck, T., *et al.* (2009). H_2O at the Phoenix landing site. Science *325*, 58–61.

Squyres, S.W., Grotzinger, J.P., Arvidson, R.E., Bell, J.F. 3rd, Calvin, W., Christensen, P.R., Clark, B.C., Crisp, J.A., Farrand, W.H., Herkenhoff, K.E., *et al.* (2004). In situ evidence for an ancient aqueous environment in Meridiani Planum, Mars. Science *306*, 1709–1714.

Taconi, K.A., Zappi, M.E., French, W.T., and Brown, L.R. (2008). Methanogenesis under acidic pH conditions in a semy-continuous reactor system. Bioresour. Technol. *99*, 8075–8081.

Webster, C.R., Mahaffy, P.R., Atreya, S.K., Flesch, G.J., Mischna, M.A., Meslin, P.Y., Farley, K.A., Conrad, P.G., Christensen, L.E., Pavlov, A.A., *et al.* (2014). Mars methane detection and variability at gale crater. Science *347*, 415–417.

Index

A

M

N

O

P

Q

R

S

T

U

www.ingramcontent.com/pod-product-compliance
Ingram Content Group UK Ltd.
Pitfield, Milton Keynes, MK11 3LW, UK
UKHW062000041026
15591UKWH00016B/1098

9 781910 190333